MÖSSBAUER EFFECT
DATA INDEX

Covering the 1975 Literature

MÖSSBAUER EFFECT DATA INDEX

DATA INDEX

Covering the 1975 Literature

Edited by

John G. Stevens

and

Virginia E. Stevens

University of North Carolina at Asheville

SPRINGER SCIENCE+BUSINESS MEDIA, LLC

Library of Congress Catalog Card Number 76-146429

ISBN 978-1-4757-5902-0   ISBN 978-1-4757-5900-6 (eBook)
DOI 10.1007/978-1-4757-5900-6

©1976 Springer Science+Business Media New York
Originally published by IFI/Plenum Data Company in 1976

Foreword

The Mössbauer effect has established itself as a powerful spectroscopic method in a surprisingly short period of time. Investigations of this effect passed very quickly through the childhood stage, when every experimental result seemed to be worthy of rapid publication in the form of a letter to the editor. This phase was terminated in 1960, only two years after the appearance of the first Mössbauer paper, by Goudsmit's editorial in the April issue of *Physical Review Letters*. He wrote: "We believe that the time has come to put a damper on the influx of letters on the Mössbauer effect, to bring them into line with policy in other fields. Most of the remaining problems on line splittings, shifts and other details do not require rapid publication and should now be presented in well prepared articles."

However, even the number of "well prepared" articles continues to grow from year to year, and so does the *MEDI*, the most useful data index for the Mössbauer community.

This community is based on a common spectroscopic method applicable to many fields of research. It is this fact that makes meetings of Mössbauer people so very different from narrow, one-field-oriented, specialized conferences. Each international conference on the Mössbauer effect is an exciting adventure for its participants, making it possible to meet specialists in physics, chemistry, mineralogy, biology, archeology, and sometimes in quite unexpected disciplines, where the application of Mössbauer spectroscopy has suddenly proved fruitful.

But just these aspects of the work, so enjoyable and interesting for the rest of us, make the task of Virginia and John Stevens very difficult. Fishing for Mössbauer data throughout the scientific literature and finding applications of Mössbauer spectroscopy in articles whose titles bear no indication of any connection to the Mössbauer effect involves deep dedication and enormous amounts of work. Virginia and John have proved their dedication to the Mössbauer community over a span of many years. The present index, *MEDI-75*, is the eighth volume in the series.

The Mössbauer Data Center at the University of North Carolina continually expands its services. The "Monthly Mössbauer Reference/Data Listing" keeps those working in Mössbauer spectroscopy up to date with the most recent developments, and the Center is now capable of doing computer searches of the Mössbauer literature for the community.

We have to keep in mind, however, that in this very successful but extremely difficult activity, for which we are so deeply thankful, Virginia and John Stevens need the assistance of all those interested in the application of the Mössbauer spectroscopy in their work.

ANDRZEJ HRYNKIEWICZ

Kraków, Poland
August 1976

Acknowledgments

Over the years many people have conscientiously helped us with various tasks associated with production of the *Index.* We wish to make note of their assistance and thank them for it publicly. Some of these people have been faithful in their help for several years, for which they deserve special recognition.

Among those in this category are Drs. L. H. Bowen, B. D. Dunlap, and D. Schrooer, who proofread the data and references and in doing so demonstrated a special kind of patience and attention to detail. Other long-time assistants are Professor G. N. Belozerskii of USSR and Dr. M. Takano of Japan, both of whom have collected the more obscure literature from their countries and, in many cases, abstracted the data for us.

Once again, Dr. F. E. Obenshain and the library staff at Oak Ridge National Laboratory have continued to accept and aid us during our regular and intensive visits to their facilities. Dr. G. K. Shenoy of Argonne National Laboratory also helped us with literature retrieval. Special reference lists have been provided by Drs. L. May, W. Meisel, and T. Zemcik.

Helping us again with the technical aspects of the production of *MEDI* were Michael and Patricia Wyatt, who have been with us in Asheville almost since the beginning of our editorship. Joining us for this year was Vickie Cassada.

Finally, special thanks go to Drs. Lewis Gevantman and Stephen Rossmassler, who together have been our contact with the Office of Standard Reference Data of the National Bureau of Standards; the financial assistance of the latter is greatly appreciated.

JOHN G. STEVENS
VIRGINIA E. STEVENS

Contents

Instructions on the Use of the Mössbauer Data Index

Arrangement of the Index

Isotope Page Information

The Mössbauer data and references for each isotope are preceded by a summary sheet which includes the parent decay schemes and a simplified nuclear level scheme for the isotope, a tabulation of the Mössbauer parameters which have been derived from these properties. The experimental results listed are "selected values" from a critical review of published results. Further details are described on page 7.

Nuclear Transition

The heading on each page gives the Mössbauer isotope and the γ-ray transition energy.

Classifications of Data

Because of the large number of data in the iron and tin sections, these sections are subdivided by grouping the substances studied into several classifications (e.g., Biological Compounds, Inorganic Cyanides, Inorganic Halides, etc.).

Source

Column 1 gives the host material in which the source atoms are embedded. The "=" symbol is used to indicate a continuation of the chemical formula in the remarks column. If the authors do not indicate a source host, the symbol xx is used. When accelerator and related techniques are employed, the appropriate reaction is indicated in parentheses (i.e., n,g = n,γ reaction and CE = Coulomb excitation). See page 5 for the abbreviations used in this column.

Source Temperature

Column 2 gives the source temperature in kelvins. R indicates nominal room temperature, N nominal liquid nitrogen temperature, He nominal liquid helium temperature, and H nominal liquid hydrogen temperature. The symbol v is used to indicate a variable-temperature experiment.

Absorber

Column 3 gives the host material in which the absorber atoms are embedded. (See the paragraph on *Source* above for additional remarks.)

Absorber Temperature

Column 4 gives the absorber temperature. (See the paragraph on *Source Temperature* above for additional remarks.)

Isomer Shift

Column 5 gives the isomer (or chemical) shift of the observed spectrum in mm/s. The generally used convention is adopted that the source and absorber have relative motion toward each other for positive velocity. Since most papers do not explicitly state the adopted convention, the sign of the shift may be subject to question. Values are listed without sign if the authors do not indicate the sign. Since a number of authors do not explicitly use the + sign for positive velocity values, it should be remarked that most of the signless shift values are probably positive. If the author specifies some source but lists the shift relative to another source or to a reference absorber, then the given shift values are entered and a notation is made in the source column by listing the source and then listing, in parentheses, IS/the name of the reference material. For example, if the experimenter reports he is using a Cu source but reports his results relative to Fe, under *Source* one would see Cu(IS/Fe).

Quadrupole Splitting

For the case of a pure quadrupole interaction with a nuclear transition between spin 3/2 and

1/2 levels (e.g., Fe^{57}, Sn^{119}, Tm^{169}, etc.), the quadrupole splitting energy, QS, is listed in column 6 in mm/s, where appropriate. This quantity is the peak separation for a simple quadrupole split doublet. QS is usually equal to $\frac{1}{2}e^2qQ(1+\eta^2/3)^{\frac{1}{2}}$, where e is the electron charge, q the electric field gradient component along the symmetry axis, Q the nuclear quadrupole moment of the 3/2 state, and η the asymmetry parameter. For higher spin states or in the presence of a magnetic interaction, where the situation is more complex, the reporting of quadrupole interaction data is indicated by the symbol "$--$" in the column. If the quadrupole coupling constant, e^2qQ, has been obtained from the experiment, it is listed in the remarks column. If no units are given with the value listed in the remarks, the units are mm/s. Other units are given with the value.

Remarks

Column 7 gives additional remarks or experimental results. Considerable abbreviation is used here and the user should consult page 5 for these abbreviations. Proper punctuation and syllabization are often ignored because of the limited space. If all the information could not be fitted on one line under remarks, additional lines were used.

Reference Code

Column 8 gives a code to the bibliography. The first two numbers are the year in which the paper was published, the letter is the initial letter of the surname of the first author, and the last three numbers are an arbitrary but unique sequence number.

References

The references for each isotope appear after the data index for that isotope. Following the listing grouped by isotope, other references appear in the Topical Reference Lists, grouped under these topic code headings: ANALYS = analysis, APPLCN = application, GENRAL = general, INSTRUM = instrumentation, MISCL = miscellaneous, PROPSL = proposal, REVIEW = review, and THEORY = theory. The Addendum Reference List appears after the Topical Reference Lists. This includes articles either not located or else not available prior to the publication of *MEDI—1974*. The information from them is incorporated into the 1975 Data, Topical, and Author Lists. Following the Addendum List is the 1975 Master Reference List, with references listed by reference code in alphanumeric sequence. There is a certain amount of cross-indexing, and some references will therefore appear in more than one topic category. Abbreviations are adopted from Chemical Abstracts' *Source Index*. Titles of French and German papers are in the original language. Special symbols such as accent marks or the German umlaut are not rendered and the spelling is not transliterated (i.e., Mössbauer = MOSSBAUER). Titles of all other international papers are in English. For Soviet and other international journals for which regular translations appear, both the original and translation references are given. In some cases only the name of the journal of the English translation is given; the lack of volume and page numbers indicates the translation was not available at the time this volume went to press. Such translations may now be available; others are many months, even years, behind.

1975 Alphabetical Author Index

This section appears at the end of the 1975 Master Reference List. (Authors with papers in the Addendum List are included in this section as well.) Under the bold heading of each author's name appears an entry for each paper by that author (name in bold type), including co-authors (if any), topic, and reference code.

Abbreviations Used in the
1975 Mössbauer Effect Data Index

A-TEMP=absorber temperature
Ac=aceto
acac=acetylacetonate/acetylacetonato
ACN=acetonitrile
aly=alloy
alys=alloys
aph=alpha

BDM=butanediol dimethacrylate
BFD=bis(trifluoromethyl)propane-1,3-dionate
Bipy=bipyridine
BMT=bromobis(morphyldithiocarbamate)
BPM=1,2 bis(diphenylarsino)methane
bond=bonding
BPE=1,2bis(diphenylarsino)ethane
Bu=butyl
bz=benzyl
BZ=benzene
Bzac=benzoylacetonato
bzbz=dibenzoylmethanato
BZP=4-(2-benzothiazolinyl)2-pentanone

c-=cis
calc=calculation
CDT=cyclohexylenedinitrolotetraacetic acid
CE=Coulomb excitation
chem=chemical
CHP=cyclohexyldithiophosphinato
cmpds=compounds
comp=composition
conc=concentration
corln=correlation
Cp=cyclopentadienyl
crln=correlation
crys=crystal

das=phenylenebis(dimethylarsine
DCB=dicyanobenzene
DDQ=dichlorodicyanoquinone
DECB=N,N-diethyldithiocarbamato
decomp=decomposition
depe=1,2-bis(diethylphosphino)ethane
detn=determination
diars=o-phenylenebisdimethylarsine
diphos=C5H5Fe(CO)2SnMe3
dipy=dipyridyl
DMCB=dimethyldithiocarbamate
DMF=N,N-dimethylformamide
DMSO=dimethyl sulfoxide
dpe=Ph2PCH2CH2PPh2 (1,2-bisdi-
      phenylphioethane)
DPP=diphenylpropane-1,3-dionate
DPSO=diphenyl sulfoxide
dtc=dithiocarbamato/dithiocarbamate
DTH=2,5-dithiahexane
DTO=dithiooxalato

EAA=N,N'-ethylenebis(acetylacetone-
      imine)
EDTA=ethylenediaminetetraacetic acid/
      ethylenediaminetetraacetato
efg=electric field gradient
EFP=EtC(CH2O)3P
EHA=N,N'-ethylenebis(2-hydroxyaceto-
      phenone imine)
elec=electronic
en=ethylenediamine
EQ=e2qQ(quadrupole coupling constant)
Et=ethyl
eta=asymmetry parameter
ETU=ethylenethiourea
exp=experiment
expl=experimental

f=recoil free fraction
fa=recoil-free fraction of absorber
fs=recoil-free fraction of source

g1=nuclear g factor of the Mossbauer state
gam=gamma
GGL=glycylglycinato
GK=Gol'danskii-Karaygin
GMI=glyoxal-bis-N-methylimine

HA=external magnetic field
Hb=deoxyhemoglobin
He=liquid helium temperature(4.2 K)
HEDTA=hydroxyethylethylenediamine
      triacetate
hex=hexagonal
HFA=hexafluoroacetylacetonato
HPP=1,1,1,5,5,k-hexafluoropentane-2,4-dionato
HI=internal magnetic field
HL=half life
HMPT=hexamethylphosphorictriamine
HPA=3-(o-hydroxyphenylamino)crotonophenon
HPB=2-(o-hydroxyphenyl)benzothiazoline
HPP=N-(2-hydroxyphenyl)-2-pyridincar-
      boxaldimine
HPT=hexaphenyl-1,4,7,10-tetraphosphadecane
HQ=8-hydroxyquinoline
HSI=N(2-hydroxyphenyl)salicyladimine
HTA=hexamethlene triamine ammonium
Hx=hexyl

ident=identification
info=information
INT=intensity
IQL=isoquinoline
IS=isomer shift
Iz=imidazole

magn=magnetic/magnetization
Mb=myoglobin
MDT=meso-2,12-dimethyl-3,7,11,17-tetraaza-
      bicyclo{11.3.1}hepta-{17},13,15-triene
Me=methyl
MG=nuclear magnetic moment of the ground
      state
MLN=malononitrile
MM=nuclear magnetic moment of the
      Mossbauer state
nmc=membered macrocycles
MNT=maleonitrildithiolato
mono=monoclinic
MPZ=methyl pyrazinium

N=liquid nitrogen temperature
NAQ=beta-naphthoquinoline
NB=nitrobenzol
nm=nuclear magneton
NPY=1,8-naphthyridine
NTA=nitrilotriacetate

obsvn=observation
Oct=octyl
OTI=ortho-tolyl isocyanides
ox=oxinate/oxine
Ox=8-hydroxyquinoline

PAO=2-pyidyl aldehye oxime
PAP=pivalamidophenyl
PBD=1-phenylbutane-1,3-dionate
PBI=2-(2-pyridyl)benzimidazole
Pc=phthalocyanino
PCA=2-pyidylcarboxylic acid
pcs=partial chemical shifts
PDA=2,6-pyridyldicarboxylic acid
PDC=pyrrolidylcarbothioates
PDM=3-(2-(1,10-phenanthrolyl))-5,6-
      dimethyl-1,2,4-triazine
PDP=3-(2-(1,10-phenanthrolyl))-5,6-
      diphenyl-1,2,4-triazine
PFC=potassium ferrocyanide
Ph=phenyl
phen=1,10-phenanthroline
pic=picoline
PIC=picolylamine
pip=piperidine
PMT=2-pyridylmethanol
POPLTN=population
POR=porphyrin
pqs=partial quadrupole splittings
PPB=pheophorbide
PPY=pheophytin
Pr=propyl
PTC=piperidyldithiocarbamato
PTD=pentane-2,4-dionate
PTS=para-tolyl isocyanides
Py=pyridine
PYD=pyridone
pyz=pyrazine

QM=nuclear quadrupole moment of the
      Mossbauer state
QS=quadrupole splitting

R=room temperature
REF=reference
relax=relaxation
RG=ratio of nuclear g factors
rhom=rhombohedral
RM=ratio of nuclear magnetic
      moments(excited to ground)
RQ=ratio of nuclear quadrupole
      moments(excited to ground)
rxn=reaction

S-TEMP=source temperature(Kelvin)
Sal H=salicyladimine
Salen=N,N'-ethylenebis(salicylidereimine)
SCATT=scattering
SFC=sodium ferrocyanide
SHIFT=isomer shift
SNP=sodium nitroprusside
SNT=succinonitrile
SS=stainless steel
struc=structure

t-=trans
T=tesla
T=temperature
TBD=trifluoromethylbutane-2,4-dionate
TC=Curie temperature
TCA=trichloroacetate
TCAA=trichloroacetetic acid
TCB=tetracyanobenzene
TCE=tetracyanoethylene
temp=temperature
TFA=trifuoroacetylacetonato
TFD=1,1,1-trifluoropentane-2,4-dionate
TFP=1,1,1-trifluoro-4-phenylbutane-2,4-dione
THF=tetrahydrofuran
thsa=thiosemicarbozone of salicyl-
      aldehyde
THT=tetrahydrotiophene
THU=thiourea
TM=Mossbauer temperature
TME=tetramethylethylenediamine
TMI=trimethylisopropoxysilane
TMSO=tetramethylene sulfoxide
TMTU=tetramethylthiourea
To=tolyl
TOA=tri-n-octylamine
TPD=thiopyridone
TPI=tetra(-o-pivalamidophenyl)porphyrin
TPL=tropolonate
TPP=tetraphenylporphinato
trans=transition
trew=2,2',2''-triaminotriethylamine
TTF=tetrathiafulvalenium
TU=thiourea

v=variable
Vin=vinyl

W=width
w/=with

Isotope Page Information

The following information is presented on each Isotope Page: the (simplified) decay scheme (the Mössbauer transitions are shown with broader lines); the γ-ray energies [E_γ] and relative intensities [I_{parent}] (where available); the half-life [$t_{1/2}$] and the total internal conversion coefficient [α_T] of the Mössbauer transition; the natural isotopic abundance [IA] of the isotope; the magnetic moments [μ (ground state dipole moment), μ^* (Mössbauer state dipole moment)] and quadrupole moments [Q (ground state quadrupole moment), Q^* (Mössbauer state quadrupole moment)] of the levels between which the transition occurs; the ratio of nuclear moments [R_μ (dipole moment ratio of Mössbauer state to ground state), R_Q (quadrupole moment ratio of Mössbauer state to ground state)] ; and special references and notes, if any. R_μ and R_Q values are given in preference to μ^* and Q^* values. All μ values have been corrected for diamagnetic shielding, using Kopfermann's calculations.[1] If the "Mössbauer level" is depopulated by more than one transition, the fraction of decays branching through the Mössbauer transition is indicated [ρ] in percent (e.g., see ^{161}Dy). When experimental values for the total internal conversion coefficient are not available, theoretical values are determined from a computer program of Hager and Seltzer.[2] When more than one measured value is considered a weighted mean is used.

The listed Mössbauer parameters and energy conversions were computer calculated from the adopted measured parameters using the equations given in *Mössbauer Effect Data Index 1958–1965*.[3] Their uncertainties are propagated from the measured values by standard statistical procedures. When more than one value is available the reported number is the weighted mean. Values used for the physical constants are as follows:

$$h = 6.626196(30) \times 10^{-34} \text{ J·s,}[4]$$
$$c = 2.997924562(11) \times 10^8 \text{ m/s,}[5]$$
$$e = 1.6021917(70) \times 10^{-19} \text{ C.}[4]$$

The energies of the K_{α_1} x-rays were obtained from Table 12 in the *Table of Isotopes*.[6] Also obtained from this reference were some of the simplified energy level diagrams and decay schemes. Extensive use was made of the Nuclear Data Sheets, compiled by the Nuclear Data Group at Oak Ridge National Laboratory (Academic Press, New York). Reference code numbers are used for any reference in any volume of the *Mössbauer Effect Data Index*.

[1] H. Kopfermann, *Nuclear Moments* (translated by E.E. Schneider) (Academic Press, New York, 1958).

[2] R.S. Hager and E.C. Seltzer, *Nuclear Data Tables* A4, 1 (1968).

[3] A.H. Muir, Jr., K.J. Ando, and H.M. Coogan, *Mössbauer Effect Data Index 1958–1965* (Interscience Publishers, New York, 1966), p. xvi.

[4] B.N. Taylor, W.H. Parker, and D.N. Langenberg, *Rev. Mod. Phys.* 41, 375 (1969).

[5] K.M. Evenson, J.S. Wells, F.R. Peterson, B.L. Danielson, G.W. Day, R.L. Barger, and J.L. Hall, *Phys. Rev. Lett.* 29, 1346 (1972).

[6] C.M. Lederer, J.M. Hollander, and I. Perlman, *Table of Isotopes*, Sixth Edition (John Wiley & Sons, New York, 1967), p. 570.

Equipment, Sources, and Supplies for Mössbauer Spectroscopy

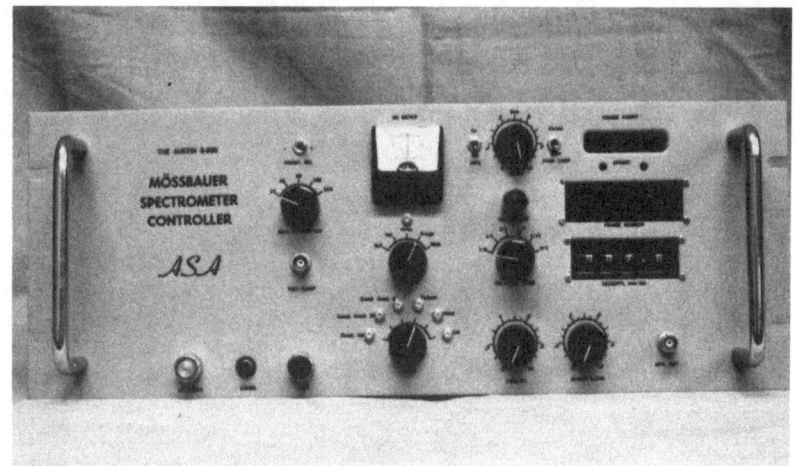

THE STANDARD OF EXCELLENCE!

S-600 Multi-Option Controller
offers all standard options,
plus laser velocity control
to maintain velocity span
accurately. Velocity to
600 mm/s. Fast electronics
for faster data collection.
Rugged K3 linear motor.

NEW! S-700 LOW COST SPECTROMETER.

This low-cost package offers constant velocity
to 10 mm/s, constant acceleration to 100 mm/s*
full data acquisition electronics, and a scaler
for point-by-point plotting. In 3 dual-width
NIM bins. Write for prices and specifications.
We don't know of any commercial spectrometer
for less money, and this one performs!

MICROPROCESSORS?   We're fascinated! Multi-channel scaling is a snap for these
small wonders. And, so cheap! If you've strong opinions on what you'd like to see
in a microprocessor-based multi-channel scaler, let us hear it.

WE ALSO PROVIDE: Mössbauer drive and motor kits - - for the adventurous.
Cryogenics, source and absorber to 4.2 K.
Vacuum furnace, absorber only to 1000 K.
Back Scatter detector, dual, for electrons or X-rays.
Mössbauer sources.
Consultation, Services, and an R & D capability to put
Mössbauer techniques to work on your special problem.
We'd like to send you more information! Just tell us where!

Write or call:
AUSTIN SCIENCE ASSOCIATES INC.
5902 WEST BEE CAVES ROAD
AUSTIN, TEXAS 78746
PHONE: 512 327-1297

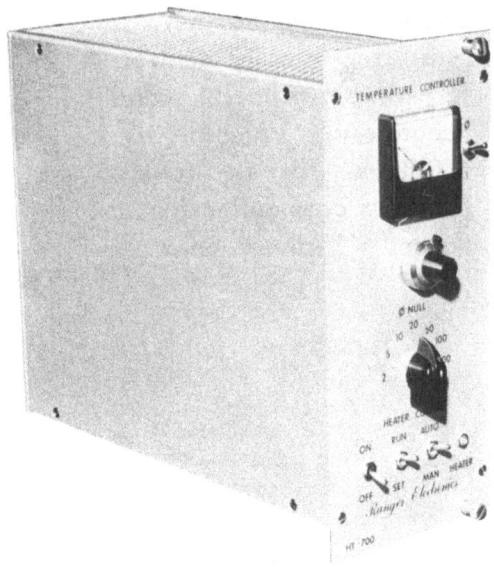
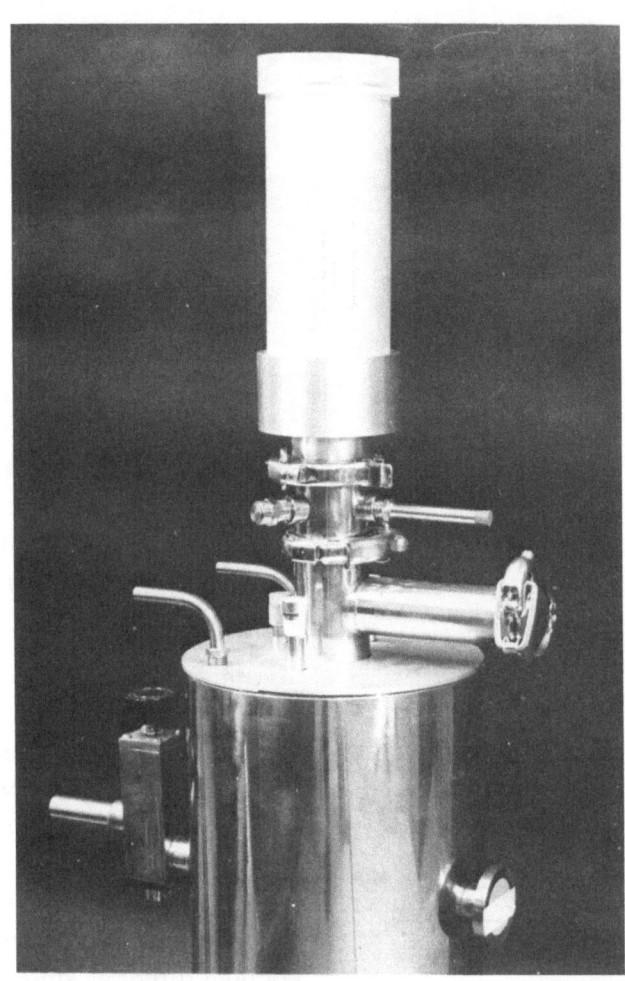

Mössbauer research
Corporation instrumentation

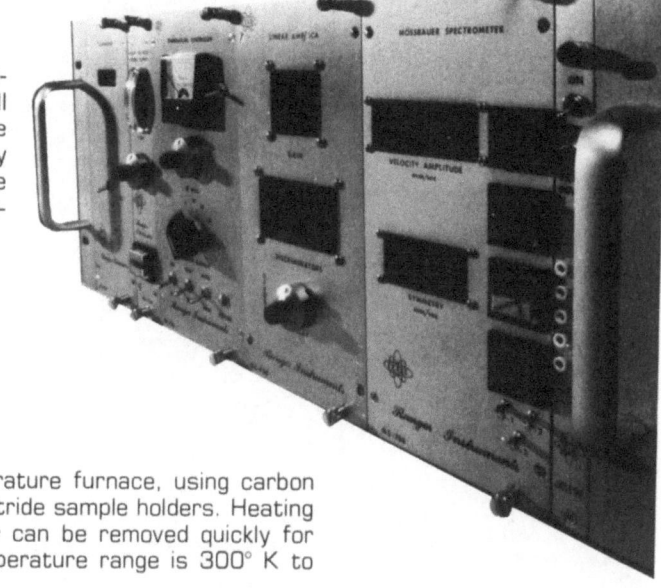

Electronic Control Unit: ▶

This fourth generation Mössbauer Spectrometer introduces a new level of flexibility and performance over the full velocity range of (0-600 mm/sec) for all available Mössbauer isotopes, and combines several previously separate functions in a single unit. Multiplexing is available to obtain simultaneously a Mössbauer spectrum and absolute velocity calibration using a Moire interferometer.

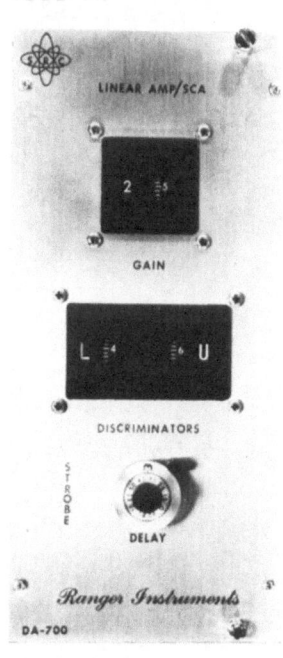

◀ **High Vacuum Furnace:**

A miniaturized high temperature furnace, using carbon disk heaters, and Boron Nitride sample holders. Heating element and sample holder can be removed quickly for sample changing. The temperature range is 300° K to 1000° K.

Conversion Electron Detector (with electron screen) ▼

The Conversion Electron Detector has been designed specifically for Backscattering (reflection) Mössbauer Spectroscopy. The conversion Detectors count the 7.3 and 5.6 keV electrons generated by the internal conversion process after the Mössbauer gamma ray is resonantly absorbed. Since the detector is not sensitive to gamma rays, the signal to noise is very good. SD-100, SD-200 or SD-300 with built-in preamp.

High Speed Data Acquisition System: ▲

The High Speed Data Acquisition System features a very high speed linear amplifier and single channel analyzer. The DA-700 makes it possible to use very high activity Mössbauer sources, and reduce the data accumulation time for a Mössbauer Spectrum by a factor of five.

For brochure and price information write
Ranger Engineering Corporation,
3132 Bryan, Fort Worth, Texas 76110
or phone (817) 921-5176

technology in our products..

Ranger Engineering Corporation

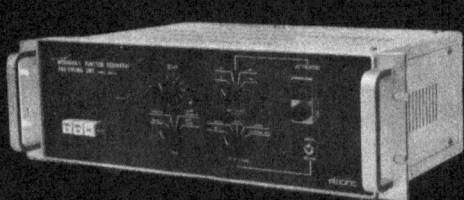
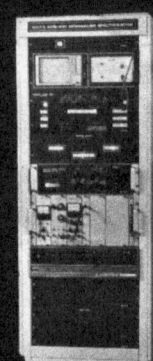

Teach Research

The Spectrometer

The Ealing Mössbauer Spectrometer is a constant-velocity, manually operated system designed to give the teaching laboratory a quality instrument at a manageable price.

You can select a complete system or buy components as required. Choose from the Basic Mössbauer Spectrometer consisting of a drive amplifier, transducer and a mounting bench, on up to a complete system with an amplifier-pulse analyzer, proportional counter, ^{57}Co source, sample absorbers, and digital counters.

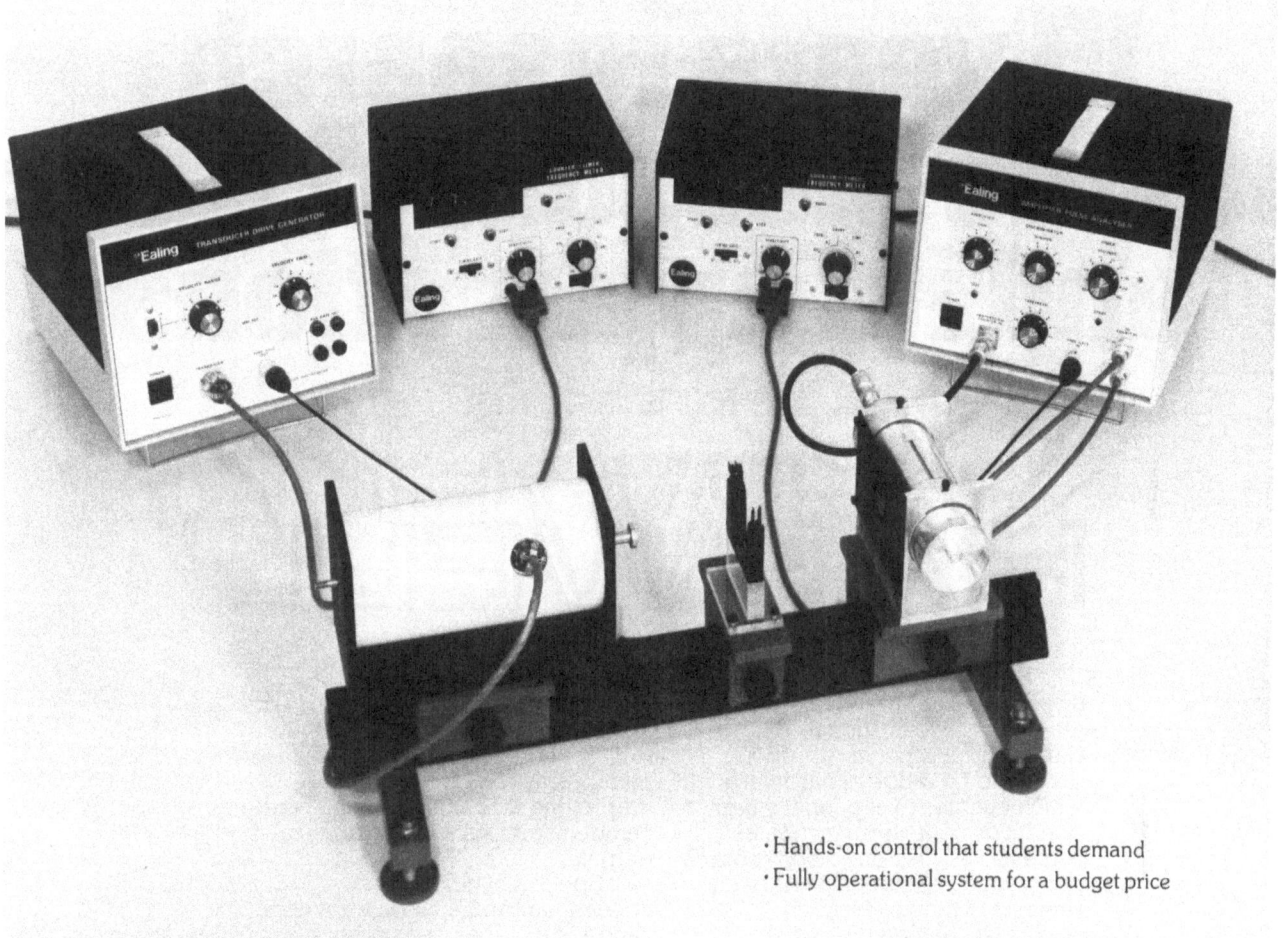

· Hands-on control that students demand
· Fully operational system for a budget price

The Components

The drive amplifier techniques are well-proven and optimized for the transducer. Velocity is highly linear and repeatable.

For accessibility and positioning flexibility, the mechanical system is mounted on an Ealing Triangular Optical Bench. This mounting technique permits easy adjustment of source and detector spacing for optimum signal strength. Yet it is simple and rigid for repeatable, vibration-free performance.

The Amplifier-Pulse Analyzer is an engineering feat for something less than rack-mount size. With gain, threshold and window

controls, an adjustable high voltage supply, and a gating timer for repeatable sample periods, it's truly a lot of technology and performance for the price.

The Company

Ealing is the leading supplier of quality physics teaching, optics and life science apparatus. Over 3,500 products are listed in the new 864-page, 1976–1977 Ealing Catalog.

A complete description of the Ealing Mössbauer System is included in this catalog. Call or write for your free copy.

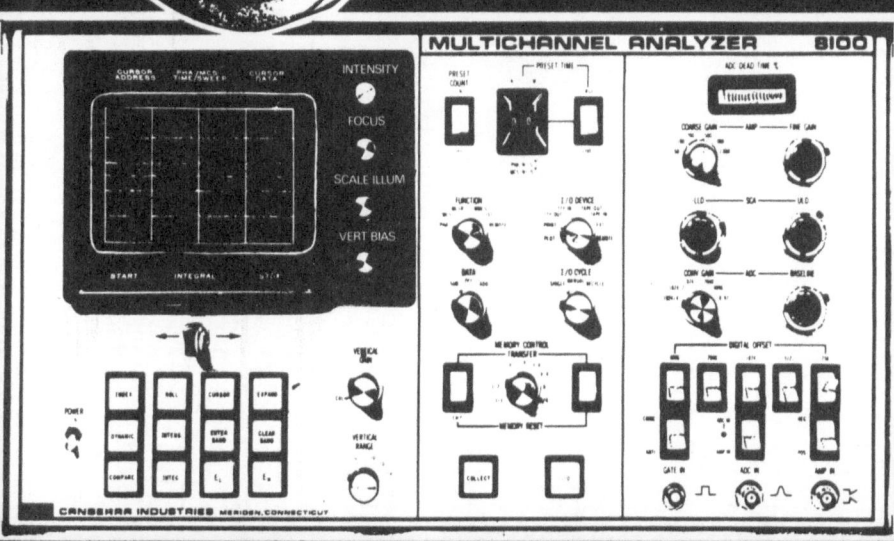

Tables for
Mössbauer Spectroscopy

Condensed Table for Mössbauer Transition Properties

ISOTOPE		ISOTOPE ABUNDANCE (%)	GAMMA ENERGY (KEV)	HALF LIFE (NS)	SPIN OF EXCITED LEVEL	SPIN OF GROUND LEVEL	MOSSBAUER CROSS-SECTION 10(-20)(CM2)	MOSSBAUER LINEWIDTH (MM/SEC)	RECOIL ENERGY 10(-3)(EV)
POTASSIUM	40	0.012	29.5600	4.2500	3.00	4.00	28.655	2.17747	11.7259
IRON	57	2.140	14.4130	97.8100	1.50	0.50	255.754	0.19405	1.9563
IRON	57	2.140	136.4785	8.7000	2.50	0.50	34.505	0.23039	175.4081
NICKEL	61	1.190	67.4080	5.2700	2.50	1.50	71.160	0.77006	39.9843
ZINC	67	4.110	93.3170	9150.0000	0.50	2.50	4.955	0.00032	69.7658
GERMANIUM	73	7.760	13.2630	2953.0000	2.50	4.50	0.761	0.00698	1.2935
GERMANIUM	73	7.760	68.7520	1.8600	3.50	4.50	22.877	2.13918	34.7571
KRYPTON	83	11.550	9.4000	147.0000	3.50	4.50	107.530	0.19797	0.5714
TECHNETIUM	99	0.000	140.5110	0.2370	3.50	4.50	8.621	8.21462	107.0487
RUTHENIUM	99	12.720	89.3600	20.5000	1.50	2.50	8.042	0.14933	43.2959
RUTHENIUM	101	17.070	127.2200	0.5810	1.50	2.50	8.688	3.70096	86.0172
TIN	117	7.610	158.5300	0.2770	1.50	0.50	16.799	6.22952	115.3010
TIN	119	8.580	23.8710	17.7500	3.50	0.50	140.313	0.64562	2.5703
ANTIMONY	121	57.250	37.1500	3.5000	3.50	2.50	19.534	2.10387	6.1225
TELLURIUM	125	6.990	35.4600	1.4810	1.50	0.50	26.563	5.20898	5.3996
IODINE	127	100.000	57.6000	1.9100	3.50	2.50	20.570	2.48651	14.0229
IODINE	129	0.000	27.7700	16.8000	2.50	3.50	39.007	0.58636	3.2089
XENON	129	26.440	39.5800	1.0100	1.50	0.50	23.485	6.84304	6.5187
XENON	131	21.180	80.1830	0.5000	0.50	1.50	7.403	6.82329	26.3445
CESIUM	133	100.000	80.9970	6.3130	2.50	3.50	10.283	0.53498	26.4778
BARIUM	133	0.000	12.2900	8.1000	1.50	0.50	29.185	2.74795	0.6096
LANTHANUM	139	99.911	165.8500	1.5000	2.50	3.50	5.282	1.09961	106.2214
PRAESEODYMIU	141	100.000	145.4200	1.8500	3.50	2.50	10.566	1.01683	80.5054
NEODYMIUM	145	8.300	67.2500	29.4000	1.50	3.50	3.810	0.13836	16.7422
NEODYMIUM	145	8.300	72.5000	0.7200	2.50	3.50	5.917	5.24053	19.4583
PROMETHIU?	145	0.000	61.2500	2.6200	3.50	2.50	11.719	1.70466	13.8880
PROMETHIUM	147	0.000	91.0300	2.5700	2.50	3.50	6.920	1.16931	30.2585
SAMARIUM	147	14.970	122.1000	0.8000	2.50	3.50	6.154	2.80053	54.4390
SAMARIUM	149	13.830	22.4940	7.1200	2.50	3.50	7.111	1.70805	1.8228
SAMARIUM	151	0.000	65.8300	20.0000	9.90	9.90	8.352	0.20777	15.4052
SAMARIUM	152	26.720	121.7800	1.4100	2.00	0.00	38.012	1.59313	52.3726
SAMARIUM	153	0.000	35.8420	2.0000	1.50	1.50	146.499	3.81614	4.5070
SAMARIUM	154	22.710	81.9900	3.0000	2.00	0.00	30.078	1.11215	23.4313
EUROPIUM	151	47.820	21.5320	9.7000	3.50	2.50	23.771	1.30975	1.6481

23

Condensed Table for Mossbauer Transition Properties

ISOTOPE		ISOTOPE ABUNDANCE (%)	GAMMA ENERGY (KEV)	HALF LIFE (NS)	SPIN OF EXCITED LEVEL	SPIN OF GROUND LEVEL	MOSSBAUER CROSS-SECTION 10(-20)(CM2)	MOSSBAUER LINEWIDTH (MM/SEC)	RECOIL ENERGY 10(-3)(EV)
EUROPIUM	153	52.180	83.3652	0.8200	3.50	2.50	9.738	4.00173	24.3823
EUROPIUM	153	52.180	97.4283	0.2100	2.50	2.50	17.970	13.37032	33.3023
EUROPIUM	153	52.180	103.1774	3.9000	1.50	2.50	5.456	0.67982	37.3486
GADOLINIUM	154	2.150	123.0700	1.1700	2.00	0.00	36.712	1.89980	52.7934
GADOLINIUM	155	14.730	60.0100	0.1340	2.50	1.50	10.484	34.01870	12.4713
GADOLINIUM	155	14.730	86.5452	6.3300	2.50	1.50	33.921	0.49934	25.9388
GADOLINIUM	155	14.730	105.3080	1.1680	1.50	1.50	16.459	2.22404	38.4049
GADOLINIUM	156	20.470	88.9656	2.1700	2.00	0.00	31.350	1.41698	27.2343
GADOLINIUM	157	15.680	54.5400	0.1870	2.50	1.50	9.586	26.82189	10.1701
GADOLINIUM	157	15.680	64.0000	460.0000	2.50	1.50	23.195	0.00929	14.0041
GADOLINIUM	158	24.870	79.5100	2.5400	2.00	0.00	27.565	1.35454	21.4774
GADOLINIUM	160	21.900	75.2600	2.7000	2.00	0.00	25.650	1.34623	19.0022
TERBIUM	159	100.000	57.9950	0.1050	2.50	1.50	10.532	44.92274	11.3548
DYSPROSIUM	160	2.290	86.7880	2.0370	2.00	0.00	29.422	1.54737	25.2694
DYSPROSIUM	161	18.880	25.6550	28.2000	2.50	2.50	95.313	0.37812	2.1944
DYSPROSIUM	161	18.880	43.8300	0.7800	3.50	2.50	31.919	0.00166	6.4049
DYSPROSIUM	161	18.880	74.5770	3.3100	1.50	2.50	6.754	0.10819	18.5430
DYSPROSIUM	162	25.530	80.6500	2.2700	2.00	0.00	121.336	1.49422	21.5521
DYSPROSIUM	164	28.180	73.3920	2.3900	2.00	0.00	22.049	1.55955	17.6299
HOLMIUM	165	100.000	94.6990	0.0222	4.50	3.50	8.277	130.12110	29.1745
ERBIUM	164	1.560	91.3900	1.4700	2.00	0.00	28.275	2.03625	27.3369
ERBIUM	166	33.410	80.5570	1.8700	2.00	0.00	23.771	1.81594	20.9843
ERBIUM	167	22.940	79.3219	0.1190	4.50	3.50	7.211	28.98051	20.2239
ERBIUM	168	27.070	79.7998	1.8800	2.00	0.00	23.600	1.82342	20.3465
ERBIUM	170	14.880	79.3100	1.9000	2.00	0.00	23.347	1.81537	19.8611
THULIUM	169	100.000	8.4010	4.0000	1.50	0.50	25.774	8.14058	0.2242
YTTERBIUM	170	3.030	84.2529	1.6080	2.00	0.00	19.042	2.01918	22.4139
YTTERBIUM	171	14.310	66.7190	0.8700	1.50	0.50	7.852	4.71278	13.9733
YTTERBIUM	171	14.310	75.8750	1.6400	2.50	0.50	14.439	2.19838	18.0716
YTTERBIUM	172	21.820	78.6900	1.8000	2.00	0.00	20.795	1.93132	19.3244
YTTERBIUM	174	31.840	76.4692	1.7600	2.00	0.00	20.096	2.03258	18.0393
YTTERBIUM	176	12.730	82.1300	2.0000	2.00	0.00	22.473	1.66538	20.5725
LUTETIUM	175	97.410	113.8030	0.1000	4.50	3.50	6.728	24.03767	39.7251
HAFNIUM	176	5.200	88.3610	1.3900	2.00	0.00	22.839	2.22726	23.8125

Condensed Table for Mossbauer Transition Properties

ISOTOPE		ISOTOPE ABUNDANCE (%)	GAMMA ENERGY (KEV)	HALF LIFE (NS)	SPIN OF EXCITED LEVEL	SPIN OF GROUND LEVEL	MOSSBAUER CROSS-SECTION 10(-20)(CM2)	MOSSBAUER LINEWIDTH (MM/SEC)	RECOIL ENERGY 10(-3)(EV)
HAFNIUM	177	18.500	112.9700	0.5000	4.50	3.50	5.991	4.84298	38.7034
HAFNIUM	178	27.140	93.1740	1.4950	2.00	0.00	25.162	1.96386	26.1798
HAFNIUM	180	35.240	93.3320	1.5000	2.00	0.00	24.594	1.95400	25.9767
TANTALUM	181	99.988	6.2380	6800.0000	4.50	3.50	167.218	0.00645	0.1154
TANTALUM	181	99.988	136.2500	0.0400	4.50	3.50	5.969	50.19374	55.0542
TUNGSTEN	180	0.140	103.7000	1.2700	2.00	0.00	25.621	2.07713	32.0687
TUNGSTEN	182	26.410	100.1040	1.3100	2.00	0.00	25.170	2.08604	29.5548
TUNGSTEN	183	14.400	46.4837	0.1840	1.50	0.50	5.523	31.98362	6.3379
TUNGSTEN	183	14.400	99.0788	0.6880	2.50	0.50	8.178	4.01307	28.7943
TUNGSTEN	184	30.640	111.2130	1.2800	2.00	0.00	27.398	1.92168	36.0819
TUNGSTEN	186	28.410	122.3000	1.0100	2.00	0.00	31.456	2.21462	43.1654
RHENIUM	187	62.930	134.2400	0.0100	3.50	2.50	5.404	203.78121	51.7271
OSMIUM	186	1.590	137.1570	0.8400	2.00	0.00	28.396	2.37437	54.2899
OSMIUM	188	13.300	155.0320	0.6950	2.00	0.00	27.965	2.53887	68.6248
OSMIUM	189	16.100	36.2200	0.5000	0.50	1.50	1.151	15.10524	3.7259
OSMIUM	189	16.100	69.5900	1.6400	2.50	1.50	8.420	2.39693	13.7540
OSMIUM	189	16.100	95.2300	0.2300	1.50	1.50	0.561	12.48948	25.7562
OSMIUM	190	26.400	186.9000	0.4700	2.00	0.00	24.662	3.11415	98.6872
IRIDIUM	191	37.300	82.3980	4.0200	0.50	1.50	1.540	0.82585	19.0808
IRIDIUM	191	37.300	129.4000	0.0890	2.50	1.50	5.649	23.75318	47.0578
IRIDIUM	193	62.700	73.0390	6.3000	0.50	1.50	3.057	0.59450	14.8371
IRIDIUM	193	62.700	138.9470	0.0800	2.50	1.50	5.831	24.60973	53.6954
PLATINUM	195	33.800	98.8570	0.1700	1.50	0.50	6.106	16.27758	26.9015
PLATINUM	195	33.800	129.7350	0.6200	2.50	0.50	7.426	3.40093	46.3314
GOLD	197	100.000	77.3450	1.8790	0.50	1.50	3.858	1.88229	16.3003
MERCURY	201	13.220	32.1900	0.2000	0.50	0.50	0.948	42.49082	2.7672
THORIUM	232	0.000	49.3690	0.3450	2.00	0.00	1.568	16.06100	5.6392
PROTOACTINIU	231	0.000	84.2400	41.0000	1.50	1.50	4.751	0.07920	16.4900
URANIUM	234	0.060	43.4910	0.2660	2.00	0.00	0.828	23.64640	4.3389
URANIUM	236	0.000	45.2420	0.2350	2.00	0.00	0.706	25.72980	4.6555
URANIUM	238	99.270	44.9150	0.2250	2.00	0.00	0.917	27.06900	4.5499
NEPTUNIUM	237	0.000	59.5370	68.3000	2.50	2.50	30.604	0.06727	8.0283
PLUTONIUM	239	0.000	57.2600	0.1010	2.50	0.50	0.770	47.30133	7.3638
AMERICIUM	243	0.000	84.0000	2.3400	2.50	2.50	25.872	1.39172	15.5865

Isotope	E_γ (keV)	$\dfrac{\Delta \langle r^2 \rangle}{\langle r^2 \rangle}$ (10^{-4})	$\Delta \langle r^2 \rangle$ $(10^{-3}\,fm^2)$	α^{+} $(a_0^3\,mm/s)$	Reference
^{57}Fe	14.41			0.35	75D046
		-11.6			75D047
				-0.19±0.01	75M005
				-0.242 ± 0.039	75T011
		-17.6 ± 1.8			75T012
^{119}Sn	23.87	3.5 ± 0.7			75D047
		3.2 ± 0.4			75R010
		3.5 ± 0.6			74R040
^{125}Te	35.46		3.1 ± 0.4		75M048
		1.3	2.0 ± 0.4		75C008
		2.3	3.4 ± 0.7		75C008
^{141}Pr	145.42	149.			75C008
^{145}Nd	72.50	21.9			75C008
^{149}Sm	22.49	13.0			75C008
^{151}Eu	21.53	206.			75C008
^{155}Gd	88.97	88			75C008
^{161}Dy	25.66	95			75C008
		44			75C008
			5.6		75B092
^{170}Yb	84.25	12			75C008
^{182}W	103.65	-0.16 ± 0.12			75B104

+ Isomer shift calibration constant

Isotope	Energy Level (keV)	I	Magnetic Moment (nuclear magnetons)	Quadrupole Moment $(10^{-28} m^2)$	Reference
^{57}Fe	14.41	$\frac{1}{2}$	$g_0 = 3.919\,(4)$ mm/s		75W004
		$\frac{3}{2}$	$g_1 = 2.238\,(2)$ mm/s		75W004
^{99}Ru	0	$\frac{5}{2}$		> 0	75D068
	89.36	$\frac{3}{2}$		> 0	75D068
^{170}Yb	84.25	2		$-1.61\,(12)$	75B018
			0.668 (4)		75B038

Nuclear Moment Ratio (Excited to Ground) Data

Isotope	E_γ (keV)	I_{ex}	I_{gd}	Ratio of Nuclear Magnetic Moments	Ratio of Nuclear Quadrupole Moments	Reference
^{83}Kr	9.40	$\frac{7}{2}$	$\frac{9}{2}$		1.99 (5)	75K026
^{99}Ru	89.36	$\frac{3}{2}$	$\frac{5}{2}$		$+2.94\,(8)$	75D068
					$+3.06\,(10)$	75D068
^{125}Te	35.46	$\frac{3}{2}$	$\frac{1}{2}$	$-0.681\,(4)$		75B061
^{155}Gd	86.54	$\frac{5}{2}$	$\frac{3}{2}$		$-0.090\,(29)$	75K012
^{180,182}W				$\dfrac{Q(^{180}W(103.65\,keV))}{Q(^{182}W(100.10\,keV))} = 0.976\,(30)$		75B104
^{197}Au	77.34	$\frac{1}{2}$	$\frac{3}{2}$	2.91 (2)		75P021
				2.87 (3)		75P021

Mössbauer Periodic Table

1975 Data Index

$^{197}_{79}$Au (77.4 keV)

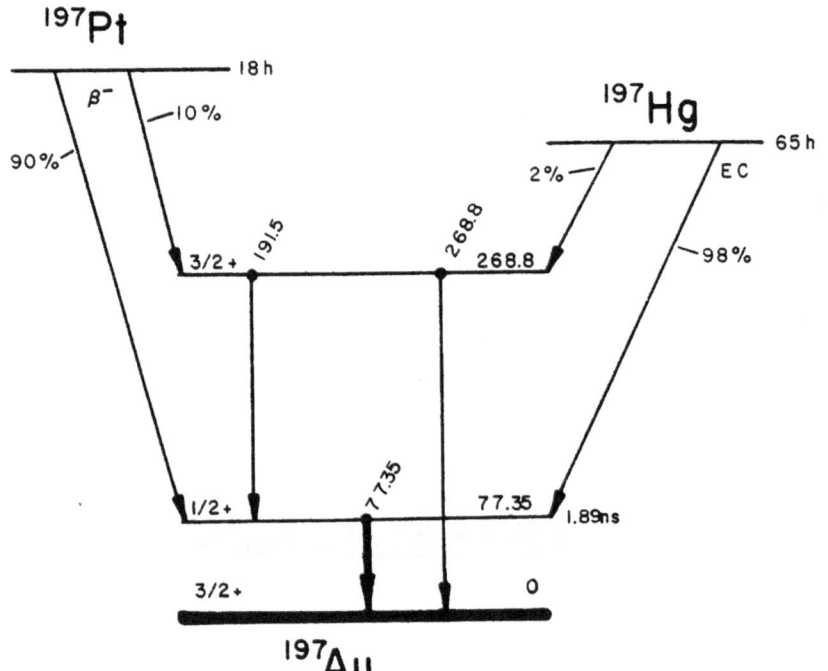

Measured Properties

E_γ = 77.345 ± 0.008 keV (1)

$t_{\frac{1}{2}}$ = 1.879 ± 0.010 ns (2),(3),(4),(5)

α_T = 4.30 ± 0.07 (4)

IA = 100%

μ = +0.1448 ± 0.0007 nm (6)

R_μ = 2.875 ± 0.022 (7),(8)

Q = +0.594 ± 0.010 b (9)

Derived Parameters

σ_o = 3.86 (5) × 10^{-20} cm^2

Γ = 2.428 (13) × 10^{-7} eV

W_o = 1.882 (10) mm/s

E_r = 1.6300 (2) × 10^{-2} eV

Energy Conversions

1 mm/s = 62.382 (6) MHz

1 mm/s = 2.5800 (3) × 10^{-7} eV

Energies and Intensities

	E_γ	I_{Pt}	I_{Hg}
γ_M :	77.35 keV	330	34.
γ_1 :	191.5 keV	100	1.
γ_2 :	269. keV	7	0.06
K_{α_1} :	68.8 keV	70	0.8

(1) I. Marklund and B. Linström, Nucl. Phys. 40, 329 (1963).
(2) 69S15.
(3) D.K. Gupta and G.N. Rao, Nucl. Phys. A182, 669 (1972).
(4) 71E003.
(5) F.J. Lynch, Phys. Rev. C7, 2160 (1973).
(6) H. Dahman and S. Panselin, Z. Phys. 200, 456 (1967).
(7) 68C019.
(8) 70P025.
(9) A.G. Blachman, D.A. Landman, and A. Lurio, Phys. Rev. 161, 60 (1967).

[Other data from M.B. Lewis, Nucl. Data Sheets B7, 129 (1972)]

SOURCE	S-TEMP	ABSORBER	A-TEMP	SHIFT	QS	REMARKS	REF
Pt		AgAu(CN)4	4.2	--	--	electron irradiation	75B088
Pt		AgAu(CN)4	4.2	4.20	7.50		75B088
Pt	4.2	Al2Au	v	+5.967		Debye T=192 K, fa=.219(T=4.2K)	75T002
Fe(Pt)	4.2	Au	4.2			detn of E2/M1 mixing ratio	75P021
Pt		Au	4.2	-1.13			75B088
Pt		Au	4.2				75K011
xx		Au	v			4.2<T<60 K, fa vs particle size	75V022
Pt	4.2	Au	v	-1.223			75T002
Pt		AuEr	4.2	+7.10		study of phase transitions	75K011
Pt		AuEr	4.2	+7.48	4.93	study of phase transitions	75K011
Pt	4.2	AuGa2	v	+4.450		Debye T=140 K, fa=.134(T=4.2K)	75T002
Pt		AuHo	4.2	+7.05		study of phase transitions	75K011
Pt		AuHo	4.2	+7.46	4.87	study of phase transitions	75K011
Pt	4.2	AuIn2	v	+3.489		Debye T=152 K, fa=.157(T=4.2K)	75T002
Pt		AuLu	4.2	+6.66		study of phase transitions	75K011
Pt	4.2	AuSb2	v	+2.288	1.5	Debye T=152 K, fa=.151(T=4.2K)	75T002
Pt		AuSc	4.2	+6.21		study of phase transitions	75K011
Pt		AuTm	4.2	+6.51		study of phase transitions	75K011
Pt		AuTm	4.2	+7.58	4.36	study of phase transitions	75K011
Pt		AuYb	4.2	+7.35		study of phase transitions	75K011
xx		(Bu4N)2Au(MNT)2	v			T=1.6 & 4.2 K, susceptibility study	75V035
Pt		KAu(CN)2			--	efg<0, single crystal, Gol'danskii-Karyagin effect	75P021
Pt		KAu(CN)2	4.2	3.07	9.80		75B088
Pt		KAu(CN)4-H2O	4.2	--	--	electron irradiation	75B088
Pt		KAu(CN)4-H2O	4.2	2.97	9.60	sample heated to 200 C	75B088
Pt		KAu(CN)4-H2O	4.2	4.75	6.91		75B088
xx		(bz)2SAuBr2				fa data	75V028

CODE	TOPIC	REFERENCE

75B088   AU-197   E Baggio-Saitovitch, U Wagner, F E Wagner, and J Danon, Mossbauer Studies of the Reduction of Complex Gold Cyanides by Electron Irradiation, in "Int Conf Mossbauer Spectrosc, Proc," Vol 1, pp 223-4 (See 75H027)

75K011   AU-197   C W Kimball, A E Dwight, G M Kalvius, B D Dunlap, and M V Nevitt, Phys Rev B 12,819-23(1975), Low-temperature Phase Transition and Isomer-shift Systematics in Intermediate Phases of Rare-earth-gold Compounds

75P021   AU-197   H Prosser, F E Wagner, G Wortmann, G M Kalvius, and R Wappling, Hyperfine Interac 1,25-32(1975), Mossbauer Determination of the E2/M1 Mixing Ratio of the 77 keV Transition in 197Au and of the Sign of the Electric Field Gradient in KAu(CN)2

75R004   AU-197   L D Roberts, The Mossbauer Effect for 197Au, in "Mossbauer Effect Data Index, Covering the 1973 Literature," edited by J G Stevens and V E Stevens (Plenum Publishing Corp, New York, 1975), pp 349-91

75T002   AU-197   J O Thomson, F E Obenshain, P G Huray, J C Love, and J Burton, Phys Rev B 11,1835-9(1975), Mossbauer Measurements with 197Au in AuCl2, AuGa2, AuIn2, and AuSb2

75V022   AU-197   M P A Viegers, J C H Van Eijkeren, M M Van Deventer, and J M Trooster, Mossbauer Fraction of Gold Microcrystals, in "Int Conf Mossbauer Spectrosc, Proc," Vol 1, p 201 (See 75H027)

75V028   AU-197   M P A Viegers, A W P G Peters Rit, and J M Trooster, Mossbauer Fraction of a Au(I)-Au(III) Complex, in "Int Conf Mossbauer Spectrosc, Proc," Vol 1, pp 287-8 (See 75H027)

75V035   AU-197   H Van Kempen, J A A J Pereboom, and M P A Viegers, Suscep-
                  tibility of the Magnetic Gold Complex (Au(II)(mnt)2)
                  Between 9 mK and 100 K, in "14th International Conference
                  on Low Temperature Physics, Proceedings: Volume 4. Tech-
                  niques and Special Topics" (Otaniemi, Finland, 1975),
                  edited by M Krusius and M Vuorio (North-Holland Publishing
                  Co, Amsterdam/American Elsevier Publishing Co, New York,
                  1975), pp 372-5

$^{133}_{55}$Cs (81.0 keV)

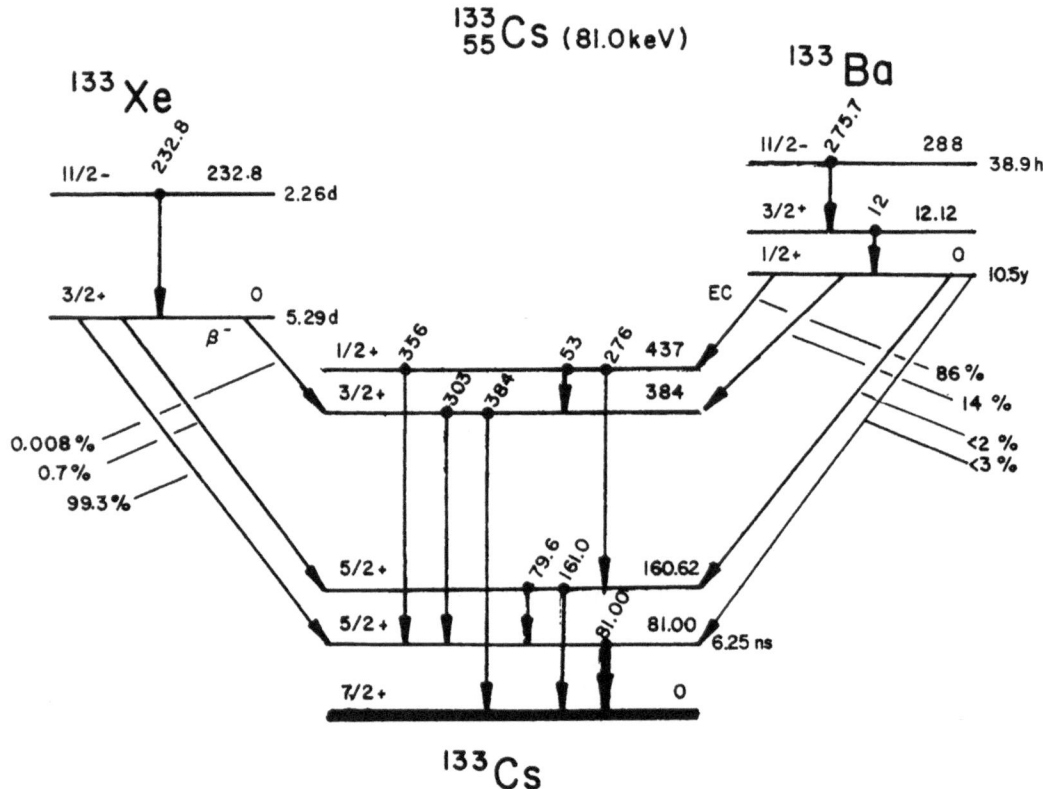

Measured Properties

E_γ = 80.997 ± 0.006 keV (1)

$t_{\frac{1}{2}}$ = 6.313 ± 0.021 ns (2), (3), (4), (5)

α_T = 1.72 ± 0.03 (6)

IA = 100%

μ = −2.5786 ± 0.0008 nm (7)

R_μ = +1.335 ± 0.008 (8)

Q = −0.0030 ± 0.0007 b (9), (10)

Derived Parameters

σ_0 = 1.028 (11) × 10^{-19} cm^2

Γ = 7.227 (24) × 10^{-8} eV

W_0 = 0.5350 (18) mm/s

E_r = 2.6478 (3) × 10^{-2} eV

Energy Conversions

1 mm/s = 65.328 (5) MHz

1 mm/s = 2.7018 (2) × 10^{-7} eV

Energy and Intensities

	E_γ	I_{Xe}(11)	I_{Ba}(12)
γ_M :	80.99 keV	932.	52.6
γ_1 :	54. keV		3.5
γ_2 :	80. keV	16.	3.9
γ_3 :	161. keV	1.7	1.2
γ_4 :	276. keV		11.4
γ_5 :	302 keV		30.2
γ_6 :	356 keV		100.
γ_7 :	382 keV	0.6	14.4
$K_{\alpha1}$:	31.0 keV		161.

(1) J.E. Thun, S. Thornkuist, K. BondeNielson, H. Snellman, F. Falk, and A. Mocoroa, Nucl. Phys. 88, 289 (1966).
(2) D. Bloess, A. Krusche, and F. Münnich, Z. Phys. 192, 358 (1966).
(3) D.K. Gupta and G.N. Rao, Nucl. Phys. A182, 669 (1972)
(4) E. Bodenstadt, H.J. Körner, and E. Matthias, Nucl. Phys. 11, 584 (1959).
(5) K.G. Valivaara, A. Marelius, and J. Kozyczkowski, Phys. Scr. 2, 19 (1970).
(6) W.G. Winn and D.G. Sarantites, Phys. Rev. C 1, 215 (1970).
(7) C.W. White, W.M. Hughes, G.S. Hayne, and H.G. Robinson, Phys. Rev. A 7, 1178 (1973)
(8) 68C008.
(9) R.M. Sternheimer and R.F. Peierls, Phys. Rev. A 3, 837 (1971).
(10) S. Rydberg and S. Svanberg, Phys. Scr. 5, 209 (1972).
(11) A. Alexander and J.P. Lau, Nucl. Phys. A121, 612 (1968).
(12) W.-D. Schmidt-Ott and R.W. Fink, Z. Phys. 249, 286 (1972).

[Other data from E.A. Henry, Nucl. Data Sheets 11, 495 (1974).]

SOURCE	S-TEMP	ABSORBER	A-TEMP	SHIFT	QS	REMARKS	REF
Fe	4.2	xx	4.2			implanted sources	74R042
Fe(Xe)	v	CsCl		—		Xe implanted impurities, fs	75R003

CODE	TOPIC	REFERENCE
74R042	CS-133	S R Reintsema, S A Drentje, and H De Waard, New Results on the Location of Xenon Impurities Implanted in Iron Derived from Mossbauer Spectra, in "Hyperfine Interactions Studied in Nuclear Reactions and Decay: Contributed Papers" (Conference, Uppsala, Sweden, 1974), edited by E Karlsson and R Wappling (Upplands Grafiska, Uppsala, 1974), pp 74-5
75R003	CS-133	S R Reintsema, S A Drentje, P J Schurer, and H De Waard, Radiat Eff 24,145-54(1975), Lattice Location of Xenon Impurities Implanted in Iron Derived from Mossbauer Effect Measurements

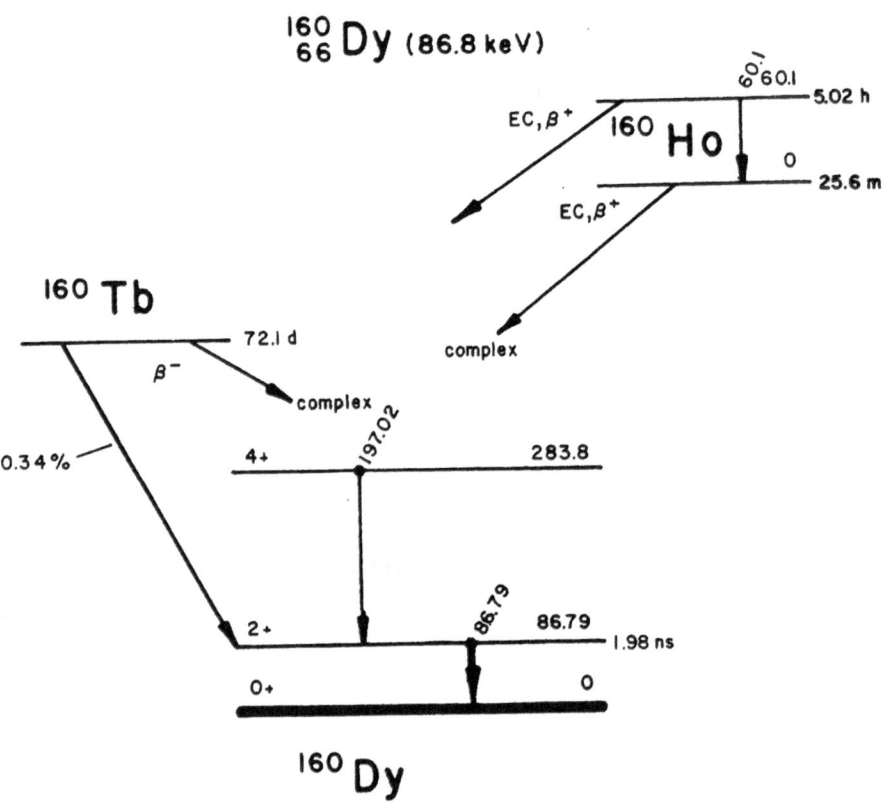

$^{160}_{66}$ Dy (86.8 keV)

^{160} Ho

^{160} Tb

^{160} Dy

Measured Properties

E_γ = 86.788 ± 0.002 keV (1)

$t_{\frac{1}{2}}$ = 2.037± 0.016 ns (2),(3),(4),(5)

α_T = 4.52 ± 0.11 (6)

IA = 2.29%

μ^* = +0.724 ± 0.018 (3),(7)

Q^* = 1.76 ± 0.39 b (8)

Derived Parameters

σ_o = 2.94 (6) x 10^{-19} cm^2

Γ = 2.240 (18) x 10^{-7} eV

W_o = 1.547 (12) mm/s

E_r = 2.52694 (8) x 10^{-2} eV

Energies and Intensities

	E_γ	I_{Tb}	I_{Ho} (9)
γ_M :	86.8 keV	13.4(5)	91.0(8)
γ_1 :	197. keV	5.24 (12)	10.0
γ_2 :	216. keV	4.02 (10)	0.018 (5)
γ_3 :	298. keV	27.4 (8)	0.066 (10)
K_{α_1} :	46.0 keV		

Energy Conversions

1 mm/s = 69.9986 (17) MHz

1 mm/s = 2.89494 (7) eV

(1) M. A. Ludington, J. J. Reddy, M. L. Weidenbeck, D. J. McMillan, and J. H. Hamilton, Nucl. Phys. A119, 398 (1968).

(2) H. Abou-Leila, A. Abd. El-Haliem, and S. M. Darwish, Nucl. Phys. A175, 663 (1971).

(3) G. Günther, G. Strube, U. Wehmann, W. Engels, H. Blumberg, H. Luig, R. M. Leider, E. Bodenstadt, and H. J. Körner, Z. Phys. 183, 472 (1965).

(4) Th. J. de Boer, E. W. ten Napel, and J. Blok, Physica 29, 1013 (1963).

(5) P. C. Lopiparo, R. L. Rasera, M. E. Caspari, Nucl. Phys. A178, 577 (1972).

(6) R. S. Dingus, W. L. Talber, Jr., and M. G. Stewart, Nucl. Phys. 83, 545 (1966).

(7) I. Ben-Zvi, P. Gilad, M. B. Goldberg, G. Goldring, K. H. Speidel, and A. Sprinzak, Nucl. Phys. A151, 401 (1970).

(8) H. F. Wagner and J. Lange, Z. Phys. 238, 35 (1970).

(9) M. P. Avotina, E. P. Grigor'ev, B. S. Dzhelepov, A. V. Zolotavin, and V. O. Sergeev, Bull. Acad. Sci. USSR, Phys. Ser. 30, 542 (1967).

[Other data from J. K. Tuli, Nucl. Data Sheets 12, 477 (1974)]

SOURCE	S-TEMP	ABSORBER	A-TEMP	SHIFT	QS	REMARKS	REF
Tb	4.2	Dy.4Sc.6H2	4.2			W=4.6 mm/s	75S082
Tb.05Dy.35Y.6H2	4.2	Dy.4Sc.6H2	4.2			W=5.1 mm/s	75S082
Tb.08Y.92H2	4.2	Dy.4Sc.6H2	4.2				75S082
Tb.08Y.92H2	1.4	Dy.4Sc.6H2	4.2				75S082
Tb.1Er.9H2	4.2	Dy.4Sc.6H2	4.2			W=5.2 mm/s	75S082
Tb.08Y.92H2	30	DyH2	20			W=8.1 mm/s	75S082

CODE	TOPIC	REFERENCE

75S082   DY-160   J Stohr and J D Cashion, Phys Rev B 12,4805-11(1975),
160Dy and 166Er Mossbauer Studies of Concentrated and Diluted Rare-earth Dihydrides: Single-line Compounds and
Crystal-field Effects

^{161}Dy (25.6 keV, 43.8 keV, 74.6 keV)

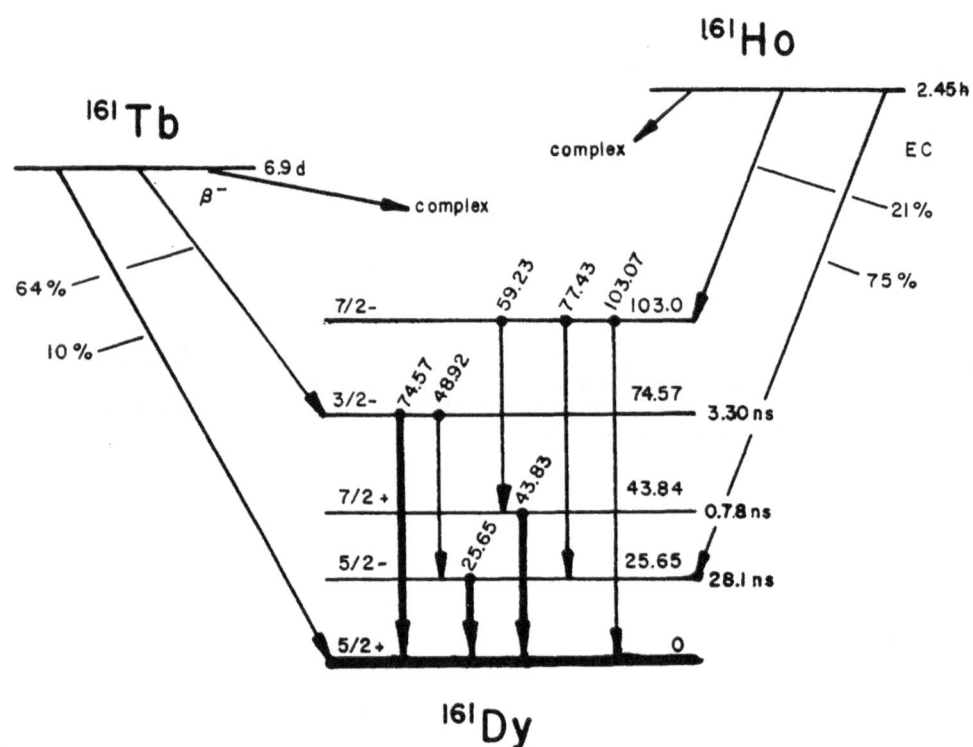

Measured Properties ($E_\gamma = 25.7$ keV)

E_γ = 25.655 ± 0.003 keV (1)

$t_{\frac{1}{2}}$ = 28.2 ± 0.9 ns (2),(3)

α_T = $2.9^{+0.3}_{-0.2}$ (4)

IA= = 18.88%

μ = -0.479 ± 0.005 nm (5)

R_μ = -1.2368 ± 0.0014 (6),(7),(8),(9),(10)

Q = $+2.35 \pm 0.16$ b$^+$ (11),(12)

R_Q = $+0.9996 \pm 0.0021$ (7),(8),(9),(10)

Derived Parameters ($E_\gamma = 25.7$ keV)

σ_0 = $9.5(6) \times 10^{-19}$ cm^2

Γ = $1.62(5) \times 10^{-8}$ eV

W_0 = $0.378(12)$ mm/s

E_r = $2.1944(4) \times 10^{-3}$ eV

Energy Conversions ($E_\gamma = 25.7$ keV)

1 mm/s = $20.692(2)$ MHz

1 mm/s = $8.5576(10) \times 10^{-8}$ eV

Measured Properties ($E_\gamma = 43.8$ keV)

E_γ = 43.83 ± 0.01 keV (13)

$t_{\frac{1}{2}}$ = 0.78 ± 0.06 ns (14)

α_T = 4.32 (15)

R_μ = $+0.293 \pm 0.010$ (15)

R_Q = 0.21 ± 0.05 (15)

Derived Parameters ($E_\gamma = 43.8$ keV)

σ_0 = $3.19(3) \times 10^{-19}$ cm^2

Γ = $5.8(4) \times 10^{-7}$ eV

W_0 = $8.0(6)$ mm/s

E_r = $6.405(2) \times 10^{-3}$ eV

Energy Conversions ($E_\gamma = 43.8$ keV)

1 mm/s = $35.351(8)$ MHz

1 mm/s = $1.4620(3) \times 10^{-8}$ eV

Measured Properties ($E_\gamma = 74.6\,\text{keV}$)

E_γ = $74.577 \pm 0.009\,\text{keV}$ (1)

$t_{\frac{1}{2}}$ = $3.31 \pm 0.06\,\text{ns}$ (16), (17)

α_T = 0.65 (1)

ρ = 38%

R_μ = $+0.840 \pm 0.007$ (18), (19), (20)

R_Q = $+0.59 \pm 0.03$ (18), (19), (20)

Derived Parameters ($E_\gamma = 74.6\,\text{keV}$)

σ_o = $6.8\,(9) \times 10^{-20}\,\text{cm}^2$

Γ = $1.378\,(25) \times 10^{-7}\,\text{eV}$

W_o = $1.108\,(20)\,\text{mm/s}$

E_r = $1.854\,(3) \times 10^{-2}\,\text{eV}$

Energies and Intensities

	E_γ (3)	I_{Tb}	I_{Ho}
γ_{M1}:	25.65 keV	23	105.
γ_{M2}:	43.84 keV		0.4
γ_{M3}:	74.58 keV	9	
γ_1 :	48.92 keV	19	0.3
γ_2 :	59.23 keV		4.
γ_3 :	77.42 keV		10.
γ_4 :	103.07 keV		12.
K_{α_1} :	46.0 keV		

Energy Conversions ($E_\gamma = 74.6\,\text{keV}$)

1 mm/s = 60.150 (7) MHz

1 mm/s = $2.4876\,(3) \times 10^{-7}\,\text{eV}$

[+] Sternheimer corrected

(1)   R. T. Brockmeier and J. D. Rogers, Nucl. Phys. 67, 428 (1965).
(2)   R. B. Begzhanov and Kh. M. Sadykov, Sov. J. Nucl. Phys. 10, 254 (1970).
(3)   M. Vetter, Z. Phys. 225, 336 (1969).
(4)   P. Steiner and G. Weyer, Z. Phys. 248, 370 (1971).
(5)   J. Ferch, W. Dankwort, and H. Gebauer, Phys. Lett. 49A, 287 (1974).
(6)   70K037.
(7)   71C051.
(8)   71N006.
(9)   71B043.
(10) 71G023.
(11) B. Elbeck, K. O. Nielsen, and M. C. Olesen, Phys. Rev. 108, 406 (1957).
(12) W. Ebenhöh, V. J. Ehlers, and J. Ferch, Z. Phys. 200, 84 (1967).
(13) E. L. Chupp, J. W. M. DuMond, F. J. Gordon, R. C. Jopson, and H. Mark, Phys. Rev. 112, 518 (1958).
(14) D. Ashery, N. Bahcall, G. Goldring, A. Sprinzak, and Y. Wolfson, Nucl. Phys. A101, 51 (1967).
(15) 73S009.
(16) A. Bäcklin, S. G. Malmskog, and H. Solhod, Ark. Fys. 34, 495 (1967).
(17) V. Berg and S. G. Malmskog, Nucl. Phys. A135, 401 (1969).
(18) 68H041.
(19) 68K006.
(20) 68C031.

[Other data from Table of Isotopes, 6th Edition (1967)]

SOURCE	S-TEMP	ABSORBER	A-TEMP	SHIFT	QS	REMARKS	REF
BaTbO3(Gd)	293	xx					75Y004
BaTiO3(Gd)	293	xx		--			75Y004
Fe(Dy)	v	xx			--	ion implantation study	75N021
Ni(Dy)	v	xx			--	ion implantation study	75N021
GdF3(Tb)		Co3Dy	4.2		--	HI data, EQ=120 mm/s	75C041
GdF3(Tb)		Co3Dy	77				75C041
GdF3(Tb)		Co7Dy2	4.2		--	HI data, EQ=100 mm/s	75C041
xx(IS/DyF3)		CuDy		--		comparison of theory & expl IS	75B092
Tb	R	DyCo3	4.2		--	comparison of efg data from Mossbauer & point charge model	75Y003
Fe	v	DyF3			--	5<T<363 K, spin relaxation	74W028
xx		Dy(FexNi(1-x))3				electronic structure info	75C033
xx		DyFe2	v		--	distribution of HI & QS values	75F032
xx(IS/DyF3)		DyRh		--		comparison of theory & expl IS	75B092
xx(IS/DyF3)		DyZn		--		comparison of theory & expl IS	75B092
bone		Dy2O3				study of rare earth uptake	74K065

CODE	TOPIC	REFERENCE

74K065   DY-161   C Kellershohn, J N Rimbert, F Soubirou, and C Hubert, Bone Uptake of Rare Earths Observed by Mossbauer Effect, in "Magnetic Resonance and Related Phenomena, Proceedings of AMPERE Congress, 18th" (Nottingham, 1974), edited by P S Allen, E R Andrew, and C A Bates (North-Holland Publishing Co, Amsterdam, 1974), pp 289-90

74W028   DY-161   H P Wit and L Niesen, Spin Relaxation Phenomena in Mossbauer Spectra of Dysprosium Impurities Implanted in Iron, in "Hyperfine Interactions Studied in Nuclear Reactions and Decay: Contributed Papers" (Conference, Uppsala, Sweden, 1974), edited by E Karlsson and R Wappling (Upplands Grafiska, Uppsala, 1974), pp 246-7

75B092   DY-161   M Belakhovsky, Solid State Commun 17,349-52(1975), Contact Charge Density in Trivalent Dysprosium Intermetallics Through APW Calculations

75C008   DY-161   J D Cashion, M A Coulthard, and D B Prowse, J Phys C 8, 1267-75(1975), Mossbauer Isomer Shifts in Rare Earth Compounds: II. Nuclear Charge Radius and Electron Density Studies

75C033   DY-161   G Crecelius and H Maletta, Electronic Structure and Transferred Hyperfine Interactions in the Series Dy(Fe(x)Ni(1-x))3, in "Int Conf Mossbauer Spectrosc, Proc," Vol 1, pp 149-50 (See 75H027)

75C041   DY-161   J M D Coey, J Chappert, J P Rebouillat, and T S Wang, 161Dy Mossbauer Spectra of an Amorphous Dy-Co Alloy, in "Int Conf Mossbauer Spectrosc, Proc," Vol 1, pp 347-8 (See 75H027)

75F032   DY-161   D W Forester, R Abbundi, R Segnan, and D Sweger, Magnetic Hyperfine Structure in Amorphous DyFe2, in "AIP Conference Proceedings-No 24, Magnetism and Magnetic Materials-1974" (20th Annual Conference, San Francisco), edited by C D Graham, Jr, G H Lander, and J J Rhyne (American Institute of Physics, New York, 1975), pp 115-6

75N021   DY-161   L Niesen, H P Wit, P J Kikkert, and H De Waard, Lattice Location of Rare-earth Impurities Implanted in Ferromagnetic Metals Derived from Mossbauer Experiments, in "Int Conf Mossbauer Spectrosc, Proc," Vol 1, pp 207-8 (See 75H027)

75S096   DY-161   J Sivardiere and M Blume, Hyperfine Interac 1,283-94(1975), Paramagnetic Mossbauer Spectra of 161Dy(Gamma 6 or Gamma 7) and 166Er(Gamma 8) in Cubic Symmetry: Influence of Relaxation

CODE	TOPIC	REFERENCE

75Y003   DY-161   J K Yakinthos and J Chappert, Solid State Commun 17,979-81 (1975), Crystal Field and Mossbauer Spectroscopy of the Intermetallic Compound DyCo3

75Y004   DY-161   V K Yarmarkin, B A Shustrov, and A V Motorny, Mossbauer Effect on Dy161, Sb121, and Fe57 Nuclei in BaTiO3 Ceramics, in "Int Conf Mossbauer Spectrosc, Proc," Vol 1, pp 325-6 (See 75H027)

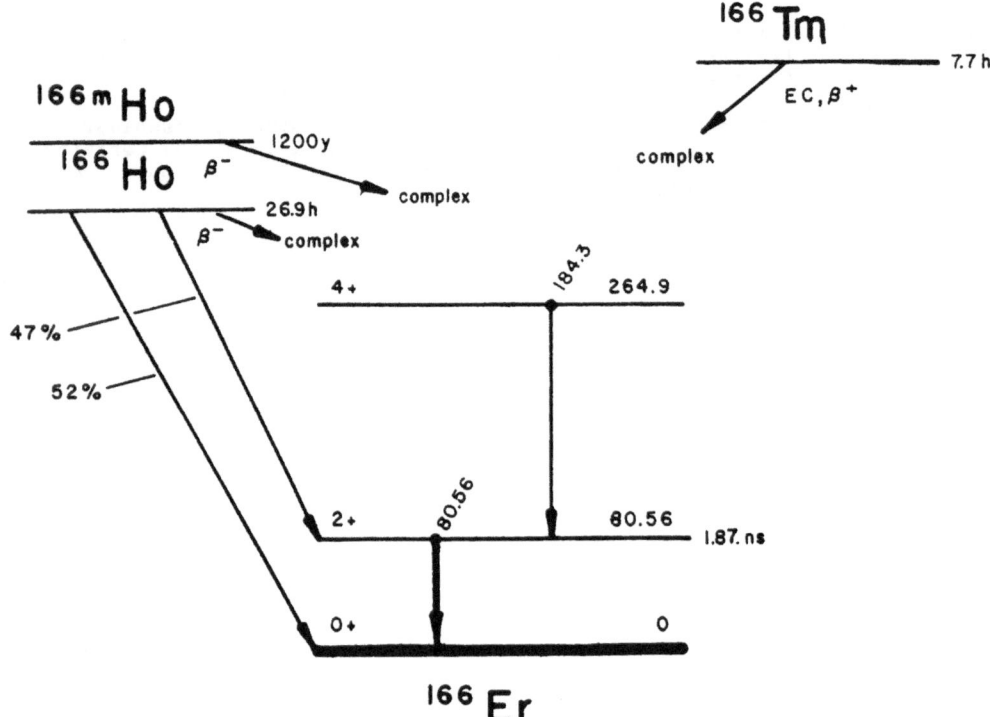

$^{166}_{68}$Er (80.6 keV)

^{166}Er

Measured Properties

E_γ = 80.557 ± 0.004 keV (1)

$t_{\frac{1}{2}}$ = 1.87 ± 0.03 ns (2)

α_T = 6.93 ± 0.03 (3)

IA = 33.41 %

μ = + 0.629 ± 0.010 nm (4),(5),(6)

Q = -1.59 ± 0.15 b (5),(7)

Derived Parameters

σ_o = 2.377 (9) × 10^{-19} cm^2

Γ = 2.44 (4) × 10^{-7} eV

W_o = 1.82 (3) mm/s

E_r = 2.09843 (15) × 10^{-2} eV

Energies and Intensities

	E_γ	I_{mHo}(8)	I_{Tm}(9)
γ_M :	80.57 keV	16.8	540(30)
γ_1 :	184.4 keV	100	877(25)
γ_2 :	215.2 keV	4.1(1)	280(10)
γ_3 :	280.5 keV	39.6(12)	14.6(4)
K_{α_1} :	49.13 keV		

Energy Conversions

1 mm/s = 64.973 (3) MHz

1 mm/s = 2.68709 (13) × 10^{-7} eV

(1) I. Marklund and B. Lingström, Nucl. Phys. 40, 329 (1963).
(2) J. D. Kurfess and R. P. Scharenberg, Phys. Rev. 161, 1185 (1967).
(3) F. W. N. DeBoer, P. F. A. Goudsmit, and B. J. Meijer, Z. Phys. 260, 75 (1973).
(4) 64D02.
(5) 64C02.
(6) 68M011.
(7) 65H02.
(8) E. W. A. Lingeman, F. W. N. DeBoer, and B. J. Meijer, Nucl. Instrum. Methods 118, 609 (1974).
(9) J. Adam, M. Honusek, K. Ya. Gromov, J. Frana, Czech. J. Phys. B25, 504 (1975).

[Other data from A. Buym, Nucl. Data Sheets 14, 471 (1975)]

SOURCE	S-TEMP	ABSORBER	A-TEMP	SHIFT	QS	REMARKS	REF
Fe(Er)	4.2	xx			--	ion implantation study	75N021
Ni(Dy)	4.2	xx			--	ion implantation study	75N021
Fe2Ho	4.2	Al3Er	V			magnetization measurements	75W010
Al-Ho alloy		AuEr	V			4.2<T<80 K HI=8300 kOe, relax	75K011
Ho	4.2	ErH2	4.2			W=5.9 mm/s	75S082
Ho.1Er.9H2	4.2	ErH2	4.2			W=8.5 mm/s	75S082
Ho.15Y.85H2	1.4	ErH2	4.2			W=8.5 mm/s	75S082
Ho.15Y.85H2	4.2	ErH2	4.2			W=8.0 mm/s	75S082
Ho.4Y.6H2	4.2	ErH2	4.2			W=7.7 mm/s	75S082
Fe(Ho)		HeAl2Er	N		--	EQ=11.8 mm/s, HI=7.68 MOe	74N020
Ho		Y	V		--	1.4<T<10 K, study of hyperfine interactions and relax effects	75S079

CODE	TOPIC	REFERENCE
74N020	ER-166	L Niesen and P J Kikkert, Mossbauer Effect of Er Impurities Implanted in Iron, in "Hyperfine Interactions Studied in Nuclear Reactions and Decay: Contributed Papers" (Conference, Uppsala, Sweden, 1974), edited by E Karlsson and R Wappling (Upplands Grafiska, Uppsala, 1974), pp 160-1
75K011	ER-166	C W Kimball, A E Dwight, G M Kalvius, B D Dunlap, and M V Nevitt, Phys Rev B 12,819-23(1975), Low-temperature Phase Transition and Isomer-shift Systematics in Intermediate Phases of Rare-earth-gold Compounds
75N021	ER-166	L Niesen, H P Wit, P J Kikkert, and H De Waard, Lattice Location of Rare-earth Impurities Implanted in Ferromagnetic Metals Derived from Mossbauer Experiments, in "Int Conf Mossbauer Spectrosc, Proc," Vol 1, pp 207-8 (See 75H027)
75S079	ER-166	J Stohr and W Wagner, J Phys F 5,812-21(1975), Hyperfine Interactions and Relaxation Effects of Er Impurities in Y Studied by Mossbauer Spectroscopy
75S080	ER-166	J Stohr, J D Cashion, and W Wagner, J Phys F 5,1417-25 (1975), Mossbauer Spectroscopy of Rare Earth Impurities in Hexagonal Host Metals: ZrEr
75S082	ER-166	J Stohr and J D Cashion, Phys Rev B 12,4805-11(1975), 160Dy and 166Er Mossbauer Studies of Concentrated and Diluted Rare-earth Dihydrides: Single-line Compounds and Crystal-field Effects
75S095	ER-166	J Sivardiere, M Blume, and M J Clauser, Hyperfine Interac 1,227-50(1975), Magnetic Relaxation and Paramagentic Mossbauer Spectra: Influence of the Off-diagonal Hyperfine Coupling
75S096	ER-166	J Sivardiere and M Blume, Hyperfine Interac 1,283-94(1975), Paramganetic Mossbauer Spectra of 161Dy(Gamma 6 or Gamma 7) and 166Er(Gamma 8) in Cubic Symmetry: Influence of Relaxation
75W010	ER-166	K Weber, Magnetization Measurements and Mossbauer 166Er Polarimetry on the Metamagnetic Phase Transitions in Hexagonal ErAl3 Single Crystals, in "5th Int Conf Mossbauer Spec, Proc," Part 3, pp 504-10 (See 75H017)

$^{167}_{68}$Er (79.3 keV)

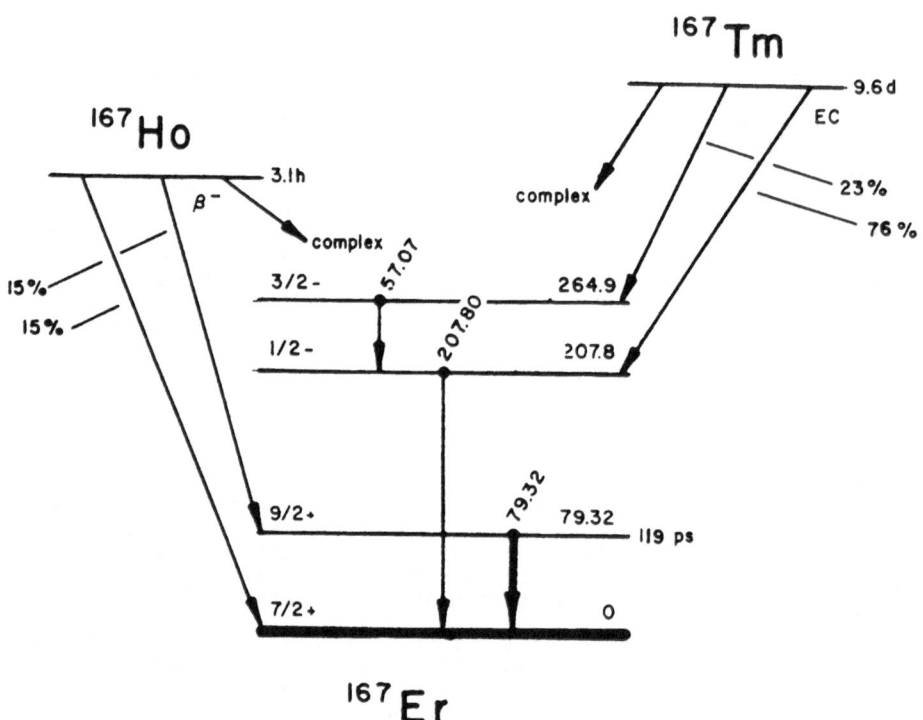

Measured Properties

E_γ = 79.3219 ± 0.0013 keV (1)

$t_{\frac{1}{2}}$ = 119 ± 9 ps (2)

α_T = 5.74 (3)

IA = 22.94

μ = 0.564 ± 0.002 nm (4)

Q = 2.827 ± 0.012 b (4)

Derived Parameters

σ_o = 7.21 (5) × 10^{-20} cm^2

Γ = 3.8 (2) × 10^{-6} eV

W_o = 29.0 (22) mm/s

E_r = 2.02239 (5) × 10^{-2} eV

Energies and Intensities

	E_γ	I_{Ho} (5)
γ_M :	79.3 keV	38 (9)
γ_1 :	83.5 keV	27 (6)
γ_2 :	207.8 keV	88 (6)
K_{α_1} :	49.13 keV	

Energy Conversions

1 mm/s = 63.9768 (11) MHz

1 mm/s = 2.64589 (4) × 10^{-7} eV

(1) H. R. Koch, Z. Phys. 187, 450 (1965).
(2) Y. Dar, J. Gerber, A. Macher, and J. P. Vivien, Nucl. Phys. A171, 575 (1971).
(3) Theory using mixing ratio from D. Ashery, A. E. Blaugrund, and R. Kalish, Nucl. Phys. 76, 336 (1966).
(4) K. F. Smith and P. J. Unsworth, Proc. Phys. Soc. (London) 86, 1249 (1965).
(5) L. Funke, et al. Nucl. Phys. A118, 97 (1968).

[Other data from B. Harmatz, Nucl. Data Sheets 17, 143 (1976)]

SOURCE	S-TEMP	ABSORBER	A-TEMP	SHIFT	QS	REMARKS	REF
Er203(CE)	xx					fs experiment	75C024

CODE	TOPIC	REFERENCE
75C024	ER-167	C L Chien and J C Walker, Studies of Temperature Spikes in Solids Using the Mossbauer Effect Following Coulomb Excitation, in "5th Int Conf Mossbauer Spec, Proc," Part 3, pp 560-3 (See 75H017)

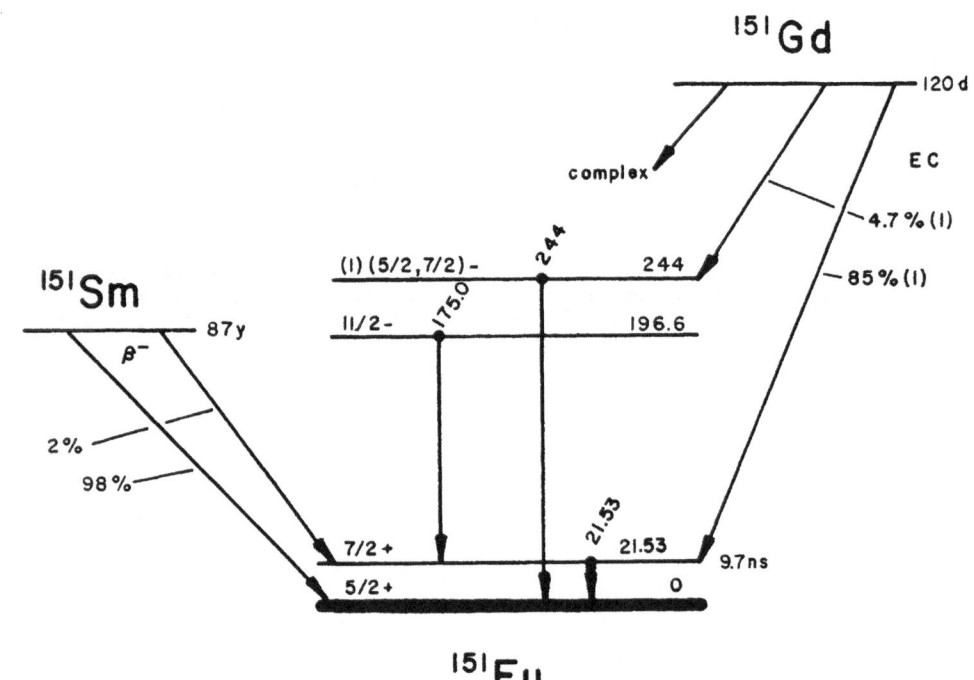

$^{151}_{63}\text{Eu}$ (21.5 keV)

Measured Properties

E_γ = 21.532 ± 0.005 keV (2), (3)

$t_{\frac{1}{2}}$ = 9.7 ± 0.3 ns (4), (5), (6)

α_T = 28.60 ± 0.15 (theory) (2)

IA = 47.82 %

μ = +3.4649 ± 0.0023 nm (7), (8)

R_μ = 0.7465 ± 0.0006 (9)

Q = 1.14 ± 0.05 b (10), (11), (12)

R_Q = +1.31 ± 0.02 (9), (13)

Energy Conversions

1 mm/s = 17.367 (4) MHz

1 mm/s = 7.1823 (17) × 10^{-8} eV

Derived Parameters

σ_o = 2.377 (12) × 10^{-19} cm^2

Γ = 4.70 (15) × 10^{-8} eV

W_o = 1.31 (4) mm/s

E_r = 1.6481 (5) × 10^{-3} eV

Energies and Intensities

	E_γ	I_{Sm}	I_{Gd}
γ_M :	21.64 keV	100	0.94
γ_1 :	175. keV		8.8
γ_2 :	244. keV		1.5
K_{α_1} :	41.5 keV		

(1) E. P. Grigor'ev at al., Bull. Acad. Sci. USSR, Phys. Ser. 32, 723 (1969).
(2) S. Antman, H. Pettersson, Z. Zehler, and I. Adam, Z. Phys. 237, 285 (1970).
(3) B. Gehrke, Document ANCR-1088, p. 392.
(4) D. J. Horen, H. H. Bolotin, and W. H. Kelly, Bull. Amer. Phys. Soc. 8, 127 (1963).
(5) O. C. Kistner, A. C. Li, and S. Monaro, Phys. Rev. 132, 1733 (1963).
(6) J. Kownacki, J. Ludziejewski, M. Moszynski, and K. Stryczniewicz, Acta Phys. Pol. B1, 173 (1970).
(7) L. Evans, P. G. H. Sandars, and G. K. Woodgate, Proc. Roy. Soc. (London) A289, 114 (1965).
(8) H. Kiefte and J. S. M. Harvey, Can. J. Phys. 48, 562 (1970).
(9) 73C045.
(10) R. Winkler, Phys. Lett. 16, 156 (1965).
(11) W. Müller, A. Steudel, and H. Walther, Z. Phys. 183, 303 (1965).
(12) G. Guthöhrlein, Z. Phys. 214, 332 (1968).
(13) 70D009.
(14) 70K008.
(15) 74C027.

[Other data from Table of Isotopes, 6th Edition (1967)]

SOURCE	S-TEMP	ABSORBER	A-TEMP	SHIFT	QS	REMARKS	REF
XX		Cu5Eu	V	--	--	Eu is divalent, magn struc.	75V013
Sm2O3(IS/Eu2O3)		Eu.1WO3	R	-1.3		bonding and structure info	75P006
Sm2O3(IS/Eu2O3)		Eu.18WO3	R	-.85		bonding and structure info	75P006
Sm2O3		EuCoO3	V	--			74J017
Sm2O3		Eu(DPP)4(pip)	V	--	--	bonding information	75L002
Sm2O3		Eu(DPP)5OH	V	--	--	bonding information	75L002
Fe(Gd)	20	EuF3		--		hf field & exchange coupling	74C040
Sm2O3	4.2	Eu(H)	4.2	--		IS vs hydrogen concentration	75M013
Sm2O3		Eu(HFP)4(pip)	V	--	--	bonding information	75L002
Sm2O3		Eu(HFP)3-2H2O	V	--	--	bonding information	75L002
SmF3(IS/Eu2O3)	R	EuIn3	N	-11.8		HI=230 kOe	75G033
SmFe(IS/Eu2O3)	R	EuIn3	V	-11.8		HI=230 kOe(T=0 K), 2 K<T<78 K	75G027
XX		EuIr(2-x)Ptx	V	--		electronic structure info	75B074
Sm2O3		Eu(PBD)4(pip)	V	--	--	bonding information	75L002
Sm2O3		Eu(PTD)3-3H2O	V	--	--	bonding information	75L002
SmF3(IS/Eu2O3)	R	EuPb3	N	-11.9		HI=245 kOe	75G033
SmFe(IS/Eu2O3)	R	EuPb3	V	-11.9		HI=245 kOe(T=0 K), 2 K<T<78 K	75G027
XX		EuRh(2-x)Alx	V	--		electronic structure info	75B074
XX		EuRh(2-x)Alx	V	--		electronic structure info	75B074
XX		EuS				study of the effect of lattice defects on the hf interactions in thin films	75M042
Sm2O3	300	EuS	4.2		--	study of lattice defects in thin films	75M043
XX		EuS	V			low temperature thermometry	75M073
Sm2O3 ·		Eu(TFP)4(pip)	V	--	--	bonding information	75L002
Sm2O3		Eu(TFP)3-2H2O	V	--	--	bonding information	75L002
EuF3(Sm)	15	EuZn2	V	--		.8 K<T<100 K, HI=238 kOe(0 K), metamagnetic substance	75B040
EuF3(Sm)	15	EuZn5	V	--		.8 K<T<100 K, HI=292 kOe(0 K)	75B040
EuF3(Sm)	15	EuZn13	V	--		.8 K<T<100 K, HI=283 kOe(0 K)	75B040
Sm2O3(IS/EuF3)		EuxWO3	R	--			74P054
XX		Eu+2 on Y zeo=	R			=lite	75S050
XX		Eu+2 on mor=	R			=denite	75S050
SmF3(IS/Eu2O3)		Eu3S4	V	--		2.2<T<295 K, HI=323 kOe(0 K), Curie temperature=4.25 K	75G037

CODE	TOPIC	REFERENCE

74C040   EU-151   R L Cohen, G Beyer, and B I Deutch, Origins of hf Field and Exchange Coupling of Eu Substitutional Impurities in Iron, in "Hyperfine Interactions Studied in Nuclear Reactions and Decay: Contributed Papers" (Conference, Uppsala, Sweden, 1974), edited by E Karlsson and R Wappling (Upplands Grafiska, Uppsala, 1974), pp 158-9

74J017   EU-151   V G Jadhao, G N Rao, R M Singru, D Bahadur, and C N R Rao, Mossbauer Spectroscopic Studies of EuCoO3 Employing 57Fe and 151Eu Gamma Rays, in "Proceedings of the Nuclear Physics and Solid State Physics Symposium" (Bombay, 1974) (Department of Atomic Energy, Bombay, 1974), Vol 17C, pp 361-4

74P054   EU-151   E Polaczkowa and A Polaczek, Pol Akad Nauk, Oddzial Krakowie, Pr Kom Ceram, Ceram 21,29-44(1974), Recent Studies on the Cubic Tungsten Bronzes of Sodium and Rare Earth Metals: Structural Problems and Magnetic Properties

75B002   EU-151   C M P Barton and N N Greenwood, Europium-151 Mossbauer Spectroscopy, in "Mossbauer Effect Data Index, Covering the 1973 Literature," edited by J G Stevens and V E Stevens (Plenum Publishing Corp, New York, 1975), pp 395-445

75B040   EU-151   K H J Buschow, W J Huiskamp, H T Le Fever, P J Steenwijk, and R C Thiel, J Phys F 5,1625-36(1975), Mossbauer Effect and Magnetic Properties of Some Eu-Zn Compounds

CODE	TOPIC	REFERENCE

75B074 EU-151 E R Bauminger, I Felner, D Levron, I Nowik, and S Ofer, Interconfiguration Fluctuations in Metallic Rare Earth Compounds, in "Int Conf Mossbauer Spectrosc, Proc," Vol 1, pp 69-70 (See 75H027)

75C008 EU-151 J D Cashion, M A Coulthard, and D B Prowse, J Phys C 8, 1267-75(1975), Mossbauer Isomer Shifts in Rare Earth Compounds: II. Nuclear Charge Radius and Electron Density Studies

75G027 EU-151 E A Gorlich, H U Hrynkiewicz, R Kmiec, K Latka, K Tomala, A Czopnik, and N Iliew, Phys Status Solidi A 30,K17-20 (1975), Mossbauer Effect Studies in EuIn3 and EuPb3

75G033 EU-151 E A Gorlich, H U Hrynkiewicz, K Latka, K Tomala, and R Kmiec, Mossbauer Effect Studies in EuIn3 and EuPb3, in "5th Int Conf Mossbauer Spec, Proc," Part 3, pp 555-6 (See 75H017)

75G037 EU-151 E A Gorlich, H U Hrynkiewicz, K Latka, K Tomala, and R Kmiec, The Mossbauer Effect in Eu3S4 at Low Temperatures, in "5th Int Conf Mossbauer Spec, Proc," Part 3, pp 557-60 (See 75H017)

75L002 EU-151 S J Lyle and A D Witts, J Chem Soc, Dalton Trans 185-8 (1975), Mossbauer Spectroscopic Investigation of Some Europium(III) Diketonates

75M013 EU-151 M Meyer, J M Friedt, L Iannarella, and J Danon, Solid State Commun 17,585-7(1975), Mossbauer Study of 151Eu in Hydrogen-loaded Palladium

75M042 EU-151 H Maletta and G Crecelius, Lattice Defects and Its Influence on Hyperfine Interactions on EuS Evaporated Thin Films, in "Int Conf Mossbauer Spectrosc, Proc," Vol 1, pp 235-6 (See 75H027)

75M043 EU-151 H Maletta and G Crecelius, Appl Phys 8,241-4(1975), Lattice Defects and Its Influence on Hyperfine Interactions on EuS Evaporated Thin Films

75M073 EU-151 H Maletta and G K Shenoy, Low Temperatrue Mossbauer Therometry, in "14th International Conference on Low Temperature Physics, Proceedings: Volume 4. Techniques and Topics" (Otaniemi, Finland, 1975), edited by M Krusius and M Vuorio (North-Holland Publishing Co, Amsterdam/American Elsevier Publishing Co, New York, 1975), pp 68-71

75P006 EU-151 A Polaczek, Rocz Chem 49,1191-6(1975), Isomer Shift in the Mossbauer Spectrum of 151Eu in the Cubic Eu(x)WO3

75S050 EU-151 E A Samuel and W N Delgass, J Chem Phys 62,1590-2(1975), The Hyperfine Structure of Eu2+ Ions in Zeolites

75V013 EU-151 F J Van Steenwijk, H T Le Fever, R C Thiel, and K H J Buschow, Physica 79B,604-9(1975), Mossbauer Effect and Magnetic Properties of EuCu5

$^{153}_{63}$Eu (83.4 keV, 97.4 keV, 103.2 keV)

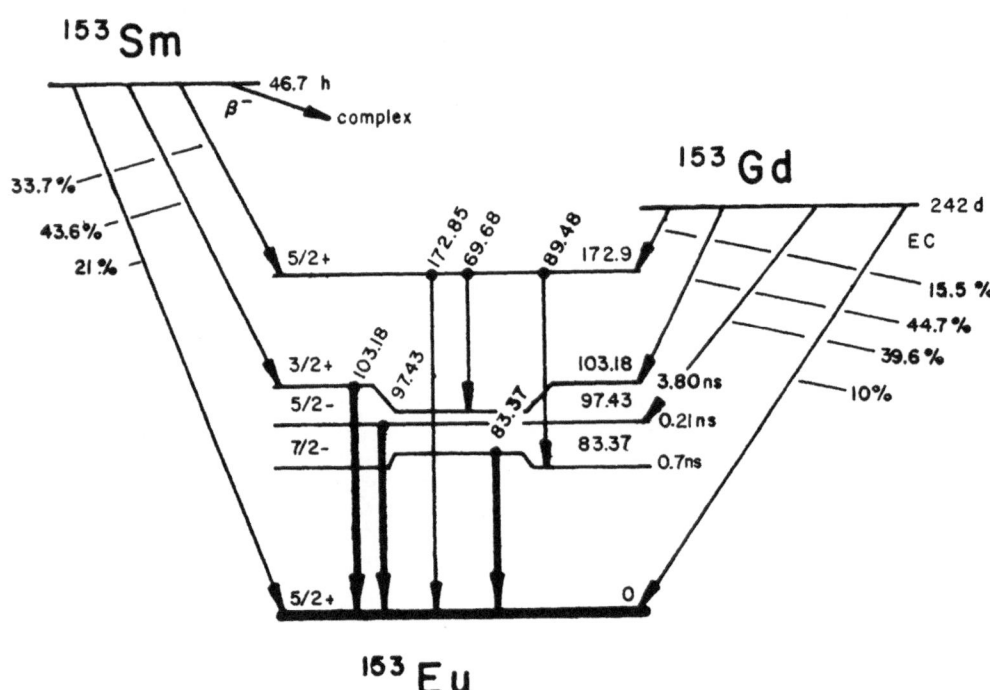

Measured Properties ($E_\gamma = 83.4$ keV)

E_γ = 83.3652 ± 0.0015 keV (1)

$t_{\frac{1}{2}}$ = 0.82 ± 0.07 ns (2)

α_T = 3.82 (theory) (3)

IA = 52.18%

μ = +1.5249 ± 0.0007 nm (4)

R_μ = +1.18 ± 0.04 (5)

Q = +2.90 ± 0.12 b (6),(7),(8)

Derived Parameters ($E_\gamma = 83.4$ keV)

σ_o = 9.74 (10) × 10^{-20} cm^2

Γ = 5.6 (5) × 10^{-7} eV

W_o = 4.0 (3) mm/s

E_r = 2.43823 (6) × 10^{-2} eV

Energy Conversions ($E_\gamma = 83.4$ keV)

1 mm/s = 67.2379 (13) MHz

1 mm/s = 2.78076 (5) × 10^{-7} eV

Measured Properties ($E_\gamma = 97.4$ keV)

E_γ = 97.4283 ± 0.0007 keV (1)

$t_{\frac{1}{2}}$ = 0.21 ± 0.02 ns (9)

α_T = 0.42 (10)

ρ = 99.4%

R_μ = 0.520 ± 0.003 (9)

Derived Parameters ($E_\gamma = 97.4$ keV)

σ_o = 1.80 (7) × 10^{-19} cm^2

Γ = 2.17 (21) × 10^{-6} eV

W_o = 13.4 (13) mm/s

E_r = 3.33023 (4) × 10^{-2} eV

Energy Conversions ($E_\gamma = 97.4$ keV)

1 mm/s = 78.5805 (7) MHz

1 mm/s = 3.24986 (2) × 10^{-7} eV

Measured Properties (E_γ = 103.2 keV)

E_γ = 103.1774 ± 0.0007 keV (1)

$t_{\frac{1}{2}}$ = 3.9 ± 0.2 ns (2)

α_T = 1.78 (theory) (9)

ρ = 99.4 %

R_μ = + 1.336 ± 0.003 (10),(11)

R_Q = 0.520 (3) (11)

Derived Parameters (E_γ = 103.2 keV)

σ_o = 5.58 (12) × 10^{-20} cm^2

Γ = 1.17 (6) × 10^{-7} eV

W_o = 0.68 (3) mm/s

E_r = 3.73486 (5) × 10^{-2} eV

Energy Conversions (E_γ = 103.2 keV)

1 mm/s = 83.2174 (8) MHz

1 mm/s = 3.44163 (23) × 10^{-7} eV

Energies and Intensities

	E_γ	$I_{Sm}^{(1)}$	I_{Gd}
γ_{M1} :	83.4 keV	0.14	1
γ_{M2} :	97.43 keV	0.71	137
γ_{M3} :	103.18 keV	28	100
γ_1 :	14.0 keV		
γ_2 :	19.8 keV		
γ_3 :	69.7 keV	4.0	10
γ_4 :	75.4 keV	0.6	
γ_5 :	89.5 keV	0.13	0.3
γ_6 :	172.9 keV	0.08	< 0.6
K_{α_1} :	41.5 keV		

(1)   K. Mühlbauer, Z. Phys. 230, 18 (1970).
(2)   R. L. Graham and J. S. Geiger, Bull. Am. Phys. Soc. 11, 369 (1966).
(3)   Y. Patin, – Thesis
(4)   L. Evans, P. G. H. Sandars, and G. K. Woodgate, Proc. Roy. Soc. (London) A289, 114 (1965).
(5)   69R25.
(6)   H. Kiefte and J. S. M. Harvey, Can. J. Phys. 48, 562 (1970).
(7)   R. Winkler, Phys. Lett. 16, 156 (1965).
(8)   W. Müller, A. Steudel, and H. Walther, Z. Phys. 183, 303 (1965).
(9)   D. Krpic, J. Milanovic, I. Bikit, and R. Stepic, Z. Phys. 253, 71 (1972).
(10)   G. Guthöhrlein, Z. Phys. 214, 332 (1968).
(11)   73A023.
(12)   73C045.

[Other data from L. A. Kroger and C. W. Reich, Nucl. Data Sheets 10, 429 (1973)]

SOURCE	S-TEMP	ABSORBER	A-TEMP	SHIFT	QS	REMARKS	REF
Dy(Sm)Cl3		Eu2O3	4.2			source and scatterer exp	75M054
Er(Sm)Cl3		Eu2O3	4.2			source and scatterer exp	75M054
Eu(Sm)Cl3		Eu2O3	4.2			source and scatterer exp	75M054
Gd(Sm)Cl3		Eu2O3	4.2			source and scatterer exp	75M054
Ho(Sm)Cl3		Eu2O3	4.2			source and scatterer exp	75M054
La(Sm)Cl3		Eu2O3	4.2			source and scatterer exp	75M054
Nd(Sm)Cl3		Eu2O3	4.2			source and scatterer exp	75M054
Pr(Sm)Cl3		Eu2O3	4.2			source and scatterer exp	75M054
SmCl3		Eu2O3	4.2			source and scatterer exp	75M054
Tb(Sm)Cl3		Eu2O3	4.2			source and scatterer exp	75M054
Tm(Sm)Cl3		Eu2O3	4.2			source and scatterer exp	75M054
Yb(Sm)Cl3		Eu2O3	4.2			source and scatterer exp	75M054

CODE	TOPIC	REFERENCE
75M054	EU-153	D Mihaila-Tarabasanu, U Wagner, F E Wagner, and G M Kalvius, Production and Stability of 153Eu Species Observed after the beta- Decay of 153Sm in Anhydrous Rare Earth Chlorides, in "Int Conf Mossbauer Spectrosc, Proc," Vol 1, pp 291-2 (See 75H027)

$^{57}_{26}Fe$ (14.4 keV, 136.5 keV)

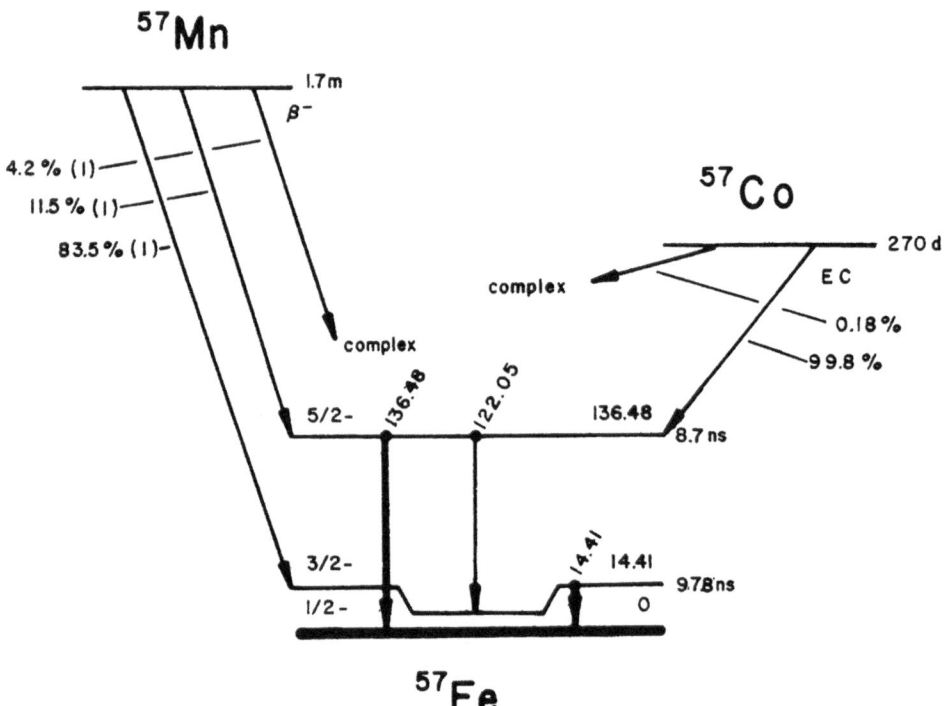

Measured Properties (E_γ = 14.4 keV)

E_γ = 14.41303 ± 0.00008 keV (3)

$t_{\frac{1}{2}}$ = 97.81 ± 0.14 ns

α_T = 8.21 ± 0.12 (4),(5),(6)

IA = 2.14 %

μ = +0.090604 ± 0.000009 nm (7)

R_μ = −1.7142 ± 0.0004 (8),(9),(10),(11),(12), (13),(14),(15),(16),(17)

Q = +0.21 ± 0.01 b$^+$ (18),(19),(20),(21),(22),(23)

Derived Parameters (E_γ = 14.4 keV)

σ_o = 2.56 (3) x 10^{-18} cm^2

Γ = 4.665 (7) x 10^{-9} eV

W_o = 0.1940 (3) mm/s

E_r = 1.956275 (25) x 10^{-3} eV

Energy Conversions (E_γ = 14.4 keV)

1 mm/s = 11.6248 (1) MHz

1 mm/s = 4.80766 (3) x 10^{-8} eV

Measured Properties (E_γ = 136.5 keV)

E_γ = 136.4785 ± 0.0035 keV (24)

$t_{\frac{1}{2}}$ = 8.7 ± 0.4 ns

α_T = 0.142 ± 0.016 (25)

ρ = 12 %

R_μ = −10.8 ± 1.1 (26)

Derived Parameters (E_γ = 136.5 keV)

σ_o = 3.45 (5) x 10^{-19} cm^2

Γ = 5.24 (24) x 10^{-8} eV

W_o = 0.230 (11) mm/s

E_r = 1.75408 (7) x 10^{-1} eV

Energy Conversions (E_γ = 136.5 keV)

1 mm/s = 110.076 (3) MHz

1 mm/s = 4.55243 (12) x 10^{-7} eV

Energies and Intensities

	E_γ	I_{Co}
γ :	14.41 keV	
γ :	136.5 keV	12
γ :	122.0 keV	88
K_{α_1} :	6.40 keV	

$^+$Sternheimer corrected

(1)   T. E. Ward, P. H. Pile, and P. K. Koroda, Nucl. Phys. A134, 60 (1969).
(2)   J. F. Emery, S. A. Reynolds, E. I. Wyatt, and G. I. Gleason, Nucl. Sci. Eng. 48, 319 (1972).
(3)   J. A. Bearden, Phys. Rev. 137, B455 (1965).
(4)   70J006.
(5)   W. Rubinson and K. P. Gopinathan, Phys. Rev. 170, 969 (1968).
(6)   H. U. Freund and J. C. McGeorge, Z. Phys. 238, 6 (1970).
(7)   A. Schwenk, Phys. Lett. 31A, 513 (1970).
(8)   62P02.
(9)   65K02.
(10)   65P03.
(11)   67C025.
(12)   67K030.
(13)   67S025.
(14)   69F001.
(15)   70L027.
(16)   70S028.
(17)   71V013.
(18)   69C04.
(19)   69I04.
(20)   70R011.
(21)   71B064.
(22)   71F001.
(23)   71M055.
(24)   U. Heim and O. W. B. Schult, Z. Naturforsch. A 27, 1861 (1972).
(25)   R. G. Albridge and D. C. Hall, Bull. Amer. Phys. Soc. 10 , 244 (1965).
(26)   G. D. Sprouse and S.S. Hanna, Nucl. Phys. A137, 658 (1969).

[Other data from J. Rapaport, Nucl. Data Sheets, B3, (3,4), 103 (1969)]

USEFUL INFORMATION FOR ^{57}Fe MÖSSBAUER SPECTROSCOPY*

Isomer-Shift Scale for ^{57}Fe(14.41 keV) in Reference Materials

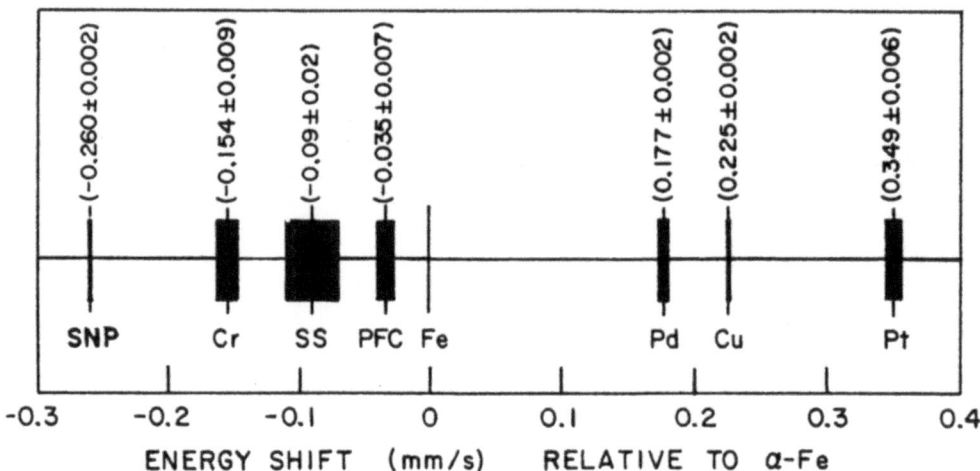

Notes: Sources and absorbers at room temperature
SNP = sodium nitroprusside [$Na_2(CN)_5NO \cdot 2H_2O$]
SS = stainless steel
PFC = potassium ferrocyanide [$K_4(CN)_6 \cdot 3H_2O$]
α-Fe is the normal form of metallic iron at room temperature.
1 mm/sec = 11.6248 ± 0.000 MHz
See (62P02) for the temperature shift of α-Fe.

The following example illustrates the use of this figure.

According to the figure, the excitation energy from the ground state to the first excited state of Fe^{57} is less when the Fe^{57} is in stainless steel than when it is in copper. In velocity units this energy difference is 0.225 + 0.09 = 0.315 mm/s. Thus the isomer shift of Fe^{57} in stainless steel relative to Fe^{57} in copper is -0.315 mm/s, or, conversely, the isomer shift of Fe^{57} in copper relative to Fe^{57} in stainless steel is +0.315 mm/s.

Splittings for ^{57}Fe(14.41 keV) in Reference Materials

α-Iron

$\Delta g_0 / \Delta T = -6.5 \times 10^{-4}$ mm/s-deg

$\Delta g_1 / \Delta T = -3.7 \times 10^{-4}$ mm/s-deg (for room temperatures)

$3g_1/g_0 = -1.7132 \pm 0.0004$

$g_0 = 3.9156 \pm 0.0017$ mm/s

$g_1 = -2.2363 \pm 0.0007$ mm/s

Sodium Nitroprusside (SNP)

$\Delta(\frac{1}{2}e^2qQ) = 10^{-4}$ mm/s at 298 K

$\frac{1}{2}e^2qQ = 1.7034 \pm 0.0014$ mm/s at 298 K

*Information from J. G. Stevens and R. S. Preston, in Mössbauer Effect Data Index, Covering the 1970 Literature, edited by J. G. Stevens and V. E. Stevens (IFI/Plenum, New York, 1972), p. 16.

FINE ARTS

SOURCE	S-TEMP	ABSORBER	A-TEMP	SHIFT	QS	REMARKS	REF
xx		Chinese =		--	--	glazed ceramics, ident of chemical species in the glazes	75H003
xx(IS/Fe)		Greek Etruscan=	v	--	--	=pottery, 4.2<T<300 K	75L006
xx		ancient Egyp=				=tian black ware, study of the origin of the color	75E009
Cr		ancient Egyp=				=tian pottery, study of baking	75E008
xx		ancient pottery		--	--		75S084
xx		pottery				study of the iron oxide transformations in fired clay	75S044

BIOLOGICAL COMPOUNDS

SOURCE	S-TEMP	ABSORBER	A-TEMP	SHIFT	QS	REMARKS	REF
xx		albumin	v			83 K<T<286 K, study of lattice dynamics	75F010
Pd(IS/Fe)		anhydrohemoglo=	v	--	--	=bin, study of the alpha and beta subunits	75C003
Cu		anhydrous Hb	v	--	--	study of alpha(SH) and beta(-SH) subunits	75P001
xx(IS/Fe)		bicarbonate-=	4.2	.46		=free ferric transferrin complex, no binding, HA	75T005
xx(IS/Fe)		bicarbonate-=	78	.47	.72	=free ferric transferrin complex, no binding	72T005
xx		catalase	v		--	T=77 & 273 K	75K059
xx(IS/SNP)		alpha-chimo=		--	--	=tripson	75R024
Cr(IS/Fe)		complex of pro=		+.17	2.0	=tohemin IX w/ Z-Gly-L-His-Gly OBzl	75L017
Cr(IS/Fe)		complex of pro=		--	--	=tohemin IX w/ Z-LLeu-L-HisOMe	75L017
Fe		D-tryptophan				circularly polarized gam sourc	75K063
Fe		D-tyrosine				circularly polarized gam sourc	75K063
xx		enterobactin					74S122
Pd(IS/Fe)		enterobactin	v	+.28	--	electronic structure info, HA	75S009
Rh	4.2	Et4NFe(S2-o=	4.2			=xyl)2, structure information	75L025
Cr	295	Fe + erythro=	78	--	--	=cyte membranes, study of the reaction	75R030
Cr(IS/SNP)	295	Fe + glutathi=	78	--	--	=one, study of the chemical interaction as a function of pH	75R029
xx(IS/SNP)		(Fe(L-alanine)=	295	.68	.51	=2H2O)3O(ClO4)7	75T021
xx(IS/SNP)		(Fe(L-isoleu=	298	.69	.47	=cine)2H2O)3O(ClO4)7	75T021
xx(IS/SNP)		(Fe(L-leucine)=	298	.68	.52	=2H2O)3O(ClO4)7	75T021
xx(IS/SNP)		(Fe(L-valine)2=	298	.68	.56	=H2O)3O(ClO4)7	75T021
xx		Fe(adenine,=	v			=2,2'dipy)SO4O3H2O	75A020
xx		Fe(adenine)2=	v			=SO4-2H2O, 77 K<T<300 K	75A020
xx		Fe(glutathione=	v			=),2,2'dipyOHSO4-2H2O	75A020
xx		(Fe(glutathi=	v			=one OH)(H2O)2)2(SO4)2-6H2O	75A020
xx		Fe(glutathione=	v			=)(SO4)-4H2O, 77 K<T<300 K	75A020
xx(IS/SNP)		(Fe(glycine)2=	298	.67	.39	=H2O)3O(ClO4)7	75T021
xx		(Fe(guanine OH=	v			=))2(SO4)2H2O-2EtOH	75A020
xx		(Fe(guanine,=	v			=2,2'dipy)(OH))-SO4-H2O	75A020
Cu		ferredoxin of =			--	=bacterial clostridium	75P015
xx		ferric hemin			--	theoretical determination of quadrupole splitting	75S017
xx		ferric hemin				discussion of previous results	75D073
xx(IS/Fe)		Fe(+3)-albumin	4.2	.46		HI=448 kOe	75T005
xx(IS/Fe)		Fe(+3)-albumin	78	.47	.74		75T005
Pd(IS/Fe)	4.2	fungus Phycomy=	v	--	--	=ces, study of chemical states of Fe, 1.5 K<T<22 K	75S020
xx(IS/310 SS)		Hb		--	--	molecular orbital calculations	74K066
xx(IS/Fe)		Hb	4.2	.93		calculated electron densities	75T011
xx(IS/Fe)		Hb-CO	4.2	.27		calculated electron densities	75T011
xx		hemin				effects of off diagonal hyperfine interaction	75D025

SOURCE	S-TEMP	ABSORBER	A-TEMP	SHIFT	QS	REMARKS	REF
xx(IS/Fe)		lecithin lipo=	V	--	--	=some-FeCl3, 4.2 K<T<180 K	75K030
Cr		lepidocrocite	R	--	--	study of the dehydration mechanism	74L047
Fe		L-tryptophan				circularly polarized gam sourc	75K063
Fe		L-tyrosine				circularly polarized gam sourc	75K063
Cu(IS/Fe)		lyophilized =	4.2	.28	2.18	=cytochrome c3	75O005
Cu(IS/Fe)		lyophilized =	77	.30	2.04	=cytochrome c3	75O005
Cu(IS/Fe)		lyophilized =	300	.18	2.02	=cytochrome c3	75O005
xx(IS/Fe)		Mb	4.2	.93		calculated electron densities	75T011
xx(IS/Fe)		Mb-CO	4.2	.27		calculated electron densities	75T011
Rh(IS/Fe)		MoFe protein	V	--	--	2 active centers, 4 Fe sites	75M083
xx		mycobactin-P					74S122
xx		myoglobin			--	single crystal sample	75T014
xx(IS/310 SS)		NO-HbO2		--	--	molecular orbital calculations	74K066
Rh		oxidized Bacil=	V	--	--	=lus stearothermophilus fer-redoxin, 4.2 K<T<195 K, HA	75M023
xx		oxidized ferre=	4.2	-.13	--	=doxin, electronic struc info	75A045
xx		oxycytochrome =			--	=P-450, discussion of previous data, electronic struc info	75L015
xx		oxyhemoglobin			--	discussion of previous data, electronic structure info	75L015
xx(IS/310 SS)		oxyhemoglobin		--	--	molecular orbital calculations	74K066
Rh(IS/Fe)	R	oxyhemoglobin	4.2	.24	2.24		75S081
Rh(IS/Fe)	R	oxyhemoglobin	77	.26	2.19		75S081
Rh(IS/Fe)	R	oxyhemoglobin	195	.20	1.89		75S081
xx		oxyhemoglobin	R	--	--	study of the effect of low intensity radiation	75D076
D-tryptophan		PFC				polarized gamma experiment	75K060
D-tyrosin		PFC				polarized gamma experiment	75K060
L-tryptophan		PFC				polarized gamma experiment	75K060
L-tyrosin		PFC				polarized gamma experiment	75K060
Rh(IS/Fe)		rat lung	V				75R007
xx		reduced	250				75A045
Rh		reduced Bacil=	V	--	--	=lus stearothermophilus fer-redoxin, 4.2 K<T<195 K, HA	75M023
xx(IS/Fe)		reduced chloro=	V	--	--	=peroxidase, HA, struc info, 1.6 K<T<180 K	75C019
Rh(IS/Fe)		reduced cyto=	V			=chrome P-450, HA, struc info, 1.6 K<T<200 K	75C018
xx		reduced ferred=	4.2	--	--	=oxin, electronic struc info	75A045
xx		reduced ferred=	4.2	--	---	=oxin, electronic struc info	75A045
xx(IS/Fe)		reduced horse=	V	--	--	=radish peroxidase, HA, struc information, T=4.2 K and 170 K	75C019
xx(IS/Fe)		sickle cell de=	V	--	--	=oxyhemoglobin, HA, struc info T=4.2 K and 180 K	75C019
Pd		yeast aconitase	V	--	--	bonding info, T=4.2 & 78 K	75S037

INORGANIC CYANIDES

SOURCE	S-TEMP	ABSORBER	A-TEMP	SHIFT	QS	REMARKS	REF
xx		Berlin blue				pyrolysis study	75G082
Cu(IS/SNP)		Cs3Fe(CN)6		.26	.50	crln of QS w/ IR frequency	75G001
Cu(IS/SNP)		H3Fe(CN)6		.20	.20	crln of QS w/ IR frequency	75G001
xx		K3(Co,Fe)(CN)6				discussion of previous data	75S074
xx		K3Fe(CN)6		.04		study of the spin structure	75H042
Cu(IS/SNP)		K3Fe(CN)6		.13	.26	crln of QS w/ IR frequency	75G001
Pd		K3Fe(CN)6		-.306		comparison with ESCA	75J003
Cr(IS/Fe)		K3Fe(CN)6	108	-.062	.45		75B031
Pd		K3Fe(CN)6	R			thermal decomposition study	75R017
Pd		K4Fe(CN)6		-.193		comparison with ESCA	75J003
Cu(IS/SNP)		Li3Fe(CN)6-4H2O		.15	.29	crln of QS w/ IR frequency	75G001
Pd		Li3Fe(CN)6-4H2O	R			thermal decomposition study	75R017
Cu(IS/SNP)		Na3Fe(CN)6-H2O		.22	.38	crln of QS w/ IR frequency	75G001
Pd		Na3Fe(CN)6-H2O	R			thermal decomposition study	75R017
xx		Ni3(Fe(CN)6)2	4.2			HI=+270 kOe	75C037
xx		Ni3(Fe(CN)6)2	4.2			HI>0	75C029

SOURCE	S-TEMP	ABSORBER	A-TEMP	SHIFT	QS	REMARKS	REF
Pd		PFC				use of a computerized piezoe-lectric spectrometer	75S041
Pt	R	PFC	R			Fourier analysis	75R005
Be		PFC	v	-.13	.58	proposed Mossbauer-perturbed angular crln exp for detn of the sign of efg	75P025
Pd		Rb3Fe(CN)6	R			thermal decomposition study	75R017
Cu(IS/SNP)		Rb3Fe(CN)6-2H2O		.24	.43	crln of QS w/ IR frequency	75G001
Cr(IS/Fe)		SFC	108	-.024			75B031
xx		SNP					75K076
Fe		SNP			--	detn of efg using a polarized source	75G065

FROZEN SOLUTIONS

SOURCE	S-TEMP	ABSORBER	A-TEMP	SHIFT	QS	REMARKS	REF
Pd		Bu4NFe(EDTA) =	78	.82	1.70	=in H2O, pH=9.0, struc info	75S099
Pt		(Bu4P)2Fe(CN)5=	80	-.603	1.972	=NO in Acetone	75P042
Pt		(Bu4P)2Fe(CN)5=	80	-.518	1.943	=NO in AcCl	75P042
Pt		(Bu4P)2Fe(CN)5=	80	-.562	1.991	=NO in ACN	75P042
Pt		(Bu4P)2Fe(CN)5=	80	-.569	1.972	=NO in Py	75P042
Pt		(Bu4P)2Fe(CN)5=	80	-.541	1.961	=NO in DMF	75P042
Pt		(Bu4P)2Fe(CN)5=	80	-.524	1.946	=NO in NB	75P042
Pd		Et4NFe(EDTA) =	78	.82	1.65	=in H2O, pH=9.0, struc info	75S099
xx		Fe(III) in H3P=				=O4	75M055
Pd		Fe(CDT) in H2O	78	--	--	structure information	75S099
xx		Fe(ClO4)2 in =	v			=H2O, 110 K<T<270 K	75P005
Cr(IS/Fe)		Fe(ClO4)2 in =	v		--	=H2O, 95 K<T<240 K	75B031
Cr(IS/Fe)		Fe(ClO4)3 in =	v		--	=H2O, 95 K<T<240 K	75B031
Cu		Fe(ClO4)3 in =	v	--	--	=H2O, 77<T<173 K	75C026
xx		FeCl2 in H2O	v			110 K<T<270 K	75P005
Cr(IS/Fe)		FeCl2 in H2O	v		--	95 K<T<240 K	75B031
Rh		FeCl2 in glyce=	v			=rol, high pressure experiment	75C001
xx		FeCl2(H2O)4				study of dynamic and static distortions	75I004
xx		FeCl3 in BDM	v			study of solid state rxn	75G077
xx		FeCl3 in H2O	v			W vs FeCl3 concentration, spin-spin relaxation	75M058
Pd		Fe(EDTA) in H2O	78	--	--	structure information	75S099
Pd		Fe(H2O)5OFe(H2=	100	--	--	=O)5 (+4) in H2O, study of hydrolysis products of Fe(III)	75K062
xx		Fe(H2O)6(+2)				study of dynamic and static distortions	75I004
Pd		Fe(HEDTA) in =	78	--	--	=H2O, structure information	75S099
xx		Fe(NO3)3 in H2O	N			pH=0, study of spin relaxation	75S073
Cu		Fe(NO3)3 in H2O	v	--	--	77<T<193 K	75C026
Pd		Fe(NTA) in H2O	78	--	--	structure information	75S099
Pd		FeSnF6 in HF				structure information	75S098
xx		Fe+2 in H3PO4				study of diffusion of Fe ions in a cold liquid	75R039
xx		Fe(+2) in pro=			--	=pane-(1,2)diol, study in vi-cinity of glass transition	75L019
Pt		K3Fe(CN)6 in =		--	--	=H2O + Bu3PCH2PhCl	75P042
Pt		K3Fe(CN)6 in =		-.41	.62	=H2O	75P042
Pt		K3Fe(CN)6 in =		-.40	--	=H2O + glycerol	75P042
Cr(IS/Fe)		K3Fe(CN)6 in =	108	-.066	.81	=H2O + glycerol	75B031
Cr(IS/Fe)		K3Fe(CN)6 in =	108	-.060	.49	=H2O	75B031
Pt		Na3Fe(CN)6 in =		-.48	1.08	=H2O + glycerol	75P042
Pt		Na3Fe(CN)6 in =		-.44	1.10	=H2O + Bu3PCH2PhCl	75P042
Pt		Na3Fe(CN)6 in =		-.44	.97	=H2O	75P042
Cr(IS/Fe)		Na4Fe(CN)6-10=	108	-.019		=H2O in H2O + glycerol	75B031
Cr(IS/Fe)		Na4Fe(CN)6-10=	108	-.021		=H2O in H2O	75B031
Pd		Pr4NFe(EDTA) =	78	.79	1.67	=in H2O, pH=9.0, struc info	75S099
CoCl2 in ace=		SS		--	--	=tone, no correlation found	75A033
CoCl2 in DMSO		SS		--	--	no correlation found	75A033
CoCl2 in DMF		SS		--	--	no correlation found	75A033

SOURCE	S-TEMP	ABSORBER	A-TEMP	SHIFT	QS	REMARKS	REF
CoCl2 in aceto=		SS		--	--	=nitrile, no corln found	75A033
xx		SnBr4 in THF		--		solvation study	75V029
xx		SnBr4 in ace=		--		=tic anhydride,solvation study	75V029
xx		SnBr4 in ace=		--		=tonitrile, solvation study	75V029
xx		SnBr4 in ace=		--		=tone, solvation study	75V029

GLASSES AND AMORPHOUS SUBSTANCES

SOURCE	S-TEMP	ABSORBER	A-TEMP	SHIFT	QS	REMARKS	REF
Pd		CaO-SiO2-Fe2O3		--	--		73N027
xx		CoO-Fe2O3-B2O3				study of magnetic properties	75B099
xx		Fe2O3-Na2O3-=		--	--	=SiO2	75P009
Cu		Fe2O3-8B2O3	v			400<T<700 K, activation energy	74J016
xx		Fe75P15C10				rf induced sideband observed, amorphous magnetic material	75C058
Cu		Na2O-B2O3-Fe2O3	R	--	--	structure change around 20-25 mole% Na2O2	75R008
Cu		xNa2O-6Fe2O3-=		---	--	=(94-x)B2O3, structure info	74R041
xx		9B2O3-2Al2O3-=		--	--	=2Li2O, IS, QS, & W vs Fe2O3 concentration	75E014

INORGANIC HALIDES

SOURCE	S-TEMP	ABSORBER	A-TEMP	SHIFT	QS	REMARKS	REF
xx(IS/Fe)		AlFeF5-2H2O	80	1.47	3.01		75S049
xx(IS/Fe)		AlFeF5-2H2O	195	1.40	2.91		75S049
xx(IS/Fe)		AlFeF5-2H2O	293	1.35	2.40		75S049
Cu(IS/Fe)	293	AlFeF5-7H2O		1.38	3.63	structure information	75S010
Cu(IS/Fe)	293	AlFeF5-7H2O		1.24	3.27	structure information	75S010
Cu(IS/Fe)	293	AlFeF5-7H2O		1.30	3.51	structure information	75S010
Cu(IS/Fe)	293	AlFeF5-7H2O	195	1.30	3.51	bonding and structure info	75S063
Cu(IS/Fe)	293	AlFeF5-7H2O	293	1.24	3.27	bonding and structure info	75S063
Cu(IS/Fe)	293	AlFeF6-7H2O	80	1.38	3.63	bonding and structure info	75S063
Pd(IS/SS)		BaFeF5	300	+.58	.70		75A024
xx		CoSiF6-6H2O	v		--	study of elec/magn properties	75V041
Pd		CsFeCl3	300	1.086	1.505		75L026
xx		CsNiCl3	v			study of the magnetic behavior	75M079
Pd		Cs2FeCl4	77	.991	3.042		75L026
Pd		Cs2FeCl4	300	.876	1.276		75L026
Pd		Cs3FeCl5	300	.889	.505		75L026
Pd		DNA + FeCl2	300	--	---		75G018
xx		Fe.3Zn.7SiF6-6=	v		--	=H2O, elec & magn properties	75V041
xx		FeCl2	4.2			description given of the magneto-elastic coupling	75N022
Pd		FeCl2	77	1.182	3.085		75G018
Cr(IS/SNP)		FeCl2(N2H2)2	4.2	1.38	2.12		75S106
Cr(IS/SNP)		FeCl2(N2H2)2	295	1.25	1.81		75S106
Cu		FeCl3	R	.44			74I016
Pd		FeCl3	300	.230	0		75G018
xx		FeCl2-4H2O	4.2		3.08	HA=40 kG, eta=.3	75S026
Rh	4.2	FeCl2-4H2O	4.2			study of magn induced texture	75N002
Fe		FeCl2-4H2O	R		--	polarized source exp, detn of efg tensor, eta=.40, fa=.28	75G023
Pd		FeCl2-4H2O	300	1.070	2.930		75G018
xx		FeCl2-4H2O	v		--	efg determination, HA	75Z012
Cu		FeCl3-1.7H2O	R	.44			74I016
Cu		FeCl3-3H2O	R	.38	.36		74I016
Cu		FeCl3-6H2O	R	.45	.83		74I016
xx(IS/Fe)		FeCoF5-2H2O	80	.57	.52		75S049
xx(IS/Fe)		FeCoF5-2H2O	293	.48	.55		75S049

SOURCE	S-TEMP	ABSORBER	A-TEMP	SHIFT	QS	REMARKS	REF
Cu (IS/Fe)	293	FeCoF5-7H2O	80	.53	.58	bonding and structure info	75S063
Cu (IS/Fe)	293	FeCoF5-7H2O	293	.44	.44	bonding and structure info	75S063
Pd (IS/SS)		FeF2	300	+1.30	2.70		75A024
Cr		FeF3				study of the effect of pressure on the magnetic field	75N029
Pd (IS/SS)		FeF3	300	+.58	0	HI=408 kG	75A024
Cr		FeF3	v			study of the pressure dependence of the Neel temperature	75N028
Cu		FeF3	v			study of small angle Mossbauer scattering near magnetic phase transitions	75S018
xx		FeI2				study of phase transitions due to changes in HA	75D036
xx		FeI2	v		--	1.5 K<T<40 K, study of magn phase transitions	75D012
xx		Fe(ND3)6(BF4)2	v		--	study of isotope effect	75B012
xx		Fe(ND3)6Cl2	v		--	study of isotope effect	75B012
xx		Fe(ND3)6SiF6			--	study of isotope effect	75B012
xx		Fe(NH3)6(BF4)2	v		--		75B012
xx		Fe(NH3)6(BF4)2	v		--	study of phase transition	75A008
xx		Fe(NH3)6Br2	v		--	study of phase transition	75A008
xx		Fe(NH3)6Cl2	v		--	study of phase transition	75A008
xx		Fe(NH3)6Cl2	v		--		75B012
xx		Fe(NH3)6I2	v		--	study of phase transition	75A008
xx		Fe(NH3)6SiF6			--		75B012
xx		Fe(NO)2Br		--	--		75L024
xx		Fe(NO)2Cl		--	--		75L024
xx (IS/Fe)		FeNiF5-2H2O	80	.53	.50		75S049
xx (IS/Fe)		FeNiF5-2H2O	293	.44	.43		75S049
Cu (IS/Fe)	293	FeNiF5-7H2O	80	.53	.53	bonding and structure info	75S063
Cu (IS/Fe)	293	FeNiF5-7H2O	293	.42	.52	bonding and structure info	75S063
Rh		FeSiF6	R		--	HA=75 kOe, efg<0	75C011
xx		FeSiF6-6H2O	R			discussion of asymmetry parameter	75C034
xx		FeSiF6-6H2O	v		--	study of elec/magn properties	75V041
Pd		FeSnF6				structure information	75S098
Pd		FeSnF6-6H2O				structure information	75S098
xx (IS/Fe)		FeZnF5-2H2O	80	.55	.55		75S049
xx (IS/Fe)		FeZnF5-2H2O	195	.51	.58		75S049
xx (IS/Fe)		FeZnF5-2H2O	293	.46	.59		75S049
Cu (IS/Fe)	293	FeZnF5-7H2O		.53	.52	structure information	75S010
Cu (IS/Fe)	293	FeZnF5-7H2O		.48	.47	structure information	75S010
Cu (IS/Fe)	293	FeZnF5-7H2O		.41	.47	structure information	75S010
Cu (IS/Fe)	293	FeZnF5-7H2O	80	.53	.52	bonding and structure info	75S063
Cu (IS/Fe)	293	FeZnF5-7H2O	195	.48	.47	bonding and structure info	75S063
Cu (IS/Fe)	293	FeZnF5-7H2O	293	.41	.47	bonding and structure info	75S063
xx (IS/Fe)		Fe2F5-2H2O		--	--		75S049
Cu (IS/Fe)	293	Fe2F5-7H2O	80	--	--	structure information	75S010
Cu (IS/Fe)	293	Fe2F5-7H2O	195	--	--	structure information	75S010
Cu (IS/Fe)	293	Fe2F5-7H2O	293	--	--	structure information	75S010
Cu (IS/Fe)		Fe2F5-7H2O	v	--	--	T=80, 195, & 293 K, 2 Fe sites	75S063
Pd		Fe4N3F3	293	.384	.745		75M078
Pd		K2FeCl4	77	--	--		75L026
Pd		K2FeCl4	300	--	--		75L026
Pd		K3FeF6		.241		comparison with ESCA	75J003
Cr (IS/SNP)		KCoF3	v			Neel T=118K, Debye T=350 K	75R038
Pd		KFeCl3	300	1.098	2.374		75L026
xx		KFeCl3	v			1.5<T<300 K, relaxation study	75P044
xx		KFeCl3	v	--	--	4.5<T<360 K	75P032
Cu	R	KFeCl3	v	--	--	QS vs temperature, 1.5<T<360 K	75P045
xx (IS/Fe)		KFeF3	v	--	--	4.2 K<T<300 K, Neel T=112 K, HI data	75I003
xx (IS/Fe)		KFeF3	v	--	--	single crystal study, magnetic structure info, 4.2<T<300 K	75I009
Cr (IS/SNP)		KFeF3	v			Neel T=110K, Debye T=350 K	75R038
xx		KMgF3(Fe+1)		--		calculated shifts	74L046
xx		KMgF3(Fe+2)		--		calculated shifts	74L046
xx		KMgF3(Fe+3)		--		calculated shifts	74L046
Cr (IS/SNP)		KMnF3	v			Neel T=86 K, Debye T=350 K	75R038
Cr (IS/SNP)		KNiF3	v			Neel T=245K, Debye T=350 K	75R038
xx		MgF2 + Fe	v	--	--	IS vs T, 1.2<T<300 K, HA, amorphous magnet	75S046

SOURCE	S-TEMP	ABSORBER	A-TEMP	SHIFT	QS	REMARKS	REF
xx		MgSiF6-6H2O	v		--	study of elec/magn properties	75V041
xx		MnSiF6-6H2O	v		--	study of elec/magn properties	75V041
Pd		(NH4)3FeF6		.247		comparison with ESCA	75J003
Pd		NOFeCl4	77	.33	.36	bonding and structure info	75A032
xx		NaCl				ion implanted absorber	75S058
xx		Na2MnFeF7				structure information	75P046
xx		NiSiF6-6H2O	v		--	study of elec/magn properties	75V041
Pd		RbFeBr3	v		--	1.54<T<400 K, QS vs T	75L004
Pd		RbFeCl3	v		--	4.2 K<T<460 K	75E001
xx		ZnSiF6-6H2O	v		--	study of elec/magn properties	75V041
Pd		adenosine + Fe=	300	.190	.670	=C13	75G018
Pd		adenosine + Fe=	v	--	--	=C12, T=77 and 300 K	75G018
Pd		cytidin + FeCl3	300	.170	.730		75G018
Pd		guanosine + Fe=	300	.210	.660	=C13	75G018
Pd		guanosine + Fe=	v	--	--	=C12, T=77 and 300 K	75G018
Pd		thymidine + Fe=	300	.140	0	=C13	75G018

IRRADIATION EXPERIMENTS

SOURCE	S-TEMP	ABSORBER	A-TEMP	SHIFT	QS	REMARKS	REF
Ag	4.2					electron irradiation, study of the formation of a defect	75M041
Al	v	xx				electron irradiation, study of the kinetics of defects	75M040
Al	v	xx			--	study of trapped atoms, Debye temperature data, 4.2K<T<140 K	75V012
Al	v	xx (IS/Fe)		--		obsvn of defects created by neutron irradiation	75M027
Cu		Au-Fe alloy				14-MeV neutron irradiation, formation of clusters	75B005
Pd		DNA + FeCl2	300	--	--	UV irradiation	75G018
xx		alpha-Fe				effect of laser irradiation	75D040
Cu		FeC2O4-2H2O	R			radiolysis study	75K031
xx		FeCl2-4H2O	R			study of the effects of proton irradiaton	75K053
Pt		Fe-Ni alloys	R	--		effect of defects on distribution of atoms	75R020
Cu		FeOx2(OH)2				study of the effects of thermal and radiolytical treatment	75F022
Pt		FeSO4-7H2O				study of effects of proton irradiation	75K038
Pt		FeSO4-7H2O	R	--	--	proton irradiation	75K015
Pt		K3Fe(CN)6				study of effects of proton irradiation	75K038
Pt		K3Fe(CN)6	R	--	--	proton irradiation	75K015
Pt		K3Fe(CN)6-3H2O				study of effects of proton irradiation	75K038
Pt		K4Fe(CN)6-3H2O	R	--		proton irradiation	75K015
Pt		NH4Fe(SO4)2-12=	R	--		=H2O, study of the effect of proton irradiation	75K013
Pt		(NH4)2Fe(SO4)2=	R	--	--	=-6H2O, study of the effect of proton irradiation	75K013
Pd		adenosine + Fe=	300	.160	.650	=C13, UV irradiation	75G018
Pd		adenosine + Fe=	300	1.210	2.420	=C12, UV irradiation	75G018
Cr		biotites		--	--	study of the affect of radiation from uranium	75I006
Pd		cytidin + FeCl3	300	.170	.670	UV irradiation	75G018
Pd		guanosine + Fe=	300	.190	.640	=C13, UV irradiation	75G018
Pd		guanosine + Fe=	300	--	--	=C12, UV irradiation	75G018
xx		haematite				study of phase transition caused by gamma irradiation	75A050
xx		magnetite				study of phase transition caused by gamma irradiation	75A050
xx		wustite				study of phase transition caused by gamma irradiation	75A050

METALS AND ALLOYS

SOURCE	S-TEMP	ABSORBER	A-TEMP	SHIFT	QS	REMARKS	REF
xx		Ag				HI data, HA	75T017
xx		Ag	v			HI data	75S100
Pd	R	AgPd	v			1.6<T<10.2 K, spin-glass behavior	75L029
xx		Al				ion implanted absorber	75S058
xx		Al				ion implantation, study of the effect of different dose	75S061
xx		Al		--	--		75S088
xx		Al				ion implantation	75S101
Pd		Al				study of impurity-vacancy bond	75W017
xx		Al-C-Fe alloys				study of phase transformations	75V011
xx		Al-Fe alloys				detn of short range order	75B051
xx		Al-Fe alloys		--		study of IS vs number of Fe atoms in first correlation sphere	74L078
xx (IS/Fe)	R	Al-Fe alloys		--	--	identification of different Fe environment	75N010
xx		Al-Fe alloys	5				75S060
Pd (IS/Fe)		Al-Fe alloys	v	--	--	T=80 & 300 K	75J009
xx		Al-Fe-Ho alloys	v				75D005
xx		AlFeHo	v				75D005
xx (IS/Fe)	R	AlNi3	4.2			HI=238 kOe	75D014
xx (IS/Fe)	R	AlNi3	R	--	--		75D014
xx		Al3FeLu2	v				75D005
Xx (IS/Fe)		Armco steel	N	.49	.66		75K056
xx		Au				ion implantation study	75S062
Pd		Au		--		HI data	75V010
xx		Au	v			.003<T<.36 K, HI vs T	75B075
xx		Au	v			HI data	75S100
xx (IS/Fe)		alpha-Be		--	--	ion implantation	75B064
xx		Be				ion implantation	75S101
xx		Be				ion implanted absorber	75S058
Pd		Be		--		HI data	75V010
Pd		Be	v	--	--	80 K<T<1123 K, investigation of electronic properties	75J002
Pd		Ce(CoxFe(1-x))2	v			Curie temp vs composition	75L001
xx		Co				ion implantation study	75S062
xx		Co-Cr-Fe alloy	R	--		spinodal decomposition	75T022
xx		CoFe	v	--		4.2<T<1000 K, Debye T data	75C031
Pd		Co-Fe-Ni alloys				study of ordering and decomposition	75M087
xx		Co-Fe-V alloy				external rf field	75B007
Pd		CoNi3		--		study of atomic rearrangement	75Z002
xx		Co(2-x)FexP(=				=1-y)AsY, band model proposal	75W013
xx		Cr				ion implantation study	75S062
xx		Cr-Fe alloys				HI vs Cr concentration study	75D033
xx		Cr-Fe alloys		--		study of HI distribution	75D032
xx		Cr-Fe alloys	R	--		IS vs Cr concentration	75D035
Pd		Cr-Fe alloys	R			study of the effect of neighboring chromium atoms on HI	75D031
xx		Cr-Fe alloys	v			80<T<488 K	75H012
xx		Cr-Fe-Mn alloys		--	--	influence of plastic deformation on the Mossbauer effect	74M065
Pt		Cr-Fe-Ni alloys				study of the effect of alloying with Cr	75B143
Pt		Cr-Fe-Ni alloys					75B057
Cr		Cr-Fe-Ni-C alys			--	study of deformed martensite	75M036
Rh	N	Cr(1-x)Fe	v	--		.2<x<.3, superparamagnetic	75L030
xx		Cr(2-x)FexP(=				=1-y)AsY, band model proposal	75W013
xx (IS/Fe)		Cu		--	--	ion implantation	75B064
Pd		Cu		--		HI data	75V010
xx		Cu	v			HI data	75S100
xx		Cu	v	--		295 K<T<780 K, thermal shift	75W005
xx		Cu-Ti-Zn alloy				corrosion study	72M076
xx		Dy(FexAl(1-x))2				magnetic structure information	75G041
Pd		Dy(FexAl(1-2))2	77			HI data	75G014
Pd		Dy(FexAl(1-2))2	300			HI data	75G014
xx		Dy(FexNi(1-x))2	4.2			HI vs composition	75B134

SOURCE	S-TEMP	ABSORBER	A-TEMP	SHIFT	QS	REMARKS	REF
xx		Dy(FexNi(1-x))2	v	--		HI data	75B085
Cu		Dy(FexNi(1-x))2	v	--		HI & IS vs T & composition	74B094
xx		Dy(FexNi(1-x))3				electronic structure info	75C033
Pd		Dy(Fe(1-x)Cox)3	v	--		T=77 K,300K, and 715K HI data, structure information	75A015
xx		Dy(Fe(1-x)Mnx)2					75W023
xx		DyFe2	v		--	distribution of HI & QS values	75F032
Cu		DyFe2	v		--	77 K<T<633 K, Curie T=633 K	75B027
xx		DyFe2-ZrCo2		--			75W023
Pd	R	DyFe3	77	--		detn of direction of magnetism	75A007
xx		DyFe3	v			3<T<800 K, magnetic info	75V020
xx		DyxFe(1-x)	R			HI vs composition	75H041
xx		Er(Fe(1-x)Mnx)2					75W023
Cu		ErFe2	v	--		77 K<T<576 K, Curie T=576 K	75B027
xx		ErFe2-ZrCo2		--			75W023
Pd	R	ErFe3	77	--		detn of direction of magnetism	75A007
xx		ErFe3	v			3<T<800 K, magnetic info	75V020
xx		ErFe3	v	--	--	HI and QS vs T, 4.2 K<T<700 K	75A014
Rh(IS/SNP)	295	ErFe3	v	--		magnetic structure information	75V024
xx		Er2Fe17	v			magnetic structure info	74G061
xx		alpha-Fe	--	--		study of the influence of uniaxial stress	75K049
Pd	295	alpha-Fe	295			g0=3.919 mm/s, g1=2.238 mm/s	75W004
xx		alpha-Fe	v			discussion of previous data	75K034
xx		epsilon-Fe	v			search for magnetic ordering at high pressures down to temperatures of 2.2 K	75K052
xx		Fe				ion implantation study	75S062
Pd		Fe				measurement of the spin contact density	75B127
Pd		Fe				use of a computerized piezoe-lectric spectrometer	75S041
xx		Fe	v			thin film, HI vs T, 4<T<650 T	75S057
Pd		Fe	v			T=4.2, 77, & 295 K, thin films	75L012
xx		Fe(V)	v			94 K<T<1089 K, Curie T & HI	75V005
xx		chi-Fe carbide	v	--		study of local magnetic properties during phase trans-formations	75A019
xx		epsilon-Fe =	v	--		=carbide, study of local magn properties during phase trans-formations	75A019
xx(IS/Fe)	R	gamma-Fe in Cu	v	--			75W015
xx(IS/Fe)		Fe on Cu	v	+.03		thin film, 1.8<T<295 K	75K044
xx		Fe(Al)	v			HI data, Curie T vs Al conc	75V005
Cr		Fe(C) alloys				structure information	75G007
Cr		Fe-C alloys		--	--	deformation study	75G069
xx(IS/Fe)		Fe-C martensite	v	--	--	study of the transformation of eta carbide	75G017
Pt		Fe-C steels	78			structure information	75G079
xx		Fe(Co)	v			HI data, Curie T vs Co conc	75V005
Cr		Fe-Co(C) alys				structure information	75G007
xx		Fe(Cr)	v			HI data, Curie T vs Cr conc	75V005
xx		Fe-Gd alloys	v			3.6 K<T<820 K, magn struc info	75V021
Pd		Fe-Ge alloys	300	--	--		75J012
Pt		Fe-Mn-C steels	78			structure information	75G079
xx		Fe(Mn)	v			HI data, Curie T vs Mn conc	75V005
Pd		Fe-Mn alloy	v				75Z017
xx		Fe(Mn.67Ni.33)=	v			=(1-x), Neel Temperature study Neel T of gamma Fe is 67 K<T< 90 K	75E005
xx		Fe-Mo alloys		--		study of ageing	75G080
Pd		Fe-Mo alloys	v	--	--	study of nitrogenation	75H032
xx		Fe-Mo-Ni-C aly				HI data	75B078
xx		Fe-N austenite	183			fa data	75D022
Pd		gam-Fe-Ni alys	R			study of atomic order	75P012
Cr		gam-Fe-Ni alys	v	--		study of magnetic properties	75C017
xx		Fe-Ni alloys				high pressure experiment	75B050
xx		Fe-Ni alloys					75B023
xx		Fe-Ni alloys				discussion of previous data	75M022
xx		Fe-Ni alloys				external rf field	75B007
xx		Fe-Ni alloys				structure information	74G066
Pd		Fe-Ni alloys				HI data	75Z010

64

SOURCE	S-TEMP	ABSORBER	A-TEMP	SHIFT	QS	REMARKS	REF
xx		Fe-Ni alloy	R			study of magn inhomogeneity	75A040
xx		Fe-Ni alloy	R			HI vs annealing times from 1000 C	75A049
Pt		Fe-Ni alloy	R			study of the effect of deformation temperature	75R019
xx		FeNi	v	--		4.2<T<1000 K, Debye T data	75C031
xx		Fe-Ni alloys	v			structure information	75J007
xx		Fe-Ni alloys	v			4.2<T<700 K. HI data	74M063
xx		Fe(Ni)	v			HI data, Curie T vs Ni conc	75V005
Cr		Fe-Ni alloys	v	--	--	T=1093 K & 1323 K, struc info	74R038
Cr		Fe-Ni alloys	v			study of phase transitions	75M077
Pd		Fe-Ni alloy	v			288 K<T<723 K	75F008
Pd(IS/Fe)		Fe-Ni alloys	v	--	--	T=80 & 300 K	75J009
SS		Fe-Ni alloys	v			4.2<T<600 K, superparamagnetic regions	74M064
xx		Fe-Ni-C alloys				structure information	74G066
Pd		Fe-Ni-C alloy	22	--		determination of magnetic hyperfine field distribution	75G038
Pd		Fe-Ni-C marten=				=site, HI vs carbon and nickel concentrations	75L010
Pt		Fe-Ni-C steels	78			structure information	75G079
xx		Fe-Ni-Ti alloys				structure information	74G066
xx		(Fe,Ni)2Mo				structure information	75C063
xx		FeNi3				study of disorder-order transformation	75F023
xx		FeNi3				study of short range order	75P027
xx		FeNi3	v			study of order-disorder trans, 4<T<870 K	75D009
Pd		Fe-Pd alloys	v			HI vs Fe concentration	75T019
xx(IS/Fe)		Fe-Ru alloy	v	--	--	4.2<T<873 K, anomalous behavior of low T QS in HA	75W027
xx		FeSb	v		--	discussion of previous data	75G050
Cu(IS/SNP)		Fe-Se alloys	300	--	--	system is paramagnetic	75L020
xx		Fe(Si)	v			HI data, Curie T vs Si conc	75V005
Cu		Fe-Si steels	R	--		study of internal magn fields	74B093
xx(IS/Fe)		FeSn	295	.41	.56		75H009
xx(IS/Fe)		FeSn	395	.34	.54		75H009
Pd	R	FeSn	v	--		20<T<395 K, spin struc info, Neel temperature=368 K	75H007
xx		FeSn2	v			fa vs temp, 300 K<T<730 K	75N009
Pd(IS/Fe)		FeTi		-.145		calibration constant detn	75M005
xx		Fe(Ti)	v			HI data, Curie T vs Ti conc	75V005
Pd		Fe-Ti alloys	v	--	--	study of nitrogenation	75H032
xx		Fe-W alloys		--		study of ageing	75G080
Pd	R	FeY	77	--		detn of direction of magnetism	75A007
Rh(IS/Fe)		FeZn13	v	--	--	4.2<T<295 K	75W014
xx		FexGd(1-x)	R			HI vs composition	75H041
xx		FexHo(1-x)	R			HI vs composition	75H041
xx		FexTb(1-x)	R			HI vs composition	75H041
xx		Fe(1-x)CoxAl		--		IS vs composition, bond info	74R039
xx		Fe(1-x)NixAl		--		IS vs composition, bond info	74R039
Pt		Fe(1+x)Sb	v	--	--	10<T<625 K, Debye T=766 K, Neel T=211 K	75R027
Cu		Fe2Ho	v		--	77 K<T<596 K, Curie T=596 K	75B027
Cr	295	Fe2TbxY(-x)			--	self-induced magnetostatic distortion	75P052
Cu		Fe2Tm	v		--	77 K<T<563 K, Curie T=563 K	75B027
Pd		Fe2U-Co2U			--	study of magnetic properties	75P051
Cu		Fe2Y	v		--	77 K<T<533 K, Curie T=533 K	75B027
Pt		Fe3(Al,Ge) alys				study of the ordering	74G060
Pd	R	Fe3Gd	77	--		detn of direction of magnetism	75A007
xx		Fe3Gd	v			3<T<800 K, magnetic info	75V020
Pd	R	Fe3Ho	77	--		detn of direction of magnetism	75A007
xx		Fe3Ho	v			3<T<800 K, magnetic info	75V020
xx		Fe3Ho	v	--	--	QS vs temperature, 4.2 K<T<700	75A014
Cu(IS/SNP)		Fe3Sb2	300	.68	.18	is paramagnetic	75L020
Pd	R	Fe3Tb	77	--		detn of direction of magnetism	75A007
xx		Fe3Tb	v			3<T<800 K, magnetic info	75V020
xx		Fe3Tb	v	--	--	HI and QS vs T, 4.2 K<T<700 K	75A014
xx		Fe3Th	v			3<T<800 K, magnetic info	75V020
Pt	R	Fe3Y	80	--	--	4 Fe sites	75F003
Pt	R	Fe3Y	295	--	--	4 Fe sites	75F003

SOURCE	S-TEMP	ABSORBER	A-TEMP	SHIFT	QS	REMARKS	REF
xx		Fe3Y	v			3<T<800 K, magnetic info	75V020
xx		Fe3Y	v	--	--	HI and QS vs T, 4.2 I<T<700 K	75A014
Rh(IS/SNP)	295	Fe3Y	v	--		magnetic structure information	75V024
xx		Fe(3-x)MnxAl					75W023
xx(IS/SNP)		Fe5Th	v	--	--	3<T<700 K	75G045
Cu(IS/SNP)		Fe7Nd	300	.06	.06	HI=165 kOe	75L020
Cu(IS/SNP)		Fe7Nd	340	.24	.74		75L020
Cu(IS/SNP)		Fe7Pr	270	.2			75L020
Cu(IS/SNP)		Fe7Pr	300	.84	.57		75L020
xx		Fe17Tb2	v			magnetic structure info	74G061
xx		Fe17Yb2	v			magnetic structure info	74G061
xx		Gd(Fe(1-x)Mnx)2					75W023
xx		GdFe2-ZrCo2		--			75W023
xx		Ge				ion implantation study,2 peaks	75S062
Pt		Ho.6Tb.4Fe2	v			77<T<200 K, study of spin rotation	75D075
xx		Ho(FexNi(1-x))2	4.2			HI vs composition	75B134
xx		Ho(FexNi(1-x))2	v	--		HI data	75B085
xx		Ho(Fe(1-x)Mnx)2					75W023
xx		HoFe2-ZrCo2		--			75W023
xx		HoxEr(1-x)Fe2	v			spin reorientation studies	75A052
xx		In				ion implantation study,2 peaks	75S062
xx(IS/Fe)		alpha-Mn		--	--	ion implantation	75B064
xx		Mn				ion implantation study,2 peaks	75S062
Pd		Mn steels				study of the influence of plastic deformation on phase transformation	75S085
xx		MnxNi(1-x)	v			study of the distribution of magnetic hyperfine fields	74R035
xx		Mn(2-x)FexP(=				=1-y)AsY, band model proposal	75W013
xx		Mn95Cu5	v			4.2<T<500 K, HA up to 50 kOe	69E022
xx		Mo				ion implantation study	75S062
xx		Nb				ion implantation study	75S062
xx		Ni				ion implantation study	75S062
xx		Ni		--		corln of IS & positron annihilation	74D056
xx		Ni	v			critical coeffeicient beta	75H034
xx		Ni	v			measurement of the critical coefficient beta	75H033
Cu		Ni	v			study of small angle Mossbauer scattering near magnetic phase transitions	75S018
Pd		Ni	v			study of relaxation near Curie point	75K086
Pd		NixBh(1-x)	v			.05<T<11 K, HA up to 5.6 T	75S065
xx		Ni(2-x)FexP(=				=1-y)AsY, band model proposal	75W013
xx		Pb				ion implantation study,2 peaks	75S062
xx		Pd				ion implantation study	75S062
xx		Pd		--		corln of IS & positron annihilation	74D056
Pd		Pd	v			study of lattice dynamics, 150 K<T<700 K	75P017
xx		Pt				ion implantation study	75S062
xx		Pt		--		corln of IS & positron annihilation	74D056
Fe52Ni48	v	SS				4.2 K<T<750 K, HI vs temp, Curie temperature=735 K	75B054
xx		Sc				ion implantation study,2 peaks	75S062
xx		Se				ion implantation study,2 peaks	75S062
xx		Sn				ion implantation study,2 peaks	75S062
xx		Tb(Fe(1-x)Mnx)2					75W023
xx		Th(Co(1-x)Fex)5	78			HI data	75B086
xx		Ti				ion implantation study	75S062
Pd		Ti	v			study of lattice dynamics, 150 K<T<700 K	75P017
xx		V				ion implantation study	75S062
xx		V				fa vs pressure	75M061
xx		W				ion implantation study	75S062
xx		Y				ion implantation study,2 peaks	75S062
Pd		Y2(Fe(1-x)Cox)=	77	--	--	=17, HI vs comp, 4 sites	75S024
Pd		Y2(Fe(1-x)Cox)=	300	--	--	=17, HI vs comp, 4 sites	75S024
xx		Y(FexCo(1-x)2=	v			=(MgCu-type), HA, study of magnetic properties	74S126

SOURCE	S-TEMP	ABSORBER	A-TEMP	SHIFT	QS	REMARKS	REF
xx		Y(Fe(1-x)Cox)2					75W023
Pd		Y(Fe(1-x)Cox)3	v	--		T=77 K, 300K, and 715K HI data, structure information	75A015
xx		Zn				ion implantation study, 2 peaks	75S062
xx		ZrFe2-ZrCo2		--			75W023
Pt		armco-Fe				study of stress	75F017
Pt		invar				study of redistribution of carbon atoms in invar	75Z021
xx		manganese steel				study of embrittlement kinetics	75B144
xx		maraging steel		--		study of age hardening	75U003
Pt		martensite				study of isothermal & athermal reactions	75G083
xx		steel				detn of short range order	75B116
xx		steel				study of structural changes during aging	74S123
Rh		steel		--	--	study of the effect of strains on the Mossbauer parameters	75M011
Cr		steel EI69				carbide phases formed under gamma irradiation	75V051
xx		theta-Fe car=	v	--		=bide, study of local magnetic properties during phase transformations	75A019

MISCELLANEOUS EXPERIMENTS

SOURCE	S-TEMP	ABSORBER	A-TEMP	SHIFT	QS	REMARKS	REF
xx		xx				study of the effect of metal working on the Mossbauer effec	75P013
Fe	.05	xx				NMR on polarized nuclei	75C007
xx (IS/Fe)		Armco foils	R	--	--	study of the initial stage of corrosion	75K056
Pd		Co-Mo on Al2O3				study of hydrodesulfurization catalysts	75T027
xx		Cp2Fe on SiO2	R	.5	1.78	adsorption study	74M060
xx		Fe				study of the effect of rf fields as a function of Fe foil thickness & rf intensity	75K045
xx		Fe				study of the effect of rf fields as a function of coatings of Zn, Si, & adhesive	75K046
xx		Fe				evidence for surface sensitive magnetic anisotropy	75D051
xx		Fe				thickness effects for polarization experiments	75W028
Pt		Fe				FM sidebands produced by rf magnetic field	75K077
xx		Fe	v			32.8 A thick film, T=4,77,&296	75G036
xx (IS/SNP)		Fe + propene		--	--	study of catalyst reaction	75M018
xx		Fe + silicomo=				=lybdenum blue	74V040
xx (IS/Fe)		Fe Urushibara=	298	--		= catalyst	75E027
xx		Fe acetilace=				=tonate, liquid crystal study	75G052
Cu		Fe clusters in=	v	--	--	= in zeolite holes	75S091
xx		Fe electrode				study of processes which take place during cycling in alkaline solutions	75G046
Cu		Fe foil				study of oxidation	74O014
Cr		Fe on Al				conversion electron spectrometer	75G043
xx		Fe on cation =		--	--	=exchange resin Dowex 50x8	75P039
Cu		Fe oxides				study of early formation of oxide film, kinetic data, depth selection	75S034
Cu		Fe treated w/ =				=H3PO4, surface study	75S005
Cu		Fe treated w/ =				=O2, surface study	75S005
Cu (IS/SNP)		FeCl3 in H2O	300	--	--	study of the products formed by boiling solution	75K024

SOURCE	S-TEMP	ABSORBER	A-TEMP	SHIFT	QS	REMARKS	REF
xx		Fe-Co-Mo cata=		--		=lyst, study of the process of charge transfer	75M071
Cr		Fe-Cr catalyst				study of structure changes	75M072
Cr		Fe-Cr-Cu cata=				=lyst, study of struc changes	75M072
Cr		Fe-Cu-Mn cata=				=lyst, study of struc changes	75M072
Cr		Fe-MgO				catalyst study, reaction w/ H2	75B041
xx		Fe-MgO catalyst				catalyst study-hydrogen activation	75B106
xx(IS/SNP)		Fe-Mo catalyst		--	--		75M057
xx		FeNi				rf modulation study	75B072
Pd(IS/SS)		Fe-Ni aly + F2	300	--	--	identification of products	75A024
xx		FeNi3				rf modulation study	75B072
Cu		FeSiF6-6H2O				compact linearly polarized source	75V001
Pd		Fe(acac)3-Co(=				=acac)3-AlEt3 in PhMe, catalyst study	75K074
Pd		Fe+2 on amor=				=phous aluminosilicate, study of the reduction proucess	75H045
Pd		Fe+2 on silica=				=gel, study of the reduction process	75H045
Pd		Fe+2 on zeo=				=lite Y study of the reduction process	75H045
Cr		alpha-Fe2O3	R			interference between different transitions of magn hyperfine struc in nuclear diffraction	75S042
Cr		Fe2O3				study of gas anodizing for Fe	75G073
Cr		Fe2O3 + MgO				analysis of a chemical rxn	75V044
Pt		Fe2O3 + SiO2				study of the affect of absorbants of NH3, propylamine, pyridine and BF3	75H016
Cu(IS/SNP)		Fe2O3-Cr2O3-K2O				study of catalyst used for the dehydrogenation of ethylbezene	75P037
Pd		Fe+3 colloidal=	R		--	=solution, thermal treatment	75C054
Cu(IS/SNP)	R	Fe+3 gels	R			study of thermal aging	75K025
xx		Fe3O4 + CO				study of kinetics of this reaction	75K084
xx		Fe3O4 in gly=	V			=cerine, study of Brownian motion	75K002
Cu(IS/298)	298	MgO + Fe		--	--	micro particles of Fe on MnO2	75B058
SS		Mn + Fe(OH)3		--	--	study of Fe(OH)3 coprecipitation with Mn	75S004
Cr(IS/SNP)		Na A-type Zeo=	V	--	--	=lite, T=4.2, 78, and 293 K	75B096
Pd(IS/SS)		NiF2 + FeF3	300	--	--	identification of products	75A024
xx		NiFe2O4 with =	V			=organic coating, HA data	75B048
Cu		PFC				time filter experiment	75K040
Pd		PFC				time dependence Mossbauer resonance scattering	75D027
Rh		PFC				time dependence Mossbauer resonance scattering	75D027
SS		SFC				study of line broadening & fa	75K068
xx		SS				no rf sidebands observed	75C035
xx		SS on Ag				no rf sidebands observed	75C035
xx		SS on Cu				no rf sidebands observed	75C035
xx		SS on Ni				rf sidebands shown to be an acoustic effect	75C035
Cr		310 SS-SiO2	V			separation of elastic and inelastic scatterings in SiO2	75A027
Pt		air sample	V	--	--	identification of alpha-Fe2O3, T=78 K and 300 K	75D020
xx		air samples	R			corln of Fe in air with meteor activity	75D058
Rh		carbon steels				study of oxidation on surface using the conversion electrons	75T018
Cr(IS/Fe)		catalysts	V			study of ammonia synthesis catalysts, 300<T<473 K	75M033
xx		cellulose				study of Fe(II) and Fe(III) bonding to cellulose	75J004
xx		cholesteri me=				=sophases	75G051
Cu	4.2	clay	4.2	--	--	study of transformations induced in clay by firing	75S078
xx		colloidal par=	V			=ticles, fa	75O012
xx		fossils				study of the fossilization process of plants and animals	75E021

SOURCE	S-TEMP	ABSORBER	A-TEMP	SHIFT	QS	REMARKS	REF
Rh		glycerol	v			scattering experiment, Debye temp=97 K, glass-transition at 185 K	75C005
xx		graphite + Fe=		--	--	=Cl3, study of the structure of layered graphite compounds using shock waves	75K080
xx		ion exchange =		--	--	=resin	75P036
xx		ion exchange =		--	--	=resin	75M055
Pd		iron ore				rapid phase analysis of Fe ore	75R037
xx		magn recording=				= tape	75S076
Pt		magnetite =		--	--	=films on MgO, study of the deposition	73B100
xx (IS/SS)		medicinal cal=	R	--	--	=cinated	75M060
xx		nematic meso=				=phases	75G051
Pd		oxide films on=				= low-carbon steel	75V045
xx		passive films		--			74E021
xx		pentaerythritol				diffraction exp, separation of elastic from inelastic parts of the scattered radiation	75B069
xx		permalloy				application of rf field causes collapse of 6 line spectrum	75K047
xx		silicomolybic =		--	--	=blue	75V018
xx		silicotungstic=		--	--	= blue	75V018
Cr		silicotungstic=		--	--	= acid-base metal polymers	75D029
xx (IS/Fe)		spinel catalyst		--			75M056
xx		steel				study of high-temperature oxidation	75S028
Cu		steel + H2SO4	R	--	--	study of the action of an organic inhibitor on corrosion	75B142
xx		steels	v			corrosion studies	75V030

MISCELLANEOUS INORGANIC COMPOUNDS

SOURCE	S-TEMP	ABSORBER	A-TEMP	SHIFT	QS	REMARKS	REF
xx		Al(NO3)3-9H2O				single crystal	75A038
Cr	R	Al(NO3)3-9H2O				single crystal, HA	75A022
Cu (IS/Fe)		BaFe2Se3	300	.47	.71	electronic structure info	75R001
Cu (IS/Fe)	R	Ba2FeSe3	78	.76	2.72	electronic structure info	75R001
Cu (IS/Fe)	R	Ba2FeSe3	300	.66	2.56	electronic structure info	75R001
Cu (IS/Fe)	300	Co(1-x)FexB	4.2	--	--	bonding information	75D002
xx (IS/Fe)	R	Co(1-x)FexSi	v		--		75M008
xx		(Cr,Fe)xC	R				75K048
Cu (IS/Fe)	300	Cr(1-x)FexB	4.2	--	--	bonding information	75D002
Cr		epsilon-Fe =	v	--		=nitride, T=80 and 300 K,HI	75M025
Cu (IS/Fe)	300	alpha-FeB	4.2	.35	.4	bonding information	75D002
Cu (IS/Fe)	300	beta-FeB	4.2	.39	--	bonding information	75D002
xx (IS/Fe)		FeC2O4-2H2O	80	1.29	2.12		75C053
xx (IS/Fe)		FeC2O4-2H2O	190	1.27	2.03		75C053
SS		FeC2O4-2H2O	R	1.35	1.70	2 forms with same results	75A011
xx (IS/Fe)		FeC2O4-2H2O	297	1.17	1.72		75C053
xx (IS/Fe)		FeC2O4-2H2O	v	--	--	thermal decomposition study	75C053
Pd		FeC2O4-2H2O	v			study of thermal decomposition in an oxidizing atmosphere, 80<T<297 K	75C025
xx		FeCO3	v			4.2<T<208 K	75S067
xx (IS/Fe)		FeCO3	v	--	--		75N001
Rh	v	FeCO3 in C	v			study of magn induced texture	75N002
xx		Fe(ClO4)2(H2O)6	v		--	pseudohexagonal to monoclinic phase transition, discussion of previous data	75C020
Pd	R	FeCr2Se4	v		--	77.5 K<T<300 K HI and eta data, Neel temperature=218 K	75H019
xx (IS/Fe)		FeGe	295	.28	.60		75H009
xx (IS/Fe)		FeGe	400	.21	.57		75H009
xx		FeGe	v	--	--	3 Fe sublattices, 4.2<T<368 K	75M035
Pd		FeGe	v		--	10<T<430 K, Neel T=398 K	75H008

SOURCE	S-TEMP	ABSORBER	A-TEMP	SHIFT	QS	REMARKS	REF
xx (IS/Fe)		FeGe (cubic)	v		--	Neel T=275.5 K	75E011
xx		FeGe (hex)	v	--	--	10<T<390 K, detn of hyperfine parameters	74H053
Cu		FeGe2	4.2	.32		0 kOe<HI<80 kOe	75B029
Cu		FeGe2	R			0 kOe<HI<80 kOe	75B029
Pd		FeH2N2		-.058		comparison with ESCA	75J003
xx		Fe (NH3) 6 (CLO4) 2	v		--	study of phase transition	75A008
xx (IS/SFC)		Fe (NH3) 6 (NO3) 2	v	--	--	phase trans study, 4.2<T<295	75A013
xx (IS/SFC)		Fe (NH3) 6 (SCN) 2	v	--	--	phase trans study, 4.2<T<295	75A013
xx		Fe (NO) 2		--	--		75L024
Pd		Fe (NO) 2S2O3K		-.099	1.309	comparison with ESCA	75J003
Pd		Fe (NO3) 3-1.5N2=	77	--	--	=O4, thermal decomp study	75A032
xx		Fe (NO3) 3-9H2O	78			HA=13 kOe, single crystal, spin-spin relaxation	75M028
xx (IS/Fe)	R	Fe (NO3) 3-9H2O	300	--		spin relaxation	75L003
xx (IS/Fe)		Fe (OH) 3 gel	4.2	.45		HI=459 kOe	75T005
xx (IS/Fe)		Fe (OH) 3 gel	75	.46	.75		75T005
Pd		FeONO3	77	.47	1.03		75A032
SS (IS/Fe)		FeSex	v	--	--	T=80&295K, study of effects of thermal treatment	75B031
xx		Fe (Si)				external rf field	75B007
xx		FeSi		--	--	study of the effect of tetragonal and trigonal distortions	75C055
Pd		Fe-Si alloy				study of domain structure	75E004
xx		FeSi	v		--	discussion of previous data	75G050
xx (IS/Fe)	R	FeSi	v				75H008
Pd		FeSi.132	v			20<T<710 K,HI vs temperature	75Z011
Cr		FeSix	v	--		study of electronic structure, HI vs temperature	75B025
xx (IS/Fe)		FeSi2 (Al)		--	--	bonding information	75B026
xx (IS/Fe)		FeSi2 (Co)		--	--	bonding information	75B026
xx		FeTex	v	--	--	T=N & R	74R036
xx		Fe-Xe		--			75S077
Pd		FexPd (80-x) Si20	R	--	--		75S070
xx		FexPd (80-x) Si20	v	--	--	2<T<300 K	75C040
Cr		Fe (1-x) CoxB	v	--	--	HI data	75T001
xx		Fe (1-x) CoxGe		--			75L021
Cr		(Fe (1-x) Cox) 2B	v	--	--	HI data	75T001
xx		Fe (1-x) CrxGe		--			75L021
Cu		(Fe (1-x) Mnx) 2B	v			HI vs composition, temperature=4, 100, and 300 K	75S047
xx		beta-Fe (1+y) Ge	77	--	--	structure information	75D044
Pd		epsilon-Fe2N		.235	.237		75L018
Pd		zeta-Fe2N		.253	.250		75L018
Pd		Fe2O (NO3) 4	77	.50	.93		75A032
Pd	R	Fe2P	v	--	--	first order magnetic transition at 214.5 K, 15 K<T<900 K	75W003
SS		Fe (2-x) Gi		--	--	HI, IS and QS vs composition	75O002
Cu		Fe (2-x) MnxAs	v		--	HI vs temperature, 77<T<615 K	75G047
Cu		Fe3C	v			study of small angle Mossbauer scattering near magnetic phase transitions	75S018
xx (IS/Fe)		Fe3Ge	v	--	--	T=5 & 295 K, structure info	75N020
xx		Fe3Se4				discussion of previous data	75G050
Pd		Fe5PB2	v	--	--	80<T<900 K	75H010
xx		GaAs	v			HI data	75K051
xx		GaAs	v	--	--	80<T<300 K, electronic state information	75I014
xx		GaSb	v			HI data	75K051
Pd		GdFe2Si2	300			study of magnetic ordering	75F020
xx		InSb	v			HI data	75K051
Cu		KFeCr (CN) 6		.85	1.71	Fe is Fe+2	75M084
Cr		LaCrO3	v			study of the angular dependence of the superexchange interaction	75O003
Cu (IS/SNP)		La (OH) 3		--	--	study of thermal treatment	75S068
Cu (IS/Fe)	300	Mn (1-x) FexB	4.2	--	--	bonding information	75D002
Cr		Mn4Si7				electronic structure info	75N027
Pd		NOFe (NO3) 4	77	.49	.51		75A032
Cu		NbHx	v	--		electronic structure info	75A002
Cr		NdCrO3	v			study of the angular dependence of the superexchange interaction	75O003

SOURCE	S-TEMP	ABSORBER	A-TEMP	SHIFT	QS	REMARKS	REF
Cu (IS/SNP)		Nd (OH) 3		--	--	study of thermal treatment	75S068
Cu (IS/Fe)	300	Ni (1-x) FexB	4.2	--	--	bonding information	75D002
Cu (IS/SNP)		Pr (OH) 3		--	--	study of thermal treatment	75S068
Cu (IS/Fe)	300	Ti (1-x) FexB	4.2	--	--	bonding information	75D002
Cu (IS/Fe)	300	V (1-x) FexB	4.2	--	--	bonding information	75D002
Cr		YCrO3	v			study of the angular dependence of the superexchange interaction	750003
xx		boracites				detn of energy level splitting	75P035
Pd (IS/Fe)		eta-FexGe	N	+.3		HI=150 kOe	75E007
Pd (IS/Fe)		eta-FexGe	R	.328	.423		75E007
xx		lamellar cmpds=	v	--	--	= of graphite, T=80 & 300 K, bonding information	75V032

ORGANIC COMPOUNDS

SOURCE	S-TEMP	ABSORBER	A-TEMP	SHIFT	QS	REMARKS	REF
xx		(BMT) FeBr	v		.056	1.4<T<4.2 K, relaxation effect	75G070
Pd		Bu4NFeCl4				comparison with ESCA	75J003
xx (IS/SNP)		C3H4Fe (CO) 3	N	.34	1.68	bonding and structure info	75P024
xx (IS/SNP)		C3H5Fe (CO) 4	N	.34	.87	bonding and structure info	75P024
xx (IS/SNP)		C3H6ClFe (CO) 3	N	.37	1.42	bonding and structure info	75P024
xx (IS/SNP)		C4H6Fe (CO) 3	N	.35	1.78	bonding and structure info	75P024
xx (IS/SNP)		C4H6Fe (CO) 4	N	.29	.93	bonding and structure info	75P024
xx (IS/SNP)		C4H8ClFe (CO) 3	N	.37	1.49	bonding and structure info	75P024
xx (IS/Fe)	R	C5H4CH2CH2COC5=	77	.48	2.20	=H4Fe, structure information	75A031
xx (IS/Fe)	R	(C5H4CH2) 2CH2Fe	77	.49	2.30	structure information	75A031
xx (IS/Fe)	R	(C5H4CHO) 2Fe	77	.51	2.16	structure information	75A031
xx (IS/Fe)	R	(C5H4COOH) 2Fe	77	.49	2.16	structure information	75A031
xx (IS/Fe)	R	C5H4SO2NHC5H4Fe	77	.49	2.33	structure information	75A031
xx (IS/SNP)		C5H8Fe (CO) 3	N	.37	1.88	bonding and structure info	75P024
xx (IS/SNP)		C5H9Fe (CO) 4	N	.35	1.05	bonding and structure info	75P024
xx (IS/SNP)		C5H9Fe (CO) 4	N	.35	1.06	bonding and structure info	75P024
xx (IS/SNP)		C5H10ClFe (CO) 3	N	.33	1.58	bonding and structure info	75P024
xx		C6H6	v			study of phases by resonance Rayleigh scattering	75G054
xx (IS/SNP)		C6H10Fe (CO) 4	N	.29	1.00	bonding and structure info	75P024
Cu	293	(CO) 4FePBu (Si=	80	-.29	2.45	=Me3) 2, bonding information	75E024
Cu	293	(CO) 4FePBu (Sn=	80	-.29	2.50	=Me3) 2, bonding information	75E024
Cu	293	(CO) 4FeP (Bu) 2=	80	-.28	2.64	=SiMe3, bonding information	75E024
Cu	293	(CO) 4FePBu2Sn=	80	-.28	2.52	=Me3, bonding information	75E024
Cu	293	(CO) 4FePBu3	80	-.30	2.82	bonding information	75E024
Cu	293	(CO) 4FeP (SiMe3=	80	-.27	2.65	=) 3, bonding information	75E024
Cu	293	(CO) 4FeP (SnMe3=	80	-.28	2.59	=) 3, bonding information	75E024
xx		C12FePc (-1)		.35	2.15	bonding information	75M047
Cu (IS/SNP)	298	C13SnFeC6CpP (O=	298	.36	1.77	=Ph) 3, bonding information	75B043
Cu (IS/SNP)	298	C13SnFeC6CpP (O=	298	.35	1.79	=Et) 3, bonding information	75B043
Cu (IS/SNP)	298	C13SnFe (CO) 2Cp	298	.32	1.82	bonding information	75B043
Cu (IS/SNP)	298	C13SnFeCOCpPEt3	298	.40	1.84	bonding information	75B043
Cu (IS/SNP)	298	C13SnFeCOCpPPh3	298	.41	1.80	bonding information	75B043
Cu (IS/SNP)	298	C13SnFeCp (P (O=	298	.44	1.80	=Ph) 3) 2, bonding information	75B043
xx (IS/Fe)	R	CpFeC5H4CHOHPh	77	.51	2.41	structure information	75A031
Pd		Cp2Fe			.274	comparison with ESCA	75J003
xx (IS/Fe)	R	Cp2Fe	77	.52	2.41	structure information	75A031
xx (IS/SNP)		EtC3H4Fe (CO) 4	N	.35	1.02	bonding and structure info	75P024
Pd		EtNFeBr4			.054	comparison with ESCA	75J003
Pd		EtNFeCl4	77	.31		bonding and structure info	75A032
Pd		EtNH3FeCl4	77	.30		bonding and structure info	75A032
Pd		EtNH3Fe (NO3) 4	77	.60		bonding and structure info	75A032
Pd		Et2NH2FeCl4	77	.31		bonding and structure info	75A032
Pd		Et2NH2Fe (NO3) 4	77	.61		bonding and structure info	75A032
Pd		Et3NHFeCl4	77	.31		bonding and structure info	75A032
Pd		Et3NHFe (NO3) 4	77	.57		bonding and structure info	75A032
xx (IS/Fe)	295	Et4NFeCo3 (CO) 12	80	.05	.14	structure information	75C028
Pd		Et4NFe (NO3) 4	77	.54		bonding and structure info	75A032
Cu		Fe (II) trisdi=		--	--	=imine complex, T=4.2 & 295 K, electronic structure info	75B095

71

SOURCE	S-TEMP	ABSORBER	A-TEMP	SHIFT	QS	REMARKS	REF
xx (IS/Fe)		Fe(IV) complex=	v	--	--	= with dithiolate ligands, 1.4<T<300 K	75P033
Cr		Fe polyazopor=	v	--	--	=phinse, T=4.2, 90, and 300 K	75S027
xx (IS/Fe)		Fe(Ac(ox))2OH	R	.33	.93	electronic structure info	75S003
xx (IS/Fe)		Fe(Ac(ox))3	R	.32	.85	electronic structure info	75S003
xx (IS/Fe)		Fe(Ac(ox))2-2=	R	1.01	2.50	=H2O, electronic struc info	75S003
xx		Fe(Bipy)2(CN)2=	v		--	=ClO4, electronic struc info	75M004
xx		Fe(Bipy)2(CN)2=	v		--	=NO3, electronic struc info	75M004
Cu		FeBipy3Cl2-5H2O	v	--	--	thermal decomposition study	75S015
Pd		FeBipy3SO4		.135		comparison with ESCA	75J003
xx (IS/SNP)		Fe(C5H5NO)6(=	v	--	--	=ClO4)2, 9 K<T<295 K, bonding and structure information	75S013
Cu (IS/SNP)		Fe(C5H5NO)6(=	v	--	--	=ClO4)2, 8 K<T<334, electronic structure information	75S014
Pt (IS/SNP)		Fe(C6H11NC)3(=	N	.20	.38	=PPh(OEt)2)3(ClO4)2	75D048
Cr (IS/SNP)	R	Fe(CHP)2Br	298	+.62	.48	bonding and structure info	75S012
Cr (IS/SNP)	R	Fe(CHP)2Cl	298	+.61	.49	bonding and structure info	75S012
Cr (IS/SNP)	R	Fe(CHP)2I	298	+.65	.48	bonding and structure info	75S012
xx		Fe(CMCB)3ClO4	v		--	QS vs temp(calcd and expl)	75G004
Cu (IS/SNP)	R	Fe(CNMe)4(CNH=	110	.16	.37	=Me)2NMe(PF6)2, bonding info	75B045
Cu (IS/SNP)	R	Fe(CNMe)4(CNMe=	110	.16	.58	=)2NH2CNHMe(PF6)2, bond info	75B045
Cu (IS/SNP)	R	Fe(CNMe)4(CN)h=	110	.14	.67	=Me)2N2H2(PF6)2, bonding info	75B045
Cu (IS/SNP)	R	Fe(CNMe)4(CN)h=	110	.17	+.68	=Me)2N2HPh(PF6)2, eta=.7	75B045
Cu (IS/SNP)	R	Fe(CNMe)5CNC4=	110	.19	.40	=H8NHMe(PF6)2, bonding info	75B045
Cu (IS/SNP)	R	Fe(CNMe)5CNH2N=	100	.17	-.36	=HMe(PF6)2, bonding info	75B045
Cu (IS/SNP)	R	Fe(CNMe)6(PF6)2	110	.18	0	bonding information	75B045
Pd (IS/Fe)	R	Fe(CO)2CpBr		.22	1.77	bonding information	75D030
Pd (IS/Fe)	R	Fe(CO)2CpCl		.22	1.89	bonding information	75D030
Pd (IS/Fe)	R	(Fe(CO)2Cp)Cr=		.13	1.71	=(CO)3CpSnCl2, bonding info	75D030
Pd (IS/Fe)	R	Fe(CO)2CpI		.20	1.86	bonding information	75D030
Pd (IS/Fe)	R	(Fe(CO)2Cp)Mo=		.12	1.54	=(CO)3CpSnCl2, bonding info	75D030
Pd (IS/Fe)	R	Fe(CO)2CpSnBr3		.10	1.80	bonding information	75D030
Pd (IS/Fe)	R	Fe(CO)2CpSnCl3		.14	1.80	bonding information	75D030
Pd (IS/Fe)	R	Fe(CO)2CpSnI3		.15	1.74	bonding information	75D030
Pd (IS/Fe)	R	Fe(CO)2CpSnPh3		.08	1.75	bonding information	75D030
Pd (IS/Fe)	R	(Fe(CO)2Cp)W=		.11	1.69	=(CO)3CpSnCl2, bonding info	75D030
Pd (IS/Fe)	R	(Fe(CO)2Cp)2Sn=		.12	1.57	=Cl2, bonding information	75D030
Pd (IS/Fe)	R	(Fe(CO)2Cp)2Sn=		.12	1.57	=(NCS)2, bonding information	75D030
Pd (IS/Fe)	R	(Fe(CO)2Cp)2Sn=		.09	1.68	=(OSOPh)2, bonding information	75D030
Pd (IS/Fe)	R	(Fe(CO)2Cp)2Sn=		.07	1.61	=(SPh)2, bonding information	75D030
Pt (IS/SNP)		Fe(CO)3(AsPh3)2		.20	3.20	bonding information	75M080
xx (IS/Fe)		Fe(CO)3CHCBrH=	80	+.061	1.730	=Fe(CO)3Br	75G048
xx (IS/Fe)		Fe(CO)3CHCBrH=	290	-.013	1.746	=Fe(CO)3Br	75G048
Pt (IS/SNP)		Fe(CO)3(CNPh)2		.16	2.03	bonding information	75M080
Pt (IS/SNP)		Fe(CO)3(PBu3)2		.15	2.31	bonding information	75M080
Pt (IS/SNP)		Fe(CO)3(P(NMe2=		.15	2.27	=3)2, bonding information	75M080
Pt (IS/SNP)		Fe(CO)3(P(OMe=		.12	2.28	=3)2, bonding information	75M080
Pt (IS/SNP)		Fe(CO)3(P(OPh)=		.15	2.61	=3)2, bonding information	75M080
Pt (IS/SNP)		Fe(CO)3(PPh3)2		.16	2.63	bonding information	75M080
Pt (IS/SNP)		Fe(CO)3(SbPh3)2		.19	3.16	bonding information	75M080
Pt (IS/SNP)		FeCl(C6H5NC)5=	N	.20	+.80	=ClO4	75D048
Pt (IS/SNP)		FeCl(C6H5NC)=	N	.34	+.54	=(PPh(OEt)2)3ClO4	75D048
Pd (IS/SNP)		t-FeCl(DCB)(=		.51	.74	=depe)2+, bonding information	75B070
Pd (IS/SNP)		t-FeCl(MLN)(=		.55	.90	=depe)2+, bonding information	75B070
Pd (IS/SNP)		t-FeClNCPh(=		.55	1.12	=depe)2+, bonding information	75B070
Pd (IS/SNP)		t-FeClNO(depe)=		.26	2.15	=2+, bonding information	75B070
xx (IS/Fe)		Fe(Cl(ox))2OH	R	.29	1.12	electronic structure info	75S003
xx (IS/Fe)		Fe(Cl(ox))3	R	.26	1.05	electronic structure info	75S003
xx (IS/Fe)		Fe(Cl(ox))2-2=	R	1.01	2.44	=H2O, electronic struc info	75S003
Cr (IS/SNP)		FeCl2(C6H5NHN=	4.2	1.40	3.21	=H2)2	75S106
Cr (IS/SNP)		FeCl2(C6H5NHN=	77	1.38	3.20	=H2)2	75S106
Cr (IS/SNP)		FeCl2(C6H5NHN=	295	1.28	2.95	=H2)2	75S106
Cr (IS/SNP)		FeCl2(n-ClC6H4=	4.2	1.42	.02	=NHNH2)2	75S106
Cr (IS/SNP)		FeCl2(n-ClC6H4=	77	1.41	1.66	=NHNH2)2	75S106
Cr (IS/SNP)		FeCl2(n-ClC6H4=	295	1.30	1.20	=NHNH2)2	75S106
Pt (IS/SNP)		FeCl(2-MeC6H4N=	N	.32	+.44	=C)(PPh(OEt)2)3ClO4	75D048
Cr (IS/SNP)		FeCl2(n-O2NC6=	4.2	1.39	1.51	=H4NHNH2)4	75S106
Cr (IS/SNP)		FeCl2(n-O2NC6=	77	1.38	1.53	=H4NHNH2)4	75S106
Cr (IS/SNP)		FeCl2(n-O2NC6=	295	1.25	1.06	=H4NHNH2)4	75S106
Cr (IS/SNP)		FeCl2(m-ClC6H4=	4.2	1.41	.35	=NHNH2)2	75S106
Cr (IS/SNP)		FeCl2(m-ClC6H4=	77	1.40	1.84	=NHNH2)2	75S106
Cr (IS/SNP)		FeCl2(m-ClC6H4=	295	1.28	1.35	=NHNH2)2	75S106

SOURCE	S-TEMP	ABSORBER	A-TEMP	SHIFT	QS	REMARKS	REF
Cr (IS/SNP)		FeCl2(o-O2NC6=	4.2	1.45	1.45	=H4NHNH2)4	75S106
Cr (IS/SNP)		FeCl2(o-O2NC6=	77	1.43	2.54	=H4NHNH2)4	75S106
Cr (IS/SNP)		FeCl2(o-O2NC6=	295	1.28	1.95	=H4NHNH2)4	75S106
Cr		FeCl3 + Pc	v	--	--	bonding information	75S094
Cr		FeCl3 + TCB	v	--	--	bonding information	75S094
Cr		FeCl3 + TCE	v	--	--	bonding information	75S094
Pt (IS/SNP)		FeCl(4-MeC6H4N=	N	.37	.74	=C)3(PPh3)2ClO4	75D048
Pt (IS/SNP)		FeCl(4-MeC6H4N=	N	.29	+.78	=C)5ClO4	75D048
Pt (IS/SNP)		FeCl(4-MeC6H4N=	N	.25	+.55	=C)2(PPh(OEt)2)3ClO4	75D048
Pt (IS/SNP)		FeCl(4-MeOC6H4=	N	.46	+.51	=NC)2(PPh(OEt)2)3ClO4	75D048
Pt (IS/SNP)		FeCl(4-NO2C6H4=	N	.38	+.62	=NC)2(PPh(OEt)2)3ClO4	75D048
Pt (IS/SNP)		FeCl2(4-MeC6H4=	N	.44	+1.54	=NC)4(violet)	75D048
Pt (IS/SNP)		FeCl2(4-MeC6H4=	N	.45	-.89	=NC)4(red)	75D048
Pt (IS/SNP)		FeCl(2,6-Me2C6=	N	.30	+.75	=H3NC)2(PPh(OEt)2)3ClO4	75D048
Pd		Fe(DECB)2Br	v	.20	--	EQ=11.6, eta=.16, HI=250 kG	75M032
Pd		Fe(DECB)2Br in=				=Fe(DECB)2Cl	75M032
xx		Fe(DECB)2Cl		.21	2.68		75M050
Pd		Fe(DECB)2Cl	v	.21	--	EQ=10.7, eta=.16, HI=335 kG	75M032
xx		Fe(DMCB)2Br		.20	2.89		75M050
Pd (I/SNP)		Fe(DMSO)6(ClO4=	R	1.45	2.61	=)2, bonding & structure info	75M069
Pd (I/SNP)		Fe(DPSO)6(ClO4=	R	1.43	2.55	=)2, bonding & structure info	75M069
Pd		Fe(GMI)3I2		-.006		comparison with ESCA	75J003
xx		Fe-HEDTA com=	v			=plexes, structure information	75D050
Pd		(Fe(HEDTA))2O	100	.26	1.67		75K062
Pd		Fe(HFA)3		.149		comparison with ESCA	75J003
Cu (IS/Fe)	295	Fe(HPT)BPh4-C=	77	+.12	2.30	=H2Cl2, electronic struc info	75K006
Cu (IS/Fe)	295	Fe(HPT)IBPh4	4.2	+.13	2.25	electronic structure info	75K006
Cu (IS/Fe)	295	Fe(HPT)IBPh4	4.2	+.10	2.30	=H2Cl2, electronic struc info	75K006
Cu (IS/Fe)	295	Fe(HPT)IBPh4	77	+.14	2.26	electronic structure info	75K006
Cu (IS/Fe)	295	Fe(HPT)IBPh4	298	+.20	1.81	electronic structure info	75K006
Cu		Fe(HQ)2		1.00	2.55		75F021
Cu		Fe(HQ)2OH		.34	.83		75F021
Cu		Fe(HQ)3		.30	.98		75F021
Cu		Fe((HQ)I-SO3H)2		1.98	1.98		75F021
Cu		Fe(HQ)I-SO3NH4=		.12	.84	=(OH)2	75F021
Cu		Fe(HQ)I-SO3NH4=		.16	.74	=OH	75F021
Cu		Fe(HQ)(OH)2		.31	.65		75F021
Cu		Fe(HQ)-SO3NH4=		.13	.68	=(OH)2	75F021
Cu		Fe((HQ)-SO3NH4=		1.40	1.40	=)2	75F021
xx (IS/Fe)	300	Fe-Iz	v	--	--	T=77 & 300 K	75W018
xx (IS/Fe)	300	Fe-Iz(Et)	4.2	.89	2.83		75M052
xx (IS/Fe)	300	Fe-Iz(Et)	300	.80	2.83		75M052
xx (IS/Fe)	300	Fe-Iz(Me)	4.2	.83	2.81		75M052
xx (IS/Fe)	300	Fe-Iz(Me)	300	.72	2.72		75M052
xx (IS/Fe)	300	Fe-Iz(Ph)	4.2	.90	3.46		75M052
xx (IS/Fe)	300	Fe-Iz(Ph)	300	.79	3.33		75M052
xx (IS/SNP)		Fe(MDT)BrPF6		1.063	3.91	structure information	75R006
xx (IS/SNP)		Fe(MDT)Br2BF4		.61	.55	structure information	75R006
xx (IS/SNP)		Fe(MDT)ClPF6		1.075	3.85	structure information	75R006
xx (IS/SNP)		Fe(MDT)Cl2BF4		.59	.66	structure information	75R006
xx (IS/SNP)		Fe(MDT)IPF6		1.051	3.77	structure information	75R006
xx (IS/SNP)		Fe(MDT)(N3)2		1.24	1.68	structure information	75R006
xx (IS/SNP)		Fe(MDT)(NCS)2		.71	.67	structure information	75R006
xx (IS/SNP)		Fe(MDT)OAcPF6		1.19	2.40	structure information	75R006
xx (IS/SNP)		Fe(MeCNOCNOHMe=	80.	58	1.64	=)2(3-CONH2Py)2, bonding info	75B122
xx (IS/SNP)		Fe(MeCNOCNOHMe=	80	.55	1.67	=)2(3-HPy)2, bonding info	75B122
xx (IS/SNP)		Fe(MeCNOCNOHMe=	196	.50	1.71	=)2(3-CONH2Py)2, bonding info	75B122
xx (IS/SNP)		Fe(MeCNOCNOHMe=	300	.45	1.60	=)2(3-CONH2Py)2, bonding info	75B122
xx (IS/SNP)		Fe(MeCNOCNOHMe=	300	.47	1.67	=)2(3-HPy)2, bonding info	75B122
Cu (IS/SNP)		Fe(Me2SO)6(=	v	--	--	=ClO4)2, 8 K<T<316, electronic structure information	75S014
Pd (IS/SNP)		t-Fe(NCMe)2(=		.53	.93	=depe)2+2, bonding information	75B070
xx (IS/Fe)		Fe(NO3(ox))3	R	.31	.82	electronic structure info	75S003
Pd		Fe(NO3)3N2O4NO=	77	.52	.45	=Fe(NO3)4	75A032
Pd		Fe(NO3)3-1.5N2=	77	.57		=O4N4O6(Fe(NO3)4)2	75A032
xx		Fe(NPY)4(ClO4)2	4.2			HA, relaxation	75Z006
Pd		Fe(NTA)-2H2O	298	.72	.52	structure information	75S099
Rh (IS/Fe)	R	FeO2(N-Me-=	v	--	--	=imid)(TPI), 4.2<T<295 K, HA	75S081
Rh (IS/Fe)	R	FeO2(THF)2TPI	4.2	.33	2.61		75S081
Rh (IS/Fe)	R	FeO2(THF)2TPI	77	.33	2.58		75S081
Rh (IS/Fe)	R	FeO2(THF)2TPI	195	.29	2.13		75S081
Rh (IS/Fe)	R	FeO2(THT)2TPI	4.2	.31	2.29		75S081

73

SOURCE	S-TEMP	ABSORBER	A-TEMP	SHIFT	QS	REMARKS	REF
Rh(IS/Fe)	R	FeO2(THT)2TPI	77	.30	2.27		75S081
Rh(IS/Fe)	R	FeO2(THT)2TPI	195	.27	1.87		75S081
xx(IS/SNP)		c-Fe(OTI)4Br2		.279		IS corln w/ binding energies	75S045
xx(IS/SNP)		t-Fe(OTI)4Br2		.369		IS corln w/ binding energies	75S045
xx(IS/SNP)		c-Fe(OTI)4Cl2		.279		IS corln w/ binding energies	75S045
xx(IS/SNP)		t-Fe(OTI)4Cl2		.382		IS corln w/ binding energies	75S045
xx(IS/SNP)		c-Fe(OTI)4I2		.278		IS corln w/ binding energies	75S045
xx(IS/SNP)		t-Fe(OTI)4I2		.351		IS corln w/ binding energies	75S045
xx(IS/Fe)		Fe(PAP)2-CHCl3	v	--	--	T=5,10,&20 K, HA=5,10,20,&40kG spin-lattice relaxation	75Z013
xx		Fe(PAP)(PQB)(1=	v	--	--	=-n-BuIz)O2, T=4.2 & 195 K	75C044
xx		Fe(PAP)(PQB)(1=	4.2	.27	.27	=-MeIz)(CO)	75C044
xx		Fe(PAP)(PQB)(1=	v	--	--	=-MeIz)O2, T=4.2,77 & 195 K	75C044
xx		Fe(PAP)(PQB)(1=	v	--	--	=-MeIz)2, T=4.2,77 & 195 K	75C044
xx		Fe(PAP)(PQB)(1=	v	--	--	=-MeIz)2, T=4.2,77 & 195 K	75C044
xx		Fe(PAP)(PQB)(=	v	--	--	=THF)2, T=4.2,77 & 195 K	75C044
xx		Fe(PAP)(PQB)(=	v	--	--	=THF)2O2, T=4.2,77 & 195 K	75C044
Cu		Fe(PBI)2Br2H2O		1.06	2.53	bonding information	74S120
Cu		Fe(PBI)2(CN)2-=		.31	.74	=2H2O, bonding information	74S120
Cu		Fe(PBI)2Cl2H2O	110	1.08	2.63	bonding information	74S120
Cu		Fe(PBI)2(N3)2-=		--	--	=H2O, bonding information	74S120
Cu		Fe(PBI)2(NCS)2		--	--	bonding information	74S120
xx		Fe(PDC)3ClO4	v		--	QS vs temp(calcd and expl)	75G004
SS		Fe(PTC)2Cl	R	.51	2.71	electronic structure info	75A041
SS		Fe(PTC)3	R	.51		electronic structure info	75A041
xx		Fe(PTC)3ClO4	v		--	QS vs temp(calcd and expl)	75G004
xx		Fe(PTC)3FeCl4	v		--	QS vs temp(calcd and expl)	75G004
xx(IS/SNP)		c-Fe(PTS)4B22		.303		IS corln w/ binding energies	75S045
xx(IS/SNP)		t-Fe(PTS)4B22		.372		IS corln w/ binding energies	75S045
xx(IS/SNP)		c-Fe(PTS)4Cl2		.285		IS corln w/ binding energies	75S045
xx(IS/SNP)		t-Fe(PTS)4Cl2		.372		IS corln w/ binding energies	75S045
xx(IS/SNP)		c-Fe(PTS)4I2		.298		IS corln w/ binding energies	75S045
xx(IS/SNP)		t-Fe(PTS)4I2		.351		IS corln w/ binding energies	75S045
xx(IS/SNP)		Fe(PTS)5Br-Br		.203		IS corln w/ binding energies	75S045
xx(IS/SNP)		Fe(PTS)5I-I		.218		IS corln w/ binding energies	75S045
xx(IS/SNP)		Fe(PTS)5I-I3		.219		IS corln w/ binding energies	75S045
Pd		FePc		.199	2.606	comparison with ESCA	75J003
Pd		FePc(beta)	v	--	--	113<T<423 K, IS,QS,W,& INT vs temperature	74D058
xx(IS/SNP)		Fe(PhCNOCNOHPh=	80	.48	1.90	=)2(3-CONEtPy)2	75B122
xx(IS/SNP)		Fe(PhCNOCNOHPh=	80	.49	1.88	=)2(3-COPhPy)2,	75B122
xx(IS/SNP)		Fe(PhCNOCNOHPh=	80	.52	1.78	=)2(3-CONH2Py)2,	75B122
xx(IS/SNP)		Fe(PhCNOCNOHPh=	80	.52	1.97	=)2(3-HPy)2,	75B122
xx(IS/SNP)		Fe(PhCNOCNOHPh=	80	.52	1.66	=)2(3-CONHCH2OHPy)2, bond info	75B122
xx(IS/SNP)		Fe(PhCNOCNOHPh=	80	.52	1.92	=)2(3-CO2EtPy)2,	75B122
xx(IS/SNP)		Fe(PhCNOCNOHPh=	196	.47	1.89	=)2(3-CONH2Py)2,	75B122
xx(IS/SNP)		Fe(PhCNOCNOHPh=	300	.46	1.71	=)2(3-CONHCH2OHPy)2, bond info	75B122
xx(IS/SNP)		Fe(PhCNOCNOHPh=	300	.47	1.78	=)2(3-CONH2Py)2,	75B122
xx(IS/SNP)		Fe(PhCNOCNOHPh=	300	.44	1.89	=)2(3-CONEtPy)2	75B122
xx(IS/SNP)		Fe(PhCNOCNOHPh=	300	.48	2.00	=)2(3-HPy)2,	75B122
xx(IS/SNP)		Fe(PhCNOCNOHPh=	300	.45	1.91	=)2(3-CO2EtPy)2,	75B122
xx(IS/SNP)		Fe(PhCNOCNOHPh=	300	.46	1.87	=)2(3-COPhPy)2,	75B122
Cu(IS/SNP)		Fe(Ph2SO)6(=	v	--	--	=ClO4)2, 8 K<T<333, electronic structure information	75S014
xx(IS/Fe)	4.2	FePh2(14-mmc)=	4.2			=N4SPh structure information	75K055
xx(IS/Fe)	77	FePh2(14-mmc)=	77	-.02	3.22	=N4SPhPy, elec struc info	75K055
xx(IS/Fe)	300	FePh2(14-mmc)=	300	.04	3.60	=N4SPh, electronic struc info	75K055
xx(IS/Fe)	77	FePh2(15-mmc)=	77	.13	2.55	=N4SPh, electronic struc info	75K055
xx(IS/Fe)	4.2	FePh2(16-mmc)=	4.2			=N4SPh, HA=80 kOe, electronic structure information	75K055
xx(IS/Fe)	4.2	FePh2(16-mmc)=	4.2			=N4SCH2Ph, HA=80 kOe, electronic structure info	75K055
xx(IS/Fe)	77	FePh2(16-mmc)=	77	.26	2.28	=N4SPh, electronic struc info	75K055
xx(IS/Fe)	300	FePh2(16-mmc)=	300	.26	1.93	=N4SPh, electronic struc info	75K055
xx(IS/Fe)	300	FePh2(16-mmc)=	300	.35	.71	=N4SCH2Ph, elec struc info	75K055
xx(IS/Fe)		FePy3(tren)(=	v	--	--	=PF6)2	75H031
xx		FePy6Br2	80	--	--		75T008
xx		FePy6Br2	298	--	--		75T008
xx(IS/Fe)		Fe(S2CNBu2)2	4.2			structure information, HA	75F013
xx(IS/Fe)		Fe(S2CNBu2)2	80	1.01	3.93	structure information	75F013
xx(IS/Fe)		Fe(S2CNC4H8)2	4.2			structure information, HA	75F013

SOURCE	S-TEMP	ABSORBER	A-TEMP	SHIFT	QS	REMARKS	REF
xx(IS/Fe)		Fe(S2CNC4H8)2	80	1.01	2.27	structure information	75F013
xx(IS/Fe)		Fe(S2CNEt2)2	4.2			structure information, HA	75F013
xx(IS/Fe)		Fe(S2CNEt2)2	80	.875	4.19	structure information	75F013
xx(IS/Fe)		Fe(S2CNMe2)2	4.2			structure information, HA	75F013
xx(IS/Fe)		Fe(S2CNMe2)2	80	1.00	2.56	structure information	75F013
xx(IS/Fe)		Fe(S2CNPr2)2	4.2			structure information, HA	75F013
xx(IS/Fe)		Fe(S2CNPr2)2	80	.87	4.20	structure information	75F013
Cu(IS/Fe)	R	Fe(SPMe2NPMe2S=	300	--	--	=)2, electronic structure info	75R001
Pd	R	Fe(Sal H)(PDC)	v	--			75K020
Pd	R	Fe(Sal H)(PTC)	v	--			75K020
Pd		Fe(TFA)3		.186		comparison with ESCA	75J003
Pd(I/SNP)		Fe(TMSO)6(ClO4=	R	1.48	1.87	=)2, bonding & structure info	75M069
xx(IS/Fe)		Fe(TPP)	v	--	--	4.2<T<300 K, bonding info	75C045
Pd		Fe(acac)3		.134	.750	comparison with ESCA	75J003
xx(IS/Fe)		Fe(acac)2-2H2O		--	--	4.2<T<295 K	75M051
xx(IS/Fe)		Fe(benzene)Cp2	77	.57	1.67		75M015
xx(IS/Fe)		Fe(biphenyl)Cp	77	.51	1.45		75M015
Cu		Fe(bipy)2(NCS)3	4.2	.55		bonding and structure info	75G011
Cu		Fe(bipy)2(NCS)3	4.2	.55		bonding and structure info	75G011
Cu		Fe(bipy)2(NCS)3	77	.34		bonding and structure info	75G011
xx(IS/SNP)		Fe-cellulose		--	--	bonding and structure info	75M053
xx		Fe(das)2Br2BF4	v		--	QS vs temp(calcd and expl)	75G004
xx		Fe(das)2Cl2BF4	v		--	QS vs temp(calcd and expl)	75G004
Pd(IS/Fe)	295	Fe(diphos)Br2	77	.63		bonding information	75L022
Pd(IS/Fe)	295	Fe(diphos)Br2	293	.58	2.76	bonding information	75L022
Pd(IS/Fe)	295	Fe(diphos)2Br2	77	.93	2.68	bonding information	75L022
Pd(IS/Fe)	295	Fe(diphos)2Br2	293	.83	2.66	bonding information	75L022
Pd(IS/Fe)	295	Fe(diphos)2Cl2	77	--	--	bonding information	75L022
Pd(IS/Fe)	295	Fe(diphos)2Cl2	195	--	--	bonding information	75L022
Pd(IS/Fe)	295	Fe(diphos)2Cl2	293	--	--	bonding information	75L022
Pd(IS/Fe)	295	Fe(diphos)2(N3=	77	.38	.27	=)2, bonding information	75L022
Pd(IS/Fe)	295	Fe(diphos)2(N3=	293	.33	.35	=)2, bonding information	75L022
Pd(IS/Fe)	295	Fe(diphos)2(NC=	77	.29	.39	=S)2, bonding information	75L022
Pd(IS/Fe)	295	Fe(diphos)2(NC=	293	.26	.30	=S)2, bonding information	75L022
xx		Fe(dtc)2Br	77		.7	electronic structure info	75M049
Pd		Fe(dtc)2Br in =	v			=Fe(dtc)2Cl, study of magn ordering vs composition	75M031
xx(IS/Fe)		Fe(mesitylene)2	77	.59	1.95		75M015
xx(IS/Fe)		Fe(naphthalene=	77	.62	1.76	=)Cp	75M015
xx(IS/Fe)		Fe(ox)2OH	R	.34	1.10	electronic structure info	75S003
Pd		Fe(ox)3		.168	.815	comparison with ESCA	75J003
xx(IS/Fe)		Fe(ox)3	R	.32	.99	electronic structure info	75S003
xx(IS/Fe)		Fe(ox)2-2H2O	R	1.02	2.45	electronic structure info	75S003
Cu(IS/SNP)	R	c-Fe(phen)2(CN=	110	.43	+.63	=)2, bonding information	75B045
Cu(IS/SNP)	R	t-Fe(phen)2(CN=	110	.49	+.62	=)2, bonding information	75B045
xx		Fe(phen)2(CN)2=	v		--	=ClO4, electronic struc info	75M004
Pd		Fe(phen)2(CN)2=		-.006	.606	=H2O, comparison with ESCA	75J003
xx		Fe(phen)2(CN)2=	v		--	=NO3-4H2O, elec struc info	75M004
Cu(IS/SNP)	R	Fe(phen)2(CNH=	110	.40	1.17	=Me)2N2H2(PF6), bonding info	75B045
Cu(IS/SNP)	R	Fe(phen)2(CNMe=	110	.44	.47	=)2(PF6)2, bonding info	75B045
Pd		Fe(phen)2Cl2		.815	2.991	comparison with ESCA	75J003
Cu		Fe(phen)2(NCS)3	1.3	.46		bonding and structure info	75G011
Cu		Fe(phen)2(NCS)3	1.3	.52		bonding and structure info	75G011
Cu		Fe(phen)2(NCS)3	77	.55		bonding and structure info	75G011
Pd		(Fe(pic)2OH)2	100	.23	.78		75K062
xx(IS/SNP)		Fe-synthetic =		--	--	=fibers, bonding & struc info	75M053
xx		FexZn(1-x)(PIC=	v			=)3Cl-EtOH, 4.2<T<300 K	75E015
xx(IS/Fe)		Fe2BipyCp2	4.2	.536	1.601		75M015
xx(IS/Fe)		Fe2BipyCp2	300	.441	1.555		75M015
Cu		Fe2Bipy4OBr4	77	.51	1.60	bonding and structure info	75G011
Cu		Fe2Bipy4OBr4	300	.36	1.60	bonding and structure info	75G011
Cu		Fe2Bipy4OCl4	77	.47	1.52	bonding and structure info	75G011
Cu		Fe2Bipy4OCl4	300	.37	1.48	bonding and structure info	75G011
Cu		Fe2Bipy4O(NCS)4	4.2	.60	1.67	bonding and structure info	75G011
Cu		Fe2Bipy4O(NCS)4	77	.50	1.23	bonding and structure info	75G011
Cu		Fe2Bipy4O(=	77	.48	1.50	=NO3)2, bonding and struc info	75G011
Cu		Fe2Bipy4O(=	300	.37	1.48	=NO3)2, bonding and struc info	75G011
Cu		Fe2Bipy4O(=	77	.49	1.45	=SO4)2, bonding and struc info	75G011
Cu		Fe2Bipy4O(=	300	.41	1.31	=SO4)2, bonding and struc info	75G011
Cu		Fe2(phen)4OBr4	77	.50	1.62	bonding and structure info	75G011
Cu		Fe2(phen)4OBr4	300	.43	1.62	bonding and structure info	75G011
Cu		Fe2(phen)4OCl4	77	.46	1.70	bonding and structure info	75G011

SOURCE	S-TEMP	ABSORBER	A-TEMP	SHIFT	QS	REMARKS	REF
Cu		Fe2(phen)4OCl4	300	.36	1.65	bonding and structure info	75G011
Cu		Fe2(phen)4O(=	4.2	.81	1.52	=NCS)4, bonding and struc info	75G011
Cu		Fe2(phen)4O(=	77	.50	1.43	=NCS)4, bonding & struc info	75G011
Cu		Fe2(phen)4O(=	77	.47	1.48	=NO3)4, bonding and struc info	75G011
Cu		Fe2(phen)4O(=	300	.36	1.45	=NO3)4, bonding and struc info	75G011
Cu		Fe2(phen)4O(=	77	.52	1.60	=SO4)2, bonding and struc info	75G011
Cu		Fe2(phen)4O(=	300	.40	1.52	=SO4)2, bonding and struc info	75G011
xx(IS/Fe)		Fe2(pyrene)Cp2	4.2	.564	1.673		75M015
xx(IS/Fe)		Fe2(pyrene)Cp2	300	.477	1.667		75M015
Rh(IS/Fe)		Fe3(CO)9S2	85	.04	.56		75H039
Rh(IS/Fe)		Fe3(CO)9S2	295	-.04	.55		75H039
Rh(IS/Fe)		Fe3Cp3(CO)2SS(=	4.2	.25	1.69	=bz)	75H039
Rh(IS/Fe)		Fe3Cp3(CO)2SSBu	77	.27	1.70		75H039
Rh(IS/Fe)		Fe3Cp3(CO)2SS(=	77	.27	1.67	=bz)	75H039
Rh(IS/Fe)		Fe3Cp3(CO)2SS(=	295	.29	1.65	=bz)	75H039
Pd		Fe3(PhCOO)5(OH=	R	.70	.44	=)3PhCOO-H2O	75M082
Pd(IS/SNP)		Fe(4-ClPyO)6(=	N	1.60	3.28	=ClO4)2, bonding & struc info	75M069
Pd(IS/SNP)		Fe(4-ClPyO)6(=	R	1.41	2.74	=ClO4)2, bonding & struc info	75M069
Pd(IS/SNP)		Fe(4-HPyO)6(=	N	1.50	1.71	=ClO4)2, bonding & struc info	75M069
Pd(IS/SNP)		Fe(4-HPyO)(=	R	1.42	1.63	=ClO4)2, bonding & struc info	75M069
Pd(IS/SNP)		Fe(4-MePyO)6(=	N	1.53	3.05	=ClO4)2, bonding & struc info	75M069
Pd(IS/SNP)		Fe(4-MePyO)6(=	R	1.44	2.55	=ClO4)2, bonding & struc info	75M069
Pd(IS/SNP)		Fe(4-NO2PyO)6(=	R	1.56	3.01	=ClO4)2, bonding & struc info	75M069
Pd(IS/SNP)		Fe(4-NO2PyO)6(=	R	1.46	2.49	=ClO4)2, bonding & struc info	75M069
Pd(IS/SNP)		Fe(4-OMePyO)6(=	N	1.52	2.92	=ClO4)2, bonding & struc info	75M069
Pd(IS/SNP)		Fe(4-OMePyO)6(=	R	1.44	2.23	=ClO4)2, bonding & struc info	75M069
xx(IS/Fe)		Fe(6-MePy)Py=	v	--	--	=(tren)(PF6)2	75H031
xx(IS/Fe)		Fe(6-MePy)Py2=	v	--	--	=(tren)(PF6)2	75H031
xx(IS/Fe)		Fe(6-MePy)3=	v	--	--	=(tren)(PF6)2	75H031
Rh(IS/Fe)		Fe(2,2'-Bipy)=	v	--	--	=Cl2, structure information	75R002
xx		Fe(2,2'-dipy)	v			77 K<T<300 K	75A020
Rh(IS/Fe)		Fe(2,9-Me2-=	4.2	.93	2.76	=phen)Cl2, structure info	75R002
Rh(IS/Fe)		Fe(2,9-Me2-=	78	.91	2.77	=phen)Cl2, structure info	75R002
Rh(IS/Fe)		Fe(2,9-Me2-=	300	.80	2.66	=phen)Cl2, structure info	75R002
Rh(IS/Fe)		Fe(5,5'-Me2-2,=	78	1.06	3.72	=2'-Bipy)Cl2, structure info	75R002
Rh(IS/Fe)		Fe(5,5'-Me2-2,=	300	.93	3.51	=2'-Bipy)Cl2, structure info	75R002
xx		HFe(Bipy)(CN)4=	v		--	=-2H2O, electronic struc info	75M004
xx(IS/Fe)	295	HFeCo3(CO)8(=	80	.13	.39	=dpe)2, structure information	75C028
xx(IS/Fe)	295	HFeCo3(CO)8(=	80	.08	1.01	=dpe)2, structure information	75C028
xx(IS/Fe)	295	HFeCo3(CO)9(P=	80	.11	.10	=MePh2)3, structure info	75C028
xx(IS/Fe)	295	HFeCo3(CO)9(=	80	.11	.11	=dpe)2, structure information	75C028
xx(IS/Fe)	295	HFeCo3(CO)10(P=	80	.12	.35	=MePh2)2, structure info	75C028
xx(IS/Fe)	295	HFeCo3(CO)11P=	80	.10	.36	=MePh2, structure information	75C028
xx(IS/Fe)	295	HFeCo3(CO)12	80	.12	.08	structure information	75C028
xx		HFe(phen)(CN)4=	v		--	=-2H2O, electronic struc info	75M004
Pd		HFe3(CO)9SBu	77	--	--	2 Fe sites in ratio of 2:1	75B091
Pd		HFe3(CO)9SPr	77	--	--	2 Fe sites in ratio of 2:1	75B091
Cu(IS/SNP)	R	KFe(phen)(CN)2	110	.34	+-.61	bonding information	75B045
Cu		(MeNH3)2FeCl4	v	--	--	Neel temperature=95.5 K	75S064
Pd		(MeNH3)2FeCl4	v	--	--	20<T<300 K, Neel temp=96 K	75K058
Cu(IS/SNP)	R	Me2AsC2AsMe2(C=	80	--	--	=F2)4Fe2(CO)6, structure info	75C038
Cu(IS/SNP)	R	Me2AsC2AsMe2(C=	80	--	--	=F2)3Fe2(CO)6, structure info	75C038
Cu(IS/SNP)	R	Me2AsC2AsMe2(C=	80	--	--	=F2)2Fe2(CO)6, structure info	75C038
Pd		Me4NFe(NO3)4	77	.59		bonding and structure info	75A032
Pt(IS/Fe)	R	MnPy2Cl2	v		--	5 K<T<300 K, phase transition	75Y002
xx(IS/SNP)		NH4Fe(thsa)2	80	.52	2.92	bonding information	75T015
xx(IS/SNP)		NH4Fe(thsa)2	300	.44	2.81	bonding information	75T015
xx(IS/SNP)		NH4Fe(3-NO2=	80	.52	3.09	=thsa)2, bonding information	75T015
xx(IS/SNP)		NH4Fe(3-NO2=	300	.49	2.97	=thsa)2, bonding information	75T015
xx(IS/SNP)		NH4Fe(5-Br=	80	.59	3.02	=thsa)2, bonding information	75T015
xx(IS/SNP)		NH4Fe(5-Br=	200	--	--	=thsa)2, bonding information	75T015
xx(IS/SNP)		NH4Fe(5-Br=	300	--	--	=thsa)2, bonding information	75T015
xx(IS/SNP)		NH4Fe(5-Cl=	80	.59	3.02	=thsa)2, bonding information	75T015
xx(IS/SNP)		NH4Fe(5-Cl=	300	.48	2.98	=thsa)2, bonding information	75T015
xx(IS/SNP)		NH4Fe(5-Me=	80	--	--	=thsa)2, bonding information	75T015
xx(IS/SNP)		NH4Fe(5-Me=	300	--	--	=thsa)2, bonding information	75T015
xx(IS/SNP)		NH4Fe(5-NO2=	80	.60	3.20	=thas)2-.5H2O, bonding info	75T015
xx(IS/SNP)		NH4Fe(5-NO2=	300	.53	3.14	=thsa)2-.5H2O, bonding info	75T015
xx(IS/SNP)		NH4Fe(3,5-Cl2=	80	.59	3.05	=thsa)2-.5H2O, bonding info	75T015
xx(IS/SNP)		NH4Fe(3,5-Cl2=	200	--	--	=thsa)2-.5H2O, bonding info	75T015
xx(IS/SNP)		NH4Fe(3,5-Cl2=	300	--	--	=thsa)2-.5H2O, bonding info	75T015
Pd		NO2Fe(NO3)4	77	.57		bonding and structure info	75A032

SOURCE	S-TEMP	ABSORBER	A-TEMP	SHIFT	QS	REMARKS	REF
Pd		NaFH2O(EDTA)-2=	298	.73	.65	=H2O, structure information	75S099
xx(IS/Fe)		Na3Fe(CN)5(=	77	.0	1.10	=DMSO)-nH2O, IS correlation with UV-visible spectrum	75T009
xx(IS/Fe)		Na3Fe(CN)5Py-=	77	.08	.74	=nH2O, isomer shift corln with UV-visible spectrum	75T009
xx(IS/Fe)		Na3Fe(CN)5(=	77	.0	1.06	=TMSO)-nH2O, IS correlation with UV-visible spectrum	75T009
xx(IS/Fe)		Na4Fe(CN)5(=	77	.06	1.10	=MPZ),NH2O, IS correlation with UV-visible spectrum	75T009
xx(IS/Fe)		Na4Fe(CN)5(=	77	.08	.96	=pz)CO2-nH2O, IS correlation with UV-visible spectrum	75T009
Pt(IS/Fe)	R	NiPy2Cl2	v		--	5 K<T<300 K, phase transition	75Y002
xx(IS/SNP)		alpha-N-m-ni=		.55	.82	=trophenyl-D-glucosylamine + FeCl3	75K070
xx(IS/SNP)		beta-N-m-ni=		.55	.50	=trophenyl-D-glucosylamine + FeCl3	75K070
xx(IS/SNP)		alpha-N-o-ni=		.28	.41	=trophenyl-D-glucosylamine + FeCl3	75K070
xx(IS/SNP)		beta-N-o-ni=		.28	.31	=trophenyl-D-glucosylamine + FeCl3	75K070
xx(IS/SNP)		alpha-N-p-ni=		.42	1.00	=trophenyl-D-glucosylamine + FeCl3	75K070
xx(IS/SNP)		beta-N-p-ni=		.42	.68	=trophenyl-D-glucosylamine + FeCl3	75K070
xx(IS/SNP)		alpha-N-p-tol=		.70	2.25	=yl-D-glucosylamine + FeCl3	75K070
xx(IS/SNP)		beta-N-p-tolyl=		.70	1.52	=-D-glucosylamine + FeCl3	75K070
Pd		OFe3(CH2CNCOO)=	R	.70	.57	=6-3H2OClO4-6H2O	75M082
Pd		OFe3(PhCOO)6=	R	.71	.51	=ClO4-H2O	75M082
Cu(IS/SNP)	298	PhCl2SnFe(CO)2=	298	.30	1.68	bonding information	75B043
Cu(IS/SNP)	298	PhCl2SnFeCOCpP=	298	.40	1.73	=Ph3, bonding information	75B043
Cu(IS/SNP)	298	Ph2ClSnFe(CO)2=	298	.30	1.71	=Cp, bonding information	75B043
Cu(IS/SNP)	298	Ph2ClSnFeCOCpP=	298	.39	1.73	=Ph3, bonding information	75B043
Cu(IS/SNP)	R	Ph2PC2AsMe2(C=	80	--	--	=F2)2Fe2(CO)6, structure info	75C038
Cu(IS/SNP)	R	Ph2PC2AsMe2(C=	80	--	--	=F2)3Fe2(CO)6, structure info	75C038
Cu(IS/SNP)	R	Ph2PC2PPh2(CF2=	80	--	--	=)3Fe2(CO)6, structure info	75C038
Cu(IS/SNP)	R	Ph2PC2PPh2(CF2=	80	--	--	=)2Fe2(CO)6, structure info	75C038
Cu(IS/SNP)	298	Ph3SnFe(CO)2Cp	298	.28	1.75	bonding information	75B043
Cu(IS/SNP)	298	Ph3SnFeCOCpPPh3	298	.39	1.79	bonding information	75B043
Pd		Ph4AsFeCl4	77	.3	.2	bonding and structure info	75A032
Pd(IS/Fe)		Sn(BFD)2Fe(CO)4	77	-.18		bonding and structure info	75C048
Pd(IS/Fe)		SnBr2Fe(CO)4	77	.06		bonding and structure info	75C048
Pd(IS/Fe)		SnCl2Fe(CO)4	77	-.17		bonding and structure info	75C048
Pd(IS/Fe)		Sn(DPP)2Fe(CO)4	77	.03		bonding and structure info	75C048
Pd(IS/Fe)		Sn(PBD)2Fe(CO)4	77	.08		bonding and structure info	75C048
Pd(IS/Fe)		Sn(PTD)2Fe(CO)4	77	.11		bonding and structure info	75C048
Pd(IS/Fe)		Sn(TBD)2Fe(CO)4	77	-.04		bonding and structure info	75C048
Pd		TPP-FeCl		-.126	.408	comparison with ESCA	75J003
xx		benzohydroxam=				=ate	74S122
xx(IS/Fe)		biferricenium-=	4.2	2.951	.573	=(DDQ)2	75M024
xx(IS/Fe)		biferricenium-=	4.2	1.726	.520	=DDQ	75M024
xx(IS/Fe)		biferricenium-=	300	2.833	.476	=(DDQ)2	75M024
xx(IS/Fe)		biferricenium-=	300	1.784	.440	=DDQ	75M024
xx(IS/Fe)		biferricenium-=	4.2	1.756	.542	=I5	75M024
xx(IS/Fe)		biferricenium-=	4.2	--	--	=I3	75M024
xx(IS/Fe)		biferricenium-=	300	--	--	=I3	75M024
xx(IS/Fe)		biferricenium-=	300	1.719	.441	=I5	75M024
xx(IS/Fe)		biferricenium-=	4.2	2.950	.567	=(PF6)2	75M024
xx(IS/Fe)		biferricenium-=	4.2	1.816	.527	=PF6	75M024
xx(IS/Fe)		biferricenium-=	300	2.890	.467	=(PF6)2	75M024
xx(IS/Fe)		biferricenium-=	300	1.738	.403	=PF6	75M024
xx(IS/Fe)		biferricenium-=	4.2	--	--	=TCA-2TCAA	75M024
xx(IS/Fe)		biferricenium-=	300	--	--	=TCA-2TCAA	75M024
xx(IS/Fe)		biferrocenylene	v	--	--	T=4.2,80, and 300 K	75M024
xx(IS/SNP)		bis(1-phenyl=		.72	1.97	=borabenzene)iron	75A053
Cu		butadiene-sty=	v			=rene-4-vinylpyridine terpolymer, struc info, 4<T<300 K	75M012
Cu		butadiene-4-vi=	v			=nylpyridine copolymer, struc information, 4<T<239 K	75M012
Cu		(bzNH3)2FeCl4	v	--	--	Neel temperature=75.6 K	75S064
Pd(IS/SNP)		t-((depe)2ClFe=		.58	1.17	=)2(SNT)+2, bonding info	75B070
xx(IS/Fe)		ferricenophani=	4.2	--	--	=um-I3-.5I2	75M024
xx(IS/Fe)		ferricenophani=	300	--	--	=um-I3-.5I2	75M024

SOURCE	S-TEMP	ABSORBER	A-TEMP	SHIFT	QS	REMARKS	REF
Pd		ferriciniumpic=		.256		=rate, comparison with ESCA	75J003
xx		hydrazonium hy=				=drazinedithiocarbaminte	75B136
xx (IS/Fe)		(pip)2Fe(TPP)	v	--	--	4.2<T<300 K, bonding info	75C045
xx		polyacrylic =				=acid, study of elec relax	75I010
xx		polyvinyl-pyri=				=dine, study of elec relax	75I010
xx		poly(4 vinyl=		--	--	=pyridine)-FeCl3-6H2O	75B093
xx (IS/Fe)		(2-MeIz)Fe(TPP)	v	--	--	4.2<T<300 K, bonding info	75C045
xx		1-1'-diacetyl=			--	=ferrocene in 4-4'-bis(heptyl-oxy)azoxybenzene, discussion of previous results	74W029
xx (IS/Fe)		1,12-dimethyl =	4.2	+.522	2.183	=ferricenophanium-(DDQ)2	75M024
xx (IS/Fe)		1,12-dimethyl =	300	+.430	2.168	=ferricenophanium-(DDQ)2	75M024
xx (IS/Fe)		1,12-dimethyl =	v	--	--	=ferricenophanium-I3, 4.2 K<T<300 K	75M024
xx (IS/Fe)		1,6-diiodobi=	4.2	1.388	.532	=ferricenium-I3	75M024
xx (IS/Fe)		1,6-diiodobi=	300	1.284	.428	=ferricenium-I3	75M024

INORGANIC OXIDES

SOURCE	S-TEMP	ABSORBER	A-TEMP	SHIFT	QS	REMARKS	REF
Cr		.6CuFe2O4-(.4-=	v		--	=x)MnFe2O4-xBiFe2O(4+z), T=80,296,&715 K	75K071
xx (IS/Fe)		AgFeO2	v	.58	.80	4.2<T<300 K, bond & struc info	75T026
Pd		Al2O3	R			solubility and structure info	75V049
xx		Al2O3-Fe2O3				Morin transition	75P038
Pd		BaEu2Fe2O7	v	--	--	77 K<T<545 K, HI data	75H015
xx (IS/PFC)		BaFeO3	298	--		bonding information	74P053
xx		BaFeSi4O10		--	--	MO calc, previous data	75T003
xx		BaFe12O19		--	--	MO calc, previous data	75T003
xx		BaFe12O19			--	discussion of previous data	75B120
xx		BaFe12O19				preparation of Ba ferrite powders	74H051
xx		BaFe12O19			--	calculate QS	75B077
xx		BaFe12O19	v		--	4.2<T<875 K	75B081
xx		BaFe12O19	v	--	--	HI data, 4.2 K<T<750 K	75B055
Cu		BaFe12O19	v		--	study of trigonalbypyramidal site	75K003
Cu		BaFe12O19	v		--	4.2 K<T<870 K, bond & struc	75K018
xx		BaFe(12-x)Alx=	v			=O19, HI vs temperature for 5 sites, 80 K<T<700 K	75A016
Pd		BaInxFe(12-x)O=		--	--	=19, magnetic structure info	74B095
xx		BaInxFe(12-x)O=	v			=19, T=100 & 300 K, fa vs comp	75B119
xx		BaIn3Fe9O19			.64	study of magn transformation	75B082
xx		BaIn3Fe9O19	v			study of the magnetic struc	75B121
xx		BaLa2Fe2O7	v			study of magnetic structure	75S083
Cu (IS/SS)		BaO-2Fe2O3	v			Neel temperature=945 K, 300<T<920 K	75O010
xx		BaSrFe4O8 (hex)				structure information	75C042
xx (IS/SNP)		BaTiO3-Fe2O3					75Y004
xx		Ba-hexaferrite	R			study of formation mechanism	75M020
Cr	R	Ba2Co2Fe12O22	v			study of the behavior of the sublattice magnetizations, 78<T<700 K	75A039
xx (IS/PFC)		Ba2FeO4	78	.04		bonding information	74P053
xx (IS/PFC)		Ba2FeO4	298	.38		bonding information	74P053
Cr	R	Ba2Mg2Fe12O22	v			study of the behavior of the sublattice magnetizations, 78<T<700 K	75A039
Pd		BeAl2O4	R			solubility and structure info	75V049
Pd		BeO	R			solubility and structure info	75V049
Cr (IS/Fe)		BiFeMoOx	v	--	--	T=80 & 300 K, structure info	75N015
Cr (IS/Fe)		Bi3FeMo2Ox	v	--	--	T=80 & 300 K, structure info	75N015
SS		Bi7Ti3Fe3O21	v			behave like thin magn films	75S103
SS		Bi12Ti4NbFe5O18	v			behave like thin magn films	75S103
Cr		CaO-Fe3O4					75V014
Pd		Ca(1-x)Fe(1+x)=	v	--	--	=Ge2O6, T=77 & 298 K	75E013

SOURCE	S-TEMP	ABSORBER	A-TEMP	SHIFT	QS	REMARKS	REF
xx		Ca2Fe(2-x)Alx05				site populations data	75G030
xx		Ca2Fe(2-x)Cox05				site populations data	75G030
xx		Ca2Fe(2-x)Crx05				site populations data	75G030
xx		Ca2Fe(2-x)Gax05				site populations data	75G030
xx		Ca2Fe(2-x)Scx05				site populations data	75G030
xx		CdMn1.9Fe.104	R			Jahn-Teller distortions	75F031
xx(IS/Fe)		Chorten rust	v	--	--	80 K<T<300 K, identification of products formed	75K021
xx		CoAlFe04	296		--	HA data	75B123
xx		CoAl1.9Fe.104				Jahn-Teller distortions	75F031
xx		CoCr204	v				75T007
xx(IS/Fe)		CoFe204	R	--		catalyst study, HI data	75C014
xx		CoMn1.9Fe.104	v		--	Jahn-Teller distortions	75F031
xx(IS/Fe)		CoTi03	4.2	1.2	+.71	HI=-105 kOe	75I002
xx(IS/Fe)		CoTi03	298	1.07	.58		75I002
xx		CoxFe(1-x)Co=			--	=(1-x)Fe(1+x-a)AlaO4, HA	75B123
Cr		CoxMn(2.9-x)Fe=	R		--	=.104, Jahn-Teller distortions	75F031
xx		Co3B7013Cl	v		--	structure info, 77 K<T<770 K	75Z007
Pd	295	Cr203-Fe203	v		--	study of magnetic ordering	75B003
Pd		Cr203-V203	v	--	--	study of metal-semiconductor transition	75S069
xx(IS/Fe)		CuFe02	v	.50	.66	4.2<T<300 K, bond & struc info	75T026
xx(IS/Fe)		CuFe204	R	--		catalyst study, HI data	75C014
xx		CuxFe(1-x)Cu=				=(1-x)Fe(1+x)04, study of aging	75K085
Cr		DyCr03	296	.51	.21	superexchange interaction	75M081
Cr	R	DyFe03	v	--			75N030
Cu		Dy3Al4.9Fe.1012	80	--	--	2 Fe sites	75F004
Cu		Dy3Al4.9Fe.1012	298	--	--	2 Fe sites	75F004
Cr		ErCr03	296	.49	.18	superexchange interaction	75M081
Cu		Er3Al4.9Fe.1012	80	--	--	2 Fe sites	75F004
Cu		Er3Al4.9Fe.1012	298	--	--	2 Fe sites	75F004
xx		FeCr204	v			HI=150 kOe(T=0 K)	75T007
xx		FeFe(3-x)Alx04	R				75M044
xx(IS/310 SS)		FeO		--	--	molecular orbital calculations	74K066
xx		Fe06(-10)				MO calc, previous data	75T003
xx		Fe06(-9)				MO calc, previous data	75T003
xx(IS/Fe)		alpha-FeOOH				study of effect of particle size on Mossbauer parameters	75K021
xx(IS/Fe)		alpha-FeOOH		.49	.12	HI=505 kOe	75K021
xx(IS/Fe)		alpha-FeOOH	300	.38	.14	HI=379 kOe	75K021
xx		alpha-FeOOH	v	--	--	study of phase transition due to mechanical activation, T=room & liquid nitrogen	75K037
xx		beta-FeOOH				HI data, microcrystals	74V041
xx(IS/Fe)		beta-FeOOH		.51	.07	HI=473 kOe	75K021
xx(IS/Fe)		beta-FeOOH		.40	.32		75K021
xx		beta-FeOOH	v	--	--	study of phase transition due to mechanical activation, T=room & liquid nitrogen	75K037
xx		delta-FeOOH				HI data, microcrystals	74V041
xx		gamma-FeOOH			--		75M085
xx(IS/Fe)		gamma-FeOOH		.38	.27		75K021
xx(IS/Fe)		gamma-FeOOH		.49	.29		75K021
Xx(IS/Fe)		FeOOH	N	.46	.72		75K056
Xx(IS/Fe)		FeOOH	R	.36	.66		75K056
xx		alpha-FeOOH-=	v			=nH20, Neel T vs H20 conc	75G053
xx		beta-FeOOh	v		--	study of the dependence of the Morin transition on the defect structure	75P053
Pd		FeOx				phase analysis after heat treatment	74G062
Pd(IS/Fe)	295	beta-FeOx(OH)=	v	--	--	=(3-2x), study of structures and dehydration mechanism	75H002
xx(IS/Fe)		FeTi03	4.2	1.23	+1.43	HI=-47 kOe	75I002
xx(IS/Fe)		FeTi03	298	1.07	.65		75I002
xx		FeV204	v			HI=110 kOe(T=0 K)	75T007
Cu(IS/310 SS)		FeV205	v		--	T=193 K and 298 K, study of cation distribution	75G032
xx		FexV(1-x)O(2+y)	v			4.2<T<700 K, phase transition	75P047
Pd(IS/SS)		FexV205	v	--	--	T=78 & 295 K, bonding info	75B076
xx(IS/SNP)		Fe2(MoO4)3	v	--	--	Debye temperature=250 K	75Z019

79

SOURCE	S-TEMP	ABSORBER	A-TEMP	SHIFT	QS	REMARKS	REF
xx		alpha-Fe2O3				Mossbauer magn diffraction	75S011
Cr		alpha-Fe2O3				study of Bragg reflections- (888) and (999)	75A028
Pd		alpha-Fe2O3				study of reduction of hematite in hydrogen at 535 C	75G012
Pd(IS/SS)		alpha-Fe2O3	300	+.50	.20	HI=515 kG	75A024
xx		alpha-Fe2O3	v			4.2 K<T<300 K, study of the magnetic properties of small particles, Morin temperatures	75K023
xx		alpha-Fe2O3	v		--	4.2<T<298 K, surface studies of fine particles	75V017
Cu		alpha-Fe2O3	v			selective excitation double Mossbauer technique	75B053
xx		gamma-Fe2O3	300			magnetic properties of ultra-fine particles	74K069
xx		gamma-Fe2O3	v			study of magnetic properties of ultra-fine particles	75K069
Pd		Fe2O3				measurement of the spin contact density	75B127
SS		Fe2O3	v			study of ferromagnetic to paramagnetic phase transistion	73D047
xx		alpha-Fe2O3 in=				= Ca-V-Fe garnet, quantitative analysis	75M019
xx		aph-Fe2O3-Al2O3	v		--	Morin transition	75D053
xx		Fe2O3-Al2O3	v		--	study of spin reorientation	75S089
xx		Fe2O3-MgO				study of reaction kinetics	75V033
Cr	295	alpha-Fe2O3-Sn	v	--		80 K<T<1020 K, HI data	75R011
Pd		Fe2O3-Ti					74V037
xx(IS/Fe)	80	Fe2Te3O9	4.2	.512	-.128	bonding and structure info	75T023
xx		Fe3B7O13Cl	v		--	structure info, 77 K<T<770 K	75Z007
xx		Fe3BO6		--	--	magnetic structure information	75B094
xx		Fe3O4		--	--	study of low temperature phase	75M045
xx		Fe3O4				study of phase transition	75E016
xx		Fe3O4	R	--		spin and charge density oscil- -lations in the B sublattice	75N005
xx		Fe3O4	v			T=4,77, and 296 K	75G036
xx		Fe3O4	v			study of the Verwey transition	75T028
xx		Fe3O4	v			multi-stage electronic trans, 100 K<T< K	75B059
Pd		Fe3O4	v			T=293 & 793 K	75K082
xx		Fe3O4-CaO				study of reaction kinetics	75V033
xx		Fe3O4 (Zn)					75E016
xx		Fe(3-x)AlxO4				site population determination	75D041
xx		Fe(3-x)AlxO4		--			75D006
Pt		Fe(3-x)AlxO4		--	--	HI data, 2 Fe sites	75D015
Pt		Fe(3-x)AlxO4	v	--		77<T<647 K, determination of cation distribution	75D070
Pt		Fe(3-x)CuxO4	77	--		study of cation distribution	75S016
Pt		Fe(3-x)CuxO4	R	--		study of cation distribution	75S016
Pd		Fe(3+x)O4				measure of deviation from stoichiometry	75V026
xx		Fe(3-x)SnxO4	78			study of phase transition	75S072
Cr		GdCrO3	296	.52	.21	superexchange interaction	75M081
Cu		Gd3Al4.9Fe.1O12	80	--	--	2 Fe sites	75F004
Cu		Gd3Al4.9Fe.1O12	298	--	--	2 Fe sites	75F004
Cu		Ho3Al4.9Fe.1O12	80	--	--	2 Fe sites	75F004
Cu		Ho3Al4.9Fe.1O12	298	--	--	2 Fe sites	75F004
Cr		LaCrO3	296	.51		superexchange interaction	75M081
Cr		LaNdCr2O6	296	.52		superexchange interaction	75M081
xx		Li.1Fe2.9O4	R	--		spin and charge density oscil- -lations in the B sublattice	75N005
xx		LiNbO3		--	--	study of impurity centers	75R036
Pd(IS/Fe)	v	LiNbO3		--	--	4.2 K<T<295 K	75K028
Pd		Lix(Fe,Mn,Ga)=				=5x08, HI data	75B141
xx		Li2O-5Fe2O3					72S098
Cr		LuCrO3	296	.51	.21	superexchange interaction	75M081
xx		Mn-Zn ferrites				study of magnetic structure	75B117
xx		MgCr2O4	v			HI=131 kOe(T=0 K)	75T007
Pd		MgFe2O4	R			quenched samples, HI data	75D061
xx		MgMn1.9Fe.1O4	v		--	Jahn-Teller distortions	75F031
xx		MgO	v			study of surface, catalytic, and magnetic properties of small iron particles	75D057

SOURCE	S-TEMP	ABSORBER	A-TEMP	SHIFT	QS	REMARKS	REF
xx (IS/Fe)		MgTiO3	4.2	1.21	1.06		75I002
xx (IS/Fe)		MgTiO3	298	1.07	.50		75I002
xx (IS/SNP)		Mg (1-x) FexGeO3	v	--	--	T=77 & 298 K	74N021
xx (IS/Fe)		MnTiO3-I	4.2	1.25	+1.60	HI=-34 kOe	75I002
xx (IS/Fe)		MnTiO3-I	298	1.10	.80		75I002
xx (IS/Fe)		MnTiO3-II	4.2	1.2	-1.9	HI=-85 kOe	75I002
xx (IS/Fe)		MnTiO3-II	298	1.07	1.30		75I002
Cu		Mn (1-x) ZnxFe2O4	4.2	--		HA=50 and 90 kOe	75M002
xx		Mn (1-x) ZnxFe2O4	v		--		75K078
xx		Mn3B7O13Cl	v		--	structure info, 77 K<T<770 K	75Z007
xx		Mn3B7O13Cl	v		--	83 K<T<723 K	75B035
xx		Mn (3-x) FexO4	v		--	Jahn-Teller distortions	75F031
Cu		(Na,Ca) (2-3) (=	v			=Fe,Mg,Al)5Si8O22(OH,F)2, study of the magnetic ordering	75B131
xx (IS/Fe)		alpha-NaFeO2	v	.41	.42	4.2<T<300 K, bond & struc info	75T026
Pd		NaFeO2	v	--	--	study of phase transition, 300 K<T<1403 K	75V007
xx (IS/SNP)		NaFexAl (1-x) O2	v	--	--	T=77 & 300 K	75B080
xx		Nd.95Ca.05Fe=				=.95Sn.05O3, T=78 & 300 K	74L049
Cr		NdCrO3	296	.52		superexchange interaction	75M081
xx		Ni.1Fe2.9O4	R	--		spin and charge density oscil--lations in the B sublattice	75N005
Cu (IS/Fe)		Ni.8Co.2Fe (2-x=	300	--		=) VxO4, study of magnetic properties and magnetic annealing	75H030
xx		NiCr2O4	v			HI=439 kOe (T=0 K)	75T007
Pd		NiFe2O4		--			75K082
Pt		NiFe2O (4-x) Sx			--	Curie temperatures	75B034
SS		NiFe (2-x) CrO4	v		--	4.2<T<858 K, study of exchange interaction	75G068
xx		NiFe (2-x) CrxO4	77	--		IS & Debye temp vs composition	75N012
xx		NiFe (2-x) CrxO4	293			IS & Debye temp vs composition	75N012
xx (IS/Fe)		NiTiO3	4.2	1.2	+.42	HI=-91 kOe	75I002
xx (IS/Fe)		NiTiO3	298	1.06	.37		75I002
xx		Ni (1-x) CdxFe2O4				Debye T decreases with increase in Cd concentration	75E012
Cr		SmCrO3	296	.55		superexchange interaction	75M081
xx		Sn.1Fe2.9O4	R	--		spin and charge density oscil--lations in the B sublattice	75N005
Pd		SrEu2Fe2O7	v	--	--	77 K<T<545 K, HI data	75H015
xx (IS/Fe)		SrFexRu (1-x) =	v	--	--	=O (3-y), perovskite structure, T=4.2, 77, and 300 K	75G009
xx		SrFe (12-x) Gax=	v			=O19, HI vs temperature for 5 sites, 80 K<T<700 K	75A016
Cu		SrO- (6-x) Fe2O3=	N	--	--	=-xAl2O3, HI data	75F018
xx		SrTiO3				elastic and inelastic scatt	75D045
Cu		TbFe (1-x) CrxO3	v			study of the effect of nearest neighbor ions on HI at Fe	75N003
Cu		Tb3Al4.9Fe.1O12	80	--	--	2 Fe sites	75F004
Cu		Tb3Al4.9Fe.1O12	298	--	--	2 Fe sites	75F004
Pd		Ti2O3-V2O3	v	--	--	study of metal-semiconductor transition	75S069
Cu		Tm3Al4.9Fe.1O12	80	--	--	2 Fe sites	75F004
Cu		Tm3Al4.9Fe.1O12	298	--	--	2 Fe sites	75F004
xx (IS/Fe)		V (1-x) FexO2		--	--	study of metal-non metal trans	74B100
xx		(V (1-x) Tix) 2O3	v			Neel temperature vs Ti conc	75M017
Cu		Y3Al4.9Fe.1O12	298	--	--	2 Fe sites	75F004
Pd		Y3Co.5xTi.5x=				=Fe (5-x) O12, bonding and structure information	75K033
xx		Y3Fe.3Ga4.4Si=				=.3O12, structure information	75H043
xx		Y3Fe (2-x) Scx=				=Fe3O12, HA data	75P029
xx		Y3Fe4.6Si.4O12				structure information	75H043
Cr		Y3Fe5O12				study of the effect of pressure on the magnetic field	75N029
Cr		Y3Fe5O12					75B112
Cr		Y3Fe5O12	R	--		HI vs pressure	75N011
Pd		Y3Fe5O12	v		--	5 K<T<300 K, thermal treatment	75P020
xx		Y3Fe (5-x) AlxO12			--	study of cation distribution	75F011
xx		Y3Fe (5-x) GaxO12	v			study in vicinity of Curie T	75V006
xx		Y3InxGaxFe (5-x=				=) O12, study of cation distribution	75G071
xx		Y (3-x) CaxFe (Fe=	v			= (3-x) Six) O12, 5 K<T<370 K, study of temperature dependence of spin order	75P007

SOURCE	S-TEMP	ABSORBER	A-TEMP	SHIFT	QS	REMARKS	REF
xx		Y(3-x)CaxFe(5-=	v			=x)SnxO12(x=.7&.9), some HI above Neel temperature	75L031
xx		Y(3-x)CaxFe=	v			=(5-x)ZrxO12, HI data, 80 K<T<513 K	75L008
xx		Y6FeSc3Fe6O12	4.2			HA data	75P003
xx		Y(3-2x)Ca2xFe=	v			=2xFe(3-x)VxO12, study of the effect of V+5 ions	75D016
Pd(IS/SNP)		Y(3-2x)Ca2xSbx=		--		=Fe(5-x)O12, study of the cation distribution	74G064
xx		YCa2Fe3Si2O12	4.2			HA data	75P003
Cr		YCrO3	296	.51	.21	superexchange interaction	75M081
Cr		YFe.7Mn.3O3				Morin transition	75G075
Pd		YFe.8Ga.2O3				HI data	75B079
Rh(IS/Fe)	R	YFeO3	4.2	.50	+.08	HI=55.2 T	75D026
Rh(IS/Fe)	R	YFeO3	293	.30	-.04	HI=55.2 T	75D026
Cr	R	YFeO3	v	--			75N030
xx		YFeO3(Co)				study of spin reorientation	75V043
xx		(Y,Gd)3(Fe,Al)=	v			=5O12	75L027
Pd		YIG			--	study of the local fields of iron	74B096
Cr		YbCrO3	296	.51	.24	superexchange interaction	75M081
Cu		Yb3Al4.9Fe.1O12	80	--	--	2 Fe sites	75F004
Cu		Yb3Al4.9Fe.1O12	298	--	--	2 Fe sites	75F004
xx		Zn.7Ni.3Fe2O4				HA data	75P004
Cr		Zn.8Co.2Fe2O4	293	--	--		75F030
xx		ZnAl1.9Fe.1O4				Jahn-Teller distortions	75F031
Cu(IS/Fe)		ZnCrFeO4	295	.33	.50	structure information	75S007
xx		ZnCr2O4	v			HI=130 kOe(T=0 K)	75T007
Pd		ZnFe2O4		--			75K082
Cu(IS/Fe)		ZnMnFeO4	295	.33	.52	structure information	75S007
xx		ZnMn1.9Fe.1O4	v		--	Jahn-Teller distortions	75F031
xx		ZnMn2O4	v		--	Neel T<50 K, like Mn3O4	75W006
xx		ZnxFe(3-x)O4	300	--			74S124
xx		ZnxNi(1-x)Fe2O4				analysis of localized canting model applied to spin struc	75P028
Cu(IS/Fe)		Zn2MnCrFe2O8	295	.38	.41	structure information	75S007
xx		lambda-FeOOH			--	study of superfine layers	75M085

SOURCE EXPERIMENTS

SOURCE	S-TEMP	ABSORBER	A-TEMP	SHIFT	QS	REMARKS	REF
Ag	v	PFC				30 mK<T<60 K, study of local magnetization	75S021
Ag	v	SFC				.022 K<T<4.2 K, study of dilute impurities, HA data	75T010
Ag	v	SFC				.022<T<20 K, HI data, HA	75T017
Al		xx				study of impurity-vacancy bond	75W017
Al-Fe alloys	v	SFC(IS/Fe)		--	--	T=80 & 300 K, study of the recoil event	75J009
AlNi3	R	xx(IS/Fe)	R	--	--		75D014
AlNi3	4.2	xx(IS/Fe)	R			HI=227 kOe	75D014
AlNi3(Co,Fe)		xx		--		HI, study of the magnetic transformation of the surface layer	75D077
Alpha-Sn		SnO2					74W027
BaFeO3	298	xx		.14			75L024
BaFeO3	298	xx		.27			75L024
BaFeO3	298	PFS(IS/SNP)		--		bonding information	74P053
Ba2CoO4	78	xx		.14			75L024
Ba2CoO4	298	xx		.10			75L024
Ba2CoO4	78	PFS(IS/SNP)		-.14		bonding information	74P053
Ba2CoO4	298	PFS(IS/SNP)		-.10		bonding information	74P053
Ba2FeO4	78	xx		.47			75L024
Ba2FeO4	298	xx		.38			75L024
CaO	4.2	xx			--	HA data, single crystal	75R026
Co	v					study of magnetic properties of thin films, 25<T<295 K	74K067

SOURCE	S-TEMP	ABSORBER	A-TEMP	SHIFT	QS	REMARKS	REF
Co/Fe(II)(CN)6	80	xx					75P016
Co/Fe(III)(CN)6	80	xx					75P016
Co(Br(acac))3	80	xx(IS/SNP)		--	--	bonding information	75A023
Co(Bzac)3	80	xx(IS/SNP)		--	--	bonding information	75A023
Co-Ga alloys	v					T=78 & 298 K, study of short-range magnetic ordering	74R034
Co-Ga alloys	v	PFC	293	--		T=78 & 293 K, short range magnetic ordering	75R015
Co(NH3)6Cr=		SS				=(C204)3-3H2O, radiolysis	75S002
Co(NH3)6Fe=		SS				=(C204)3-3H2O, radiolysis	75S002
Co(NH3)6Fe(CN)6		SS				radiolysis study	75S002
Co(NH3)6(NO3)3		SS				radiolysis study	75S002
(Co(NH3)6)2=		SS				=(C204)3-4H2O, radiolysis	75S002
Co(NO)2Cl	298	xx		--	--		75L024
Co(NO2(acac))3	80	xx(IS/SNP)		--	--	bonding information	75A023
CoO	v	SFC		--		297 K<T<1423 K, phase trans	75S029
Co(SCN(acac))3	80	xx(IS/SNP)		--	--	bonding information	75A023
CoSn	295	Fe	R	-.36	.54		75H009
Co(acac)3	80	xx(IS/SNP)		--	--	bonding information	75A023
Co(dipy)3ClO4	80	SNP				bonding information	75A034
Co(dipy)3(ClO4=	v	SNP				=)2,T=80 & 295 K, bonding info	75A034
Co(dipy)3Cl2-=	v	SNP				=H2O, T=80 & 295 K, bond info	75A034
Co(dipy)3Cl2-6=	v	SNP				=H2O, T=80 & 295 K, bond info	75A034
Co(dipy)3(NO3)=	v	SNP				=2-H2O, T=80 & 295K, bond info	75A034
Co(en)3(NO3)3		SS				radiolysis study	75S002
(Co(en)3)2=		SS				=(C204)3-9H2O, radiolysis	75S002
Co(phen)3(ClO4=	25	KFC	25			=)2, time differential exp	75G055
Co(1-x)FeSi	v	xx		--			75M008
Cr203(Mn)	N	310 SS		--	--	57Mn source, Cr(alpha,proton) Mn reaction	75P043
Cu		xx				relaxation times of superparamagnetic particles	75K032
Cu		xx				fs vs pressure, discussion of previous data	75M070
Cu	v	PFC				.025<T<60 K, Kondo effect	74S125
Cu		PFC	R	--	--	study of atomic short range order	75C061
ErCoO3	v	PFC		--		298<T<1075 K, bonding info	75J008
EuCoO3		PFC	v	--		78<T<1158 K	74J017
Fe	v	xx				.07<T<1 K, study of thermal contact	75B103
Fe/Co(CN)6	80	xx					75P016
FeF2	v	PFC		--	--	ion implantation, struc info	75F019
Fe(NO)2Br	298	xx		--	--		75L024
Fe-Ni alloys		SFC(IS/Fe)		--	--	T=80 & 300 K study of the recoil event	75J009
Fe(OH)3-H2O-KOH	77	xx(SNP)		--	--		75O011
Fe(OH)3-H2O-N=	77	xx(SNP)		--	--	=H4OH	75O011
Fe(OH)3-H2O-N=	77	xx(SNP)		--	--	=H4OH-NH4Cl	75O011
Fe-ge alloys	300	SFC		--	--	study of recoil damage	75J012
Fe52Ni48	v	SS				4.2<T<900 K, HI vs T	74B092
GaAs	v	xx				HI data	75K051
GaNi3	R	xx(IS/Fe)	R	--			75D014
GaSb	v	xx				HI data	75K051
Gd	v	SS	R			T=77 296 K	75B146
Ge	R	PFC		--		ion implantation study,2 peaks	75W016
Hf	R	xx					75W029
InSb	v	xx				HI data	75K051
K2CoO3	298	xx		.12			75L024
K2CoO3	298	PFS(IS/SNP)		-.16		bonding information	74P053
K3Co(CN)6	80	xx					75P016
K3Co(CN)6		SS				radiolysis study	75S002
K3Co(NO2)6		SS				radiolysis study	75S002
K3Fe(CN)6	v	PFC		--	--	ion implantation, struc info	75F019
K4Fe(CN)6	v	PFC		--	--	ion implantation, struc info	75F019
KFeO2	298	xx		.48			75L024
KFeO2	298	xx		.48			75L024
KMgF3	4.2	xx			--	HA data, single crystal	75R026
La(1-x)SrCoO3		310 SS	v			78 K<T<440 K, study of magn ordering	75B028
La(1-x)SrxCoO3	v	xx				78<T<1200 K, HI data	75B105
La(1-x)SrxCoO3	v	PFC				=t=78 & 298 K, elec struc info	74R037

SOURCE	S-TEMP	ABSORBER	A-TEMP	SHIFT	QS	REMARKS	REF
LiNbO3	v	SFC (IS/Fe)		--	--	4.2 K<T<295 K	75K028
Ln(1-x)SrxCoO3		PFC				electron hopping conduction	75R033
MgF2		xx	4.2		3.11	HI=-270 kOe	75A001
MgF2		xx	295		2.83	HI=-270 kOe	75A001
MnNi2	v	xx				disordered sample, HI data	75C032
MnxNi(1-x)	v	xx	v	--		T=4.2 & 80 K, HI data	75C057
Nb3Sn(Pd)	2	PFC				study of local fields inside a superconductor	75M014
NdCoO3	v	PFC				78<T<628 K, elec struc info	74R037
Nd(1-x)SrxCoO3		PFC				electron hopping conduction	75R033
Ni	v	PFC				spin correlation times	75K005
Ni/Co(CN)6	80	xx					75P016
Os	R	xx(IS/Fe)	R	-.125	-.153		75W029
Re	R	xx(IS/Fe)	R	-.02	-.10		75W029
Re-W alloys		xx				structure information	75B030
Rh		xx				examination of Rh sources	75D054
Ru	R	xx(IS/Fe)	R	-.065	-.143		75W029
SS		xx				study of line broadening	75D055
Si	R	PFC		--		ion implantation study,2 peaks	75W016
Si		PFC(IS/Fe)				structure information	75B087
Si		SnO2					74W027
Si (n-type)		Pd		--		diffusion study	75U002
beta-Sn	v	xx		--		FeSn2 clusters formed	75N031
Ti	R	xx(IS/Fe)	R	-.012	-.10		75W029
YCoO3	v	PFC		--		78<T<1158 K, bonding info	75J008
ZnF2		xx	4.2		3.13	HI=-275 kOe	75A001
ZnF2		xx	295		2.80	HI=-275 kOe	75A001
Zr	R	xx					75W029
asbestos		SS	999	--	--	T=1225 K	75Z003
diamond	R	PFC		--		ion implantation study,2 peaks	75W016
mica		SS	999	--	--	T=1200 K	75Z003
quartz		SS	v	--	--	T up to 1375 K	75Z003

INORGANIC SULFATES

SOURCE	S-TEMP	ABSORBER	A-TEMP	SHIFT	QS	REMARKS	REF
xx(IS/Fe)	R	CsFe(SO4)2-12=	87	--		=H2O, spin relaxation	75L003
xx(IS/Fe)	R	CsFe(SO4)2-12=	300	--		=H2O, spin relaxation	75L003
xx		Fe(N2H5)2(SO4)2	v	--	--	slow paramagnetic relaxation	75T029
xx		FeNH4(SO4)2-12=				=H2O, discussion of previous data, electronic relaxation	75B011
xx(IS/SFC)		Fe(NH3)6SO4	v	--	--	phase trans study, 4.2<T<295	75A013
Fe		Fe(NH4)2(SO4)2=			1.740	=-6H2O, polarized source exp, determination of efg tensor	75G022
xx		alpha-FeSO4	v			studies around the Neel temp	75W008
xx		alpha-FeSO4	v		--	discussion of previous data	75Z008
xx		alpha-FeSO4	v		--	electronic & magn struc info	75T025
Rh	4.2	FeSO4-H2O	4.2			study of magn induced texture	75N002
Cu(IS/Fe)		FeSO4-H2O	v	--	--	thermal decomposition study	75S048
xx		FeSO4-4H2O	v		--	7<T<300 K, eta info	75G049
Pd		FeSO4-7H2O		1.089	3.235	comparison with ESCA	75J003
Cu(IS/Fe)		FeZrO(SO4)2-8=	300	1.21	3.08	=H2O, bonding & structure info	75S008
Cr(IS/SNP)		Fe2(SO4)3-7H2O	v	--	--	study of thermal decomposition 293<T<873 K	75B019
Cr(IS/SNP)		Fe2(SO4)3-7H2O	v	--	--	study of thermal decomposition 293<T<873 K	75B020
xx(IS/Fe)	R	KFe(SO4)2-12H2O	87	.45		spin relaxation	75L003
xx(IS/Fe)	R	KFe(SO4)2-12H2O	273	.36		spin relaxation	75L003
xx(IS/Fe)	R	MeNH3Fe(SO4)2-=	103	--		=H2O, spin relaxation	75L003
xx(IS/Fe)	R	MeNH3Fe(SO4)2-=	164	--		=H2O, spin relaxation	75L003
xx		MeNH3Fe(SO4)2-=	v			=12H2O, fa v temperature, study of ferroelectric phase transformation	75D052
xx		NH4FeSO4-12H2O				study of angular distributions of resonant scattered gamma	75D037
xx(IS/Fe)	R	NH4Fe(SO4)2-12=	87	.49		=H2O, spin relaxation	75L003

84

SOURCE	S-TEMP	ABSORBER	A-TEMP	SHIFT	QS	REMARKS	REF
xx(IS/Fe)	R	NH4Fe(SO4)2-12=	298	.38		=H2O, spin relaxation	75L003
xx(IS/Fe)	R	RbFe(SO4)2-12=	87	.47		=H2O, spin relaxation	75L003
xx(IS/Fe)	R	RbFe(SO4)2-12=	300	.38		=H2O, spin relaxation	75L003
xx(IS/Fe)	R	TlFe(SO4)2-12=	87	.47		=H2O, spin relaxation	75L003
xx(IS/Fe)	R	TlFe(SO4)2-12=	220	.41		=H2O, spin relaxation	75L003

INORGANIC SULFIDES

SOURCE	S-TEMP	ABSORBER	A-TEMP	SHIFT	QS	REMARKS	REF
Cu(IS/Fe)		BaFe2S3	78	.53	.67	electronic structure info	75R001
Cu(IS/Fe)		BaFe2S3	300	.41	.60	electronic structure info	75R001
Cu(IS/Fe)	R	Ba2FeS3	78	.75	2.64	electronic structure info	75R001
Cu(IS/Fe)	R	Ba2FeS3	300	.62	2.56	electronic structure info	75R001
Cu(IS/Fe)		Ba6Fe8S15	78			electronic structure info, HI	75R001
Cu(IS/Fe)		Ba6Fe8S15	195	.54	.71	electronic structure info	75R001
Cu(IS/Fe)		Ba6Fe8S15	300	.47	.68	electronic structure info	75R001
Cu(IS/Fe)	R	Ba7Fe6S14	78			electronic structure info, HI	75R001
Cu(IS/Fe)	R	Ba7Fe6S14	300	--	--	electronic structure info	75R001
Cu(IS/Fe)	R	Ba9Fe16S32	78			electronic structure info, HI	75R001
Cu(IS/Fe)	R	Ba9Fe16S32	195			electronic structure info, HI	75R001
Cu(IS/Fe)	R	Ba9Fe16S32	300	.20	.65	electronic structure info	75R001
xx		CdCr2S4	V			HI=51 kOe(T=0 K)	75T007
Cr		Cd(1-x)FexCr2S4	V			Jahn-Teller stabilization	75S025
xx		CoCr2S4	V			HI=70 kOe(T=0 K)	75T007
Cu		CoS2	4.2	--	--	magnetic structure info	75W030
xx		CsFeS2	V			HI=136 kOe(T=13 K)	75N017
Pd(IS/Fe)	R	CuFeS2	V	--		78 K<T<820 K, Neel T=815 K	75O008
xx		CuGaS2				obsvn of valency changes	75V027
Cu		CuGaS2	V	--	--	Fe is Fe+2 and Fe+3	75V002
Pd(IS/Fe)		Fe.28TaS2	V	--	--	T=4.2 and 296 K, HI data	75E010
xx		FeCr2S4	V			HI=180 kOe(T=0 K)	75T007
Pd	R	FeCr2S4	V			structure information, T=5 & 300 K	75L016
Pd(IS/Fe)		Fe-Ni-S		--	--	study of magnetic properties	74V036
Cu		FeS		--	--	crystallographic texture	75K008
xx		FeS	V			418 K<T<458 K, single crystal	75G039
xx(IS/310 SS)		FeS2		--	--	molecular orbital calculations	74K066
xx		FeS-ZnS	V	--	--	2 Fe sites, synthetic sphalerites	75S052
xx		FeSx	V			HI data	75C050
Pt	R	Fe(1-x)CdxCr2S4	V	--	--	QS vs temperature and comp	75B125
xx		Fe(1-x)NixS	V			study of magnetic properties in vicinity of metal non-metal transition	74C038
xx		Fe(1-x)S				study of spin flop transition	75N024
Cr		Fe(1-x)S				study of ordering vacancies	75B128
xx		Fe(1-x)S	V				75N023
Cu		Fe2GeS4	V	--	--	study of magnetic and crystallographic properties	75M001
Cu(IS/Fe)		Fe7S8	V	--		HA, HI vs temp, 4 Fe sites	75G003
Cu(IS/Fe)		KFeS2	N	.27		HI=215 kG, bonding information	75T004
Cu(IS/Fe)		KFeS2	R		.53	bonding information	75T004
xx		KFeS2	V			HI=221 kOe(T=77 K)	75N017
Cu(IS/Fe)		NiS2	V			study of the magnetic structure, 4.2<T<295 K	75N032
Cu(IS/Fe)		NiS2(Co)	V			study of the magnetic structure	75N032
xx		Ni(1-x)S	V			two magnetic phase 78<T<294 K	75O009
xx		RbFeS2	V			77<T<239 K, HI=192 kOe(T=77 K)	75N017

TERRESTRIAL AND EXTRATERRESTRIAL MINERALS

SOURCE	S-TEMP	ABSORBER	A-TEMP	SHIFT	QS	REMARKS	REF
xx		alkali amphi=			--	=boles	75B084
xx		almandine	v		--	discussion of previous data, proposed model for 3d levels	75H023
xx		amposite	v			4<T<298 K, study of antiferro-magnetic order	75E022
Pd(IS/Fe)		Apollo 17 soils	v	--	--	study of the effect of heating and subsolidus reduction on lunar materials	75B132
Pd		arfvedsonite	v	--	--	site population data, 120 K<T <520 K	75A026
Cr		axinite	v	--	--	T=80 & 300 K	75A051
Pd(IS/SNP)	R	beidellite	R	--	--	2 Fe sites, site popltn data	75R014
xx		biotite				application of conversion electron Mossbauer	75T020
xx		biotite		--	--	study of the effects of proton irradiation	75K054
xx		biotite		--	--	study of the effect of thermal treatment	75V031
Cr		biotite		--	--	spectra use as a prospecting criteria for uranic minerals	75I018
xx		biotite	295	--	--	study of the effect of electron irradiation	75D038
xx		biotite	v	--	--	study of thermal decomposition 295<T<800 K	75H013
Cr		biotite	v	--	--	823<T<1173 K, study of oxida-tion processes	75I016
Cu(IS/SNP)		biotites	295	--	--	Fe+3/Fe+2 ratio data	75B042
xx(IS/Fe)		biotites	v	--	--	bonding and structure info, T=77, 298, and 405 K	75A035
Pd		carbonaceous =		--	--	=meteorites	75V042
xx		crocidolite		--	--	study of antiferromagnetic order	75E022
Cr		cryophllite	v	--	--	T=80 & 300 K	75A051
Cr		danburite	v	--	--	T=80 & 300 K	75A051
xx(IS/Fe)		epidote	v	--	--	4.2<T<300 K, structure info	75P034
xx		Fe minerals				study of age	75A055
xx(IS/Fe)		freshwater fer=		--	--	=romanganese nodules	75C051
Pd(IS/Fe)		glauconite		--	--	is a primary clay mica	75A017
Pd(IS/SNP)	R	glauconite	R	--	--	3 Fe sites, site popltn data	75R014
xx		glauconites			--	study of the effect of age on the Mossbauer parameters	75M062
xx		goethite	v			77<T<300 K	75G056
Pd(IS/SNP)		gothite	R	--	--		75R014
Cr		Gressk meteor=	N			=ite	75M026
xx		hedenbergite				study of hydrothermal synthe-sis of hedenbergite	75K022
Cu(IS/SNP)		hornblendes	295	--	--	Fe+3/Fe+2 ratio data	75B042
Pd(IS/SNP)	R	illite	R	--	--	2 Fe sites, site popltn data	75R014
Cu		ilmenite	R	--	--		75A030
310 SS		ilmenite	v			300 K<T<630 K, Debye tempera-ture data, single crystal	75B037
xx		ilvaite	v		--	Neel temp=4.2 K	75G005
xx		ilvaite	v	--		electron hopping, 4.2<T<500 K	74G068
xx(IS/SNP)		iron ores	v			T=78 & 298 K	74G063
xx		kaolinites	v			80<T<295 K, HI vs temperature	75H014
Rh		kaolinites	v	--	--	88<T<290 K	75J010
Cr(IS/SS)		Krymka meteor=		--	--	=ite, study of the various phases	75I017
xx(IS/Fe)		lake sdeiments	R	--	--	study of surface sediments from various Canadian lakes	74C041
xx		lepidocrocite					75M085
xx		lunar olivines	R			superparamagnetic clusters	75H044
Pd		Luna-16 parti=			--	=cle	75Z015
Pd		Luna-16 samples	R	--	--	identification of 5 components	75Z004
Pd		Luna-16 small =		--	--	=particle, approx .2 mm size	75Z005
Pd		Luna-20 parti=			--	=cles, identification of the mineral content	75Z015
Pd		Luna-20 sample	R	--	--	identification of 4 components	75Z004

SOURCE	S-TEMP	ABSORBER	A-TEMP	SHIFT	QS	REMARKS	REF
xx		magnetite				study at different redox stages	75F033
Cu(IS/Fe)		metamict thoro=		--	--	=-steenstrupine, struc info	75A009
Pd(IS/SNP)	R	montmorillonite	R	--	--	2 Fe sites, site popltn data	75R014
Pd(IS/Fe)		muscovite				study of the reverse pleochroism	75R022
Pd(IS/SNP)	R	nontronite	R	--	--	2 Fe sites, site popltn data	75R014
Cr		olivine		--	--	obsvn of low spin state of Fe	75E006
Cr		olivines	V	--	--	bonding information	75E023
xx		ore samples			--	Mossbauer dating of ores and evaluating thermodynamic conditions of formation	75A054
xx(IS/Fe)	298	orthopyroxene	77	--	--	study of the effect of shock intensities up to 1000 kb	75H021
Pd(IS/Fe)		pentlandite		--	--	structure information	75G067
Pd(IS/SNP)	R	palygorskite	R	--	--	3 Fe sites, site popltn data	75R014
Cr		phosphorites				samples from Indian Ocean	75G021
xx(IS/Fe)		post-glacial =	77	--	--	=lake sediment core	75C009
xx(IS/Fe)		post-glacial =	296	--	--	=lake sediment core	75C009
Th	298	pyralspite gar=	298	1.4	3.6	=net, electronic struc info	75E028
xx		pyroxenes		--	--	study of the properties of the aegirite-diopside series	74E022
xx		pyroxenes		--	--	site population data	73K053
Cr		pyrrhotite	V			study of phase transformation, 80<T<150 K	75N033
Cr(IS/Armco Fe)		pyrrhotite(hex)		--	--	structure information	75I005
Cr(IS/Armco Fe)		pyrrhotite(mon)		--	--	structure information	75I005
xx		pyrrhotites		--	--	study of magnetic phases	75H020
Pd		rhodonite	V	--	--	5Fe+2 site, T=77,295, & 495 K	75D028
Cr		Santa Catharin=	V			=a iron meteorite, T=room temp and liquid nitrogen temp	75M026
Pd(IS/SNP)	R	saponite	R	--	--	2 Fe sites, site popltn data	75R014
Pd(IS/SNP)	R	seledonite	R	--	--	3 Fe sites, site popltn data	75R014
Pt		stannite		--	--	obsvn tetragonal to cubic phase transition	75P014
Cr		talc	300	1.31	2.66		75A051
Cr		Toluca meteor=	N			=ite	75M026
xx		tourmaline	V		--	fast electron hopping	75P041
Pd		tourmaline	V		--	6 Fe sites	75B124
Cu		tourmalines				fa=.52, detn of Fe content	75P010
xx(IS/Fe)	R	troilite	R	--		HI=313 kG	75I008
Pd		troilite	V			study of phases	75Z016
xx		vermiculite	295	--	--	study of the effect of electron irradiation	75D038
Pd(IS/SNP)	R	vermiculite	R	--	--	3 Fe sites, site popltn data	75R014
Cu		vermiculite	V	--	--	Debye T=190 K, 83<T<323 K, fa data, temperature coef of isomer shift	75H026
Rh(IS/Fe)		vesuvianites		--	--	site population data	75M034
Cr		volcanic horn=	R	--	--	=blende, structure information	75M006
xx		wurtzite	V	--	--	80<T<300 K	75C052

References

To save space, the Fe-57 references, which comprise the majority of entries in the Master Reference List, have not been reproduced here. They will be found in the Master Reference List on pages 231 to 311.

$^{155}_{64}$ Gd (60.0 keV, 86.5 keV, 105.3 keV)

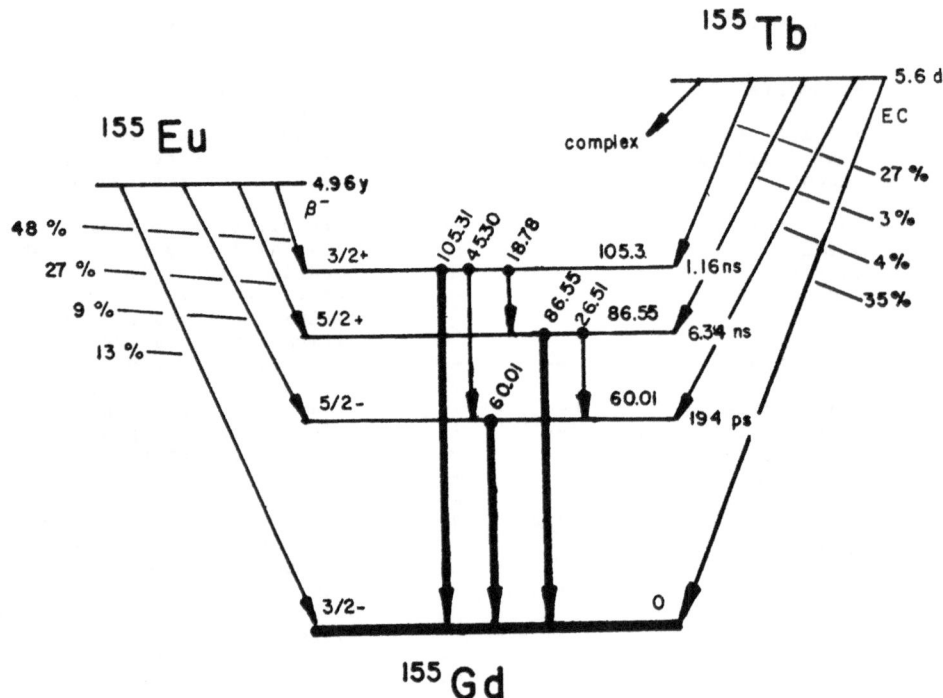

Measured Properties ($E_\gamma = 60.0$ keV)

E_γ = 60.0100 ± 0.0018 keV (1)

$t_{\frac{1}{2}}$ = 194 ± 15 ps (2)

α_T = 8.72 (3)

IA = 14.73 %

μ = −0.2584 ± 0.0005 nm (4)

Q = + 1.59 ± 0.16 b (4)

Derived Parameters ($E_\gamma = 60.0$ keV)

σ_o = 1.048 (5) × 10^{-19} cm^2

Γ = 2.35 (18) × 10^{-6} eV

W_o = 23.5 (18) mm/s

E_r = 1.24713 (5) × 10^{-2} eV

Energy Conversions ($E_\gamma = 60.0$ keV)

1 mm/s = 48.4009 (15) MHz

1 mm/s = 2.00172 (6) × 10^{-7} eV

Measured Properties ($E_\gamma = 86.5$ keV)

E_γ = 86.5452 ± 0.0033 keV (1)

$t_{\frac{1}{2}}$ = 6.33 ± 0.06 ns (5), (6), (7)

α_T = 0.43 (3)

ρ = 98%

R_μ = +2.049 ± 0.019 (8), (9)

R_Q = 0.20 ± 0.05 (8), (9)

Derived Parameters ($E_\gamma = 86.5$ keV)

σ_o = 3.39 (21) × 10^{-19} cm^2

Γ = 7.21 (7) × 10^{-8} eV

W_o = 0.499 (5) mm/s

E_r = 2.59388 (14) × 10^{-2} eV

Energy Conversions ($E_\gamma = 86.5$ keV)

1 mm/s = 69.803 (3) MHz

1 mm/s = 2.88684 (11) × 10^{-7} eV

Measured Properties (E_γ = 105.3 keV)

E_γ = 105.308 ± 0.003 keV (1)

$t_{\frac{1}{2}}$ = 1.168 ± 0.016 ns (5),(6),(7)

α_T = 0.26 (3)

ρ = 56%

R_μ = −0.55 ± 0.05 or + 1.80 ± 0.18 (8),(9)

R_Q = 0.99 ± 0.04 (8),(9)

Derived Parameters (E_γ = 105.3 keV)

σ_o = 1.64 (11) × 10^{-19} cm^2

Γ = 3.91 (5) × 10^{-7} eV

W_o = 2.22 (3) mm/s

E_r = 3.84049 (16) × 10^{-2} eV

Energy Conversions (E_γ = 105.3 keV)

1 mm/s = 84.9358 (25) MHz

1 mm/s = 3.51270 (10) × 10^{-7} eV

Energies and Intensities

	E_γ	I_{Eu}(3),(10)	I_{Tb}(11)
γ_{M1}:	60.0 keV	54	3.8
γ_{M2}:	86.5 keV	1480	100
γ_{M3}:	105.3 keV	1000	67.9
γ_1:	19. keV	2	
γ_2:	26.5 keV	15	1
γ_3:	45.3 keV	58	2.8
K_{α_1}:	43.00 keV	516	

(1) D. E. Raeside, Nucl. Instrum. Methods 87, 7 (1970).
(2) 73A002.
(3) R. A. Meyer and J. W. T. Meadows, Nucl. Phys. A132, 177 (1969).
(4) P. J. Unsworth, Proc. Phys. Soc. London (At. Mol. Phys.), Ser. 2 2, 122 (1969).
(5) A. Krusche, D. Bloess, and F. Münnich, Z. Phys. 192, 490 (1967).
(6) E. V. Lan'ko and G. S. Dombrovskaya, Izv. Akad. Nauk SSSR, Ser. Fiz. 33, 2083 (1969).
(7) H. Bakhru, Phys. Rev. C 3, 1706 (1971).
(8) 74A030.
(9) 74B091.
(10) F. F. Tomblin and P. H. Barrett, in Hyperfine Structure and Nuclear Radiations, edited by E. Matthias and D. A. Shirley (American Elsevier Publishing Co., New York, 1968), p. 245.
(11) P. Alexander, Nucl. Phys. A108, 145 (1968).

[Other data from L. A. Kroger and C. W. Reich, Nucl. Data Sheets 15, 409 (1975)]

SOURCE	S-TEMP	ABSORBER	A-TEMP	SHIFT	QS	REMARKS	REF
Al2La(Sm)	4.2	AgGd	4.2	-.307		bonding and structure info	75R023
Al2La(Sm)	4.2	CuGd	4.2	-.375		bonding and structure info	75R023
Pd3Sm	4.1	Gd			--	study of hyperfine interaction	75B022
xx(IS/GdF3)		Gd	4.2	--			75L009
Al2La(Sm)	4.2	GdAg(1-x)Inx	4.2	--		bonding and structure info	75R023
Al2La(Sm)	4.2	GdAg(1-x)Pdx	4.2			bonding and structure info	75R023
Al2La(Sm)	4.2	GdCu(1-x)Agx	4.2	--		bonding and structure info	75R023
Al2La(Sm)	4.2	GdCu(1-x)Gax	4.2	--		bonding and structure info	75R023
xx(IS/GdF3)	He	GdHx	v	--		HI data, results agree with anionic model of metal hydrides	75L009
Pd3Sm	4.1	Gd(Lu)			--	study of hyperfine interaction	75B022
Pd3Sm	4.1	Gd(Sc)			--	study of hyperfine interaction	75B022
Pd3Sm	4.1	Gd(Y)			--	study of hyperfine interaction	75B022
Al2La(Sm)	4.2	GdZn	4.2	-.180		bonding and structure info	75R023
Pd3Sm	4.1	Gd(1-x)Tb			--	study of hyperfine interaction	75B022
Pd3Sm	4.1	Gd(1-x)Tmx			--	study of hyperfine interaction	75B022
Sm2(Eu)Sn2O7		Sc(Gd)H		+.093		RQ=-.090(29), Goldanski-Karayagin effect	75K012

CODE	TOPIC	REFERENCE
75B022	GD-155	E R Bauminger, A Diamant, I Felner, I Nowik, and S Ofer, Phys Rev Lett 34,962-5(1975), Anisotropic Hyperfine Interactions in Gadolinium Metal
75C008	GD-155	J D Cashion, M A Coulthard, and D B Prowse, J Phys C 8, 1267-75(1975), Mossbauer Isomer Shifts in Rare Earth Compounds: II. Nuclear Charge Radius and Electron Density Studies
75K012	GD-155	R Kmiec, K Latka, T Matlak, K Ruebenbauer, and K Tomala, Phys Status Solidi B 68,K125-8(1975), Quadrupole Moment Ratio for the 86.5 keV Transition in 155Gd and the Goldanskii-Karyagin Effect for Sm2Sn2O7
75L009	GD-155	S J Lyle, P T Walsh, A D Witts, and J W Ross, J Chem Soc, Dalton Trans 1406-9(1975), Mossbauer Spectroscopic Study of the Gadolinium-Hydrogen System
75N016	GD-155	D J Newman and D C Price, J Phys C 8,2985-91(1975), Determination of the Electrostatic Contributions to Lanthanide Quadrupolar Crystal Fields
75R023	GD-155	J W Ross and J Sigalas, J Phys F 5,1973-80(1975), Isomer Shifts in Gadolinium Compounds Having the Caesium Chloride Structure

$^{73}_{32}$ Ge (13.3 keV, 68.8 keV)

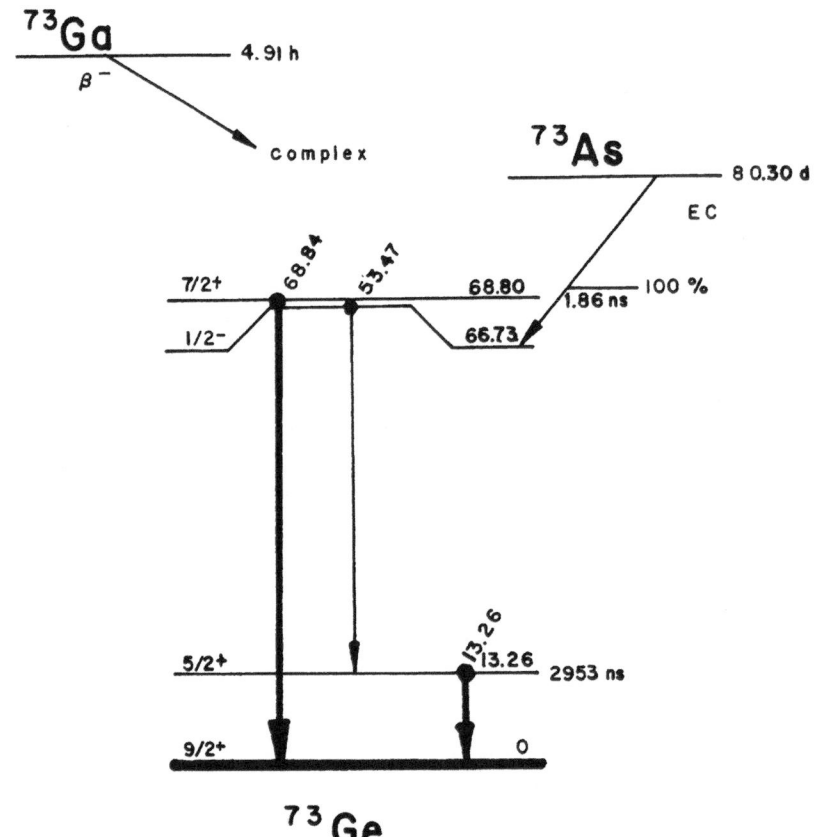

Measured Properties (E$_\gamma$ = 13.3 keV)

E$_\gamma$ = 13.263 ± 0.015 keV (1)

t$_{\frac{1}{2}}$ = 2953 ± 23 ns (2),(3),(4),(5)

α_T = 1095 ± 55 (3)

IA = 7.76%

μ = −0.87919 ± 0.00012 nm (6)

R$_\mu$ = +0.107 ± 0.003 (7)

Q = −0.18 ± 0.03 b (8)

Derived Parameters (E$_\gamma$ = 13.3 keV)

σ_o = 7.6 (4) × 10^{-21} cm^2

Γ = 1.545 (12) × 10^{-10} eV

W$_o$ = 0.00698 (6) mm/s

E$_r$ = 1.2935 (21) × 10^{-3} eV

Energy Conversions (E$_\gamma$ = 13.3 keV)

1 mm/s = 10.697 (12) MHz

1 mm/s = 4.424 (5) × 10^{-8} eV

Measured Properties (E$_\gamma$ = 68.8 keV)

E$_\gamma$ = 68.752 ± 0.007 keV (9)

t$_{\frac{1}{2}}$ = 1.86 ± 0.10 ns (10)

α_T = 0.81 ± 0.22 (11)

Derived Parameters (E$_\gamma$ = 68.8 keV)

σ_o = 2.3 (3) × 10^{-19} cm^2

Γ = 2.45 (13) × 10^{-7} eV

W$_o$ = 2.14 (12) mm/s

E$_r$ = 3.4757 (5) × 10^{-2} eV

Energy Conversions (E$_\gamma$ = 68.8 keV)

1 mm/s = 55.452 (6) MHz

1 mm/s = 2.29332 (23) × 10^{-7} eV

Energies and Intensities

$$E_\gamma (6),(8),(9) \quad I_{As}(1)$$

γ_{M1}:   13.26 keV

γ_{M2}:   68.75 keV

γ_1 :   53.4 keV   111

γ_2 :   13.3 keV   1

K_{α_1} :   9.89 keV

(1) D. G. Douglas, Can. J. Phys. 47, 1815 (1969).
(2) K. P. Vester, Z. Phys. 242, 320 (1971).
(3) R. S. Raghavan, Z. Phys. 243, 441 (1971).
(4) D. G. Douglas, Can. J. Phys. 48, 930 (1970).
(5) J. Kyles, J. C. McGeorge, F. Shaikh, and J. Bryne, Nucl. Phys. A150, 143 (1970).
(6) W. Sahm and A. Schwenk, Z. Naturforsch. A 29, 1763 (1974).
(7) H. Haas, E. Ivanov, and E. Recknagel, Phys. Lett. 58B, 423 (1975).
(8) W. J. Childs and L. S. Goodman, Phys. Rev. C 1, 750 (1971).
(9) K. W. Jones and H. W. Kraner, Phys. Rev. C 4, 125 (1971).
(10) 68C029.
(11) R. Weishaupt and D. Rabenstein, Z. Phys. 251, 105 (1972).

[Other data from K. R. Alvar, Nucl. Data Sheets 13, 305 (1974)]

SOURCE	S-TEMP	ABSORBER	A-TEMP	SHIFT	QS	REMARKS	REF
Ga		Ge				source preparation	74M059
As	R	Ge	R				74R040
Ge(As)		Ge	R	--		narrowest linewidth at room T	75P030

CODE	TOPIC	REFERENCE
74M059	GE-73	K Matsui, O Konno, and S Ishino, Kakuriken Kenkyu Hokoku 7,147-55(1974), Preparation of Ga73 Mossbauer Source in Germanium Through Photonuclear Reactions
74R040	GE-73	R S Raghavan and L N Pfeiffer, Ge73: A New High-resolution Mossbauer Nuclide, in "Hyperfine Interactions Studied in Nuclear Reactions and Decay: Contributed Papers" (Conference, Uppsala, Sweden, 1974), edited by E Karlsson and R Wappling (Upplands Grafiska, Uppsala, 1974), pp 282-3
75P030	GE-73	L N Pfeiffer, R S Raghavan, C P Lichtenwalner, and A G Cullis, Phys Rev B 12,4793-804(1975), Mossbauer Effect of the 13.3-keV Transition in 73Ge

$^{178}_{72}$Hf (93.2 KeV)

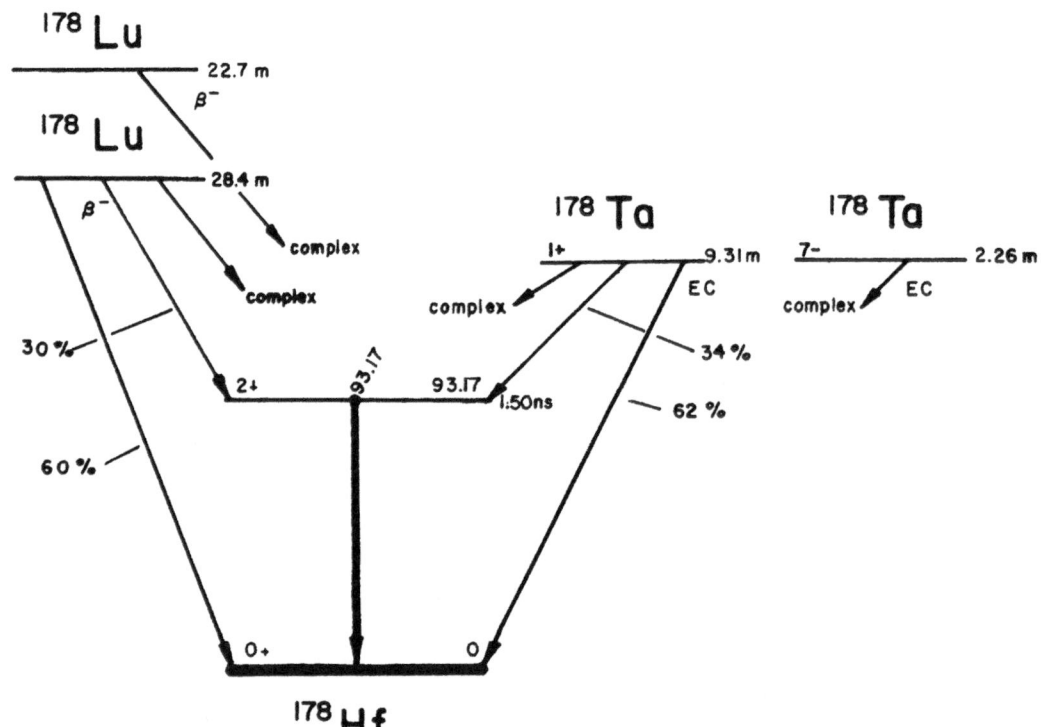

Measured Properties

E_γ = 93.174 ± 0.004 keV (1),(2)

$t_{\frac{1}{2}}$ = 1.495 ± 0.024 ns (3),(4)

α_T = 4.6 (5)

IA = 27.14%

μ^* = 0.52 ± 0.03 nm (4),(6),(7)

Q^* = -1.95 b (8),(9)

Derived Parameters

σ_o = 2.52 (22) × 10^{-19} cm^2

Γ = 3.05 (5) × 10^{-7} eV

W_o = 1.96 (3) mm/s

E_r = 2.61781 (12) × 10^{-2} eV

Energy Conversions

1 mm/s = 75.147 (2) MHz

1 mm/s = 3.10785 (11) × 10^{-7} eV

Energies and Intensities

	E_γ	$I_{178m_{Lu}}$(5)	I_{Ta}(10)
γ_M :	93.2 keV	21	1372
γ_1 :	88.8 keV	71	
γ_2 :	213. keV	106	19
K_{α_1} :	46.0 keV		

(1) T. Mori, H. Inoue, and Y. Yoshizawa, J. Phys. Soc. Jap. 38, 611 (1975).
(2) B. Fogelberg and A. Bäcklin, Nucl. Phys. A171, 353 (1971).
(3) E. Bodenstadt, H. J. Körner, E. Gerdau, J. Radeloff, K. Auerback, L. Mayer, and A. Roggenbuck, Z. Phys. 168, 103 (1962).
(4) E. Karlsson, E. Matthias, and S. Ogaza, Ark. Fys. 22, 257 (1962).
(5) T. Tomura, J. Phys. Soc. Jap. 23, 691 (1967).
(6) I. Ben-Zvi, P. Gilad, G. Goldring, P. Hillman, A. Schwarschild, and Z. Vager, Nucl. Phys. A109, 201 (1968).
(7) P. Gilad et al., Nucl. Phys. A91, 85 (1967).
(8) R. E. Snyder, J. W. Ross, and D. St. P. Bunbury, J. Phys. C (Proc. Phys. Soc.) 1, 1662 (1968).
(9) E. Gerdau, P. Steiner, and D. Steenken, in Hyperfine Structure and Nuclear Radiations, edited by E. Matthias and D. Shirley (American Elsevier Publ. Co., New York, 1968), p. 316.
(10) H. L. Nielsen, K. Wilsky, and J. Zylicz, Nucl. Phys. A93, 385 (1967).

[Other data from L. R. Greenwood, Nucl. Data Sheets 13, 549 (1974)]

SOURCE	S-TEMP	ABSORBER	A-TEMP	SHIFT	QS	REMARKS	REF
Ta(W)	4.2	Hf	4.2	--	--	.	75J001
Ta(W)	4.2	HfO2	4.2	--	--		75J001

CODE	TOPIC	REFERENCE
75J001	HF-178	C Jeandey and P Peretto, Phys Status Solidi A 28,529-37 (1975), Effet Mossbauer en Ligne et Defauts de Recul dans le Hafnium Metallique et dans l'Oxyde de Hafnium

$$^{165}_{67}\text{Ho}(94.7\text{ keV})$$

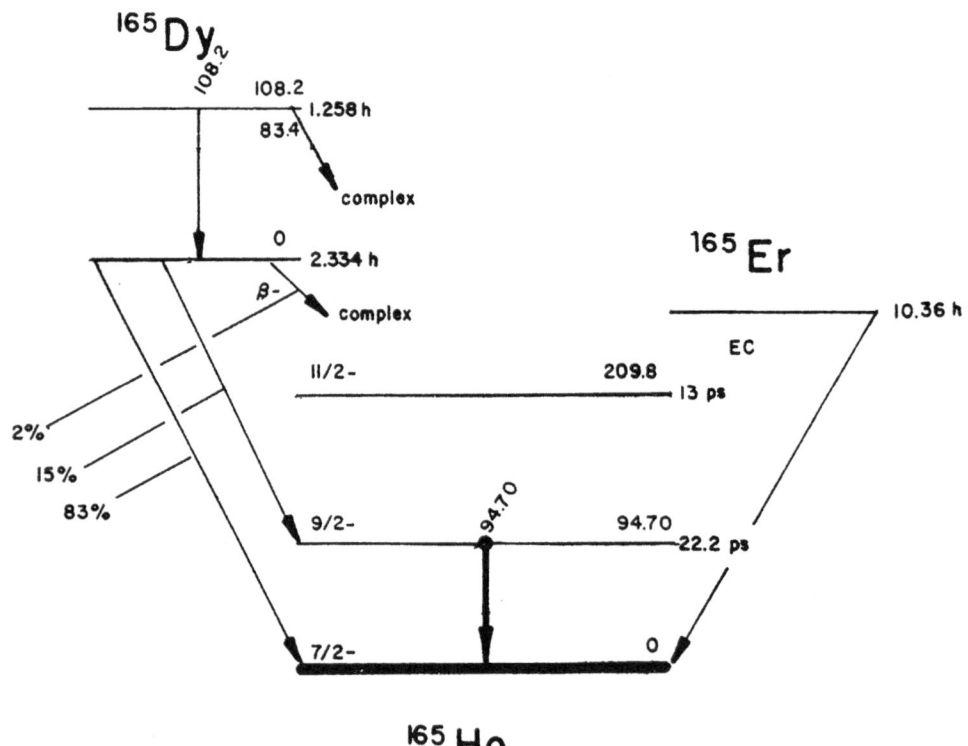

$$^{165}\text{Ho}$$

Measured Properties

E_γ = 94.699 ± 0.003 keV (1),(2)

$t_{\frac{1}{2}}$ = 22.2 ± 0.7 ps (3)

α_T = 3.12 (theory)

IA = 100%

μ = +4.12 ± 0.04 nm (4)

R_μ = 0.99 ± 0.04 (3)

Q^* = 2.6 ± 0.4 b (5),(6)

Derived Parameters

σ_o = 8.28 (10) x 10^{-20} cm^2

Γ = 2.06 (6) x 10^{-5} eV

W_o = 130 (4) mm/s

E_r = 2.91744 (13) x 10^{-7} eV

Energies and Intensities

	E_γ	I_{Dy}
γ_M :	94.70 keV	32
γ_1 :	108.2 keV	20
γ_2 :	184.3 keV	18
K_{α_1} :	47.55 keV	

Energy Conversions

1 mm/s = 3.15882 (10) x 10^{-7} eV

1 mm/s = 76.3792 (25) MHz

(1) I. Marklund and B. Lindstrom, Nucl. Phys. 40, 329 (1968).
(2) O. W. B. Schult, B. P. Maier, and U. Gruber, Z. Phys. 182, 171 (1964).
(3) 72G037.
(4) R. A. Haberstroh, T. I. Moran, and S. Penselin, Z. Phys. 252, 421 (1972).
(5) L. S. Goodman and K. Schlüpmann, Z. Phys. 178, 235 (1964).
(6) B. G. Wyboume, J. Chem. Phys. 37, 1807 (1962).

[Other data from A. Buyrn, Nucl. Data Sheets 11, 189 (1974)].

SOURCE	S-TEMP	ABSORBER	A-TEMP	SHIFT	QS	REMARKS	REF
Dy203(CE)	xx					fs experiment	75C024
Dy203(d,p)	xx					fs experiment	75C024

CODE	TOPIC	REFERENCE
75C024	HO-165	C L Chien and J C Walker, Studies of Temperature Spikes in Solids Using the Mossbauer Effect Following Coulomb Excitation, in "5th Int Conf Mossbauer Spec, Proc," Part 3, pp 560-3 (See 75H017)

$^{127}_{53}I$ (57.6 keV)

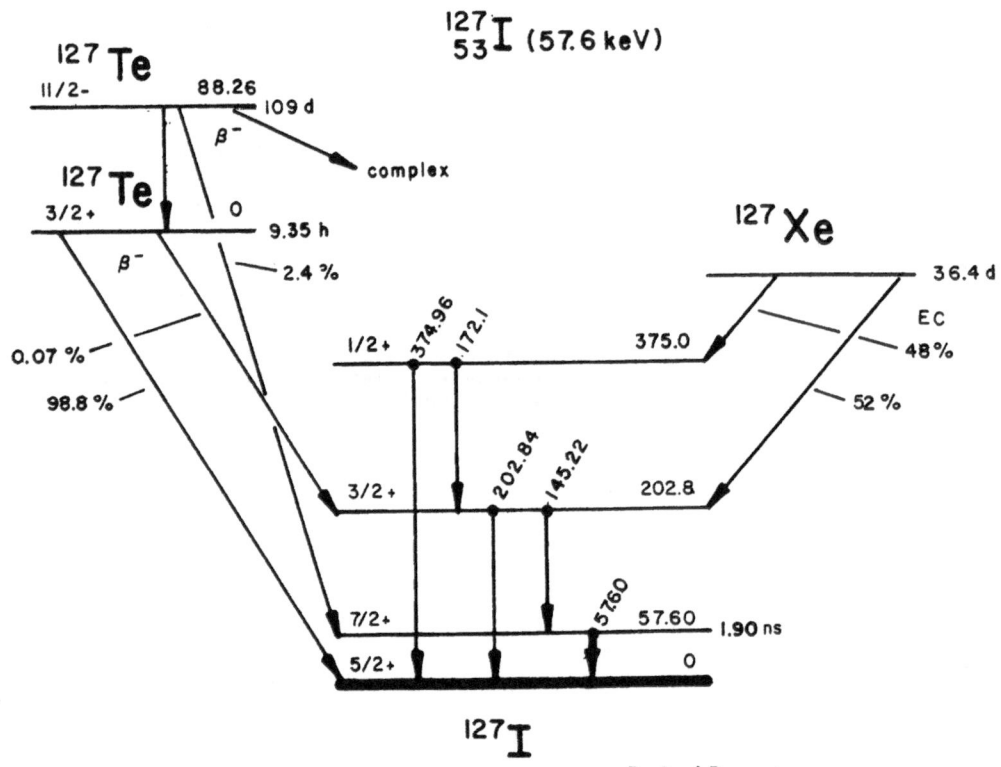

Measured Properties

E_γ = 57.60 ± 0.02 keV (1)

$t_{\frac{1}{2}}$ = 1.91 ± 0.06 ns (2), (3)

α_T = 3.78 (theory) (1)

IA = 100 %

μ = + 2.8091 ± 0.0004 nm (4), (5), (6), (7), (8)

R_μ = 0.905 ± 0.016 (9)

Q = - 0.79 ± 0.10 b (10)

R_Q = - 0.896 ± 0.002 (11)

Energy Conversions

1 mm/s = 46.457 (16) MHz

1 mm/s = 1.92133 (7) $\times 10^{-7}$ eV

Derived Parameters

σ_o = 2.057 (22) $\times 10^{-19}$ cm^2

Γ = 2.39 (8) $\times 10^{-7}$ eV

W_o = 2.49 (8) mm/s

E_r = 1.4023 (7) $\times 10^{-2}$ eV

Energies and Intensities

	E_γ	I_{Te}(12)	I_{Xe}(13)
γ_M :	57.60 keV	56.	1.3
γ_1 :	145. keV	0.3	4.2
γ_2 :	172. keV	0.03	25
γ_3 :	203. keV	5.9	68
γ_4 :	360 keV	13.6	17
K_{α_1} :	28.32 keV		

(1) J. S. Geiger, Phys. Rev. 158, 1094 (1967).
(2) A. G. Svensson, R. W. Sommerfeldt, L.-O. Norlin, and P. N. Tandon, Nucl. Phys. A68, 653 (1965).
(3) J. S. Geiger, R. L. Graham, I. Berström, and F. Brown, Nucl. Phys. A68, 352 (1965).
(4) R. V. Pound, Phys. Rev. 73, 1112 (1948); erratum: Phys. Rev. 74, 228 (1948).
(5) J. R. Zimmerman and D. Williams, Phys. Rev. 76, 350 (1949).
(6) R. E. Sheriff and D. Williams, Phys. Rev. 82, 651 (1951).
(7) H. E. Walchi, R. Livingston, and G. Herbert, Phys. Rev. 82, 92 (1951).
(8) E. Yasaitis and B. Smaller, Phys. Rev. 82, 750 (1951).
(9) 72W016.
(10) H. Stroke – Private communication (see G. H. Fuller and V. W. Cohen, Nucl. Data Tables A5, 433 (1969) – reference 59St46).
(11) 64P05.
(12) K. E. Apt, W. B. Walters, and G. E. Gordon, Nucl. Phys A152, 344 (1970).
(13) R. Colle and R. Kishore, Phys. Rev. C 9, 981 (1974).

[Other data available from R. L. Auble, Nucl. Data Sheets B8, 77 (1972)]

SOURCE	S-TEMP	ABSORBER	A-TEMP	SHIFT	QS	REMARKS	REF
xx		AgGaTe2			--		75M065
xx		AgInTe2			--		75M065
xx		CdAl2Te4			--		75M065
xx		CdGa2Te4			--		75M065
xx		CdIn2Te4			--		75M065
xx		CdTe			--		75M065
xx		CuGaTe2			--		75M065
xx		CuInTe2			--		75M065
xx		Ga2Te3			--		75M065
xx(IS/ZnTe)		I2Cl6			--	EQ=+3060 MHz	75M066
xx		In2Te3			--		75M065
xx(IS/ZnTe)		KICl2			--	EQ=-3189 MHz	75M066
xx(IS/ZnTe)		KICl4			--	EQ=+3094 MHz	75M066
ZnTe	20	L-thyroxine	20	-.32	--	EQ(I-127)=-1959 MHz, eta=.05, bonding information	750007
xx(IS/ZnTe)		PhICl2			--	EQ=+2525 MHz	75M066
xx(IS/ZnTe)		PhIO			--	EQ=+2345 MHz	75M066
xx		ZnAl2Te4			--		75M065
xx		ZnGa2Te4			--		75M065
xx		ZnIn2Te4			--		75M065
xx		ZnTe			--		75M065
ZnTe	20	3-iodo-L-tyro=	20	-.25	--	=sine, EQ(I-127)=-1992 MHz, eta=.22, bonding information	750007
ZnTe	20	3,5-diiodo-L-=	20	-.34	--	=tyrosine, EQ(I-127)=-1894 MHz eta=.09, bonding information	750007
ZnTe	20	5,3'-triiodo-L=	20	-.23	--	=thyronine, EQ(I-127)=-1902 MHz, eta=.07, bonding info	750007

CODE	TOPIC	REFERENCE
75M065	I-127	K V Makaryunas, E K Makaryuene, A K Dragunas, and M L Bal'chyuene, Electric Field Gradients at the Nuclei of Tellurium and Impurity Iodine in the A(1)B(2)Te2 and A(2)B(3)Te4 Crystals, in "5th Int Conf Mossbauer Spectrosc, Proc," Vol 3, pp 529-30 (See 75H017) (In Russian)
75M066	I-127	L I Molkanov, I M Band, M B Trzhaskovskaya, Yu S Grushko, and A V Oleynik, Electron Valence Configurations of Iodine from the Chemical Shift of the X-ray Emission Lines and Mossbauer Effect, in "5th Int Conf Mossbauer Spectrosc, Proc," Vol 3 pp 546-9 (See 75H017) (In Russian)
750007	I-127	L W Oberley and J C Ehrhardt, J Chem Phys 63,2329-33 (1975), 127I Mossbauer Studies of Thyroid Compounds

$^{129}_{53}I$ (27.8 KeV)

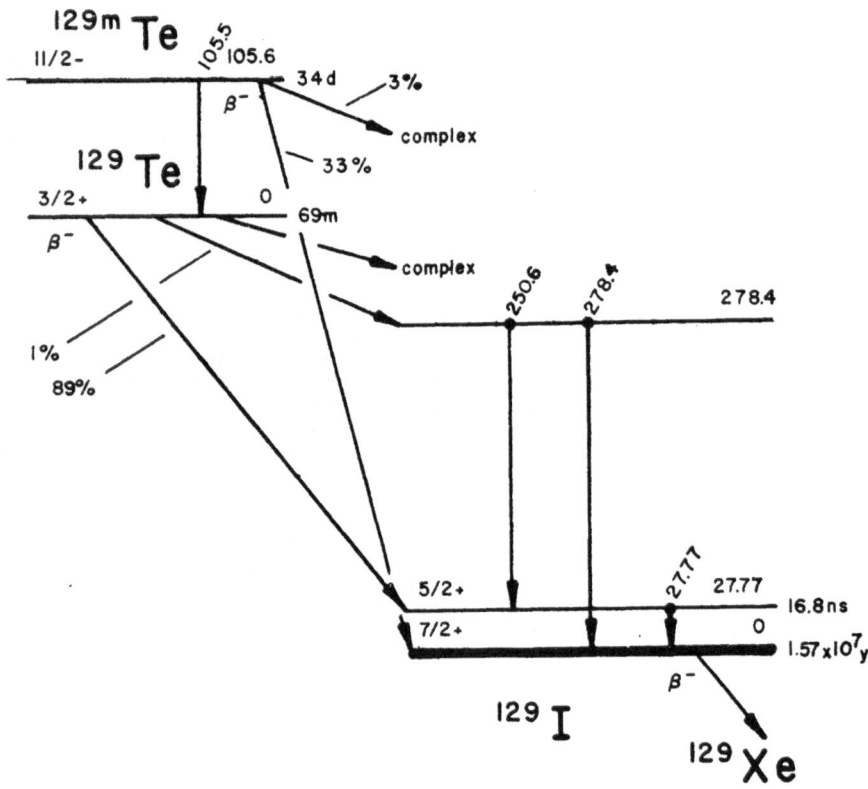

Measured Properties

E_γ = 27.77 ± 0.02 keV (1)

$t_{\frac{1}{2}}$ = 16.8 ± 0.2 ns (2)

α_T = 5.1 ± 0.3 (2),(3)

IA = 0 (radioactive)

μ = +2.6174 ± 0.0008 nm (4)

R_μ = +1.0687 ± 0.0011 (5)

Q = -0.55 ± 0.07 b (6)

R_Q = +1.2380 ± 0.0016 (7),(8)

Energy Conversions

1 mm/s = 22.398 (16) MHz

1 mm/s = 9.2631 (7) × 10^{-8} eV

Derived Parameters

σ_o = 3.90 (19) × 10^{-19} cm^2

Γ = 2.72 (3) × 10^{-8} eV

W_o = 0.586 (7) mm/s

E_r = 3.209 (3) × 10^{-3} eV

Energies and Intensities

	E_γ	I_{Te} (1)
γ_M :	27.77 keV	230
γ_1 :	105.5 keV	3.0
γ_2 :	209.0 keV	2.6
γ_3 :	250.6 keV	5.6
K_{α_1} :	28.61 keV	

(1)   W. C. Dickerson, S. D. Bloom, and L. G. Mann, Nucl. Phys. A123, 481 (1969).
(2)   R. Sanders and H. de Waard, Phys. Rev. 146, 907 (1966).
(3)   S. H. Devare and H. G. Devare, Phys. Rev. 134, B705 (1964).
(4)   H. E. Walchi, R. Livingston, and G. Herbert, Phys. Rev. 82, 97 (1951).
(5)   70D026.
(6)   R. Livingston and H. Zeldes, Phys. Rev. 90, 609 (1953).
(7)   65P08.
(8)   72R025.

[Other data from D. J. Horen, Nucl. Data Sheets B8, 123 (1972)]

SOURCE	S-TEMP	ABSORBER	A-TEMP	SHIFT	QS	REMARKS	REF
CrTe	v	CuI		-.54		EQ=85 MHz, T=85 and 300 K	75G026
Cr2Te	v	CuI		-.48		EQ=143 MHz, T=85 and 380 K	75G026
Cr2Te3	v	CuI		-.61		EQ=128 MHz, T=85 and 240 K	75G026
Fe		CuI				ion implanted source, HI data	75R028
Fe(Te)		CuI				ion implantation, HI data	75C023
Fe(Te)		CuI			--	ion implantation, EQ=39 MHz	75V016
Ni	v	CuI				ion implanted source,98<T<307	75R028
Ni(Te)		CuI				ion implantation, HI data	75C023
S(Te)	He	CuI		-.88		ion implantation, EQ=-1125 MHz	75L014
Se(Te)	He	CuI		-1.27	--	ion implantation, EQ=-1253 MHz	75L014
Te	N	CuI		-1.19	--	EQ=-404 MHZ, eta=.73	75L011
Te(Te)	He	CuI		-1.19	--	ion implantation, EQ=-404 MHz	75L014
Te	He	CuI	He		--	single crystal experiment	75L011
Te	He	CuI	He	-1.19	--	EQ=-404 MHz, eta=.73	75L011
ZnTe	4.2	I2	4.2	+.91	--	EQ=-1510 MHz, eta=.16	75J006
ZnTe	He	I2 in Acetone	He	.83	--	EQ=+2350 MHz, eta=.01	75B013
ZnTe	He	I2 in Acridine	He		--	2 I sites, bonding information	75B013
ZnTe	He	I2 in Benzene	He	.77	--	EQ=+2380 MHz, bonding info	75B013
ZnTe	He	I2 in C6H6	He	.93	--	EQ=+2170 MHz, eta=.19	75B013
ZnTe	He	I2 in EtOEt	He	--	--	2 I sites, bonding information	75B013
ZnTe	He	I2 in HTA	He		--	2 I sites, bonding information	75B013
ZnTe	He	I2 in MeOH	He	.80	--	EQ=+2320 MHz, eta=.07	75B013
ZnTe	He	I2 in Phenazine	He	.93	--	EQ=+2220 MHz, eta=.06	75B013
ZnTe	He	I2 in P-xylene	He	.99	--	EQ=+2150 MHz, eta=.18	75B013
ZnTe	He	I2 in Pyridine	He		--	2 I sites, bonding information	75B013
ZnTe	He	I2 in Toluene	He	.93	--	EQ=+2240 MHz, eta=.16	75B013
ZnTe	4.2	I2O4	4.2	--	--	2 I sites	75J006
ZnTe	4.2	IAc3	4.2	+3.60	--	EQ=+2390 MHz, eta=.12	75J006
ZnTe	4.2	IPO3	4.2	+1.91	--	EQ=-2594 MHz, eta=0	75J006
ZnTe	4.2	IPy2ClO4	4.2	+1.84	--	EQ=-2280 MHz, eta=.03	75J006
ZnTe	4.2	IPy2NO3	4.2	+1.80	--	EQ=-2270 MHz, eta=.03	75J006
ZnTe	4.2	I(THU)I	4.2	--	--	2 I sites	75J006
Te(IV) in H2O	77	KI		2.6	--	EQ=745 MHz, chem after-effects	74A049
Te(IV) in HCl	77	KI		--		chemical after-effects	74A049
Te(IV) in HNO3	77	KI		3.4	--	EQ=745 MHz, chem after-effects	74A049
(NH4)2H4TeO6	80	KI	80	-2.3	--		75L024
beta-TeO3	80	KI	80	--	--		75L024
H6TeO6	80	KI	80	-2.6	--		75L024
Na2H4TeO6	80	KI	80	-2.4	--		75L024
PbTeO6	80	KI	80	-3.2	--		75L024
Te(TM)4Cl2-2H2O	80	KI	80	+3.7	--		75L024
TeO2	80	KI	80	+2.1	--		75L024
xx		KI	v			fa, discussion of previous exp	75S053
ZnTe	4.2	NaI	4.2	-.45			75J006
xx		NaI	v			fa, discussion of previous exp	75S053
ZnTe	v	NiI2	v	--	--	T=4.2 & 77 K, magn struc detn	75P018
xx		RbI	v			fa, discussion of previous exp	75S053

CODE	TOPIC	REFERENCE

74A049 I-129 N E Ablesimov, S I Bondarevskii, I S Kirin, and V A Tarasov, Radiokhimiya 16,919-22(1974)/Sov Radiochem 16, 895-7(1975), Post Effects of beta-Decay in Frozen Tellurium-containing Solutions

74H055 I-129 R H Herber, Failure of the Point-charge Model for Quadrupole Hyperfine Interactions in Organo-tin Compounds, in "Hyperfine Interactions Studied in Nuclear Reactions and Decay: Contributed Papers" (Conference, Uppsala, Sweden, 1974), edited by E Karlsson and R Wappling (Upplands Grafiska, Uppsala, 1974), pp 234-5

75B013 I-129 S Bukshpan, C Goldstein, T Sonnino, L May, and M Pasternak, J Chem Phys 62,2606-9(1975), Molecular Complexes of I2: A Mossbauer Effect Study

75B014 I-129 S Bukshpan, M Pasternak, and T Sonnino, J Chem Phys 62, 2916-7(1975), Mossbauer Effect Results for I2, IBr, and ICl in Different Chemical States

CODE	TOPIC	REFERENCE
75C023	I-129	R Coussement, G Dumont, G Langouche, H Pattyn, M Rots, K P Schmidt, and M Van Rossum, Implantation of 129mTe in Fe and Ni Foils and Determination of the Magnetic Hyperfine Fields, in "5th Int Conf Mossbauer Spec, Proc," Part 3, pp 549-52 (See 75H017)
75D001	I-129	H De Waard, 129I Mossbauer Spectroscopy, in "Mossbauer Effect Data Index, Covering the 1973 Literature," edited by J G Stevens and V E Stevens (Plenum Publishing Corp, New York, 1975), pp 447-94
75G026	I-129	J Granot and S Bukshpan, J Phys C 8,1435-42(1975), Mossbauer Effect Measurements in Ferromagnetic $Cr(x)Te(y)$ Compounds
75G081	I-129	R K Gupta, Phys Rev B 12,4452-9(1975), Mean-square Amplitudes of Vibration for Ionic Crystals
75J006	I-129	C H W Jones, J Chem Phys 62,4343-9(1975), 129I Mossbauer Studies of Iodine in the +1 and +3 Oxidation States
75L011	I-129	G Langouche, M Van Rossum, K P Schmidt, and R Coussement, The Quadrupole Interaction of 125Te and 129I in Polycrystalline Te and in Te Single Crystals, in "5th Int Conf Mossbauer Spec, Proc," Part 3, pp 531-6 (See 75H017)
75L014	I-129	G Langouche, P Boolchand, M Van Rossum, and R Coussement, Nuclear Quadrupole Interaction of 129I in Crystalline Te, Se and S, in "Int Conf Mossbauer Spectrosc, Proc," Vol 1 pp 203-4 (See 75H027)
75L024	I-129	R A Lebedev, Yu D Perfil'ev, L A Kulikov, M I Afanasov, A M Babeshkin, and A N Nesmeyanov, Gamma-resonance Spectroscopy Investigation of the Chemical Subsequences of the Isomeric Transition, Electron Capture, Beta-minus-decay in Solid Substances, in "5th Int Conf Mossbauer Spectrosc, Proc," Vol 3, pp 537-45 (See 75H017) (In Russian)
75P018	I-129	M Pasternak, S Bukshpan, and T Sonnino, Solid State Commun 16,871-2(1975), Magnetic Structure Determination of NiI2 by 129I Mossbauer Effect
75R028	I-129	S R Reintsema, H De Waard, and S A Drentje, Mossbauer Effect Studies on Sources of 129mTe Implanted in Iron and Nickel, in "Int Conf Mossbauer Spectrosc, Proc," Vol 1, pp 525-6 (See 75H027)
75S053	I-129	M Sneh and B Dayal, Phys Status Solidi B 70,341-6(1975), Recoilless Fraction of Iodine Ion in RbI, NaI, and KI by the Shell Model
75V016	I-129	M Van Rossum, G Langouche, K P Schmidt, and R Coussement, Evidence of a Quadrupole Interaction of 129I Implanted in Iron Foils, in "5th Int Conf Mossbauer Spec, Proc," Part 3, pp 552-4 (See 75H017)

$^{193}_{77}$ Ir (73.0keV, 138.9 keV)

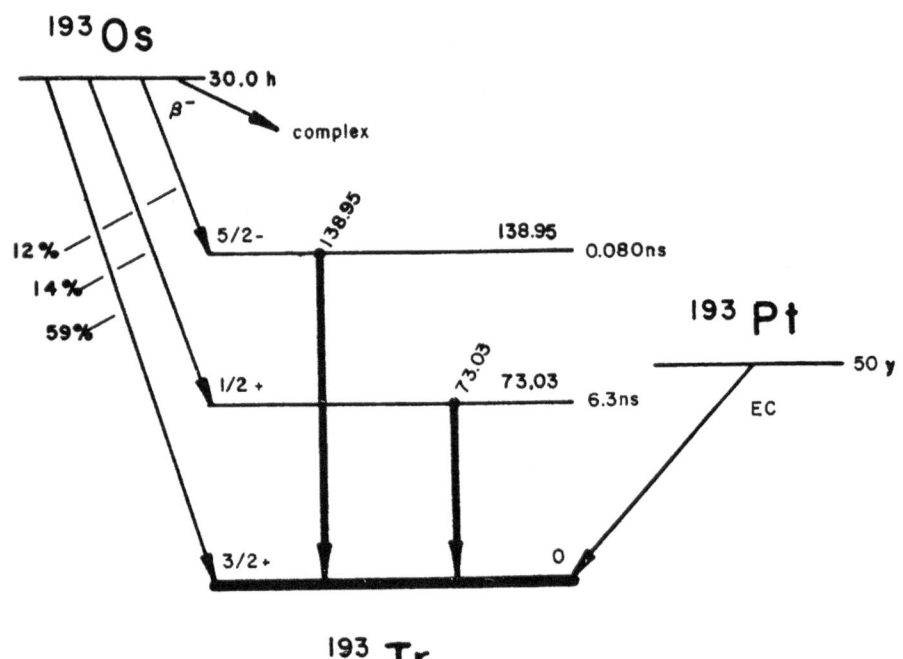

Measured Properties (E_γ = 73.0keV)

E_γ = 73.039 ± 0.011 keV (1), (2)

$t_{\frac{1}{2}}$ = 6.3 ± 0.2 ns (3)

α_T = 6.5 ± 0.2 (4)

IA = 62.7%

μ = + 0.1583 ± 0.0006 nm (5)

R_μ = 2.958 ± 0.006 (6)

Q = + 0.70 ± 0.18 b (7)

Derived Parameters (E_γ = 73.0keV)

σ_o = 3.06 (8) x 10^{-20} cm^2

Γ = 7.2 (2) x 10^{-8} eV

W_o = 0.595 (19) mm/s

E_r = 1.4837 (3) x 10^{-2} eV

Measured Properties (E_γ = 138.9keV)

E_γ = 138.947 ± 0.008 keV (8)

$t_{\frac{1}{2}}$ = 80 ± 2 ps (9)

α_T = 2.26 ± 0.03 (4)

Derived Parameters (E_γ = 138.9keV)

σ_o = 5.83 (5) x 10^{-20} cm^2

Γ = 5.70 (14) x 10^{-6} eV

W_o = 24.6 (6) mm/s

E_r = 5.3695 (4) x 10^{-2} eV

Energy Conversion ($E_\gamma = 73.0\,\text{keV}$)

1 mm/s = 58.909 (9) MHz

1 mm/s = 2.4363 (4) $\times 10^{-7}$ eV

Energy Conversion ($E_\gamma = 138.9\,\text{keV}$)

1 mm/s = 112.067 (6) MHz

1 mm/s = 4.6348 (3) $\times 10^{-7}$ eV

Energies and Intensities

	E_γ	$I_{Os}(10)$
γ_{M1}:	73.0 keV	50
γ_{M2}:	138.9 keV	100
γ_1 :	107. keV	13
K_{α_1} :	64.9 keV	

(1) R. H. Price and M. W. Johns, Nucl. Phys. A187, 641 (1972).
(2) M. A. Ludington and D. E. Raeside, Nucl. Instrum. Methods 94, 195 (1971).
(3) J. Lindskog, K.-G. Valivaara, Z. Awwad, S.-E. Hägglund, A. Marelius, and J. Phil, Nucl. Phys. A137,
(4) C. R. Cothern, H. J. Hennecke, J. C. Manthuruthil, and R. C. Lange, Phys. Rev. 182, 1286 (1969).
(5) A. Narath, Phys. Rev. 165, 506 (1968).
(6) 68W020.
(7) S. Büttgenbach, M. Hershel, G. Meisel, E. Schrödl, W. White, and W. J. Childs, Z. Phys. 263, 341 (19
(8) D. E. Raeside, Nucl. Instrum. Methods 87, 7 (1970).
(9) 69P07.
(10) V. Berg, S. G. Malmskog, and A. Bäcklin, Nucl. Phys. A143, 177 (1970).

[Other data from M. B. Lewis, Nucl. Data Sheets B8, 389 (1972)]

SOURCE	S-TEMP	ABSORBER	A-TEMP	SHIFT	QS	REMARKS	REF
V(Os)		CeIr2		+.91	3.85	study of magnetic structure	74T033
V(Os)		DyIr2		+.90	3.56	study of magnetic structure	74T033
Os		Fe70IrxPt(30-x)	V	—	--	IS, W, and QS vs Ir conc	75K009
V(Os)		GdIr2		+.87	3.58	study of magnetic structure	74T033
V(Os)		HoIr2		+.89	3.56	study of magnetic structure	74T033
Nb-Os alloy		Ir		-.84		a single line source, W=.64	75D018
Nb-Os aly(IS/Ir		IrCNCl2(PMe2Ph=	He	-.30	4.83	=)3, bonding & structure info	75W021
Nb-Os aly(IS/Ir		Ir(CO)(PPh3)2Cl	He	-.06	6.66	bonding information	75W020
Nb-Os aly(IS/Ir		Ir(CO)(PPh3)2F	He	+.28	7.22	bonding information	75W020
Nb-Os aly(IS/Ir		Ir(CO)(PPh3)2I	He	-.02	6.65	bonding information	75W020
Nb-Os aly(IS/Ir		Ir(CO)(PPh3)2N=	He	-.22	7.90	=CMePF6, bonding info	75W020
Nb-Os aly(IS/Ir		Ir(CO)(PPh3)2=	He	+.06	7.26	=O2CCF3, bonding info	75W020
Nb-Os aly(IS/Ir		Ir(CO)(PPh3)2OH	He	+.28	7.17	bonding information	75W020
Nb-Os aly(IS/Ir		Ir(CO)(PPh3)2P=	He	.00	8.31	=F6, bonding information	75W020
Nb-Os aly(IS/Ir		Ir(CO)(PPh3)2SH	He	+.15	7.31	bonding information	75W020
Nb-Os aly(IS/Ir		IrCl2Py(PMe2Ph)	He	-.80	4.01	=)3PF6, bond & struc info	75W021
Nb-Os aly(IS/Ir		IrCl3-4H2O		-2.11		bonding and structure info	75W021
Nb-Os aly(IS/Ir		IrHCl2(PMe2Ph)3	He	-.59	6.10	bonding and structure info	75W021
Nb-Os aly(IS/Ir		Ir(NO)2(PPh3)=	He	-.12	3.56	=2PF6, bonding information	75W020
Nb-Os aly(IS/Ir		Ir(NO)(PPh3)2=	He	+.73	7.12	=OHPF6, bonding info	75W020
Nb-Os aly(IS/Ir		trans-Ir(dpe)2=	He	-.82	3.48	=Cl2Cl, bond & struc info	75W021
Nb-Os aly(IS/Ir		cis-Ir(dpe)2H2=	He	+.44	1.92	=ClO4, bond & struc info	75W021
Nb-Os aly(IS/Ir		trans-Ir(dpe)2=	He	-.12	1.01	=HClPF6, bond & struc info	75W021
Nb-Os aly(IS/Ir		trans-Ir(dpe)2=	He	-.09	.98	=HIPF6, bond & struc info	75W021
Nb-Os aly(IS/Ir		cis-Ir(dpe)2O2=	He	-.26	1.49	=PF6, bonding & struc info	75W021
Nb-Os aly(IS/Ir		Ir(dpe)2PF6	He	-.4	8.8	bonding information	75W020
Nb-Os aly(IS/Ir		Ir(pMePh2)4PF6	He	-.25	7.80	bonding information	75W020
V(Os)		Ir2La		+.78	3.86	study of magnetic structure	74T033
V(Os)		Ir2Nd		+.87	3.82	study of magnetic structure	74T033
V(Os)		Ir2Pr		+.86	3.79	study of magnetic structure	74T033
V(Os)		Ir2Tb		+.88	3.62	study of magnetic structure	74T033
V(Os)		Ir2Yb		+.96	3.56	study of magnetic structure	74T033
Os(IS/Ir)		KIrO3	V			T=4.2 & 77 K, structure info	75H029
Nb-Os aly(IS/Ir		PhMe2HAs-trans-	He	-1.60	2.70	IrCl4(AsMe2Ph)2	75W021
Nb-Os aly(IS/Ir		PhMe2HP-trans-=	He	-1.31	5.85	=IrCl4(PMe2Ph)2	75W021
Nb-Os aly(IS/Ir		fac-IrCl3(PMe2=	He	-.58		=Ph)3, bonding & struc info	75W021
Nb-Os aly(IS/Ir		mer-IrCl3(AsMe=	He	-1.46	1.06	=2Ph)3, bond & struc info	75W021
Nb-Os aly(IS/Ir		mer-IrCl3(PMe2=	He	-.81	3.81	=Ph)3, bonding & struc info	75W021
Nb-Os aly(IS/Ir		mer-IrCl3(SMe2=	He	-2.36	1.85	=)3, bonding & structure info	75W021

CODE	TOPIC	REFERENCE

74T033   IR-193   G Tanner, F E Wagner, G M Kalvius, G K Shenoy, and K H J Buschow, Mossbauer Studies of (RE)Ir2 Intermetallic Compounds, in "Magnetic Resonance and Related Phenomena, Proceedings of AMPERE Congress, 18th" (Nottingham, 1974), edited by P S Allen, E R Andrew, and C A Bates (North-Holland Publishing Co, Amsterdam, 1974), pp 87-8

75D018   IR-193   G J Davies, A G Maddock and A F Williams, J Chem Soc, Chem Commun 264(1975), A Single-line Source for 193Iridium Mossbauer Spectroscopy

75H029   IR-193   R Hoppe and K Claes, J Less-Common Metals 43,129-42(1975), Uber Oxoiridate: Zur Kenntnis von KIrO3

75K009   IR-193   M Kanashiro, M Nishi, N Kunitomi, and H Sakai, J Phys Soc Jpn 38,897(1975), Mossbauer Effect of Ir193 in Fe-Pt-Ir Alloys

75W020   IR-193   A F Williams, S Bhaduri, and A G Maddock, J Chem Soc, Dalton Trans 1958-62(1975), Mossbauer Spectroscopy of Iridium Compounds. Part II. Some Iridium(I) Compounds

75W021   IR-193   A F Williams, G C H Jones, and A G Maddock, J Chem Soc, Dalton Trans 1952-7(1975), Mossbauer Spectroscopy of Iridium Compounds. Part I. Some Iridium(III) Complexes

$$\frac{83}{36}\text{Kr} \ (9.4\,\text{keV})$$

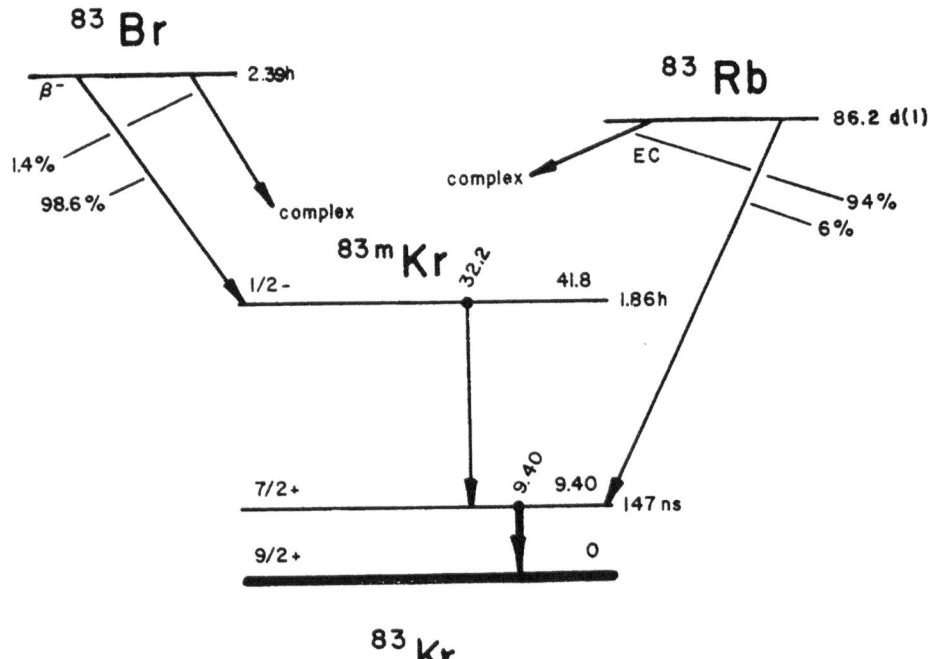

Measured Properties

E_γ = 9.40 ± 0.01 keV (1)

$t_{\frac{1}{2}}$ = 147 ± 4 ns (2)

α_T = 19.6 ± 0.7 (3)

IA = 11.55 %

μ = -0.9703 ± 0.0002 nm (3)

R_μ = +0.971 ± 0.002 (4)

Q = 0.253 ± 0.007 b (5), (6)

R_Q = 1.70 ± 0.07 (7)

Energy Conversions

1 mm/s = 7.582 (8) MHz

1 mm/s = 3.136 (3) × 10^{-8} eV

Derived Parameters

σ_0 = 1.08 (4) × 10^{-18} cm^2

Γ = 3.10 (8) × 10^{-9} eV

W_0 = 0.198 (5) mm/s

E_r = 5.714 (9) × 10^{-4} eV

Energies

γ_M :   9.3 keV

γ_1 :   32. keV

γ_2 :   530. keV

K_{α_1} :   12.65 keV

(1)   B. Kolk, F. Pleiter, and W. Heeringa, Nucl. Phys. A194, 614 (1972).
(2)   S. L. Ruby, Y. Hazony, and M. Pasternak, Phys. Rev. 129, 826 (1963).
(3)   D. Brinkman, Phys. Lett. 27A, 466 (1968).
(4)   69C08.
(5)   H. Kuiper, Z. Phys. 165, 402 (1964).
(6)   W. L. Faust and L. Y. Chow, Chin. Phys. Rev. 129, 1214 (1963).
(7)   66R009.

[Other data from D. C. Kocher, Nucl. Data Sheets 15, 169 (1975)]

SOURCE	S-TEMP	ABSORBER	A-TEMP	SHIFT	QS	REMARKS	REF
Kr		Kr				fs is 40% less than fa	75K041
RbCl		Kr-HQ	4.2			line broadening	75K026
RbHF2		Kr-HQ	4.2			line broadening	75K026
ZnSe(Kr)	v	Kr-HQ	4.2	--		T=4.2, 92, 112 K, fs data, RQ=1.99	75K026

CODE	TOPIC	REFERENCE
75K026	KR-83	B Kolk, Phys Rev B 12,1620-5(1975), Mossbauer Investigations of the Quadrupole Interaction of 83Kr in Various Hosts
75K041	KR-83	B Kolk, Phys Rev B 12,4695-701(1975), Mossbauer Studies on 83Kr Aftereffects in Solid Krypton

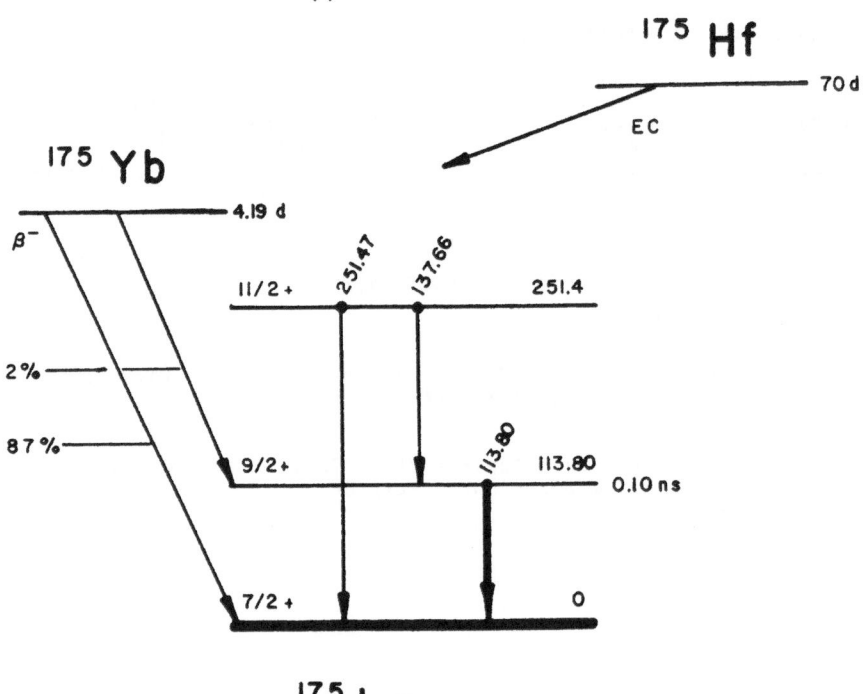

$^{175}_{71}$ Lu (113.8 keV)

Measured Properties

E_γ = 113.803 ± 0.004 keV (1)

$t_{\frac{1}{2}}$ = 100 ± 4 ps (2),(3),(4),(5),(6)

α_T = 2.51

IA = 97.41%

μ = +2.216 ± 0.013 nm (7),(8)

R_μ = +1.92 ± 0.12 (9),(10)

Q = +5.68 ± 0.06 b (11)

Derived Parameters

σ_o = 7.42 (12) × 10^{-20} cm^2

Γ = 4.56 (18) × 10^{-6} eV

W_o = 24 (10) mm/s

E_r = 3.973 (2) × 10^{-2} eV

Energies and Intensities

	E_γ	I_{Yb}
γ_M :	113.81 keV	29.4 (9)
γ_1 :	137.7 keV	1.8 (1)
γ_2 :	144.9 keV	5.1 (3)
γ_3 :	251.4 keV	1.3 (1)
K_{α_1} :	54.07 keV	

Energy Conversions

1 mm/s = 3.79606 (13) × 10^{-7} eV

1 mm/s = 91.787 (3) MHz

(1) J. D. Reierson, G. C. Nelson, and E. N. Hatch, Nucl. Phys. A153, 109 (1970).
(2) A. E. Blaugrund, Y. Dar, and G. Goldring, Phys. Rev. 120, 1328 (1960).
(3) Y. Dar, J. Gerber, A. Macher, and J. P. Vivien, Nucl. Phys. A171, 575 (1971).
(4) K. M. M. S. Ayyangar, V. Lakshminarayana, and S. Jnanananda, Indian J. Pure Appl. Phys. 3, 408 (1965).
(5) E. E. Berlovich, Yu. K. Gusev, V. V. Il'in, and M. K. Nikitin, Sov. Phys.-JETP 16, 1144 (1963).
(6) P. R. Rougny, J. J. Samuel, and A. Sarazin, J. Phys. (Paris) 26, 63 (1965).
(7) A. H. Reddoch and G. J. Ritter, Phys. Rev. 126, 1493 (1962).
(8) V. Kaufman and J. Sugar, J. Opt. Soc. Am. 61, 1693 (1971).
(9) E. Karlsson, E. Matthias, A. G. Svenson, and K. Johansson, Nucl. Phys. 64, 8 (1965).
(10) R. Wäppling, E. Karlsson, and K. Johansson, Phys. Status Solidi 32, 151 (1969).
(11) G. J. Ritter, Phys. Rev. 126, 240 (1962)

[Other data taken from M. M. Minor, Nucl. Data Sheets 18, 331 (1976)]

SOURCE	S-TEMP	ABSORBER	A-TEMP	SHIFT	QS	REMARKS	REF
Lu203(CE)	xx					fs experiment	75C024

CODE	TOPIC	REFERENCE
75C024	LU-175	C L Chien and J C Walker, Studies of Temperature Spikes in Solids Using the Mossbauer Effect Following Coulomb Excitation, in "5th Int Conf Mossbauer Spec, Proc," Part 3, pp 560-3 (See 75H017)

$$^{145}_{60}\mathbf{Nd}\ (67.2\,keV,\ 72.5\,keV)$$

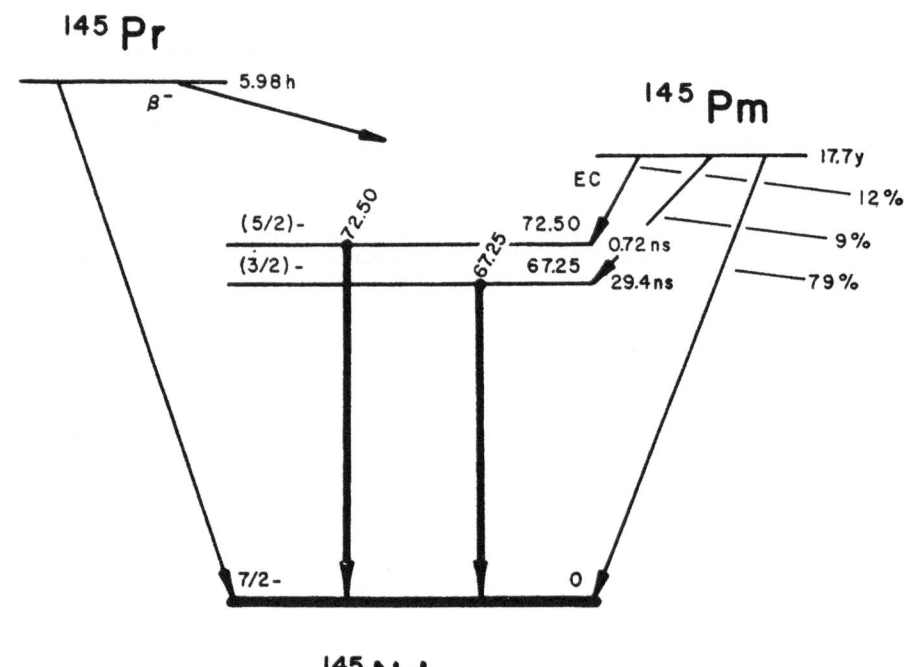

$$^{145}\mathbf{Nd}$$

Measured Properties (E_γ = 67.2 keV)

E_γ = 67.25 ± 0.02 keV (1)

$t_{\frac{1}{2}}$ = 29.4 ± 0.1 ns (2)

α_T = 6.1 ± 0.3 (3)

IA = 8.30%

μ = −0.654 ± 0.004 nm (4)

Q = −0.254 ± 0.005 b (4)

Derived Parameters (E_γ = 67.2 keV)

σ_0 = 3.81 (16) × 10^{-20} cm^2

Γ = 1.55 (5) × 10^{-8} eV

W_0 = 0.138 (5) mm/s

E_r = 1.6742 (7) × 10^{-2} eV

Energy Conversion (E_γ = 67.2 keV)

1 mm/s = 54.240 (16) MHz

1 mm/s = 2.2432 (7) × 10^{-7} eV

Measured Properties (E_γ = 72.5 keV)

E_γ = 72.50 ± 0.02 keV (1)

$t_{\frac{1}{2}}$ = 0.72 ± 0.05 ns (2)

α_T = 4.9 ± 0.4 (3)

R_μ = +0.489 ± 0.004 (1)

Derived Parameters (E_γ = 72.5 keV)

σ_0 = 5.9 (4) × 10^{-20} cm^2

Γ = 6.3 (4) × 10^{-7} eV

W_0 = 5.2 (4) mm/s

E_r = 1.9458 (8) × 10^{-2} eV

Energy Conversion (E_γ = 72.5 keV)

1 mm/s = 58.475 (16) MHz

1 mm/s = 2.4183 (7) × 10^{-7} eV

Energies and Intensities

	E_γ	I_{Pm}
γ_{M1}:	67.25 keV	10
γ_{M2}:	72.50 keV	23
K_{α_1}:	37.36 keV	

(1)   70K042.
(2)   B. Myslek, Z. Sujkowski, and A. Zglinski, Institute for Nuclear Research, Swierk, Poland (referenced in 70L04.
(3)   A. R. Brosi, B. H. Ketelle, H. C. Thomas, and R. J. Kerr, Phys. Rev. 113, 239 (1959).
(4)   K. F. Smith and P. J. Unsworth, Proc. Phys. Soc. (London) 86, 1251 (1965).

[Other data from T. W. Burrows, Nucl. Data Sheets 12, 203 (1974).]

CODE   TOPIC   REFERENCE

75C008   ND-145   J D Cashion, M A Coulthard, and D B Prowse, J Phys C 8,
                  1267-75 (1975), Mossbauer Isomer Shifts in Rare Earth Com-
                  pounds: II. Nuclear Charge Radius and Electron Density
                  Studies

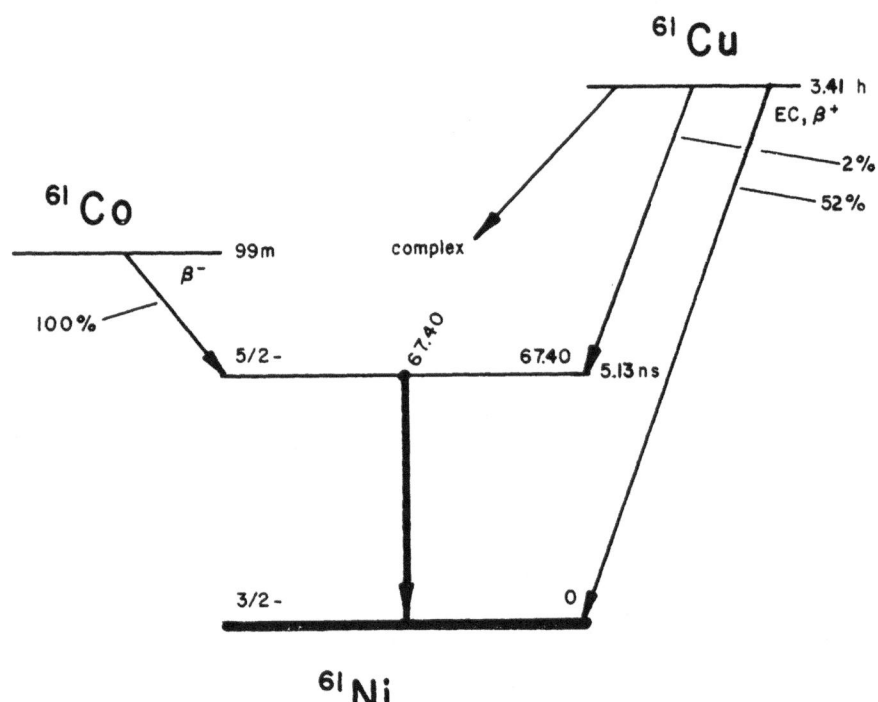

$^{61}_{28}$Ni (67.4 keV)

Measured Properties

E_γ = 67.403 ± 0.01 keV (1), (2)

$t_{\frac{1}{2}}$ = 5.27 ± 0.09 ns (3)

α_T = 0.135 (theory)

IA = 1.19%

μ = 0.7498 ± 0.0001 nm (4)

R_μ = −0.637 ± 0.011 (5), (6), (7)

Q = +0.162 ± 0.015 b (8)

R_Q = −1.21 ± 0.13 (6)

Energy Conversions

1 mm/s = 54.368 (6) MHz

1 mm/s = 2.24849 (23) × 10^{-7} eV

Derived Parameters

σ_o = 7.12 (3) × 10^{-19} cm^2

Γ = 8.66 (15) × 10^{-8} eV

W_o = 0.770 (13) mm/s

E_r = 3.9984 (6) × 10^{-2} eV

Energies

E_γ

γ_M :   67.40 keV

K_{α_1} :   7.48 keV

(1) R. Beraud, I. Berkes, J. Daniere, M. Levy, G. Marest, and R. Roughby, Nucl. Phys. A99, 577 (1967).
(2) R. J. Gehrke, – Document, ANCR-1088, p. 392 (1972).
(3) S. Roodbergen, – Thesis, Vrije University, Amsterdam (1974).
(4) L. E. Drain, Phys. Lett. 11, 114 (1964).
(5) 68E001.
(6) 71G050.
(7) 71L011.
(8) W. J. Childs and L. S. Goodman, Phys. Rev. 170, 136 (1968).

[Other data from J. Vervier, Nucl. Data Sheets B2, 81 (1968)]

SOURCE	S-TEMP	ABSORBER	A-TEMP	SHIFT	QS	REMARKS	REF
XX		Mn-Ni alloys	V		--	4.2<T<180 K, HI data	75F016
XX		NiS(2-x)Sex			--	study of magnetic structure and phase transitions	75C030
XX		NiS(2-x)Sex				existence of an antiferromagnetic metallic phase	75G064

CODE	TOPIC	REFERENCE
75C030	NI-61	G Czjzek, J Fink, H Schmidt, F Gautier, G Krill, M F Lapierre, P Panissod, and C Robert, An Investigation of Magnetic Structures and Phase Transitions in NiS(2-x)Se(x) by 61Ni-Mossbauer Spectroscopy, in "Int Conf Mossbauer Spectrosc, Proc," Vol 1, pp 81-2 (See 75H027)
75F016	NI-61	J Fink, G Czjzek, and H Schmidt, Magnetic Hyperfine Interactions at 61Ni in Ni-Mn Alloys, in "Int Conf Mossbauer Spectrosc, Proc," Vol 1, pp 79-80 (See 75H027)
75G064	NI-61	F Gautier, G Krill, M F Lapierre, P Panissod, C Robert, G Czjzek, J Fink, and H Schmidt, Phys Lett 53A,31-3(1975), Existence of an Antiferromagnetic Metallic Phase (AFM) in the NiS(2-x)Sex System with Pyrite Structure

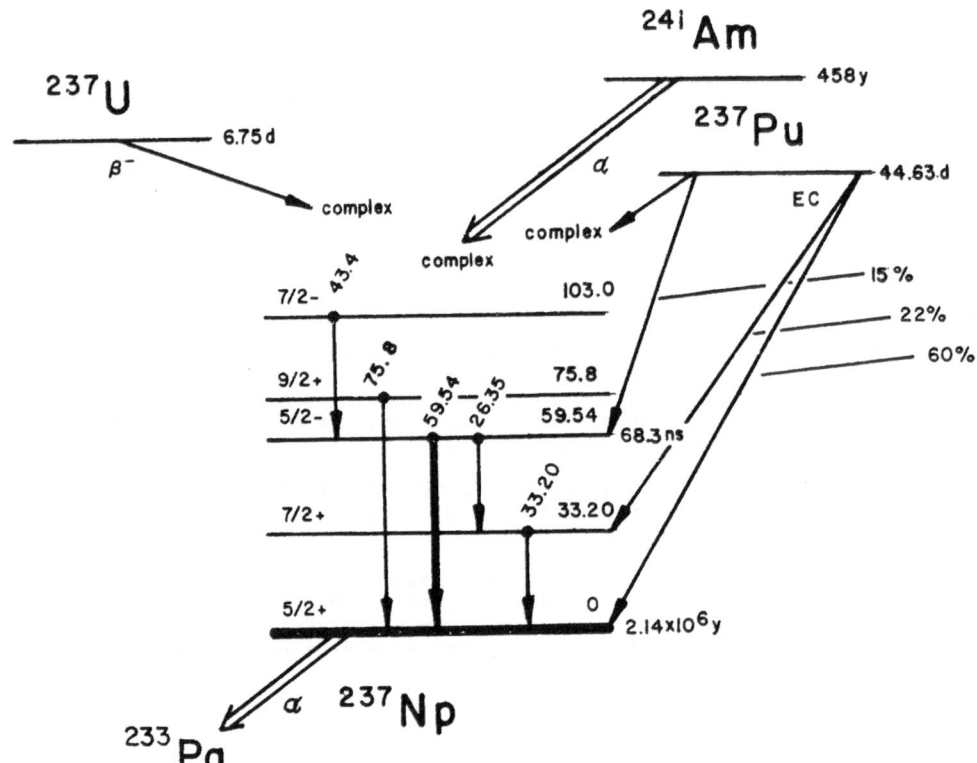

$^{237}_{93}$Np (59.5 keV)

Measured Properties

E_γ = 59.537 ± 0.001 keV (1),(2)

$t_{\frac{1}{2}}$ = 68.3 ± 0.2 ns (3)

α_T = 1.12 ± 0.12 (4)

ρ = 94 %

IA ≈ 0 (radioactive)

μ = +2.5 ± 0.3 nm (5),(6),(7)

R_μ = + 0.535 ± 0.004 (8),(9)

Q = + 4.1 ± 0.7 b (10)

R_Q = + 0.99 ± 0.01 (8)

Energy Conversion

1 mm/s = 48.0194 (9) MHz

1 mm/s = 1.98594 (3) × 10^{-7} eV

Derived Parameters

σ_o = 3.06 (24) × 10^{-19} cm^2

Γ = 6.680 (20) × 10^{-9} eV

W_o = 0.06727 (20) mm/s

E_r = 8.02826 (20) × 10^{-3} eV

Energies and Intensities

		E_γ	I_{Am} (11)
γ_M	:	59.54 keV	35.3
γ_1	:	26.35 keV	2.5
γ_2	:	33.20 keV	0.11
γ_3	:	43.4 keV	0.07
K_{α_1}	:	101.1 keV	

(1) R. W. Jewell, W. John, R. Massey, and B. G. Saunders, Nucl. Instrum. Methods 62, 68 (1968).
(2) G. C. Nelson and B. G. Saunders, Nucl. Instrum. Methods 84, 90 (1970).
(3) G. H. Miller, P. Dillard, M. Eckhause, and R. E. Welsh, Nucl. Instrum. Methods 104, 11 (1972).
(4) R. K. Garg, S. D. Chauhan, S. L. Gupta, and N. K. Saha, Z. Phys. 244, 312 (1971).
(5) C. A. Hutchinson, Jr. and B. Weinstock, J. Chem. Phys. 32, 56 (1960).
(6) J. C. Eisenstein and M. H. L. Pryce, J. Res. Nat. Bur. Stand., Sect. A69, 217 (1965).
(7) W. B. Lewis, J. B. Mann, D. A. Libermann, and D. T. Cromer, J. Chem. Phys. 53, 809 (1970).
(8) 68P001.
(9) 68D002.
(10) 69D024.

[Other data from Y. A. Ellis, Nucl. Data Sheets B6, 539 (1971)]

SOURCE	S-TEMP	ABSORBER	A-TEMP	SHIFT	QS	REMARKS	REF
xx		Co2Np				study of magnetic properties	75A012
xx(IS/Al2Np)		Co2Np	4.2	-22	--	EQ=5.0x10(-7) eV, HI=980 kOe	75A004
Th(Am)(IS/Al2Np	4.2	Cs2Np(NO3)6	4.2	-.40	--	EQ=-13 mm/s	75C002
Th(Am)(IS/Al2Np	4.2	(Et4N)2Np(NO3)6	4.2	-.33	--	EQ=+2 mm/s	75C002
xx		Fe2Np				study of magnetic properties	75A012
xx(IS/Al2Np)		Fe2Np	4.2	-23	--	EQ=2.9x10(-6) eV, HI=1670 kOe	75A004
xx(IS/Al2Np)		Mn2Np	4.2	-24	--	EQ=1.7x10(-6) eV, HI=400 kOe	75A004
xx		Ni2Np				study of magnetic properties	75A012
xx(IS/Al2Np)		Ni2Np	4.2	-17	--	EQ=1.0x10(-6) eV, HI=2350 kOe	75A004
Th(Am)(IS/Al2Np	4.2	Np(IV) in 20% =		-6.4		=TOA-benzene extract from 12M HCl	75C002
Th(Am)(IS/Al2Np	4.2	Np(IV) in 20% =	4.2	-.28	--	=TOA-benzene extract from 8M HNO3, EQ=+17 mm/s	75C002
Th(Am)(IS/Al2Np	4.2	Np(IV) on =		-6.4		=Dowex 1-X8 from 12M HCl	75C002
Th(Am)(IS/Al2Np	4.2	Np(IV) on =		-1.9		=Dowex 50W-8 from 1.0M HNO3	75C002
Th(Am)(IS/Al2Np	4.2	Np(IV) on =		-2.3		=Dowex 50W-8 from .3M HNO3	75C002
Th(Am)(IS/Al2Np	4.2	Np(IV) on =		-2.3		=Dowex 50W-8 from 1.0M HCl	75C002
Th(Am)(IS/Al2Np	4.2	Np(IV) on =		-2.5		=Dowex 50W-8 from 4.0M HNO3	75C002
Th(Am)(IS/Al2Np	4.2	Np(IV) on =	4.2	-.28	--	=Dowex 1-X8 from 8M HNO3, EQ=+17 mm/s	75C002
AmO2		NpO2				scattering experiment	75F028
NpO2		NpO2				scattering experiment	75F028
UO2	78	NpO2			-.93	source made by bremsstrahlung	740015
NpO2	77	NpO2	77				75F029
PuO2	77	NpO2	77				75F029
ThO2	77	NpO2	77				75F029
UO2.05	78	NpO2	78	--	--	study of annealing effects	75M076
AmO2	77	NpO2	150			scattering experiment	75F029
AmO2	77	NpO2	150			scattering experiment	75F029
xx		NpOs2	1.6			HI and HA data	75D010
xx		NpOs2	4.2			HI and HA data	75D010
xx		NpOs(2-x)Rux	v			study of magnetic properties	75A003
xx		NpS	v			4.2<T<25 K, HI data	74L045

CODE	TOPIC	REFERENCE
74L045	NP-237	D J Lam, B D Dunlap, A R Harvey, M H Mueller, A T Aldred, I Nowik, and G H Lander, The Magnetic Properties of Monosulphides of Neptunium and Plutonium, in "Proceedings of the International Conference on Magnetism ICM-73" (Moscow, 1973) ("Nauka," Moscow, 1974), Vol VI, pp 74-8
740015	NP-237	A Ohkawa, T Shoji, and K Matsui, Kakurikin Kenkyu Hokoku 7, 140-6(1974), Np237 Mossbauer Probe Embedded in Uranium Dioxide by Means of U238 (gamma/n)U237
75A003	NP-237	A T Aldred, D J Lam, A R Harvey, and B D Dunlap, Phys Rev B 11,1169-75(1975), Magnetic Properties of Neptunium Laves Phases: NpOs(2-x)Ru(x) Pseudobinary System
75A004	NP-237	A T Aldred, B D Dunlap, D J Lam, G H Lander, M H Mueller, and I Nowik, Phys Rev B 11,530-44(1975), Magnetic Properties of Neptunium Laves Phases: NpMn2, NpFe2, NpCo2, and NpNi2
75A012	NP-237	A T Aldred, B D Dunlap, D J Lam, G H Lander, M H Mueller, and I Nowik, Magnetic Properties of the Neptunium Laves Phases: NpMn2, NpFe2, NpCo2, NpNi2, in "AIP Conference Proceedings-No 24, Magnetism and Magnetic Materials-1974" (20th Annual Conference, San Francisco), edited by C D Graham, Jr, G H Lander, and J J Rhyne (American Institute of Physics, New York, 1975), pp 347-8
75C002	NP-237	C A Clausen, III and J A Stone, J Inorg Nucl Chem 37, 261-4(1975), Mossbauer Spectra of 237Np in Ion Exchange Resins and in a Solvent Extractant

CODE	TOPIC	REFERENCE
75C062	NP-237	A A Chaikhorskii, Radiokhimiya 17,910-8(1975)/Sov Phys-Radiochem, Results of Studying Neptunium Compounds by a Mossbauer Spectroscopic Method
75D010	NP-237	B D Dunlap, A T Aldred, D J Lam, and G R Davidson, High-field Susceptibility in Ferromagnetic NpOs2, in "AIP Conference Proceedings-No 24, Magnetism and Magnetic Materials-1974" (20th Annual Conference, San Francisco), edited by C D Graham, Jr, G H Lander, and J J Rhyne (American Institute of Physics, New York, 1975), pp 351-2
75F028	NP-237	V M Filin, V I Gol'danskii, and V S Nefedov, Mossbauer Scattering at 237Np, in "5th Int Conf Mossbauer Spectrosc, Proc," Vol 3, pp 563-8 (In Russian)
75F029	NP-237	V M Filin, V I Gol'danskii, and V S Nefedov, Mossbauer Effect Investigation of the State of the 237Np Atom Formed after the Alpha-decay of 241Am, in "5th Int Conf Mossbauer Spectrosc, Proc," Vol 3, pp 568-72 (See 75H017) (In Russian)
75M076	NP-237	K Matsui, A Ohkawa, and T Shoji, Kakuriken Kenkyu Hokoku 7,141-54(1975)8 Annealing Effects in Recoil-damaged Uranium Dioxide Studied by Np237 59.54 keV Mossbauer Effect

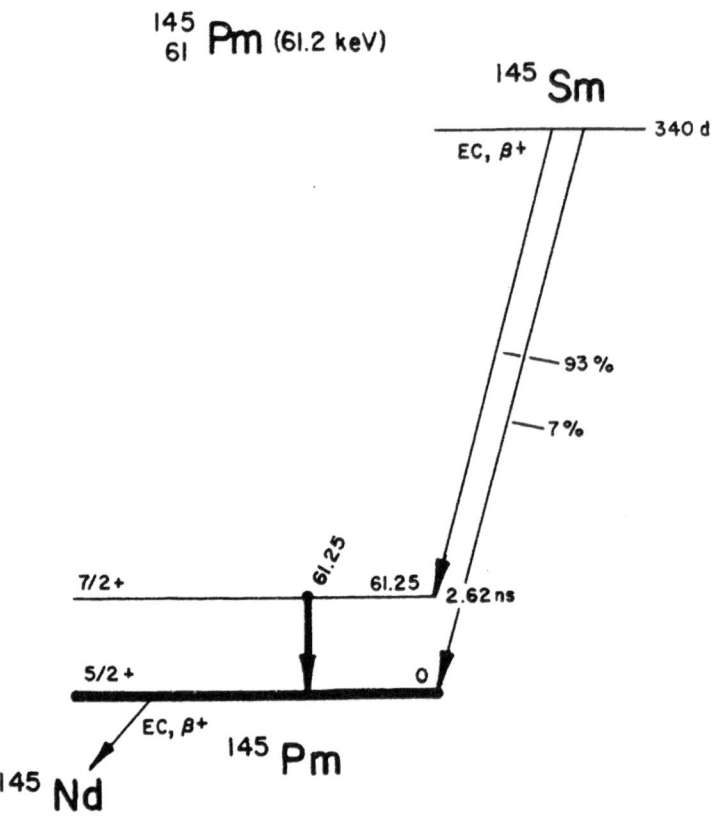

$^{145}_{61}$ **Pm** (61.2 keV)

^{145} **Sm**

340 d

EC, β^+

93 %

7 %

7/2+   61.25   61.25   2.62 ns

5/2+   0

EC, β^+

^{145} **Pm**

^{145} **Nd**

Measured Properties

E_γ = 61.25 ± 0.05 keV (1)

$t_{\frac{1}{2}}$ = 2.62 ± 0.08 ns

α_T = 6.42

IA = radioactive

Energy Conversions

1 mm/s = 49.40 (4) MHz

1 mm/s = 2.0431 (17) × 10^{-7} eV

Derived Parameters

σ_o = 1.172 (8) × 10^{-19} cm^2

Γ = 1.74 (5) × 10^{-7} eV

W_o = 1.70 (5) mm/s

E_r = 1.3888 (16) × 10^{-2} eV

(1)   A. Bäcklin and S. G. Malmskog, Ark. Fysik <u>34</u>, 531 (1967).

[Other data from T. W. Burrows, Nucl. Data Sheets <u>12</u>, 203 (1974)]

SOURCE	S-TEMP	ABSORBER	A-TEMP	SHIFT	QS	REMARKS	REF
Sm203		Pm203				first reported observation of the Mossbauer effect for this transition, W=2.2 mm/s	75B104

CODE	TOPIC	REFERENCE
75B104	PM-145	H Bokemeyer, K Wohlfahrt, E Kankeleit, and D Eckardt, Z Phys A 274,305-18(1975), Mossbauer Conversion Spectroscopy: Measurements on the First Excited States of 180,182W and 145Pm

$${}^{141}_{59}\text{Pr} \ (145.2\,\text{keV})$$

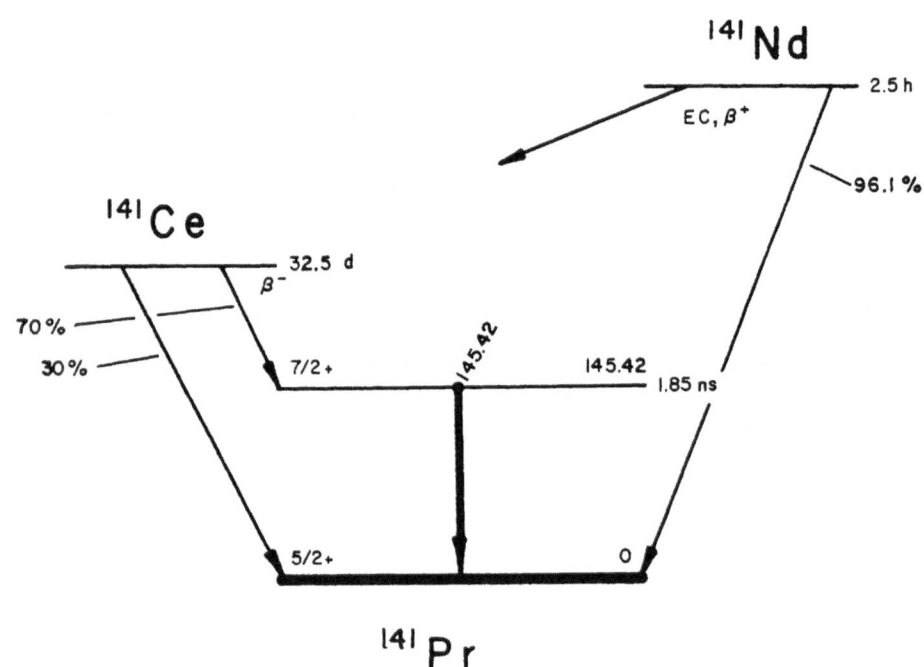

$${}^{141}\text{Pr}$$

Measured Properties

E_γ = 145.42 ± 0.21 keV (1),(2)

$t_{\frac{1}{2}}$ = 1.85 ± 0.03 ns (3)

α_T = 0.46 (theory) (4)

IA = 100%

μ = + 4.162 ± 0.002 nm (5)

R_μ = + 0.69 ± 0.02 (6),(7),(8)

Q = – 0.059 ± 0.004 b (9)

Energy Conversions

1 mm/s = 117.29 (17) MHz

1 mm/s = 4.8507 (7) × 10^{-7} eV

Derived Parameters

σ_o = 1.06 (4) × 10^{-19} cm^2

Γ = 2.47 (4) × 10^{-7} eV

W_o = 1.017 (17) mm/s

E_r = 8.050 (16) × 10^{-2} eV

Energies

E_γ

γ_M : 145.2 keV

K_{α_1} : 36.03 keV

(1) D. B. Beery, W. H. Kelly, and W. C. McHarris, Phys. Rev. 171, 1283 (1968).
(2) V. S. Buttsev, K. Ya. Gromov, and V. G. Kalinnikov, Izv. Akad. Nauk SSSR, Ser. Fiz. 37, 1024 (1973).
(3) D. Bloess, A. Krusche, and F. Münnich, Z. Phys. 192 , 358 (1966).
(4) Used multipole mixing ratio from J. S. Geiger, R. L. Graham, I. Bergström, and F. Brown, Nucl. Phys. 68, 352 (1965).
(5) H. Lew, Bull. Amer. Phys. Soc. 15, 795 (1970).
(6) 71B024 .
(7) 71K043.
(8)   73G015.
(9) B. Bleany, "Quantum Electronics," In Proceedings of the Third International Congress (Paris) (Columbia University Press, New York, 1964), p. 595.

[Other data from R. L. Auble, Nucl. Data Sheets 10, 151 (1973)]

CODE	TOPIC	REFERENCE
75C008	PR-141	J D Cashion, M A Coulthard, and D B Prowse, J Phys C 8, 1267-75 (1975), Mossbauer Isomer Shifts in Rare Earth Compounds: II. Nuclear Charge Radius and Electron Density Studies

$$^{99}_{44}\text{Ru} \text{ (89.4 keV)}$$

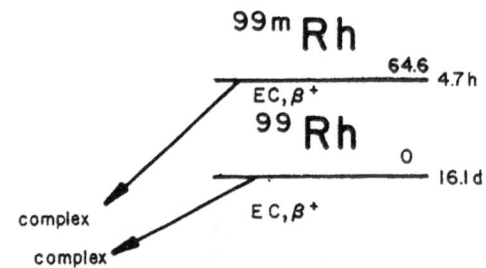

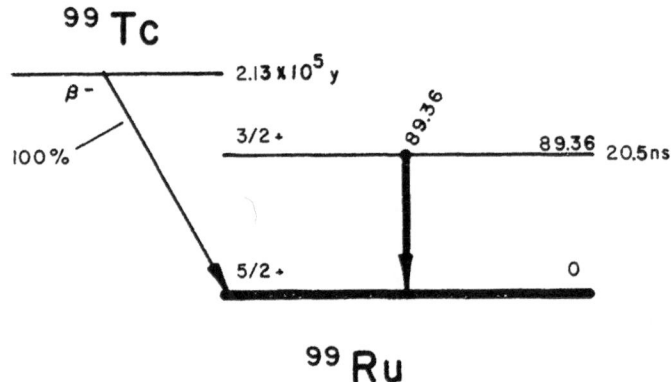

Measured Properties

E_γ = 89.36 keV (1)

$t_{\frac{1}{2}}$ = 20.5 ±0.1 ns (1)

α_T = 1.54 ± 0.10 (theory) (2)

IA = 12.72 %

μ = − 0.626 ± 0.013 nm (3),(4)

R_μ = + 0.456 ± 0.002 (3)

Q = 0.12 ± 0.03 b (5),(6)

R_Q = + 2.93 ± 0.05 (2),(6)

Energy Conversions

1 mm/s = = 72.07 (4) MHz

1 mm/s = 2.9807 (17) × 10^{-7} eV

Derived Parameters

σ_o = 8.0 (3) × 10^{-20}cm^2

Γ == 2.226 (11) × 10^{-8} eV

W_o = 0.1493 (7) mm/s

E_r = 4.330 (3) × 10^{-2} eV

Energies and Intensities

	E_γ	I_{Rh} (7)
γ_M :	89.4 keV	100
γ_1 :	119. keV	< 2
γ_2 :	175. keV	4.5
K_{α_1} :	19.28 keV	

(1)   D. K. Gupta, C. Rangacharyula, R. Singh, and G. N. Rao, Nucl. Phys. A180, 311 (1972).
(2)   75D068.
(3)   73G001.
(4)   E. Matthias, S. S. Rosenbaum, and D. A. Shirley, Phys. Rev. 139, B532 (1965)
(5)   72B055.
(6)   74G037.
(7)   N. M. Anton'eva, E. P. Grigor'eva, and L. F. Protasova, Izv. Akad. Nauk SSSR, Ser. Fiz. 34, 865 (1970).

[Other data from L. R. Medsker, Nucl. Data Sheets 12, 431 (1974)]

SOURCE	S-TEMP	ABSORBER	A-TEMP	SHIFT	QS	REMARKS	REF
Ru(Rh)(IS/Ru)		BaRuO4-H2O	4.2	+.38	.44		75C012
xx(IS/Ru)		Co2RuO4	4.2	-.567	--	RQ=+2.94	75D068
Ru(Rh)(IS/Ru)		KRuO4	4.2	+.82	.37		75C012
xx(IS/Ru)		NBu4RuC14N	4.2	+.081	--	RQ=+2.82	75D068
Ru(Rh)(IS/Ru)		Ru	4.2	.00			75C012
Ru(Rh)(IS/Ru)		Ru on SiO2	4.2			catalyst study, size effects	75C012
Ru(Rh)(IS/Ru)		Ru on eta-Al2O3	4.2			catalyst study	75C012
Ru(Rh)(IS/Ru)		Ru(CO)3Cl3	4.2	-.31			75C012
xx(IS/Ru)		beta-RuCl3	4.2	-.642	--	RQ=+3.06(10)	75D068
Ru(Rh)(IS/Ru)		RuCl3-nH2O	4.2	-.34			75C012
Ru(Rh)(IS/Ru)		Ru(NH3)5COBr	4.2	-.54			75C012
Ru(Rh)(IS/Ru)		Ru(NH3)5NOCl3-=	4.2	-.16	.34	=H2O	75C012
Ru(Rh)(IS/Ru)		Ru(NH3)6Cl2	4.2	-.72			75C012
xx(IS/Ru)		RuO2		-.23	--	EQ=-33.7 MHz	75F002
Ru(Rh)(IS/Ru)		RuO2	4.2	-.23	.51		75C012
Ru(Rh)(IS/Ru)		RuO4	4.2	+1.06			75C012
Ru(Rh)(IS/Ru)		Ru3(CO)12	4.2	-.24			75C012
xx		SrFexRu(1-x)=	4.2	--		=O(3-y), perovskite structure, HI data	75G009

CODE	TOPIC	REFERENCE
75C012	RU-99	C A Clausen, III and M L Good, J Catal 38,92-100(1975), Mossbauer Effect Studies of Supported Ruthenium Catalysts
75D068	RU-99	F M DaCosta, T C Gibb, R Greatrex, and N N Greenwood, Chem Phys Lett 36,655-7(1975), The 99Ru Mossbauer Spectrum of beta-RuCl3
75F002	RU-99	D C Foyt, M L Good, J G Cosgrove, and R L Collins, J Inorg Nucl Chem 37,1913-6(1975), Mossbauer Data Reduction: The M1-E2 Transition of 99Ru
75G009	RU-99	T C Gibb, R Greatrex, N N Greenwood, and K G Snowdon, J Solid State Chem 14,193-202(1975), A Study of the New Perovskite Solid Solution Series SrFe(x)Ru(1-x)O(3-y) by Ruthenium-99 and Iron-57 Mossbauer Spectroscopy

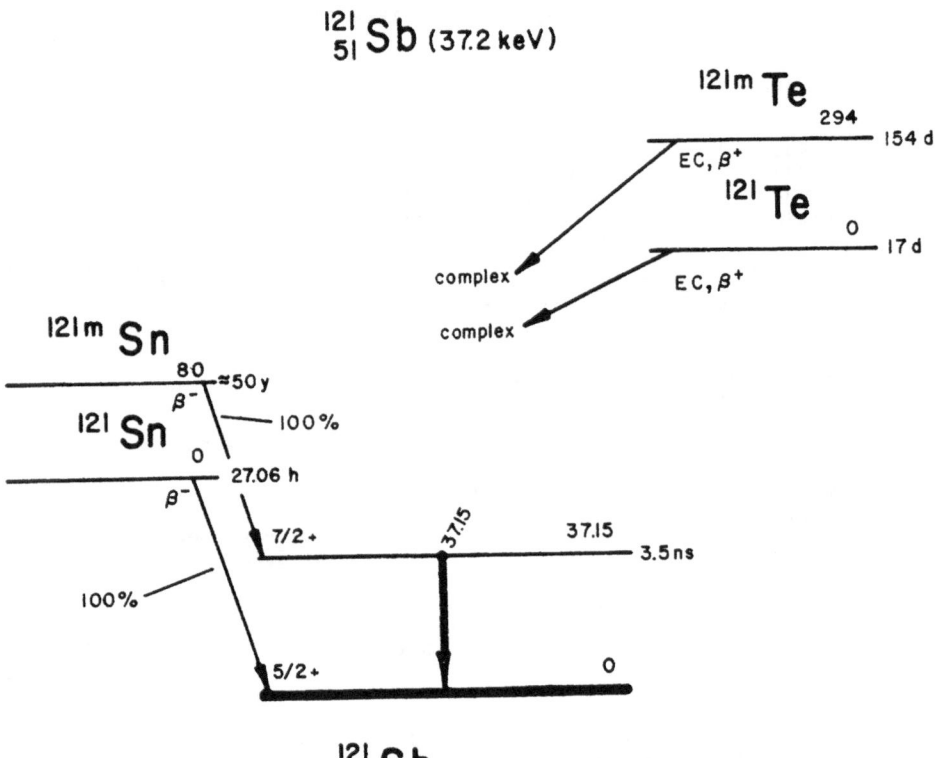

$^{121}_{51}$Sb (37.2 keV)

Measured Properties

E_γ = 37.15 ± 0.02 keV (1),(2)

$t_{\frac{1}{2}}$ = 3.5 ± 0.2 ns (1)

α_T = 11.1 (theory)

IA = 57.25%

μ = + 3.3591 ± 0.0006 nm (3),(4)

R_μ = +0.735 ± 0.009 (5),(6)

Q = − 0.28 ± 0.06 b (7)

R_Q = + 1.34 ± 0.01 (8)

Energy Conversions

1 mm/s = 29.963 (16) MHz

1 mm/s = 1.2392 (7) × 10^{-7} eV

Derived Parameters

σ_o = 1.95 (8) × 10^{-19} cm^2

Γ = 1.30 (7) × 10^{-7} eV

W_o = 2.10 (12) mm/s

E_r = 6.122 (5) × 10^{-3} eV

Energies and Intensities

	E_γ	I_{Te}
γ_M :	37.15 keV (2)	20
γ_1 :	65.6 keV	1
K_{α_1} :	26.36 keV	117

(1) Y. Y. Chu, O. C. Kistner, A. C. Li, S. Monaro, and M. L. Perlman, Phys. Rev. 133, B1361 (1964).
(2) R. E. Snyder and G. B. Beard, Nucl. Phys. A113, 581 (1968).
(3) V. W. Cohen, W. D. Knight, T. Wentink, Jr., and W. S. Koshi, Phys. Rev. 79, 191 (1950).
(4) W. G. Proctor and F. C. Yu, Phys. Rev. 81, 20 (1951).
(5) 67R012.
(6) 70I004.
(7) 67R013.
(8) 70S010.

[Other data from D. J. Horen, Nucl. Data Sheets B6, 75 (1971)]

SOURCE	S-TEMP	ABSORBER	A-TEMP	SHIFT	QS	REMARKS	REF
SnO2		BaTiO3		--			75Y004
CaSnO3	N	CaSb2O6	N	-.36			75F001
CaSnO3	N	Ca-halophos=	N	--		=phate phosphors, struc info	75F001
BaSnO3		CoSb2O6	4.2	-.16	--	EQ=2.7 mm/s	75D074
CaSnO3		Co2MnSb	77			HI=+277 kOe	75C036
BaSnO3		CrSbO4	4.2	-.19	--	EQ=7.8 mm/s	75D074
xx(IS/InSb)		CsSbClF3		-5.23	--	EQ=+16.1 mm/s, bond/struc info	75G076
CaSnO3	4.2	CsSbF6	4.2	2.63		electron transfer from cation	75D024
SnO2		CsSb2F7		-4.3	--	EQ=17 mm/s, eta=.3, struc info	75G074
xx(IS/InSb)		CsSb4F13		-5.90	--	EQ=+19.4 mm/s, bond/struc info	75G076
xx(IS/InSb)		Cs2SbCl3F2		--	--	bonding and structure info	75G076
xx(IS/InSb)		Cs2SbF5		-3.78	--	EQ=+14.8 mm/s, bond/struc info	75G076
SnO2		Cs2SbF5		-3.9	--	EQ=15 mm/s, eta=3, struc info	75G074
SnO2		CuMnSb		-7.6		electronic structure info	74T034
BaSnO3		CuSb2O6	4.2	+.13	--	EQ=1.3 mm/s	75D074
BaSnO3		FeSbO4	4.2	-.06	--	EQ=5.1 mm/s	75D074
BaSnO3		GaSbO4	4.2	+.34	--	EQ=1.5 mm/s	75D074
SnO2	78	GaSbO4	78	0	--	EQ=9.74 mm/s	75V038
beta-Sn		HSbCl6-nH2O		+9.2			75L024
SnO2		InSb		8.55			75G074
CaSnO3	4.2	InSb	4.2	-8.53			75D024
SnS2	78	InSb	78	--		nuclear decay synthesis of antimony(V) sulfide	75A018
SnO2	78	InSbO4	78	0	--	EQ=11.46 mm/s	75V038
SnO2	78	In(2-x)SbxO(=	78	0	--	=3+x), EQ=9.89 mm/s	75V038
xx(IS/InSb)		K2SbC2O4F3-H2O		-4.64	--	EQ=+15.8 mm/s, bond/struc info	75G076
xx(IS/InSb)		K2SbF5		-3.94	--	EQ=+15.0 mm/s, bond/struc info	75G076
beta-Sn		K2Sb(OH)6		+12.3			75L024
xx(IS/InSb)		K2SbSO4F3		-4.69	--	EQ=+15.3 mm/s, bond/struc info	75G076
xx(IS/InSb)		KSbClF3		-4.97	--	EQ=+17.3 mm/s, bond/struc info	75G076
CaSnO3	4.2	KSbF6	4.2	2.73		electron transfer from cation	75D024
SnO2		KSb2F7		-4.7	--	EQ=17 mm/s, eta=0, struc info	75G074
xx(IS/InSb)		KSb4F13		-5.75	--	EQ=+18.5 mm/s, bond/struc info	75G076
CaSnO3	4.2	LiSbF6	4.2	2.73		electron transfer from cation	75D024
BaSnO3	80	Me3Sb(OCOCCl3)2	8	-5.50	--	EQ=-27.5 mm/s, bond/struc info	75G002
BaSnO3	80	Me3Sb(OCOCF3)2	8	-5.5	--	EQ=-28.0 mm/s, bond/struc info	75G002
BaSnO3	80	Me3Sb(OCOCH2Br=	8	-5.24	--	=)2, EQ=-23.8 mm/s, struc info	75G002
BaSnO3	80	Me3Sb(OCOCH2Cl=	8	-5.27	--	=)2, EQ=-25.1 mm/s, struc info	75G002
BaSnO3	80	Me3Sb(OCOCH2Cl=	8	-5.3	--	=)2, EQ=-25.2 mm/s, struc info	75G002
BaSnO3	80	Me3Sb(OCOCH2F)2	8	-5.3	--	EQ=-24.4 mm/s, bond/struc info	75G002
BaSnO3	80	Me3Sb(OCOCH2NC=	8	-5.37	--	=)2, EQ=-25.8 mm/s, struc info	75G002
BaSnO3	80	Me3Sb(OCOCHBr2=	8	-5.4	--	=)2, EQ=-26.2 mm/s, struc info	75G002
BaSnO3	80	Me3Sb(OCOCHF2)2	8	-5.4	--	EQ=-26.4 mm/s, bond/struc info	75G002
BaSnO3	80	Me3Sb(OCOMe)2	8	-5.17	--	EQ=-23.3 mm/s, bond/struc info	75G002
BaSnO3		MgSb2O6	4.2	+.15	--	EQ=3.8 mm/s	75D074
CaSnO3		MnNi2Sb	77			HI=293 kOe	75C036
CaSnO3		MnRhSb	77			HI=116 kOe	75C036
xx(IS/InSb)		NH4SbClF3		-5.62	--	EQ=+17.9 mm/s, bond/struc info	75G076
CaSnO3	4.2	NH4SbF6	4.2	2.53		electron transfer from cation	75D024
xx(IS/InSb)		(NH4)2SbC2O4F3=		-5.86	--	=-2H2O, EQ=+16.9 mm/s	75G076
xx(IS/InSb)		(NH4)2SbCl3F2		--	--	EQ=+9.3 mm/s,bond/struc info	75G076
xx(IS/InSb)		(NH4)2SbF5		-3.99	--	EQ=+15.3 mm/s, bond/struc info	75G076
xx(IS/InSb)		NaSbClF3-H2O		-3.76	--	EQ=+15.8 mm/s, bond/struc info	75G076
CaSnO3	4.2	NaSbF6	4.2	2.89		electron transfer from cation	75D024
xx(IS/InSb)		NaSbNO3F3-H2O		-4.8 -	-	EQ=+15.8 mm/s, bond/struc info	75G076
SnO2		NaSb2F7		-4.8	--	EQ=17 mm/s, eta=0, struc info	75G074
xx(IS/InSb)		Na2SbSO4F3		-5.10	--	EQ=+15.8 mm/s, bond/struc info	75G076
xx		NbSb2		-9.0	--	EQ=9.0 mm/s, structure info	75B021
SnO2		NiMnSb		-7.6		HI=290 kG, elec struc info	74T034
SnO2		Ni(1-x)CuxMnSb		--		HI data, elec struc info	74T034
SnO2		PdMnSb		-7.5		HI=300 kG. elec struc info	74T034
BaSnO3	77	Ph2SbCl2(Ox)	9	-4.02	--	EQ=-13.3 mm/s, eta=.75, cis structure	75R009
BaSnO3	77	Ph2SbCl2(acac)	9	-5.39	--	EQ=+26.4 mm/s, trans struc	75R009
SnO2		PtMnSb		-7.5		HI=280 kG. elec struc info	74T034
CaSnO3		PtMnSb	77			HI=+225 kOe	75C036
xx(IS/InSb)		RbSbClF3		-4.57	--	EQ=+16.5 mm/s, bond/struc info	75G076
xx(IS/InSb)		RbSbF4		-4.02	--	EQ=+17.4 mm/s, bond/struc info	75G076
CaSnO3	4.2	RbSbF6	4.2	2.58		electron transfer from cation	75D024
SnO2		RbSb2F7		-5.3	--	EQ=18 mm/s, eta=0, struc info	75G074
xx(IS/InSb)		RbSb4F13		-5.85	--	EQ=+18.7 mm/s, bond/struc info	75G076
xx(IS/InSb)		Rb2SbCl3F2		--	--	bonding and structure info	75G076

SOURCE	S-TEMP	ABSORBER	A-TEMP	SHIFT	QS	REMARKS	REF
xx (IS/InSb)		Rb2SbF5		-3.70	--	EQ=+15.8 mm/s, bond/struc info	75G076
xx (IS/InSb)		Rb2Sb2SO4F6		-4.51	--	EQ=+16.9 mm/s, bond/struc info	75G076
BaSnO3		SbBr3	4.2	-14.42	--	EQ=+11.60 mm/s, eta=.39	75P011
BaSnO3		SbBr3(C3H2S3)2	4.2	-15.07	--	EQ=+5.89 mm/s, bonding info	75P011
BaSnO3		SbBr3(C9H6S3)2	4.2	-15.40	--	EQ=+7.00 mm/s, bonding info	75P011
BaSnO3		SbBr3(C10H8OS3=	4.2	-15.49	--	=)1.5, EQ=-7.22 mm/s	75P011
SnO2	R	SbC2O4OH	N	-16			75A025
xx		SbCl3		-15.3	--	EQ=12.8 mm/s, bonding info	75S105
BaSnO3		SbCl3	4.2	-14.37	--	EQ=+12.25 mm/s, eta=.187	75P011
BaSnO3		SbCl3(C3H2S3)2	4.2	-14.94	--	EQ=+8.12 mm/s, bonding info	75P011
BaSnO3		SbCl3(C7H4S4)2	4.2	-14.89	--	EQ=+9.62 mm/s, bonding info	75P011
BaSnO3		SbCl3(C10H8OS3=	4.2	-14.79	--	=)2, EQ=+9.75 mm/s, bond info	75P011
xx		SbCl3-MeCN		-14.4	--	EQ=18.7 mm/s, bonding info	75S105
xx		SbCl3-PhCN		-14.4	--	EQ=11.3 mm/s, bonding info	75S105
xx		SbCl3-PhCONHCO=		-15.8	--	=Bu, EQ=13.8mm/s, bonding info	75S105
xx		SbCl3-PhCONHCO=		-16.3	--	=Me EQ=15.4 mm/s, bonding info	75S105
xx		SbCl3-PhCONHCO=		-15.4	--	=Ph, EQ=13.8mm/s, bonding info	75S105
xx		SbCl3-PhCONHCO=		-15.8	--	=Pr, EQ=15.8mm/s, bonding info	75S105
xx		SbCl3-.5MeCN		-14.3	--	EQ=15.9 mm/s, bonding info	75S105
xx		SbCl3-.5PhCN		-14.1	--	EQ=14.3 mm/s, bonding info	75S105
BaSnO3	77	SbCl5-NCCl	77	-2.87	--	EQ=-7.9 mm/s	75S039
BaSnO3	77	SbCl5-NCMe	77	-2.66	--	EQ=-6.9 mm/s	75S039
xx (IS/InSb)		SbF3		-5.97	--	EQ=+19.7 mm/s, bond/struc info	75G076
BaSnO3		SbSBr	4.2	-14.9	--	EQ=10.4 mm/s, eta=.14, struc	75D008
BaSnO3		SbSI	4.2	-14.8	--	EQ=12.3 mm/s, eta=.31, struc	75D008
BaSnO3		SbSI	77	-15.1	--	EQ=10.3 mm/s, eta=.31, struc	75D008
BaSnO3		SbSeBr	4.2	-14.5	--	EQ=10.6 mm/s, eta=.0, struc	75D008
BaSnO3		SbSeI	4.2	-14.9	--	EQ=6.4 mm/s, eta=1, struc info	75D008
BaSnO3		SbTeI	4.2	-14.0	--	EQ=-12.0 mm/s, eta=.2, struc	75D008
CaSnO3	R	Sb2O3	N	-11.0			75F001
beta-Sn		Sb2O3-nH2O		.0			75L024
(NH4)2SnCl6		Sb2O5		--			75L024
beta-Sn		Sb2O5		+12.7			75L024
K2Sn(OH)6		Sb2O5		--			75L024
Sn(OH)2		Sb2O5		--			75L024
SnC2O4		Sb2O5		--			75L024
SnO2		Sb2O5		--			75L024
CaSnO3	N	Sb2O5	N	+.5			75F001
xx		Sb2Ta		-9.2	--	EQ=6.4 mm/s, structure info	75B021
xx		Sb2Ti	4.2	-10.2	--	EQ=11.8 mm/s, structure info, eta=.7	75D013
xx		Sb2V	4.2	-10.2	--	EQ=5.9 mm/s, eta=1	75D013

CODE	TOPIC	REFERENCE

74T034   SB-121   S T Tamaev, R N Kuz'min, Kh Kh Valiev, and S M Irkaev, The Effective Magnetic Fields on the 121Sb Nuclei in the Heusler-type Alloys, in "Proceedings of the International Conference of Magnetism ICM-73" (Moscow, 1973) ("Nauka," Moscow, 1974), Vol V, pp 83-7 (In Russian)

75A018   SB-121   S Ambe and F Ambe, J Chem Phys 63,4077-8(1975), Mossbauer Emission Spectrum of 121Sb after the Beta- Decay of 121mSn in SnS2: Nuclear Decay Synthesis of Antimony(V) Sulfide

75A025   SB-121   S Ambe, J Inorg Nucl Chem 37,2023(1975), Chemical Properties of Sb(III)(C2O4)OH

75B021   SB-121   L Brattas, J D Donaldson, A Kjekshus, D G Nicholson, and J T Southern, Acta Chem Scand, Ser A 29,217-9(1975), 121Sb Mossbauer Studies on NbSb2 and TaSb2

75C036   SB-121   C C M Campbell, J Phys F 5,1931-45(1975), Hyperfine Field Systematics in Heusler Alloys

75D008   SB-121   J D Donaldson, A Kjekshus, D G Nicholson, and J T Southern, Acta Chem Scand, Ser A 29,220-4(1975),121Sb Mossbauer Studies on Antimony(III) Chalcogenohalides

CODE	TOPIC	REFERENCE
75D013	SB-121	J D Donaldson, A Kjekshus, D G Nicholson, and T Rakke, J Less-Common Met 41,255-63(1975), Properties of TiSb2 and VSb2
75D024	SB-121	J P Devort and J M Friedt, Chem Phys Lett 35,423-5(1975), 121Sb Mossbauer Spectroscopy in Alkali Antimony(V) Hexa-fluorides
75D074	SB-121	J D Donaldson, A Kjekshus, D G Nicholson, and T Rakke, Acta Chem Scand, Ser A 29,803-9(1975), Properties of Sb-compounds with Rutile-like Structures
75F001	SB-121	V Fraknoy-Koros, P Gelencser, B Levay, and A Vertes, J Lumin 9,467-74(1975), Application of Mossbauer Spectroscopy to Follow the Incorporation of Antimony into Calcium-halophosphate Phosphors. I
75G002	SB-121	R G Goel, J N R Ruddick, and J R Sams, J Chem Soc, Dalton Trans 67-71(1975), Antimony-121 Mossbauer Spectroscopic Study of Bis(halogenoacetato)-trimethylantimony Derivatives
75G074	SB-121	S E Gukasyan, V P Gor'kov, L A Sadokhina, F Kh Chibirova, and V S Shpinel', Zh Strukt Khim 16,207-11(1975)/J Struct Chem (USSR) 16,191-4(1975), NGR Spectra of Certain Complex Antimony(III) Fluorides
75G076	SB-121	V P Gor'kov, R L Davidovich, G V Zimina, L A Sadokhina, F Kh Chibirova, and V S Shpinel', Koord Khim 1,561-6(1975), Mossbauer Study of Antimony(III) Complex Compounds (In Russian)
75L024	SB-121	R A Lebedev, Yu D Perfil'ev, L A Kulikov, M I Afanasov, A M Babeshkin, and A N Nesmeyanov, Gamma-resonance Spectroscopy Investigation of the Chemical Consequences of the Isomeric Transition, Electron Capture, Beta-minus-decay in Solid Substances, in "5th Int Conf Mossbauer Spectrosc, Proc," Vol 3, pp 537-45 (See 75H017) (In Russian)
75P011	SB-121	F Petillon and J E Guerchais, J Inorg Nucl Chem 37,1863-70 (1975), Complexes Soufres (Partie VII) du Titane(III), de l'Antimoine(V) et (III), du Bismuth(III) et de l'Etain(IV) Etude Mossbauer
75R009	SB-121	J N R Ruddick and J R Sams, Inorg Nucl Chem Lett 11,229-31 (1975), Verification of the -1:2 Quadrupole Coupling Constant Ratio in cis- and trans-Octahedral Diorganoantimony(I) Complexes
75S039	SB-121	G K Shenoy and J M Friedt, Methodology of the 121Sb Mossbauer Quadrupole Spectra, in "5th Int Conf Mossbauer Spec, Proc," Part 3, pp 510-3 (See 75H017)
75S105	SB-121	T N Sumarokova, E F Makarov, D Kh Kamysbaev, A Yu Aleksandrov, I I Amelin, and M I Usanovich, Izv Akad Nauk Kaz SSR, Ser Khim 25(5),9-13(1975), Mossbauer Effect in Some Complexes of Antimony(III) with Organic Substances (In Russian)
75V038	SB-121	M B Varfolomeev, M N Sotnikova, F Kh Chibirova, and V S Shpinel', Zh Neorg Khim 20,1163-6(1975)/Russ J Inorg Chem 20,655-7(1975), Some Structural Differences in Binary Oxides Formed by Indium and Gallium with Sb(V)
75Y004	SB-121	V K Yarmarkin, B A Shustrov, and A V Motorny, Mossbauer Effect on Dy161, Sb121, and Fe57 Nuclei in BaTiO3 Ceramics, in "Int Conf Mossbauer Spectrosc, Proc," Vol 1, pp 325-6 (See 75H027)

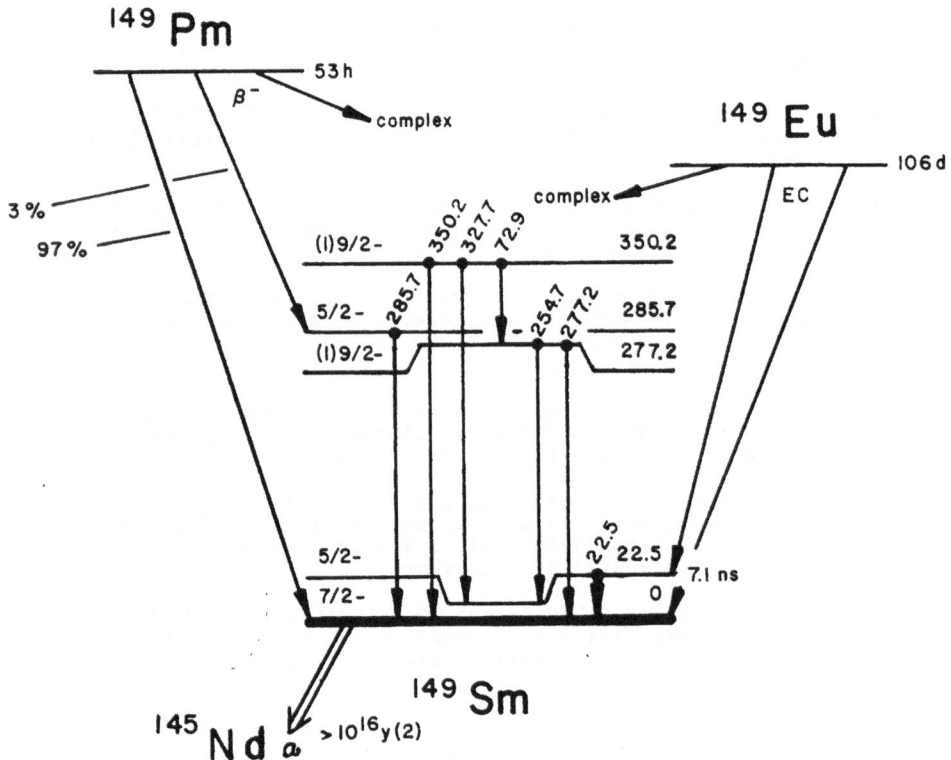

$^{149}_{62}$Sm (22.5 KeV)

Measured Properties

E_γ = 22.494 ± 0.011 keV (1)

$t_{\frac{1}{2}}$ = 7.12 ± 0.11 ns (3)

α_T ≈ 50 (theory)

IA = 13.83%

μ = −0.670 ± 0.006 nm (4)

R_μ = +0.929 ± 0.001 (5)

Q = +0.058 ± 0.006 b (4)

R_Q = 8.3 ± 2.1 (5)

Derived Parameters

σ_o = 7.1 (7) × 10^{-20} cm^2

Γ = 6.41 (10) × 10^{-8} eV

W_o = 1.708 (26) mm/s

E_r = 1.8228 (13) × 10^{-3} eV

Energy Conversions

1 mm/s = 18.142 (9) MHz

1 mm/s = 7.503 (4) × 10^{-8} eV

(1) S. Antman, H. Pettersson, Z. Zehley, and I. Adam, Z. Phys. 237, 285 (1970).
(2) V. A. Korolev, G. D. Alkhazov, A. A. Vorob'ev, A. I. Egorov, and L. M. Vasil'eva, Sov. J. Nucl. Phys. 8, 131 (1969).
(3) J. Kownacki, J. Ludziejewski, M. Moszynski, and K. Styrczniewicz, Acta Phys. Pol. B1, 173 (1970).
(4) G. K. Woodgate, Proc. Roy. Soc. (London) A293, 117 (1966).
(5) 71E009.

[Other data from Table of Isotopes, 6th Edition (1967)]

SOURCE	S-TEMP	ABSORBER	A-TEMP	SHIFT	QS	REMARKS	REF
EuF3		Sm.77Y.23S	R	-.37			75C027
EuF3		SmS	R	--		0<P<11 kbar	75C027
EuF3		SmS(semicon=	R	-.72		=ductor)	75C027
EuF3		Sm(1-x)YxS	R	--		electronic structure info	75C049

CODE	TOPIC	REFERENCE
75C008	SM-149	J D Cashion, M A Coulthard, and D B Prowse, J Phys C 8, 1267-75(1975), Mossbauer Isomer Shifts in Rare Earth Compounds: II. Nuclear Charge Radius and Electron Density Studies
75C027	SM-149	J M D Coey, S K Ghatak, and F Holtzberg, Semiconductor-metal Transition in SmS - A 149Sm Mossbauer Study, in "AIP Conference Proceedings No 24, Magnetism and Magnetic Materials-1974" (20th Annual Conference, San Francisco), edited by C D Graham, Jr, G H Lander, and J J Rhyne (American Institute of Physics, New York, 1975), pp 38-9
75C049	SM-149	J M D Coey and S K Ghatak, Electronic Configuration Fluctuations of Samarium in (Sm(1-x)Y(x))S, in "Int Conf Mossbauer Spectrosc, Proc," Vol 1, pp 369-70 (See 75H027)

$^{119}_{50}$ Sn (23.9 keV)

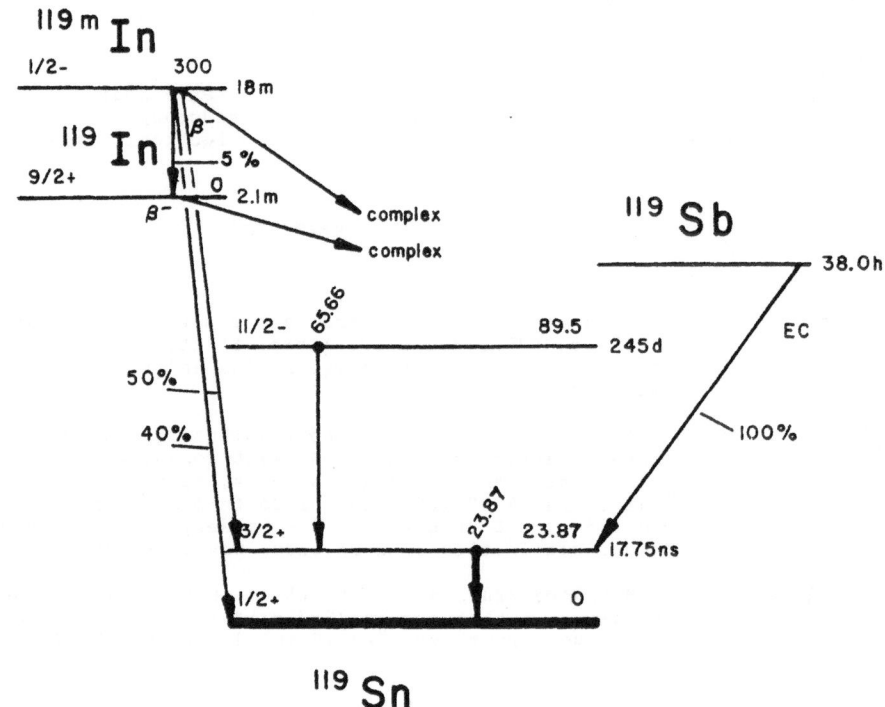

Measured Properties

E_γ = 23.871 ± 0.007 keV (1), (2)

$t_{\frac{1}{2}}$ = 17.75 ± 0.12 ns (3)

α_T = 5.12 ± 0.10 (4), (5)

IA = 8.58%

μ = −1.0461 ± 0.0003 nm (6)

R_μ = −0.605 ± 0.017 (7), (8)

Q^* = −0.064 ± 0.005 b (9), (10)

Derived Parameters

σ_0 = 1.403 (23) × 10^{-18} cm^2

Γ = 2.570 (17) × 10^{-8} eV

W_0 = 0.646 (4) mm/s

E_r = 2.5703 (11) × 10^{-3} eV

Energy Conversions

1 mm/s = 19.253 (6) MHz

1 mm/s = 7.9625 (23) × 10^{-8} eV

(1) S. Raman, P. H. Stelson, G. G. Slaughter, J. A. Harvey, T. A. Walkiewicz, G. T. Lutz, L. G. Multhauf, and K. G. Tirsell, Nucl. Phys. A206, 343 (1973).
(2) J. P. Bocquet, Y. Y. Chu, G. T. Emery, and M. L. Perlman, Phys. Rev. 167, 1117 (1968).
(3) N. Benczer-Koller and T. Fink, Nucl. Phys. A161, 123 (1971).
(4) J. P. Bocquet, Y. Y. Chu, G. T. Emery, and M. L. Perlman, Phys. Rev. Lett. 17, 809 (1966).
(5) V. O. Kostroun and B. Craseman, Phys. Rev. 174, 1535 (1968).
(6) W. G. Proctor, Phys. Rev. 79, 35 (1950).
(7) 73C046.
(8) 73W010.
(9) 72B055.
(10) 73B014.

[Other data from Table of Isotopes, 6th Edition (1967)]

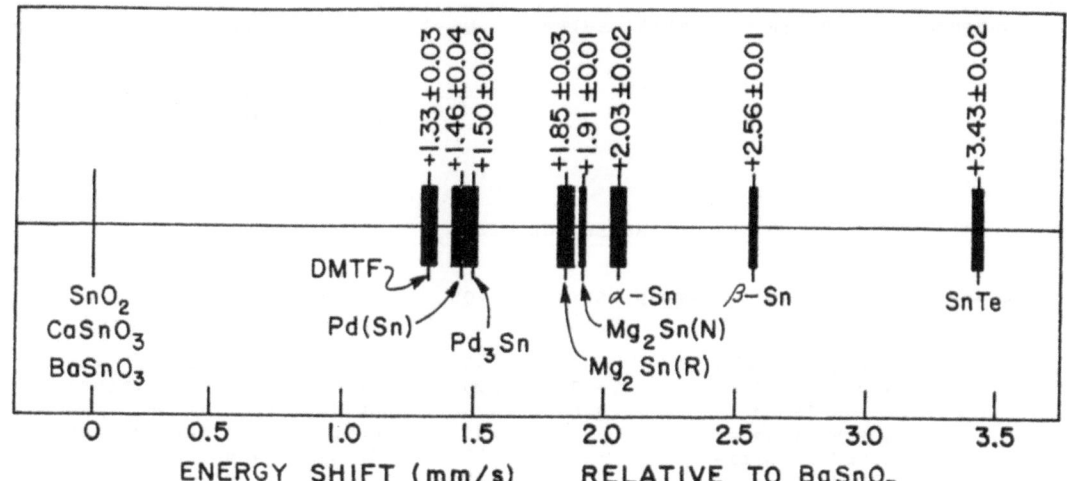

Notes: Sources and absorbers at room temperature except α-Sn which is at liquid nitrogen temperature and Mg$_2$Sn is given for both liquid nitrogen and liquid temperatures. See the pages following the ^{57}Fe isotope for a discussion on sign convention and an illustration of the use of the figure. DMTF = dimethyltin difluoride (Me$_2$SnF$_2$)

[*]Information from Mössbauer Effect Data Index–Covering the 1970 Literature, edited by J. G. Stevens and V. E. Stevens (IFI Plenum, New York, 1972), p 23.

BIOLOGICAL COMPOUNDS

SOURCE	S-TEMP	ABSORBER	A-TEMP	SHIFT	QS	REMARKS	REF
BaSnO3		ants				study of the reaction of a group of ants	75B111

FROZEN SOLUTIONS

SOURCE	S-TEMP	ABSORBER	A-TEMP	SHIFT	QS	REMARKS	REF
xx		Bu2SnH2 in hex=		-.67	.45	=ane, bonding information	75R032
xx		Bu2SnH2 in 2-=		-.71	.20	=picoline, bonding information	75R032
xx		Bu3SnH in DMF		-.67	.20	bonding information	75R032
xx		Bu3SnH in hex=		-.67	.15	=ane, bonding information	75R032
xx		Bu3SnH in pyri=		-.66	.20	=dine, bonding information	75R032
xx		Bu3SnH in 2-=		-.66	.20	=picoline, bonding information	75R032
xx		C6F5Et3SnCC in=			2.16	= HMPT	75Z022
xx		C6F5Et3SnCC in=			1.41	= C6H6	75Z022
xx		C6F5Me3Sn in =			2.10	=HMPT	75Z022
xx		C6F5Me3Sn in =			1.20	=C6H6	75Z022
xx		C6F5Me3SnCC in=			2.06	= HMPT	75Z022
xx		C6F5Me3SnCC in=			1.37	= C6H6	75Z022
BaSnO3		SnBr4 in ben=		1.22		=zene or acetic anhydride	75V003
BaSnO3		SnI4 + TMI in =	N	--		=CCl4, study of equilibrium	75V003
BaSnO3		SnI4 in CCl4	N	1.60			75V003

GLASSES AND AMORPHOUS SUBSTANCES

SOURCE	S-TEMP	ABSORBER	A-TEMP	SHIFT	QS	REMARKS	REF
SnO2		As-Ge-Se-Sn	80	--	--		75B068
CaSnO3	264	Cu-Sn	V	--		amorphous sample, fa vs temp	75B004
xx		CuxSn(1-x)	V	--		2.6<T<108 K, fa vs temperature	75B130
CaSnO3	264	Sn	V	--		amorphous sample, fa vs temp	75B004
xx		Tl2Se-2SnSe-7=		--		=As2Se	75V048
xx		silicate glass				study of the distribution of tin in diffusion layers after heat treatment	75E026
xx		silicate glass				study of the distribution of tin in diffusion layers after heat treatment	75E025
xx		5.5Tl2Se-2.5Sn=		--		=Se-2As2Se3	75V048

INORGANIC HALIDES

SOURCE	S-TEMP	ABSORBER	A-TEMP	SHIFT	QS	REMARKS	REF
SnO2		BaSnCl4		3.97	.70	shown to exist	75G025
BaSnO3		CsSnBr3	4.2	2.09		bonding information	75D049
BaSnO3		CsSnBr6	4.2	2.30		bonding information	75D049
BaSnO3	R	CsSnCl3	80	3.53	1.03	bonding information	74M061
BaSnO3		CsSn2Br5	80	1.77	.82	bonding information	75D049
BaSnO3		CsSn2ClBr4	80	1.76	.92	bonding information	75D049
BaSnO3		CsSn2Cl2Br3	80	1.74	.93	bonding information	75D049
BaSnO3		CsSn2Cl3Br2	80	1.74	.96	bonding information	75D049
BaSnO3		CsSn2Cl4Br	80	1.73	1.07	bonding information	75D049

SOURCE	S-TEMP	ABSORBER	A-TEMP	SHIFT	QS	REMARKS	REF
BaSnO3		CsSn2Cl5	80	1.72	1.00	bonding information	75D049
BaSnO3		Cs2SnCl6	80	.45		bonding information	75D056
BaSnO3		Cs2(Sn,Te)Br6	80	.84		bonding information	75D056
BaSnO3		Cs2(Sn,Te)Cl6	80	.51		bonding information	75D056
xx		Cs4Rh(SnF2(H2O=		--	--	=) 2) 2Sn4F15-4H2O, struc info	75K083
BaSnO3		FeSnF6				structure information	75S098
BaSnO3		FeSnF6-6H2O				structure information	75S098
BaSnO3		K2SnBr6	80	.75		bonding information	75D056
BaSnO3	R	K2SnCl4-H2O	80	3.68	.89	bonding information	74M061
BaSnO3		K2SnCl6	80	.45		bonding information	75D056
BaSnO3		K2(Sn,Te)Br6	80	.84		bonding information	75D056
BaSnO3		K2(Sn,Te)Cl6	80	.46		bonding information	75D056
BaSnO3	R	KSnBr3-H2O	80	3.96	.93	bonding information	74M061
BaSnO3	R	KSnCl3-H2O	80	3.73	.91	bonding information	74M061
BaSnO3	R	NH4SnBr3-H2O	80	3.92	1.01	bonding information	74M061
BaSnO3	R	NH4SnCl3-H2O	80	3.48	1.04	bonding information	74M061
BaSnO3	R	(NH4)2SnBr4-H2O	80	3.96	.98	bonding information	74M061
BaSnO3		(NH4)2SnBr6	80	.80		bonding information	75D056
BaSnO3	R	(NH4)2SnCl2Br2=	80	3.66	.70	=-H2O, bonding information	74M061
BaSnO3	R	(NH4)2SnCl4-H2O	80	3.52	.99	bonding information	74M061
BaSnO3		(NH4)2SnCl6	80	.48		bonding information	75D056
BaSnO3		(NH4)2(Sn,Te)=	80	.84		=Br6, bonding information	75D056
BaSnO3		(NH4)2(Sn,Te)=	80	.48		=Cl6, bonding information	75D056
BaSnO3		Rb2SnCl6	80	.43		bonding information	75D056
BaSnO3		Rb2(Sn,Te)Br6	80	.82		bonding information	75D056
BaSnO3		Rb2(Sn,Te)Cl6	80	.46		bonding information	75D056
BaSnO3	R	SnBr2	80	3.95		bonding information	74M061
xx(IS/BaSnO3)	300	SnBr2 in Ar	77	--	--	structure information	75S066
xx(IS/BaSnO3)	300	SnBr4 in Ar	77	--		structure information	75S066
BaSnO3	R	SnCl2	77	4.04	.60	bonding and structure info	75B090
BaSnO3	R	SnCl2	80	4.03		bonding information	74M061
xx(IS/BaSnO3)	300	SnCl2 in Ar	77	--	--	structure information	75S066
xx(IS/BaSnO3)	300	SnCl4 in Ar	77	--		structure information	75S066
BaSnO3	R	SnFe(ortho)	77	3.30	2.15	bonding and structure info	75B090
xx(IS/BaSnO3)	300	SnI2 in Ar	v	--	--	T=4.2 & 77 K, structure info	75S066
xx(IS/BaSnO3)	300	SnI2 in paraf=	77	--		=fin, Debye T=80 K	75S066
xx(IS/BaSnO3)	300	SnI4 in Ar	77	--		structure information	75S066
BaSnO3	R	Sn(SO3Cl)2	77	4.09	.97	bonding and structure info	75B090
BaSnO3	R	Sn(SO3F)2	77	4.06	.69	bonding and structure info	75B090
BaSnO3		Sn(SbF6)2		4.44			75R034
BaSnO3(IS/SnO2)		Sn4(OH)6Cl6	77	2.81	1.59		75T013

IRRADIATION EXPERIMENTS

SOURCE	S-TEMP	ABSORBER	A-TEMP	SHIFT	QS	REMARKS	REF
Si		BaSnO3		-1.85		source exp, Debye T=281 K, study of vacancy interaction	75M010
alpha-Sn	4.2	PdSn	140	--	--	neutron irradiation of source, study of defects	75V015
xx		alpha-Sn		--	--	neutron irradiation, study of defects	75V015

METALS AND ALLOYS

SOURCE	S-TEMP	ABSORBER	A-TEMP	SHIFT	QS	REMARKS	REF
xx		Bi-Sn alloys	v			fa vs Sn conc, T=77 and 293 K	75B039
xx		Cd-Sn alloys	v			fa vs Sn conc, T=77 and 293 K	75B039
CaSnO3		Co-Pd alloys				Debye temp=310 K, magnetic anomaly of the intensity	75S032
BaSnO3		CoSn	v	--	--	8<T<395 K	75H009

SOURCE	S-TEMP	ABSORBER	A-TEMP	SHIFT	QS	REMARKS	REF
BaSnO3	4.2	Co2HfSn	4.2	1.35		HI(T=0 K)=120.0 kOe	75G029
BaSnO3		Co2HfSn	77			HI=85 kOe	75C036
BaSnO3	R	Co2HfSn	468	1.31			75G029
BaSnO3		Co2HfSn	V			T=90 & 300 K, HI data	75V019
BaSnO3		Co2HfSn	V	--		T=4.2 & 468 K,HI=120kOe(T=0 K)	75G044
BaSnO3		Co2MnSn	V			T=90 & 300 K, HI data	75V019
BaSnO3	R	Co2SnTe	405	1.30			75G029
BaSnO3	4.2	Co2SnTi	4.2	1.32		HI(T=0 K)=84.5 kOe	75G029
Pd3Sn	R	Co2SnTi	4.2	--		HI & HA data	75B133
BaSnO3		Co2SnTi	77			HI=+76 kOe	75C036
CaSnO3		Co2SnTi	V			77<T<400 K, HI data, Curie T=358 K	75D072
BaSnO3	4.2	Co2SnV	4.2	1.42		HI(T=0 K)=8.4 kOe	75G029
BaSnO3	R	Co2SnV	293	1.38			75G029
BaSnO3		Co2SnZr	77			HI=63 kOe	75C036
BaSnO3		Co2TiSn	V	--		T=4.2 & 405 K, HI=84kOe(T=0 K)	75G044
BaSnO3		Co2VSn	V	--		T=4.2 & 293K	75G044
BaSnO3	4.2	Co2ZrSn	4.2	1.37		HI(T=0 K)=103.5 kOe	75G029
BaSnO3	R	Co2ZrSn	473	1.36			75G029
BaSnO3		Co2ZrSn	V	--		T=4.2 & 473 K,HI=104kOe(T=0 K)	75G044
BaSnO3		Co2ZrSn	V			T=90 & 300 K, HI data	75V019
xx		Cu-Sn alloys					75B052
CaSnO3	264	Cu-Sn alloys	V	--		fa vs temperature	75B004
xx		CuxSn(1-x)	V	--		2.6<T<108 K, fa vs temperature	75B130
BaSnO3		Cu2InMn	77			HI=+196 kOe	75C036
xx		FeSn	V	--	--	80<T<295 K, detn of hyperfine parameters	74H053
BaSnO3		FeSn	V	--	--	10<T<395 K	75H009
BaSnO3	R	FeSn	V	--		20<T<395 K, spin struc info, Neel temperature=368 K	75H007
xx		FeSn2	V			fa vs temp, 300 K<T<600 K	75N009
BaSnO3		Fe3Sn	He			polarimetry experiment	75G035
BaSnO3		InMnNi2	77			HI=109 kOe	75C036
BaSnO3	4.2	IrMnSn	4.2	1.45		HI=59.3 kOe	75G028
BaSnO3	R	IrMnSn	293	1.47			75G028
BaSnO3		MnNi2Sb	77			HI=+52 kOe	75C036
BaSnO3	78	MnPtSn	78	1.59		HI=42.9 kOe	75G028
BaSnO3	R	MnPtSn	394	1.56			75G028
BaSnO3		MnSn2	V	--	--	95 K<T<290 K	75V009
BaSnO3		Mn2Sn	V	--	--	95 K<T<290 K	75V009
BaSnO3		alpha-Nb		1.45			74G065
BaSnO3		NbSn2		2.26	2.15		74G065
BaSnO3		Nb3Sn		1.76			74G065
BaSnO3		Nb6Sn5		1.97	1.13		74G065
BaSnO3		Ni-Sn alloys		--		HI data	75E002
xx		Ni-Sn alloys	V	--		study of the critical T range	75D071
BaSnO3		Ni2HfSn	V			T=90 & 300 K, HI data	75V019
BaSnO3		Ni2MnSn	V			T=90 & 300 K, HI data	75V019
BaSnO3		Ni2ZrSn	V			T=90 & 300 K, HI data	75V019
BaSnO3		Pd	V			study of lattice dynamics, 150 K<T<700 K	75P017
BaSnO3		Sb-Sn alloy	V			fa vs temperature and Sn conc	75S031
BaSnO3		alpha-Sn		2.01			75W026
BaSnO3		beta-Sn		2.65			75B032
xx		beta-Sn	4.2		-.16		75P026
xx		beta-Sn	77		-.19		75P026
BaSnO3		beta-Sn	80	2.65		fa=.15	75V003
xx		beta-Sn	295		-.22		75P026
xx(IS/BaSnO3)	R	beta-Sn	295	2.580		lattice dynamics data	75H022
xx		Sn	V	--		2.6<T<108 K, fa vs temperature	75B130
BaSnO3		Sn	V			study of melting	75W001
CaSnO3	264	Sn	V	--		fa vs temperature	75B004
BaSnO3		Ti	V			study of lattice dynamics, 150 K<T<700 K	75P017
BaSnO3		zircaloy 2	V	--	--	77<T<670 K	75B015

MISCELLANEOUS EXPERIMENTS

SOURCE	S-TEMP	ABSORBER	A-TEMP	SHIFT	QS	REMARKS	REF
BaSnO3		CaSnO3				resonance detector	75K064
BaSnO3		Cp2Fe on SiO2 =		--	--	=with Sn, adsorption study	74M060
BaSnO3		Me4Sn on graph=	v	1.11		=ite, fa data, 78<T<300 K	75B148
xx		NaX zeolite + =				=Ga(Sn),no effect observed	75B129
BaSnO3		Pt-Sn-Al2O3				study of catalysts	75B067
xx		Pt-Sn-Al2O3 =				=catalyst, affect of hydrogen reduction	75B106
Sn(IV) complex=	v	Sn(IV)				=es on anion exchangers, T=77 and 298 K	75A021
BaSnO3		Sn evoporated =	78			=in O2, study of lattice dynamics	75A010
BaSnO3		Sn treated w/ =				=F2, surface study	75S005
BaSnO3		Sn treated w/ =				=NaOH, surface study	75S005
xx(IS/beta-Sn)		Sn.67NbSe		--		structure information	75K029
BaSnO3		SnCl4 on graph=	v	.72		=ite, fa data, 78<T<300 K	75B148
BaSnO3		SnI4 on graph=	v	1.49		=ite, fa data, 78<T<300 K	75B148
BaSnO3		SnO2			--	resonance detector	75K064
SnO2		SnO2				study of the properties of a scintillation resonance detector	74K064
BaSnO3(IS/SnO2)		SnO2 + CO	77	--	--	study of chemisorption	75T013
xx		butyl rubber				study of the acceleration of the resin vulcanization of butyl rubber	75C060
xx		butyl rubber				study of the acceleration of the resin vulcanization of butyl rubber	75C059
xx		butyl rubber				study of the acceleration of the resin vulcanization using SnCl2	75C064

MISCELLANEOUS INORGANIC COMPOUNDS

SOURCE	S-TEMP	ABSORBER	A-TEMP	SHIFT	QS	REMARKS	REF
BaSnO3	R	AgCrSnSe4		1.50			75K010
BaSnO3	R	Ag2CdSnSe4		1.47			75K010
BaSnO3	R	Ag8SnSe6					75K010
Mg2Sn		As2Se3	80	3.35	.65	electronic structure info	75S033
xx		As2Se3Snx	v	--	--	study of the glass-crystal transformation	75B097
Mg2Sn		As2Se3(glass)	80	1.78	.40	electronic structure info	75S033
BaSnO3	R	CuAgCdSnSe4		1.52			75K010
BaSnO3	R	CuCrSnSe4		1.51			75K010
xx		CuCr2Se4				HI=+490 kOe	75L007
xx(IS/SnO2)		CuCr2Se4	v			Curie temperature=374 K	75L013
BaSnO3	R	Cu2CdSnSe4		1.65			75K010
BaSnO3	R	Cu2CoSnSe4		1.77			75K010
BaSnO3	R	Cu2FeSnSe4		1.65			75K010
BaSnO3	R	Cu2HgSnSe4		1.65			75K010
BaSnO3	R	Cu2MnSnSe4		1.63			75K010
BaSnO3	R	Cu2NiSnSe		1.81	.89		75K010
BaSnO3	R	Cu2ZnSnSe4		1.67			75K010
BaSnO3	R	Cu3SnSe4		1.80	.63		75K010
Mg2Sn	80	Ge.15Te.85				study of crystal-glass trans	75N007
SnO2		In2Te3		--			75K079
Pd		LaCrO3				HI data	75O003
BaSnO3(IS/SnO2)		Li(8+4x)Sn=	v	--	--	=(1-x)P4, structure info, fa	75M003
BaSnO3		Mn4Si7				electronic structure info	75N027
BaSnO3(IS/SnO2)		NaSn(OH)3	77	2.60	2.29		75T013
Pd		NdCrO3				HI data	75O003
BaSnO3		PbSe-SnSe	298				75S036
SnO2		Pb(1-x)SnxTe		--		study of softening of phonon spectrum	75N008

SN-119, 23.9 KEV TRANSITION

SOURCE	S-TEMP	ABSORBER	A-TEMP	SHIFT	QS	REMARKS	REF
SnO2		Si	v	1.85		T=80 & 295 K,Sn is substitutional impurity	75N013
BaSnO3	R	SnMn3N	v	--		IS, W, and HI vs temperature	75N004
BaSnO3		SnP3	295	--	--		75H011
Mg2Sn		SnSe	80	3.55	.60	electronic structure info	75S033
BaSnO3		SnSe	298				75S036
CaSnO3		SnSe	v	--	--	77 K<T<300 K, fa vs temp	75O004
Mg2Sn		SnSe2	80	1.55	.40	electronic structure info	75S033
BaSnO3		SnSe2	298				75S036
BaSnO3	R	SnSe2	v			298<T<943 K, study of thermal decomposition	75K010
CaSnO3		SnTe	v	--	--	77 K<T<300 K, fa vs temp	75O004
xx		Sn2Se		--			75V048
BaSnO3		Sn3P4	295	--	--		75H011
BaSnO3		Sn4P3	295	--	--		75H011
Mg2Sn		TlAsSe2	80	1.58	.40	electronic structure info	75S033
Mg2Sn		TlAsSe2 (glass)	80	1.81	.40	electronic structure info	75S033
Pd		YCrO3				HI data	75O003

ORGANIC COMPOUNDS

SOURCE	S-TEMP	ABSORBER	A-TEMP	SHIFT	QS	REMARKS	REF
BaSnO3		AcCNNSnPh3Cl=	77	+1.34	3.12	=Et3NH, trigonal bipyramidal structure	75F006
BaSnO3 (IS/SnO2)		(BPD) 2SnCr (CO) 5	77	1.99	2.60	bonding information	75C046
BaSnO3 (IS/SnO2)		(BPD) 2SnMo (CO) 5	77	2.03	2.45	bonding information	75C046
BaSnO3 (IS/SnO2)		(BPD) 2SnW (CO) 5	77	2.13	2.56	bonding information	75C046
BaSnO3		BipyMo (CO) 3Sn=	80	1.30	+1.29	=Cl3Cl, bonding & struc info	75C013
BaSnO3		BipyMo (CO) 3Sn=	80	1.54	+2.44	=MeCl2Cl, bonding & struc info	75C013
BaSnO3		BipyMo (CO) 3Sn=	80	1.44	+2.10	=PhCl2Cl, eta=.3, bond/struc	75C013
BaSnO3		BipyMo (CO) 3Sn=	80	1.47	2.49	=Ph2ClCl, eta=.8, bond & struc	75C013
BaSnO3		BipyW (CO) 3Sn=	80	1.68	+1.67	=Cl3Cl, bonding & struc info	75C013
BaSnO3		BipyW (CO) 3SnMe=	80	1.57	2.62	=Cl2Cl, bonding & struc info	75C013
BaSnO3		BipyW (CO) 3SnPh=	80	1.48	2.21	=Cl2Cl, bonding & struc info	75C013
BaSnO3		BipyW (CO) 3Sn=	80	1.48	2.50	=SnPh2ClCl, bond & struc info	75C013
BaSnO3	R	p-BrC6 H4SnMe3	77	1.25			75K017
BaSnO3	R	Br2Sn (BZP)	80	.86	.62	structure information	75B114
BaSnO3	R	Br2Sn (HPA)	80	.62	.81	structure information	75B114
BaSnO3	R	Br2Sn (HPB)	80	.73	.83	structure information	75B114
BaSnO3	R	Br2Sn (HSI)	80	.43	.73	structure information	75B114
xx (IS/SnO2)		Br2Sn2 (Mn (CO) 5=	77	2.06	1.80	=) 4	75B049
SnO2	300	(Bu2POO) 2SnCl2	80	.00		structure information	75M075
BaSnO3	R	Bu2SnCl2	N	1.59	3.43	bonding and structure info	74M062
BaSnO3		Bu2SnCl2- (BPE)	80	1.49	4.22	bonding and structure info	75M046
BaSnO3	R	Bu2SnCl2-Bipy	N	1.58	4.09	bonding and structure info	74M062
BaSnO3		Bu2SnCl2- (dpe)	80	1.65	4.19	bonding and structure info	75M046
BaSnO3	R	Bu2SnCl2-phen	N	1.62	4.19	bonding and structure info	74M062
BaSnO3	R	Bu2SnCl2-2Py	N	1.60	4.04	bonding and structure info	74M062
BaSnO3	R	Bu2SnCl2-2QL	N	1.71	4.33	bonding and structure info	74M062
xx		Bu2Sn (EtCOC (CN=		1.57	3.67	=) 2) 2, structure information	75K075
xx		Bu2SnH2		-.72	.35	bonding information	75R032
xx		Bu2Sn (MeCOC (CN=		1.87	4.40	=) 2) 2, structure information	75K075
BaSnO3		Bu2Sn (NCS) 2-=	80	1.42	4.23	=(BPE), bonding & struc info	75M046
BaSnO3		Bu2Sn (NCS) 2-=	80	1.40	4.31	=(BPM), bonding & struc info	75M046
BaSnO3		Bu2Sn (NCS) 2 (=	80	1.47	4.48	=Bu3PO) 2, bonding & struc info	75M046
BaSnO3		Bu2Sn (NCS) 2 (=	80	1.44	4.27	=HMPT) 2, bonding & struc info	75M046
BaSnO3		Bu2Sn (NCS) 2 (=	80	1.53	4.41	=Ph3PO) 2, bonding & struc info	75M046
BaSnO3		Bu2Sn (NCS) 2 (=	80	1.47	4.16	=Ph3AsO) 2, bond & struc info	75M046
BaSnO3		Bu2Sn (NCS) 2-=	80	1.52	4.38	=(dpe), bonding & struc info	75M046
BaSnO3		(Bu2Sn (NCS) 2) 2=	80	1.52	4.02	=Bipy02, bonding & struc info	75M046
BaSnO3		Bu2Sn (SC6F5) 2	8	1.49	2.51	bonding information	75B089
BaSnO3	R	Bu2Sn (SCH2) 2		1.70	2.30	structure information	75D003
BaSnO3		Bu2Sn (SMe) 2	8	1.43	2.02	bonding information	75B089
BaSnO3		Bu2Sn (SPh) 2	8	1.46	1.91	bonding information	75B089
BaSnO3	R	Bu3Sn-Mn (CO) 5	79	1.51	.88		75O001
xx		Bu3SnH		-.61	.10	bonding information	75R032

SOURCE	S-TEMP	ABSORBER	A-TEMP	SHIFT	QS	REMARKS	REF
BaSnO3		Bu3SnSC6F5	8	1.45	2.32	bonding information	75B089
BaSnO3		Bu3SnSMe	8	1.39	1.71	bonding information	75B089
BaSnO3		Bu3SnSPh	8	1.42	2.00	bonding information	75B089
BaSnO3	R	Bu3Sn(TPL)2	v	--	--	structure info, 78<T<131 K	75R018
BaSnO3		(Bu3Sn)2CrO4	80	1.58	3.71	fa=-.302, non-polymeric struc	75S001
BaSnO3		(Bu3Sn)2SO4	80	1.54	3.87	fa=-.240, non-polymeric struc	75S001
BaSnO3		(Bu3Sn)2SeO4	80	1.69	4.09	fa=-.246, non-polymeric struc	75S001
BaSnO3		Bu4Sn	77		--	study of smectic B liquid-crystalline glass	75U001
xx		C6F5Et3SnCC			1.43		75Z022
xx		C6F5Me3SnCC			1.30		75Z022
xx		C6F5Me3SnCC			1.41		75Z022
BaSnO3		C6H6Sn(AlCl4)2=			3.93	=-C6H6, bonding information	75R034
BaSnO3		(C6H11)3SnO2CMe	77	1.69	3.27	structure information	75H040
BaSnO3		(C6H11)3SnOH	77	1.46	2.98	hydroxyl-bridged structure	75H018
BaSnO3		((C6H11)3Sn)2=	77	1.62	3.39	CO3, hydroxyl-bridged struc	75H018
BaSnO3		CF3COCNNSnPh3	77	1.07	4.54	structure information	75K027
BaSnO3		CH2CHCH2Sn(=	77	.91	1.70		75B145
BaSnO3	293	(CO)4FePBu(Sn=	80	1.36	1.22	=Me3)2, bonding information	75E024
BaSnO3	293	(CO)4FePBu2Sn=	80	1.37	1.21	=Me3, bonding information	75E024
BaSnO3	293	(CO)4FeP(SnMe3=	80	1.36	1.32	=)3, bonding information	75E024
BaSnO3	R	o-ClC6H4SnMe3	77	1.24			75K017
BaSnO3	R	m-ClC6H4SnMe3	77	1.28			75K017
SnO2	300	(Cl2EtPO)2SnCl4	80	.37	.5	structure information	75M075
SnO2	300	(Cl2MePO)2SnCl4	80	.37	.5	structure information	75M075
BaSnO3	300	(Cl2PNMe2)2Sn=	80	.56	.4	=Cl4, bonding information	75K035
SnO2	300	(Cl2PhPO)2SnCl4	80	.46	1.60	structure information	75M075
BaSnO3	R	Cl2Sn(HPA)	80	.40	.73	structure information	75B114
BaSnO3	R	Cl2Sn(HPB)	80	.56	.81	structure information	75B114
BaSnO3	R	Cl2Sn(HSI)	80	.42	.72	structure information	75B114
xx(IS/SnO2)		Cl2Sn2(Mn(CO)5=	77	2.04	1.84	=)4	75B049
BaSnO3		Cl3Ru2(CO)5Sn=	80	1.71	+1.84	=Cl3, bonding and struc info	75C013
BaSnO3	R	Cl3SnFe(CO)CpP=	80	1.95	1.91	=Et3, bonding information	75B043
BaSnO3	R	Cl3SnFe(CO)CpP=	80	1.77	1.85	=(OEt)3, bonding information	75B043
BaSnO3	R	Cl3SnFe(CO)CpP=	80	1.79	1.82	=(OPh)3, bonding information	75B043
BaSnO3	R	Cl3SnFe(CO)CpP=	80	1.86	1.89	=Ph3, bonding information	75B043
BaSnO3	R	Cl3SnFeCp(P(O=	80	1.88	1.92	=Ph)3)2, bonding information	75B043
BaSnO3		CpFe(CO)2SnPh=	80	1.70	2.84	=Cl2, bonding & structure info	75C013
BaSnO3		CpMo(CO)3SnCl3	80	1.73	1.73	bonding and structure info	75C013
BaSnO3		CpMo(CO)3SnMe=	80	1.68	2.83	=Cl2, bonding and struc info	75C013
xx		Cr(CO)4(SnCl3)=		2.19	1.77	=2(-3), bonding information	75O014
xx		Cr(CO)5SnCl3=		2.04	1.86	=(-1), bonding information	75O014
xx		Cr(CO)5SnF2Cl=		1.78	2.34	=(-1), bonding information	75O014
xx		Cr(SnCl3)6(-6)		3.44	1.10	bonding information	75O014
BaSnO3		Cs3IrCl3(SnF3)3	95	1.45	2.01	bonding information	75Y006
BaSnO3		Cs3Pt(SnCl3)5		1.73	1.90	=Cl3)5, structure information	75V047
BaSnO3		Cs3Pt(SnF3)5	95	1.46	1.78	bonding information	75Y006
BaSnO3		Cs5Rh(SnF3)6		1.26	2.20	=Cl3)5, structure information	75V047
BaSnO3		(DTH)Mo(CO)3Sn=	80	1.41	1.32	=Cl3Cl, bonding & struc info	75C013
BaSnO3		(DTH)Mo(CO)3Sn=	80	1.49	2.26	=MeCl2Cl, bonding & struc info	75C013
BaSnO3		(DTH)Mo(CO)3Sn=	80	1.50	2.20	=PhCl2Cl, bonding & struc info	75C013
BaSnO3		(DTH)W(CO)3Sn=	80	1.45	1.48	=Cl3Cl, bonding & struc info	75C013
BaSnO3		(DTH)W(CO)3Sn=	80	1.61	2.57	=MeCl2Cl, bonding & struc info	75C013
BaSnO3		(DTH)W(CO)3Sn=	80	1.55	+2.43	=PhCl2Cl, eta=.4, bond & struc	75C013
BaSnO3	R	o-EtC6H4SnMe3	77	1.27			75K017
SnO2	300	((EtO)2POSMe)2=	80	1.47	4.00	=Me2SnCl2, structure info	75M075
BaSnO3		EtOOCCH2Sn(=	77	.87	1.77		75B145
SnO2	300	(Et2ClPO)2SnCl4	80	.34	.5	structure information	75M075
SnO2	300	(Et2ClPS)2SnCl4	80	.65	.4	structure information	75M075
SnO2	300	(Et2POO)2SnCl2	80	.73		structure information	75M075
SnO2	300	(Et2POSMe)2Et=	80	1.10	2.10	=SnCl3, structure information	75M075
SnO2	300	(Et2POSMe)2Me2=	80	1.45	4.15	=SnCl2, structure information	75M075
BaSnO3	300	(Et2PSNMe2)2Sn=	80	.96		=Br4, bonding information	75K035
BaSnO3	300	(Et2PSNMe2)2Sn=	80	.66		=Cl4, bonding information	75K035
SnO2	300	(Et2PrPS)2SnCl4	80	.60		structure information	75M075
xx		Et2Sn(EtCOC(CN=		1.85	4.15	=)2)2, structure information	75K075
xx		Et2Sn(MeCOC(CN=		1.56	4.01	=)2)2, structure information	75K075
BaSnO3	R	Et2Sn(OCH2CH2O=		1.30	3.20	=)2SnEt2, structure info	75D003
BaSnO3	R	Et2Sn(OMe)2		1.20	2.60	structure information	75D003
BaSnO3	R	Et2Sn(SCH2)2		1.55	2.50	structure information	75D003
BaSnO3	R	Et2Sn(SMe)2		1.70	2.30	structure information	75D003
BaSnO3	R	Et3Sn-Mn(CO)5	79	1.58	.78	2	75O001

SOURCE	S-TEMP	ABSORBER	A-TEMP	SHIFT	QS	REMARKS	REF
BaSnO3	R	Et3Sn-Mn(CO)5P=	79	1.44	.43	=Ph3	75O001
BaSnO3		Et3Sn(PAO)	77	1.42	2.73	structure information	75H040
BaSnO3		Et3Sn(PCA)	77	1.49	3.83	structure information	75H040
BaSnO3		Et3Sn(PYD)	77	1.43	3.18	structure information	75H040
BaSnO3		Et3Sn(PYD)	77	1.42	3.32	structure information	75H040
BaSnO3	R	Et3Sn(TPL)2	v	--	--	structure info, 78<T<137 K	75R018
BaSnO3		(Et4N)2IrClCO(=	95	1.73	2.15	=SnCl3)2, bonding information	75Y006
BaSnO3		(Et4N)2IrCl3CO=	95	1.83	2.08	=(SnCl3)2, bonding information	75Y006
BaSnO3		trans-(Et4N)2=	95	1.90	2.05	=PtCl2(SnCl3)2, bonding info	75Y006
BaSnO3		cis-(Et4N)2Pt=	95	1.68	2.07	=Cl2(SnCl3)2, bonding info	75Y006
BaSnO3		(Et4N)2PtHCl(=	95	1.53	1.33	=SnLc3)2, bonding information	75Y006
BaSnO3		(Et4N)2RhCOCl(=	95	1.64	1.92	=SnLc3)2, bonding information	75Y006
BaSnO3		(Et4N)3IrCl3(=	95	1.63	1.93	=SnCl3)3, bonding information	75Y006
BaSnO3		(Et4N)3Ir(Sn=		1.87	2.00	=Cl3)5, structure information	75V047
BaSnO3		(Et4N)3Ir(Sn=	95	1.87	2.00	=Cl3)5Cl, bonding information	75Y006
BaSnO3		(Et4N)3Pd(Sn=		1.85	1.77	=Cl3)5, structure information	75V047
BaSnO3		(Et4N)3Pd(Sn=	95	1.85	1.77	=Cl3)5, bonding information	75Y006
BaSnO3		(Et4N)3Pt(Sn=		1.75	1.82	=Cl3)5, structure information	75V047
BaSnO3		(Et4N)3Pt(Sn=	95	1.76	1.83	=Cl3)5, bonding information	75Y006
BaSnO3		(Et4N)3Rh(Sn=		1.80	1.82	=Cl3)5, structure information	75V047
BaSnO3		(Et4N)4Rh(Sn=	95	1.80	1.82	=Cl3)5, bonding information	75Y006
BaSnO3		(Et4N)4Rh2Cl2(=	95	1.60	1.95	=SnLc3)4, bonding information	75Y006
BaSnO3		Et4NPtCOCl2Sn=	95	1.58	1.60	=Cl3, bonding information	75Y006
BaSnO3	R	o-FC6H4SnMe3	77	1.29			75K017
BaSnO3	R	m-FC6H6SnMe3	77	1.32			75K017
BaSnO3	R	(Fe(CO)2Cp)2Sn=	110	1.92	2.40	=Cl2, structure information	75B044
BaSnO3	R	(Fe(CO)2Cp)2Sn=	295	1.91	2.36	=Cl2, structure information	75B044
xx		Fe(CO)4SnCl2Br=		1.95	2.00	=(-1), bonding information	75O014
xx		Fe(CO)4SnCl3=		1.74	2.09	=(-1), bonding information	75O014
BaSnO3		H2IrClCO(Sn=	95	1.18	2.15	=(OH)3)2, bonding information	75Y006
BaSnO3		H2Ir(Sn(OH)3)5		1.35	2.23	=Cl3)5, structure information	75V047
BaSnO3		H2Ir(Sn(OH)3)5	95	1.35	2.23	bonding information	75Y006
xx(IS/SnO2)		H2Sn2(Mn(CO)5)4	77	1.85	.63		75B049
BaSnO3		H3Pt(SnCl3)5-3=		1.74	1.87	=C10H8N2, structure info	75V047
BaSnO3		H3Pt(SnCl3)5-3=		1.75	1.82	=PhN, structure information	75V047
BaSnO3		H3Pt(SnCl3)5-5=		1.75	2.03	=C2H6SO, structure information	75V047
BaSnO3		H3Pt(SnCl3)5-6=		1.75	1.82	=C4H9ON, structure information	75V047
BaSnO3		H3Ru(SnCl3)5-3=		1.97	1.97	=C10H8N2, structure info	75V047
BaSnO3		H3Ru(SnCl3)5-3=	95	1.97	1.97	=dipy, bonding information	75Y006
BaSnO3		H4Rh(SnCl3)5-4=		1.81	1.84	=C10H8N2, structure info	75V047
BaSnO3		H4Rh(Sn(OH)3)5		1.34	2.20	=Cl3)5, structure information	75V047
BaSnO3		H4RhSn(OH)3)5	95	1.34	2.20	bonding information	75Y006
SnO2	300	(Hx2POO)2SnCl2	80	.06		structure information	75M075
BaSnO3		ICH2Sn(acac)2I	77	.77	1.60		75B145
BaSnO3		K5Rh(C2O4)Sn=	95	1.32	1.88	=(C2O4)2)2, bonding info	75Y006
BaSnO3		K7Ir(C2O4)3Sn=	95	1.38	1.84	=(C2O4)2)2, bonding info	75Y006
BaSnO3		K7Ir(Sn(C2O4)2=		1.58	2.16	=)5, structure information	75V047
BaSnO3		K8Pt(Sn(C2O4)2=		1.57	1.99	=)5, structure information	75V047
BaSnO3		K9Rh(Sn(C2O4)2=		1.46	1.96	=)5, structure information	75V047
BaSnO3	R	o-MeC6H4SnMe3	77	1.27			75K017
BaSnO3	R	m-MeC6H4SnMe3	77	1.29			75K017
BaSnO3		MeCOCNNSnPh3	77	1.55	4.23	structure information	75K027
SnO2	300	(MeEtClP)2SnCl4	80	.69		structure information	75M075
SnO2	300	(MeMeOPO2)2Sn=	80	.02		=Cl2, structure information	75M075
SnO2	300	((MeO)2PO2)2Sn=	80	.04		=Cl2, structure information	75M075
BaSnO3		MeOCOCNNSnPh3	77	1.55	4.23	structure information	75K027
BaSnO3		MeOCOCNNSnPh3=	77	+1.34	3.12	=ClEt3NH, trigonal bipyrami-dal structure	75F006
SnO2		MeSON(C6H11)SnP	78	1.27	2.85	structure information	75W007
SnO2		MeSON(C6H11)SnB	78	--	--	structure information	75W007
SnO2		MeSON(C6H11)SnN	78	1.27	2.85	structure information	75W007
SnO2		MeSONMeSnMe3	78	1.26	2.60	structure information	75W007
xx		MeSnCl2(salen)		+3.33		structure information	75C006
BaSnO3		MeSn(acac)2I	77	.83	1.93		75B145
SnO2	300	(Me2ClPO)2SnBr4	80	.64		structure information	75M075
SnO2	300	(Me2ClPO)2SnCl4	80	.36		structure information	75M075
SnO2	300	(Me2ClPS)2SnCl4	80	.64		structure information	75M075
BaSnO3		Me2ClSn(CH2)2C=	77	1.47	3.26	=oPh, structure information	75K081
BaSnO3		Me2ClSn(CH2)2C=	77	1.54	3.70	=oMe, structure information	75K081
BaSnO3		Me2ClSn(CH2)3C=	77	1.52	3.48	=oPh, structure information	75K081
BaSnO3		Me2ClSn(CH2)3C=	77	1.51	3.66	=oMe, structure information	75K081
SnO2	300	(Me2MePO)2SnCl4	80	.28		structure information	75M075

144

SOURCE	S-TEMP	ABSORBER	A-TEMP	SHIFT	QS	REMARKS	REF
SnO2	300	(Me2OMePO)2Sn=	80	.65	.4	=Br4, structure information	75M075
SnO2	300	(Me2OMePO)2Sn=	80	.32		=Cl4, structure information	75M075
SnO2	300	(Me2OMePS)2Sn=	80	1.07		=Br4, structure information	75M075
SnO2	300	(Me2OMePS)2Sn=	80	.65		=Cl4, structure information	75M075
SnO2	300	(Me2OPCl)2BuSn=	80	1.06	2.30	=Cl3, structure information	75M075
SnO2	300	(Me2PO)2Me2Sn=	80	1.38	4.30	=Cl2, structure information	75M075
SnO2	300	Me2POClMe3SnCl	80	1.52	3.50	structure information	75M075
SnO2	300	Me2POCl2EtSnCl3	80	1.17	2.25	structure information	75M075
SnO2	300	(Me2POCl)2MeSn=	80	1.04	2.26	=Cl3, structure information	75M075
SnO2	300	(Me2POCl)2Me2=	80	1.58	4.10	=SnCl2, structure information	75M075
SnO2	300	(Me2POO)2SnCl2	80	.07		structure information	75M075
SnO2	300	Me2POOMeMe2Sn=	80	1.50	4.37	=Cl2, structure information	75M075
SnO2	300	Me2POOMeMe3SnCl	80	1.38	3.65	structure information	75M075
SnO2	300	(Me2POOMe)2Me=	80	.90	2.20	=SnCl3, structure information	75M075
SnO2	300	(Me2POSMe)2Me2=	80	1.45	4.20	=SnCl2, structure information	75M075
BaSnO3	R	Me2SnCl2	N	1.55	3.58	bonding and structure info	74M062
BaSnO3	R	Me2SnCl2	110	1.52	3.57	structure information	75B044
BaSnO3	R	Me2SnCl2	295	1.47	3.37	structure information	75B044
BaSnO3	R	Me2SnCl2-Bipy	N	1.45	4.07	bonding and structure info	74M062
BaSnO3	R	Me2SnCl2-phen	N	1.46	4.16	bonding and structure info	74M062
BaSnO3	R	Me2SnCl2-2ACR	N	1.49	4.04	bonding and structure info	74M062
BaSnO3	R	Me2SnCl2-2IQL	N	1.38	3.92	bonding and structure info	74M062
BaSnO3	R	Me2SnCl2-2NAQ	N	1.58	4.29	bonding and structure info	74M062
BaSnO3	R	Me2SnCl2-2Py	N	1.36	3.92	bonding and structure info	74M062
xx		Me2Sn(EtCOC(CN=		1.78	4.12	=)2)2, structure information	75K075
xx		Me2Sn(MeCOC(CN=		1.46	3.26	=)2)2, structure information	75K075
xx		Me2Sn(NCS)2O		1.25	3.38	structure information	75S030
BaSnO3	R	Me2SnOHNO3	110	1.28	3.67	structure information	75B044
BaSnO3	R	Me2SnOHNO3	295	1.24	3.66	structure information	75B044
xx		(Me2SnOHNO3)2		1.28	3.67	structure information	75S030
xx		Me2Sn(PhCONCN)2		1.45	2.96	structure information	75K075
BaSnO3	R	Me2Sn(SCH2)2		1.50	2.60	structure information	75D003
BaSnO3	R	Me2Sn(acac)2	110	1.16	3.93	structure information	75B044
BaSnO3	R	Me2Sn(acac)2	295	1.08	3.81	structure information	75B044
BaSnO3	R	Me3Sn-Mn(CO)4=	79	1.28		=AsPh3	75O001
BaSnO3	R	Me3Sn-Mn(CO)4P=	79	1.27		=Ph3	75O001
BaSnO3	R	Me3Sn-Mn(CO)5	79	1.33	.61		75O001
BaSnO3	R	Me3Sn(Bzac)	110	1.13	3.69	structure information	75B138
BaSnO3		(Me3SnC3H6OC6=	77	--	--	=H4HCN)2C6H4, study of smectic B liquid-crystalline glass	75U001
BaSnO3		Me3SnC6H4HCNC6=	77		--	=H4C4H9, study of smectic B liquid-crystalline glass	75U001
BaSnO3		Me3Sn(CH2)2CO=	77	1.41		=Me, structure information	75K081
BaSnO3		Me3Sn(CH2)2CO=	77	1.44		=Ph, structure information	75K081
BaSnO3		Me3Sn(CH2)3CO=	77	1.41		=Me, structure information	75K081
BaSnO3		Me3Sn(CH2)3CO=	77	1.42		=Ph, structure information	75K081
BaSnO3		Me3SnOH	77	1.23	2.82	hydroxyl-bridged structure	75H018
BaSnO3	R	Me3SnOH-Me3PbN3		1.28	3.12	polymeric structure	75B036
BaSnO3	R	Me3SnOH-Me3Sn=		1.31	3.24	=NCO, polymeric structure	75B036
BaSnO3	R	Me3SnOH-Me3Sn=		1.28	3.16	=NCS, polymeric structure	75B036
BaSnO3		Me3Sn(PAO)	77	1.33	2.93	structure information	75H040
BaSnO3		Me3Sn(PAO)	77	1.30	2.81	structure information	75H040
BaSnO3		Me3Sn(PCA)	77	1.37	3.89	structure information	75H040
BaSnO3		Me3Sn(PDA)	77	1.36	3.77	structure information	75H040
BaSnO3		Me3Sn(PYD)	77	1.30	3.10	structure information	75H040
BaSnO3		Me3Sn(PYD)	77	1.28	3.21	structure information	75H040
BaSnO3		Me3Sn(TPD)	77	1.49	2.00	structure information	75H040
BaSnO3	R	Me3Sn(TPL)2	v	--	--	structure info, 4.2<T<135 K	75H018
BaSnO3	R	Me3Sn(acac)	110	1.21	3.81	structure information	75B138
BaSnO3	R	Me3Sn(bzbz)	110	1.15	3.86	structure information	75B138
BaSnO3		Me3Sn(ox)	77	1.18	2.27	structure information	75H040
BaSnO3		(Me3Sn)2CO3	77	1.34	3.05	hydroxyl-bridged structure	75H018
BaSnO3	R	Me4HPh3SnCl2	110	1.22	3.00	structure information	75B044
BaSnO3	R	Me4HPh3SnCl2	295	1.25	3.00	structure information	75B044
BaSnO3	300	(Me4N)2Cl4Sn(=		--	--	=N3)2SnCl4, structure info	75B065
BaSnO3	300	(Me4N)2SnCl4(=		--	--	=N3)2, structure information	75B065
BaSnO3		Mn(CO)5SnPbCl2	80	1.63	2.52	bonding and structure info	75C013
xx		Mo(CO)3(SnClBr=		2.27	1.79	=)3(-3), bonding information	75O014
xx		Mo(CO)3(SnCl3)=		2.23	1.79	=2(-3), bonding information	75O014
xx		Mo(CO)5SnBr3=		2.22	1.89	=(-1), bonding information	75O014
xx		Mo(CO)5SnCl2Br=		2.18	1.98	=(-1), bonding information	75O014

SOURCE	S-TEMP	ABSORBER	A-TEMP	SHIFT	QS	REMARKS	REF
xx		Mo(CO)5SnCl3=		2.12	1.94	=(-1), bonding information	75O014
xx		Mo(CO)5SnF2Cl=		1.95	2.24	=(-1), bonding information	75O014
BaSnO3	R	o-NMe2C6H4SnMe3	77	1.30			75K017
xx(IS/SnO2)	R	NaSnPh3 in di=	N	--	--	=methoxyethane	75B016
BaSnO3		Na4Pt(C2O4)CO(=	95	1.26	1.76	=Sn(C2O4)2)2, bonding info	75Y006
BaSnO3		Na6Pt(C2O4)2Sn=	95	1.47	1.82	=(C2O4)2)2, bonding info	75Y006
BaSnO3		Na8PtSn(C2O4)2=	95	1.50	1.76	=)5, bonding info	75Y006
BaSnO3		(Nh4)3Pt(SnF3)5		1.46	2.63	=Cl3)5, structure information	75V047
BaSnO3	R	o-OEtC6H4SnMe3	77	1.22			75K017
BaSnO3	R	p-OEtC6H4SnMe3	77	1.29			75K017
BaSnO3	R	m-OMeC6H4SnMe3	77	1.26			75K017
BaSnO3	R	o-OMeC6H4SnMe3	77	1.22			75K017
BaSnO3(IS/SnO2)		(PBD)2SnCr(CO)5	77	2.02	2.07	bonding information	75C046
BaSnO3(IS/SnO2)		(PBD)2SnMo(CO)5	77	1.93	2.51	bonding information	75C046
BaSnO3(IS/SnO2)		(PBD)2SnW(CO)5	77	1.90	2.39	bonding information	75C046
BaSnO3(IS/SnO2)		(PTD)2SnCr(CO)5	77	1.81	2.28	bonding information	75C046
BaSnO3(IS/SnO2)		(PTD)2SnMo(CO)5	77	1.82	2.30	bonding information	75C046
BaSnO3(IS/SnO2)		(PTD)2SnW(CO)5	77	1.80	2.35	bonding information	75C046
xx(IS/SnO2)	R	(PhCH2)2SnCl2	N	1.63	3.0		75B016
xx(IS/SnO2)	R	(PhCH2)2SnH2	N	1.42	.0		75B016
xx(IS/SnO2)	R	(PhCH2)3SnCl	N	1.52	2.82		75B016
xx(IS/SnO2)	R	(PhCH2)3SnH	N	1.43	.0		75B016
BaSnO3	R	PhCl2SnFe(CO)=	80	1.83	3.03	=CpPEt3, bonding information	75B043
BaSnO3	R	PhCl2SnFe(CO)=	80	1.80	3.00	=CpPPh3, bonding information	75B043
BaSnO3	R	PhCl2SnFe(CO)=	80	1.77	2.83	=CpP(OPh)3, bonding info	75B043
BaSnO3	R	PhCl2SnFeCp(P(=	80	1.75	2.90	=OPh)3)2, bonding information	75B043
BaSnO3		PhSO2CNNSnPh3	77	1.07	4.54	structure information	75K027
BaSnO3	R	PhSn(TPL)3	N	.48	1.95	structure info	75R018
SnO2	300	(Ph2ClP)2SnCl4	80	.70		structure information	75M075
BaSnO3	R	Ph2ClSnFe(CO)=	80	1.57	2.69	=CpP(OPh)3, bonding info	75B043
BaSnO3	R	Ph2ClSnFe(CO)=	80	1.63	2.59	=CpPEt3, bonding information	75B043
BaSnO3	R	Ph2ClSnFe(CO)=	80	1.63	2.74	=CpPPh3, bonding information	75B043
BaSnO3	R	Ph2ClSnFeCp(P(=	80	1.65	2.71	=OPh)3)2, bonding information	75B043
BaSnO3	300	(Ph2PNMe2)2SnCl4	80	.83	.4	=Cl4, bonding information	75K035
BaSnO3	300	(Ph2PONMe2)2Sn=	80	.54	.77	=Br4, bonding information	75K035
BaSnO3	300	(Ph2PONMe2)2Sn=	80	.27	.77	=Cl4, bonding information	75K035
SnO2	300	(Ph2POO)2SnBr2	80	.16	.5	structure information	75M075
SnO2	300	(Ph2POO)2SnCl2	80	.05		structure information	75M075
BaSnO3	300	(Ph2PSNMe2)2Sn=	80	.95	.3	=Br4, bonding information	75K035
BaSnO3	300	(Ph2PSNMe2)2Sn=	80	.60	.3	=Cl4, bonding information	75K035
BaSnO3		Ph2SnCl2-(BPE)	80	1.20	3.63	bonding and structure info	75M046
BaSnO3		Ph2SnCl2-(BPE)	80	1.17	3.67	bonding and structure info	75M046
BaSnO3		Ph2SnCl2-Bipy=	80	1.31	3.67	=O2,bonding and structure info	75M046
BaSnO3		Ph2SnCl2-(DPE)	80	1.31	3.27	bonding and structure info	75M046
BaSnO3		Ph2SnCl2-(TMTU)	80	1.29	2.60	bonding and structure info	75M046
BaSnO3		Ph2SnCl2-diars	80	1.30	3.07	bonding and structure info	75M046
BaSnO3		Ph2SnCl2-(dpe)	80	1.29	4.00	bonding and structure info	75M046
BaSnO3		Ph2Sn(NCS)2(=	80	1.10	3.60	=Bu3PO)2, bond & struc info	75M046
BaSnO3		Ph2Sn(NCS)2(=	80	1.07	3.73	=HMPT)2, bond & struc info	75M046
BaSnO3		Ph2Sn(NCS)2(=	80	1.13	3.79	=Ph3PO)2, bond & struc info	75M046
BaSnO3		Ph2Sn(NCS)2(=	80	1.18	3.71	=Ph3AsO)2, bond & stru info	75M046
BaSnO3		Ph2Sn(NCS)2(=	80	1.20	3.39	=TMTU)2, bond & struc info	75M046
BaSnO3		Ph2Sn(NCS)2-=	80	1.21	4.12	=(dpe), bonding & struc info	75M046
BaSnO3		(Ph2Sn(NCS)2)2=	80	.77	2.10	=BipyO2, bonding & struc info	75M046
xx		Ph2Sn(PhCOC(CN=		--	--	=)2)2, structure information	75K075
xx		Ph2Sn(PhCONCN)2		1.08	1.96	structure information	75K075
BaSnO3		Ph3CSn(acac)2Br	77	.86	1.37		75B145
BaSnO3	R	Ph3Sn-Mn(CO)4=	79	1.38		=AsPh3	75O001
BaSnO3	R	Ph3Sn-Mn(CO)4P=	79	1.43		=Ph3	75O001
BaSnO3	R	Ph3Sn-Mn(CO)4=	79	1.40		=SbPh3	75O001
BaSnO3	R	Ph3Sn-Mn(CO)5	79	1.41			75O001
BaSnO3		Ph3Sn(Bzac)	110	1.08	2.25	structure information	75B138
CaSnO3		Ph3SnCdCl-TME	N	1.51	1.03	bonding information	75B115
BaSnO3	R	Ph3SnCl	110	1.32	2.54	structure information	75B044
BaSnO3	R	Ph3SnCl	295	1.21	2.46	structure information	75B044
xx		Ph3SnCl-PDM		1.26	3.10		75S097
xx		Ph3SnClPDP		1.36	3.19		75S097
BaSnO3	R	Ph3SnFe(CO)CpP=	80	1.51	.76	=Et3, bonding information	75B043
BaSnO3	R	Ph3SnFe(CO)CpP=	80	1.53	.66	=Ph3, bonding information	75B043
BaSnO3	R	Ph3SnFeCp(P(O=	80	1.49	.78	=Ph)3)2, bonding information	75B043
BaSnO3	R	Ph3SnMn(CO)5	110	1.35	.41	structure information	75B044
BaSnO3	R	Ph3SnMn(CO)5	295	1.23	--	structure information	75B044

SOURCE	S-TEMP	ABSORBER	A-TEMP	SHIFT	QS	REMARKS	REF
BaSnO3	R	Ph3SnOH-Ph3SnN3		1.29	3.25	polymeric structure	75B036
BaSnO3		Ph3Sn(PCA)	77	1.29	3.49	structure information	75H040
BaSnO3		Ph3Sn(PYD)	77	1.30	3.09	structure information	75H040
BaSnO3		Ph3Sn(PYD)	77	1.25	2.92	structure information	75H040
BaSnO3	R	Ph3Sn(TPL)2	v	--	--	structure info, 78<T<135 K	75R018
CaSnO3		Ph3SnZnCl-TME	N	1.57	1.20	bonding information	75B115
BaSnO3	R	Ph3Sn(acac)	110	1.09	1.38	structure information	75B138
BaSnO3	R	Ph3Sn(bzbz)	110	1.13	2.25	structure information	75B138
BaSnO3	R	Ph3Sn(bzbz)	110	1.13	2.25	structure information	75B044
BaSnO3	R	Ph3Sn(bzbz)	295	1.00	2.20	structure information	75B044
CaSnO3		(Ph3Sn)2Cd-Bipy	N	1.54	1.01	bonding information	75B115
CaSnO3		(Ph3Sn)2Cd-TME	N	1.54	1.05	bonding information	75B115
CaSnO3		(Ph3Sn)2Cd-phen	N	1.55	1.00	bonding information	75B115
BaSnO3		(Ph3Sn)2(PDA)	77	1.30	3.36	structure information	75H040
CaSnO3		(Ph3Sn)2Zn-TME	N	1.60	1.25	bonding information	75B115
Xx(IS/BaSnO3)	R	Ph4AsPh3SnN3NCS	81	1.21	2.97	structure information	75B009
Xx(IS/BaSnO3)	R	Ph4AsPh3Sn(N3)2	81	1.20	2.75	structure information	75B009
BaSnO3		(Ph4As)2B10H12=	77	1.75	3.13	=MeSnCl2, structure info	75G013
BaSnO3		(Ph4As)2B10H12=	77	3.17	1.26	=SnCl2, structure information	75G013
Xx(IS/BaSnO3)	R	(Ph4As)2Me2Sn(=	81	1.23	3.61	=N3)4, structure information	75B009
Xx(IS/BaSnO3)	R	(Ph4As)2Ph2Sn(=	81	1.10	3.72	=Nb)2(NCS)2, structure info	75B009
BaSnO3	R	Ph4Sn	110	1.21	0	structure information	75B044
BaSnO3	R	Ph4Sn	295	1.13	0	structure information	75B044
BaSnO3		PrSn(acac)2I	77	.90	2.09		75B145
SnO2	300	(Pr2POO)2SnCl2	80	.70		structure information	75M075
SnO2	300	(Pr2POSMe)2Me2=	80	1.43	4.20	=SnCl2, structure information	75M075
BaSnO3	R	Pr3Sn-Mn(CO)5	79	1.57	.85		75O001
BaSnO3		Pr3Sn(PAO)	77	1.43	2.63	structure information	75H040
BaSnO3		Pr3Sn(PCA)	77	1.49	3.88	structure information	75H040
BaSnO3		Pr3Sn(PYD)	77	1.44	3.38	structure information	75H040
BaSnO3		Pr3Sn(PYD)	77	1.41	3.22	structure information	75H040
BaSnO3		Pr3Sn(TPD)	77	1.43	1.99	structure information	75H040
BaSnO3	R	Pr3Sn(TPL)2	v	--	--	structure info, 78<T<145 K	75R018
BaSnO3(IS/SnO2)		Sn(BFD)2Fe(CO)4	77	1.37	2.03	bonding and structure info	75C048
Pd(IS/SnO2)	R	SnBr2Cr(CO)3Cp=	N	2.04	2.26	=W(CO)3Cp, eta=.95	75D030
Pd(IS/SnO2)	R	SnBr2Cr(CO)3Cp=	N	2.03	2.16	=Mo(CO)3Cp, eta=.98	75D030
Ph(IS/SnO2)	R	SnBr2(Cr(CO)3=	N	2.10	1.85	=Cp)2, pqs study, eta=1.00	75D030
BaSnO3		SnBr2(EAA)		3.73	.89	bonding and structure info	75S102
Ph(IS/SnO2)	R	SnBr2(Fe(CO)3=	N	1.85	2.13	=Cp)2, pqs study, eta=1.00	75D030
BaSnO3(IS/SnO2)		SnBr2Fe(CO)4	77	2.40	2.23	bonding and structure info	75C048
Ph(IS/SnO2)	R	SnBr2(Mn(CO)5)2	N	2.04	2.12	pqs study, eta=1.00	75D030
Ph(IS/SnO2)	R	SnBr2(Mo(CO)3=	N	2.03	2.35	=Cp)2, pqs study, eta=1.00	75D030
Ph(IS/SnO2)	R	SnBr2PhMn(CO)5	N	1.75	2.70	pqs study, eta=.91	75D030
xx(IS/BaSnO3)		SnBr2(Salen)	80	.41	.67	structure information	75P005
Ph(IS/SnO2)	R	SnBr2(W(CO)3=	N	2.01	2.42	=Cp)2, pqs study, eta=1.00	75D030
BaSnO3	R	SnBr2(salen)	N	.45	.65	structure information	75R016
Pd(IS/SnO2)	R	SnBr3Co(CO)4	N	1.49	1.30	pqs study	75D030
Pd(IS/SnO2)	R	SnBr3Cr(CO)3Cp	N	1.77	1.34	pqs study	75D030
Pd(IS/SnO2)	R	SnBr3Fe(CO)2Cp	N	1.76	1.61	pqs study	75D030
Pd(IS/SnO2)	R	SnBr3Mn(CO)5	N	1.78	1.47	pqs study	75D030
BaSnO3	R	SnBr3-Mn(CO)5	79	1.79	1.41		75O001
Pd(IS/SnO2)	R	SnBr3Mo(CO)3Cp	N	1.82	1.63	pqs study	75D030
Pd(IS/SnO2)	R	SnBr3W(CO)3Cp	N	1.80	1.62	pqs study	75D030
BaSnO3		SnBr4(C3H2S3)2	77	.994			75P011
BaSnO3		SnBr4(C5H6S3)2	77	1.013			75P011
BaSnO3		SnBr4(C7H4S4)2	77	1.098			75P011
BaSnO3		SnBr4(C9H6S3)2	77	.990			75P011
BaSnO3	R	SnBr4(EHA)	N	.63	.44	structure information	75R016
BaSnO3	R	SnBr4(Salen)	N	.55	.50	structure information	75R016
xx		SnBr3(-1)		3.71	1	bonding information	75O014
Pd(IS/SnO2)	R	SnBu3CoCO(EFP)3	N	1.38	0	pqs study	75D030
xx		SnCl2Br(-1)		3.51	1	bonding information	75O014
Pd(IS/SnO2)	R	SnCl2Cr(CO)2Cp=	N	2.10	2.46	=W(CO)3Cp, eta=1.00	75D030
Pd(IS/SnO2)	R	SnCl2Cr(CO)2Cp=	N	2.01	2.45	=Mo(CO)3Cp, eta=.98	75D030
Pd(IS/SnO2)	R	SnCl2Cr(CO)3Cp=	N	1.99	2.09	=Mo(CO)3Cp, eta=.98	75D030
Pd(IS/SnO2)	R	SnCl2Cr(CO)3Cp=	N	2.03	2.16	=Fe(CO)2Cp, eta=.93	75D030
Pd(IS/SnO2)	R	SnCl2Cr(CO)3Cp=	N	1.98	2.15	=W(CO)3Cp, eta=.96	75D030
Pd(IS/SnO2)	R	SnCl2(Cr(CO)3=	N	2.00	1.95	=Cp)2, pqs study, eta=1.00	75D030
BaSnO3		SnCl2(EAA)		3.63	1.09	bonding and structure info	75S102
Ph(IS/SnO2)	R	SnCl2Fe(CO)2Cp	N	1.95	2.41	pqs study, eta=1.00	75D030
BaSnO3(IS/SnO2)		SnCl2Fe(CO)4	77	2.01	2.09	bonding and structure info	75C048
BaSnO3		SnCl2(HPP)		3.23	1.72	bonding and structure info	75S102

SOURCE	S-TEMP	ABSORBER	A-TEMP	SHIFT	QS	REMARKS	REF
Ph(IS/SnO2)	R	SnCl2(Mn(CO)5)2	N	1.90	2.10	pqs study, eta=1.00	75D030
Pd(IS/SnO2)	R	SnCl2(Mo(CO)3=	N	2.03	2.39	=Cp)2, pqs study, eta=1.00	75D030
BaSnO3	R	SnCl2(NiCOCp)2		1.87	2.15	bonding information	75B001
BaSnO3		SnCl2(Salen)		3.29	1.67	bonding and structure info	75S102
xx(IS/BaSnO3)		SnCl2(Salen)	80	.31	.73	structure information	75P005
Pd(IS/SnO2)	R	SnCl2(W(CO)3=	N	1.96	2.55	=Cp)2, pqs study, eta=1.00	75D030
BaSnO3	R	SnCl2(salen)	N	.31	.68	structure information	75R016
BaSnO3	R	SnCl2(salphen)	N	.35	.68	structure information	75R016
BaSnO3		SnCl2(trew)		2.88	2.22	bonding and structure info	75S102
Pd(IS/SnO2)	R	SnCl3Cr(CO)3Cp	N	1.67	+1.54	pqs study, HA perturbation	75D030
Pd(IS/SnO2)	R	SnCl3Fe(CO)2Cp	N	1.71	1.81	pqs study	75D030
BaSnO3	R	SnCl3-Mn(CO)4=	79	1.66	1.74	=AsPh3	750001
BaSnO3	R	SnCl3-Mn(CO)4P=	79	1.72	1.61	=Ph3	750001
Pd(IS/SnO2)	R	SnCl3Mn(CO)5	N	1.62	1.57	pqs study	75D030
BaSnO3	R	SnCl3-Mn(CO)5	79	1.68	1.57		750001
Pd(IS/SnO2)	R	SnCl3Mo(CO)3Cp	N	1.69	1.71	pqs study	75D030
BaSnO3	R	SnCl3NiPPh3-C=		1.80	2.06	=H2Cl2, bonding information	75B001
BaSnO3	R	SnCl3NiPPh3Cp-=		1.81	1.97	=Me2CoO, bonding information	75B001
BaSnO3	R	SnCl3NiPPh3Cp		1.76	2.01	bonding information	75B001
BaSnO3	R	SnCl3Ni(PPh3)2=		3.15	1.29	=Cp, bonding information	75B001
BaSnO3	R	SnCl3Ni(PPh3)2=		3.15	1.23	=Cp-.5C6H6	75B001
BaSnO3	R	SnCl3Ni(PPh3)2=		3.11	1.29	=Cp-THF, bonding information	75B001
BaSnO3	R	SnCl3Ni(PPh3)2=		3.21	1.33	=Cp-MeOH, bonding information	75B001
BaSnO3	R	SnCl3Ni(PPh3)2=		3.19	1.24	=Cp-Me2CO, bonding information	75B001
BaSnO3	R	SnCl3Ni(PPh3)2=		3.18	1.20	=Cp-1.3CH2Cl2, bonding info	75B001
BaSnO3	R	SnCl3Ni(dppe)Cp		3.20	1.49	bonding information	75B001
Pd(IS/SnO2)	R	SnCl3W(CO)3Cp	N	1.67	1.77	pqs study	75D030
BaSnO3	R	SnCl4-Bipy	N	.46		bonding and structure info	74M062
BaSnO3		SnCl4(C2H4OC2=	95	.34	.78	=H4O)2, bonding info, fa=.55	75V040
BaSnO3		SnCl4(C2H6SO)2	95	.39	.34	bonding information, fa=.36	75V040
BaSnO3		SnCl4(C3H2S3)2	77	.773			75P011
BaSnO3		SnCl4(C4H8O)2	95	.47	1.23	bonding information, fa=.40	75V040
BaSnO3		SnCl4(C5H5N)2	95	.48		bonding information, fa=.39	75V040
BaSnO3		SnCl4(C5H6S3)2	77	.804			75P011
BaSnO3		SnCl4(C5H11N)2	95	.57		bonding information, fa=.42	75V040
BaSnO3		SnCl4(C7H7S4)2	77	.749			75P011
BaSnO3		SnCl4(C9H6S3)2	77	.683			75P011
BaSnO3		SnCl4(C10H8OS3=	77	.679		=)2	75P011
BaSnO3	R	SnCl4(EHA)	N	.33	.23	structure information	75R016
BaSnO3		SnCl4(EtOEt)2	95	.51	1.37	bonding information	75V040
BaSnO3		SnCl4(EtOH)2	95	.41	.89	bonding information	75V040
BaSnO3		SnCl4(HCONC2=	95	.36	.73	=H6)2, bonding info, fa=.39	75V040
BaSnO3		SnCl4(MeCN)2	95	.33	.64	bonding information, fa=.35	75V040
BaSnO3		SnCl4(MeOH)2	95	.37	.73	bonding information	75V040
BaSnO3		SnCl4(Me2S)2	95	.58	.40	bonding information	75V040
BaSnO3		SnCl4(P-H3CO=	95	.35	.26	=C5H4N2Ph)2, fa=.41, bond info	75V040
BaSnO3		SnCl4(PhCHNC6=	95	.18		=H4-p-NO2)2, bonding info	75V040
BaSnO3		SnCl4(PhCHNC6=	95	.31		=H4-p-Me)2, bonding info	75V040
BaSnO3		SnCl4(PhCHNPh=	95	.40	.37	=)2, fa=.43, bonding info	75V040
BaSnO3		SnCl4(PhN2Ph)2	95	.38	.48	bonding information, fa=.52	75V040
BaSnO3		SnCl4PhNPPh3	95	.31		bonding information, fa=.42	75V040
BaSnO3		SnCl4Ph3PNC6=	95	.29		=H4-p-NO2, fa=.22, bond info	75V040
BaSnO3		SnCl4Ph3PNC6=	95	.33		=H4-p-Me, fa=.22, bonding info	75V040
BaSnO3		SnCl4(PrOH)2	95	.27	.34	bonding information	75V040
BaSnO3	R	SnCl4(Salen)	N	.38	.28	structure information	75R016
BaSnO3		SnCl4(p-BrC6=	95	.34	.42	=H4N2Ph)2, fa=.48, bond info	75V040
BaSnO3		SnCl4(p-MeOC6=	95	.23		=H4CHNC6H4-p-Me)2, bond info	75V040
BaSnO3		SnCl4(p-O2NC6=	95	.25		=H4CHPh)2, bonding information	75V040
BaSnO3	R	SnCl4-phen	N	.43		bonding and structure info	74M062
xx		SnCl3(-1)		3.31	1.13	bonding information	750014
BaSnO3	R	SnCl4-2ACR	N	.27		bonding and structure info	74M062
BaSnO3	R	SnCl4-2IQL	N	.50		bonding and structure info	74M062
BaSnO3	R	SnCl4-2NAQ	N	.37	.52	bonding and structure info	74M062
BaSnO3	R	SnCl4-2Py	N	.44		bonding and structure info	74M062
BaSnO3	R	SnCl4-2QL	N	.39		bonding and structure info	74M062
BaSnO3		SnCl4(3-MeC5H4=	95	.36	.73	=N)2, bonding information	75V040
BaSnO3(IS/SnO2)		Sn(DPP)2Fe(CO)4	77	1.38	1.73	bonding and structure info	75C048
BaSnO3		Sn(HFP)	77	3.60	1.66	bonding and structure info	75E017
BaSnO3		Sn(HFP)-Bipy	77	3.15	.90	bonding and structure info	75E017
BaSnO3		Sn(HFP)Cl	77	3.56	1.71	bonding and structure info	75E017
BaSnO3		Sn(HFP)-phen	77	3.55	1.34	bonding and structure info	75E017
Ph(IS/SnO2)	R	SnI2(Cr(CO)3=	N	2.19	1.80	=Cp)2, pqs study, eta=1.00	75D030

SOURCE	S-TEMP	ABSORBER	A-TEMP	SHIFT	QS	REMARKS	REF
Ph(IS/SnO2)	R	SnI2(Fe(CO)3=	N	2.00	2.27	=Cp)2, pqs study, eta=1.00	75D030
Ph(IS/SnO2)	R	SnI2(Mo(CO)3=	N	2.09	2.06	=Cp)2, pqs study, eta=1.00	75D030
Ph(IS/SnO2)	R	SnI2(W(CO)3Cp)2	N	2.11	2.38	pqs study, eta=1.00	75D030
BaSnO3	R	SnI2(salen)	N	.55	.60	structure information	75R016
Pd(IS/SnO2)	R	SnI3Co(CO)4	N	1.71	.81	pqs study	75D030
Pd(IS/SnO2)	R	SnI3Cr(CO)3Cp	N	1.80	1.06	pqs study	75D030
Pd(IS/SnO2)	R	SnI3Fe(CO)2Cp	N	1.97	1.48	pqs study	75D030
BaSnO3	R	SnI3-Mn(CO)5	79	1.92	1.32		750001
Pd(IS/SnO2)	R	SnI3Mo(CO)3Cp	N	1.87	1.27	pqs study	75D030
Pd(IS/SnO2)	R	SnI3W(CO)3Cp	N	1.88	1.35	pqs study	75D030
xx		SnI4-(BZ)(Me3=	80	--		=SiOCHMe2, bond & struc info	75C022
xx		SnI4-(BZ)(Me3=	80	--		=SiO(CH2)2Me, bond & struc	75C022
xx		SnI4-(BZ)(Me3=	80	--		=SiOMe, bond & structure info	75C022
xx		SnI4-C6H6	80	1.60		bonding and structure info	75C022
xx		SnI4-CC14	80	1.60		bonding and structure info	75C022
BaSnO3	R	SnI4(EHA)	N	1.03	.62	structure information	75R016
xx		SnI4-MeC2H3Si(=	80	1.52		=OEt)2, bond & struc info	75C022
xx		SnI4-MePhSi(O=	80	1.52		=Et)2, bonding & struc info	75C022
xx		SnI4-(Me2C2H3=	80	--		=Si)O, bond & structure info	75C022
xx		SnI4-(Me2PhSi)=	80	.85		=2O, bonding & structure info	75C022
xx		SnI4-Me2Si(OEt=	80	1.52		=)2, bonding & structure info	75C022
xx		SnI4-Me3SiOCH=	80	--		=Me2, bonding & structure info	75C022
xx		SnI4-Me3SiO(C=	80	.90		=H2)2Me, bond & structure info	75C022
xx		SnI4-Me3SiOCMe3	80	--		bonding and structure info	75C022
xx		SnI4-Me3SiOEt	80	--		bonding and structure info	75C022
xx		SnI4-Me3SiOMe	80	--		bonding and structure info	75C022
xx		SnI4-Me3SiOSi=	80	.40		=PhMe2, bond & structure info	75C022
xx		SnI4-(Me3Si)2O	80	1.32		bonding and structure info	75C022
BaSnO3	R	SnI4(Salen)	N	.98	.59	structure information	75R016
Pd(IS/SnO2)	R	SnIzCr(CO)3Cp=	N	2.10	2.00	=Mo(CO)3Cp, eta=.97	75D030
Pd(IS/SnO2)	R	SnIzCr(CO)3Cp=	N	2.19	1.80	=W(CO)3Cp, eta=.93	75D030
BaSnO3	R	o-SnMe3C6H4Sn=	R	1.28		=Me3	75K017
BaSnO3	R	p-SnMe3C6H4Sn=	R	1.22		=Me3	75K017
Pd(IS/SnO2)	R	Sn(NCS)2(Fe(CO=	N	1.82	2.58	=)2Cp)2, pqs study	75D030
BaSnO3		Sn(O2CC6H4Cl-o=	77	3.40	1.91	=)2	75E018
BaSnO3		Sn(O2CC6H4Cl-m=	77	3.49	1.69	=)2	75E018
BaSnO3		Sn(O2CC6H4Cl-p=	77	3.35	1.83	=)2	75E018
BaSnO3		Sn(O2CC6H4Br-o=	77	3.38	1.93	=)2	75E018
BaSnO3		Sn(O2CC6H4NH2-=	77	3.26	1.79	=o)2	75E018
BaSnO3		Sn(O2CC6H4NH2-=	77	3.37	1.89	=p)2	75E018
BaSnO3		Sn(O2CC6H4CMe3=	77	3.26	2.03	=-p)2	75E018
BaSnO3		Sn(O2CC6H4NO2-=	77	3.38	2.03	=p)2	75E018
BaSnO3		Sn(O2CC6H4Me-p=	77	3.30	1.85	=)2	75E018
BaSnO3		Sn(O2CC6H4Me-o=	77	3.26	1.86	=)2	75E018
BaSnO3		Sn(O2CC6H4Me-m=	77	3.32	1.88	=)2	75E018
BaSnO3		Sn(O2CC6H5)2	77	3.46	1.75		75E018
BaSnO3		Sn(O2CC10H7)2	77	3.35	2.00		75E018
BaSnO3		Sn(O2CEt)2	77	3.43	1.83		75E018
BaSnO3		Sn(O2CMe)2	77	3.31	1.77		75E018
BaSnO3		Sn(O2CPh)2	77	3.17	1.69		75E018
BaSnO3		Sn(O2SC6H4Me-p=	77	3.69	1.69	=)2	75E018
BaSnO3		Sn(O3SC6H4NO2-=	77	3.25	2.54	=m)2	75E018
BaSnO3		Sn(OBu)2		2.87	1.97	structure informaiton	75G031
BaSnO3	77	Sn(OC4H5COCH)2	77	2.99	2.25	bonding and structure info	75C047
BaSnO3(IS/SnO2)		Sn(OC6H4Cl-p)2		3.20	1.67	structure information	75E019
BaSnO3(IS/SnO2)		Sn(OC6H4Me-m)2		3.10	1.81	structure information	75E019
BaSnO3(IS/SnO2)		Sn(OC6H4Me-o)2		3.13	1.97	structure information	75E019
BaSnO3(IS/SnO2)		Sn(OC6H4Me-p)2		3.21	1.74	structure information	75E019
BaSnO3(IS/SnO2)		Sn(OC6H4Me2-2,=		3.13	1.97	=6)2, structure information	75E019
BaSnO3(IS/SnO2)		Sn(OC6H4Me2-3,=		3.14	1.69	=5)2, structure information	75E019
BaSnO3(IS/SnO2)		Sn(OC6H4N-3)2		2.96	1.93	structure information	75E019
BaSnO3(IS/SnO2)		Sn(OC6H4NH2-p=		3.15	1.69	=)2, structure information	75E019
BaSnO3(IS/SnO2)		Sn(OC6H4NH2-m=		3.52	1.85	=)2, structure information	75E019
BaSnO3(IS/SnO2)		Sn(OC6H4NO2-p=		3.42	1.47	=)2, structure information	75E019
BaSnO3	77	Sn(OC6H7O)2	77	2.96	2.12	bonding and structure info	75C047
BaSnO3	77	Sn(OC7H5O)2	77	2.91	1.95	bonding and structure info	75C047
BaSnO3	77	Sn(OCMeCHCOBuO=	77	3.03	2.14	=)2, bonding & structure info	75C047
BaSnO3	77	Sn(OCMeCHCOEtO=	77	3.15	2.19	=)2, bonding & structure info	75C047
BaSnO3	77	Sn(OCMeCHCOMeO=	77	3.07	2.13	=)2, bonding & structure info	75C047
BaSnO3	77	Sn(OCPhCHCOMeO)2	77	3.07	1.98	bonding and structure info	75C047
BaSnO3	77	Sn(OCPhCHCPhO)2	77	3.31	1.80	bonding and structure info	75C047
BaSnO3		Sn(OEt)2		2.85	1.92	structure informaiton	75G031

SOURCE	S-TEMP	ABSORBER	A-TEMP	SHIFT	QS	REMARKS	REF
BaSnO3		Sn(OMe)Br	77	3.21	1.91		75E018
BaSnO3		Sn(OMe)Cl	77	3.00	2.28		75E018
BaSnO3		Sn(OMe)I	77	3.66	1.46		75E018
BaSnO3		Sn(OMe)2		2.80	2.02	structure information	75G031
BaSnO3		Sn(OMe)2	77	2.92	1.89		75E018
BaSnO3		Sn(OMe)2	77	2.92	1.89	bonding and structure info	75E017
Pd(IS/SnO2)	R	Sn(OSOPh)2(Fe=	N			=(CO)2Cp)2, pqs study	75D030
BaSnO3(IS/SnO2)		Sn(PBD)2Fe(CO)4	77	1.45	1.91	bonding and structure info	75C048
BaSnO3(IS/SnO2)		Sn(PTD)2Fe(CO)4	77	1.46	1.85	bonding and structure info	75C048
Ph(IS/SnO2)	R	SnPhCl2Fe(CO)2=	N	1.71	2.65	=Cp, pqs study, eta=.98	75D030
BaSnO3	R	Sn(SCH2)4		1.50	1.00	structure information	75D003
Pd(IS/SnO2)	R	Sn(SPh)2(Fe(CO=	N	1.47	1.49	=)2Cp)2, pqs study	75D030
BaSnO3(IS/SnO2)		Sn(TBD)2Fe(CO)4	77	1.47	1.49	bonding and structure info	75C048
BaSnO3		Sn(TFD)	77	3.40	1.87	bonding and structure info	75E017
BaSnO3		Sn(TFD)-Bipy	77	3.05	1.02	bonding and structure info	75E017
BaSnO3		Sn(TFD)-phen	77	3.47	1.44	bonding and structure info	75E017
BaSnO3		Sn(acac)2	77	3.10	2.02	bonding and structure info	75E017
BaSnO3		Sn2(O2CC6H4NO2=	77	--	--	=)40-THF	75E018
BaSnO3(IS/SnO2)		(TBD)2SnCr(CO)5	77	1.95	2.29	bonding information	75C046
BaSnO3(IS/SnO2)		(TBD)2SnMo(CO)5	77	1.93	2.42	bonding information	75C046
BaSnO3(IS/SnO2)		(TBD)2SnW(CO)5	77	2.00	2.31	bonding information	75C046
BaSnO3	R	(TPL)2SnCl2	N	.25		structure info	75R018
XX		To2Sn(MeCOC(CN=		--	--	=)2)2, structure information	75K075
XX		To2Sn(PhCOC(CN=		--	--	=)2)2, structure information	75K075
XX		To2Sn(PhCONCN)2		1.53	2.56	structure information	75K075
BaSnO3	R	To3Sn(TPL)2	V	--	--	structure info, 78<T<138 K	75R018
BaSnO3		Vin2Sn(NCS)2-=	80	1.10	3.83	=(BPE), bonding & struc info	75M046
BaSnO3		Vin2Sn(NCS)2-=	80	1.19	4.01	=(dpe), bonding & struc info	75M046
XX		W(CO)3(SnCl3)3=		2.10	1.87	=(-3), bonding information	75O014
XX		W(CO)5SnBr3(-1)		2.11	1.97	bonding information	75O014
XX		W(CO)5SnCl3(-1)		1.87	2.00	bonding information	75O014
XX		W(CO)5SnF2Cl=		1.80	2.32	=(-1), bonding information	75O014
XX(IS/BaSnO3)		(diphos)Mo(CO)=		--	--	=4SnCl3SnCl5OH2	75C043
XX(IS/BaSnO3)		(diphos)W(CO)4=				=SnCl3SnCl5OH2	75C043
BaSnO3		(phen)Mo(CO)3=	80	1.43	2.53	=Ph2ClCl, bonding & struc info	75C013
BaSnO3		(phen)Mo(CO)3=	80	1.54	2.53	=SnMeCl2Cl, bond & struc info	75C013
BaSnO3		(phen)Mo(CO)3=	80	1.40	2.26	=SnPhCl2Cl, bond & struc info	75C013
BaSnO3		(phen)W(CO)3Sn=	80	1.66	1.66	=Cl3Cl, bonding & struc info	75C013
BaSnO3		(phen)W(CO)3Sn=	80	1.61	2.64	=MeCl2Cl, bonding & struc info	75C013
BaSnO3		(phen)W(CO)3Sn=	80	1.44	2.32	=PhCl2Cl, bonding & stru info	75C013
SnO2		2SnCl4-(EtCONH=		.42		=CO)2Ph, structure information	75S104
SnO2		2SnCl4-(MeCONH=		.42		=CO)2Ph, structure information	75S104
SnO2		4-H-2-O2NC6H3S=	77	1.32	.76	=CH2CHSCNSnPh3	75W022
SnO2		4-H-2-O2NC6H3S=	77	1.32	.51	=CH2CHClSnPh3	75W022
SnO2		4-Me-2-O2NC6H3=	77	1.29	.46	=SCH2CHSCNSnPh3	75W022
SnO2		4-NO2-2O2NC6=	77	1.30	.55	=H3SCH2CHSCNSnPh3	75W022
BaSnO3	R	2,4-CF3C6H4Sn=	R	1.22		=Me3	75K017
BaSnO3	R	2,4-MeC6H4SnMe3	R	1.23			75K017
BaSnO3	R	2,4,6-OMeC6H4=	R	1.21		=SnMe3	75K017
BaSnO3	R	2,6-CF3C6H4Sn=	R	1.26	.90	=Me3	75K017
BaSnO3	R	2,6-ClC6H4SnMe3	R	1.27	.87		75K017
BaSnO3	R	2,6-FC6H6SnMe3	77	1.32	1.00		75K017
BaSnO3	R	2,6-MeC6H4SnMe3	R	1.33			75K017
BaSnO3	R	2,6-OEtC6H4Sn=	R	1.20		=Me3	75K017
BaSnO3	R	2,6-OMeC6H4Sn=	R	1.28		=Me3	75K017
BaSnO3	R	3,4-MeC6H4SnMe3	R	1.30			75K017
BaSnO3	R	3,5-CF3C6H4Sn=	R	1.34	.61	=Me3	75K017

INORGANIC OXIDES

SOURCE	S-TEMP	ABSORBER	A-TEMP	SHIFT	QS	REMARKS	REF
BaSnO3(IS/SnO2)		BaSn(OH)2OSn=	77	2.41	1.86	=(OH)2	75T013
BaSnO3	R	BaTiO3-BaSnO3	V			77 K<T<490 K, polymeric transformation	75S019
CaSnO3	R	CaMnO3	4			HI=-75 kOe(T=0 K)	75T016
XX		CaxY(3-x)Fe(5-=				=x)SnxO12	75N026

SOURCE	S-TEMP	ABSORBER	A-TEMP	SHIFT	QS	REMARKS	REF
CaSnO3	R	Ca(1-x)SrxMnO3	4			HI data	75T016
xx		Cd2Nb(2-2x)Sn=	v	--	--	=2xO(7-2x)F2x, 4.2 K<T<400 K, study of phase transition	75P008
BaSnO3	300	alpha-Fe2O3	80			HI=130 kG	75R025
BaSnO3	300	alpha-Fe2O3	300			HI=126 kG	75R025
BaSnO3	295	alpha-Fe2O3-Sn	v	--		80 K<T<960 K, HI data	75R011
BaSnO3	300	Fe3O4	80			HI=242 kG	75R025
BaSnO3	300	Fe3O4	300			HI=211 kG	75R025
SnO2	78	In2O3 + SnO2		0	1.2	study of the product formed	75V039
Pd		LaFeO3				HI data	75O003
xx		Nd.95Ca.05Fe=				=.95Sn.0503, T=78 & 300 K	74L049
Pd		NdFeO3				HI data	75O003
xx(IS/BaSnO3)	R	SnO	78	2.79	1.43	lattice dynamics data	75H022
xx(IS/BaSnO3)	R	SnO	295	2.95	1.43	lattice dynamics data	75H022
BaSnO3	295	SnO in N2	v			matrix isolation, study of the reactivity and diffusion	75B006
BaSnO3(IS/SnO2)		SnO-.4H2O	77	2.85	2.04		75T013
BaSnO3(IS/SnO2)		SnO(black)	77	2.71	1.45		75T013
BaSnO3	295	(SnO)n in N2	v			matrix isolation, study of the reactivity and diffusion	75B006
BaSnO3(IS/SnO2)		SnO(red)	77	2.60	2.20		75T013
BaSnO3	R	SnP2O7	v	--	--	bonding and structure info	75H005
xx		SnTe3O8				Sn is Sn(IV)	75T023
BaSnO3	R	alpha-SnWO4	77	3.48	1.15	bonding and structure info	75B090
BaSnO3	R	beta-SnWO4	77	3.53	1.38	bonding and structure info	75B090
BaSnO3	295	Sn2O2 in N2	v			matrix isolation, study of the reactivity and diffusion	75B006
BaSnO3(IS/SnO2)	R	Sn(2-x)Nb(2-y)=	v	--	--	=SnyO(7-x-y/2), 4<T<295	75B017
BaSnO3(IS/SnO2)	R	Sn(2-x)Ta(2-y)=	v	--	--	=SnyO(7-x-y/2), 4<T<295	75B017
CaSnO3	R	SrMnO3	4			HI=+20 kOe(T=0 K)	75T016
xx		Y(3-x)CaxFe(K-=	v			=x)SnxO12(x=.7&.9), some HI above Neel temperature	75L031
Pd		YFeO3				HI data	75O003
SnO2		Zn2SnO4				Sn is Sn(IV)	75I011
BaSnO3		xMg2SnO4-(1-x)=		--	--	=Mg3(BO3)2	75M088

SOURCE EXPERIMENTS

SOURCE	S-TEMP	ABSORBER	A-TEMP	SHIFT	QS	REMARKS	REF
Al	v	BaSnO3					75I015
AlZn	v	BaSnO3				eutectoid mixture, grain boundary segregation, fs data	75I015
BaSnO3	77	alpha-Sn				fs=.62	73F033
BaSnO3	293	alpha-Sn				fs=.50	73F033
BaTiO3		xx				anharmonic atomic vibrations	74P057
C		SnO2					74W027
CaS		xx		--	--	2 Sn sites	75R021
CuMgSn	293	alpha-Sn				fs=.26	73F033
CuMgSn	77	alpha-Sn				fs=.8	73F033
Cu4MgSn	77	alpha-Sn				fs=.36	73F033
Cu4MgSn	293	alpha-Sn				fs=.32	73F033
Ge		xx(IS/SnO2)		--		ion implanted, bonding info	75W026
Ge	80	Mg2Sn		--		structure of impurity site	75S093
Ge		SnO2					74W027
H6TeO6		BaSnO3		-.02	--	study of oxidation state of Sn after successive EC decays	73A046
Mg2Sn	293	alpha-Sn				fs=.29	73F033
Mg2Sn	77	alpha-Sn				fs=.72	73F033
Mg2Sn	v	alpha-Sn		--		77<T<293 K, study of different source preparation	73F033
Mg2Sn(Pb)	v	alpha-Sn					73F033
Mg2Sn(Pb)	v	alpha-Sn					73F033
Mg2Sn(Si)	v	alpha-Sn					73F033
NaCl		xx		--	--	2 Sn sites	75R021
PbTeO3		xx				anharmonic atomic vibrations	74P057
PbZrO3		xx				anharmonic atomic vibrations	74P057

SOURCE	S-TEMP	ABSORBER	A-TEMP	SHIFT	QS	REMARKS	REF
Pb(1-x)SnxTe=		beta-Sn				=(1-x)Sex, bond & struc info	75B032
Pd		xx				discussion of previous results	75P023
Sb	v	BaSnO3				91<T<300 K, fs vs temperature	75A029
SbOHC2O4	78	BaSnO3	78	--	--		75A005
Si		xx(IS/SnO2)		--		ion implanted, bonding info	75W026
Si	80	Mg2Sn		--		structure of impurity site	75S093
Si(n-type)		SnO2		--		comparison with channeling exp	75W002
alpha-Sn	4.2	PdSn(IS/BaSnO3)	240	2.04			75V015
beta-Sn	v	BaSnO3					75I015
beta-Sn	N	BaSnO3	R	2.56		nuclear radius experiment	75R010
beta-Sn	N	BaSnO3	R	2.60		nuclear radius experiment	75R010
SnCl2	N	BaSnO3	R	3.88		nuclear radius experiment	75R010
SnI4	N	BaSnO3	R	1.48		nuclear radius experiment	75R010
SnS	N	BaSnO3	R	3.31		nuclear radius experiment	75R010
SnS2	N	BaSnO3	R	1.06		nuclear radius experiment	75R010
Sn(SO4)2	N	BaSnO3	R	.06		nuclear radius experiment	75R010
Sn(SO4)2	N	BaSnO3	R	.02		nuclear radius experiment	75R010
SnSe		beta-Sn		-3.42	-.6	bonding and structure info	75B032
SnTe		beta-Sn		-3.58		bonding and structure info	75B032
Te(Sn)		beta-Sn		-3.68		bonding and structure info	75B032
Te(Sn)		HeBaSnO3	He	--	--	ion implantation, struc info	75V023
Zn	v	BaSnO3				Debye T=128 K	75I015
diamond		xx(IS/SnO2)		--		ion implanted, bonding info	75W026

INORGANIC SULFATES

SOURCE	S-TEMP	ABSORBER	A-TEMP	SHIFT	QS	REMARKS	REF
BaSnO3		SnSO4		4.3			75R034
BaSnO3(IS/SnO2)		Sn3O(OH)2SO4	77	2.57	2.00		75T013

INORGANIC SULFIDES

SOURCE	S-TEMP	ABSORBER	A-TEMP	SHIFT	QS	REMARKS	REF
BaSnO3	R	AgCrSnS4		1.29			75K010
BaSnO3	R	Ag2CdSnS4		1.28			75K010
BaSnO3	R	Ag2FeSnS4		1.31			75K010
BaSnO3	R	Ag8SnS6					75K010
Mg2Sn		As2S3	80	1.32	.40	electronic structure info	75S033
Mg2Sn		As2S3(glass)	80	1.58	.40	electronic structure info	75S033
BaSnO3	R	BaSnS3-nH2O	N	1.13	.73	bonding and structure info	75I007
BaSnO3	R	Ba2SnS3	N	1.25	.75	bonding and structure info	75I007
BaSnO3	R	Ba2SnS4	N	1.29		bonding and structure info	75I007
BaSnO3	R	Ba2SnS4-nH2O	N	1.19		bonding and structure info	75I007
BaSnO3	R	CaSnS3-nH2O	N	1.03	.62	bonding and structure info	75I007
BaSnO3	R	Ca2SnS3	N	1.01	.52	bonding and structure info	75I007
BaSnO3	R	Ca2SnS4	N	1.14		bonding and structure info	75I007
BaSnO3	R	Ca2SnS4-nH2O	N	1.28		bonding and structure info	75I007
xx		CoCr2S4				HI=-405 kOe	75L007
xx(IS/SnO2)		CoCr2S4	v			Curie temperature=206 K	75L013
BaSnO3	R	CuAgCdSnS4		1.31			75K010
BaSnO3	R	CuAgFeSnS4		1.38			75K010
BaSnO3	R	CuCrSnS4		1.25			75K010
xx		CuCr2S4				HI=+530 kOe	75L007
xx(IS/SnO2)		CuCr2S4	v			Curie temperature=344 K	75L013
BaSnO3	R	Cu2CdSnS4		1.42			75K010
BaSnO3	R	Cu2CoSnS4		1.52			75K010
BaSnO3	R	Cu2FeSnS4		1.54			75K010
BaSnO3	R	Cu2HgSnS4		1.40			75K010
BaSnO3	R	Cu2MnSnS4		1.44			75K010
xx(IS/SnO2)		Cu2MnSnS4	1.6			HI=12 kOe	74C039

SOURCE	S-TEMP	ABSORBER	A-TEMP	SHIFT	QS	REMARKS	REF
xx(IS/SnO2)		Cu2MnSnS4	4.2			HI=12 kOe	74C039
xx(IS/SnO2)		Cu2MnSnS4	78	1.48			74C039
xx(IS/SnO2)		Cu2MnSnS4	R	1.37			74C039
BaSnO3	R	Cu2NiSnS4		1.56	.90		75K010
BaSnO3	R	Cu2SnS3		1.61	.87		75K010
BaSnO3	R	Cu2SnS(3-x)Sex		--	--		75K010
BaSnO3	R	Cu2ZnSnS4		1.41			75K010
BaSnO3	R	Cu3SnS4		1.56	.64		75K010
BaSnO3	R	Cu8SnS6					75K010
xx		FeCrS4				HI=-470 kOe	75L007
xx(IS/SnO2)		FeCr2S4	v			Curie temperature=162 K	75L013
BaSnO3	R	K2SnS3	N	1.27	1.13	bonding and structure info	75I007
BaSnO3	R	K2SnS3-nH2O	N	1.34	1.15	bonding and structure info	75I007
BaSnO3	R	MgSnS3-nH2O	N	1.15	.53	bonding and structure info	75I007
BaSnO3	R	Mg2SnS3	N	1.05	.43	bonding and structure info	75I007
BaSnO3	R	Mg2SnS4	N	1.08		bonding and structure info	75I007
BaSnO3	R	Mg2SnS4-nH2O	N	1.20		bonding and structure info	75I007
BaSnO3	R	Na2SnS1.5Se1.5	N	1.50	1.05	bonding and structure info	75I007
BaSnO3	R	Na2SnS2Se	N	1.36	1.20	bonding and structure info	75I007
BaSnO3	R	Na2SnS3	N	1.37	1.08	bonding and structure info	75I007
BaSnO3	R	Na2SnS3(alpha)	N	1.15		bonding and structure info	75I007
BaSnO3	R	Na2SnS3(beta)	N	1.29	1.13	bonding and structure info	75I007
BaSnO3	R	Na2SnS3-nH2O	N	1.37	1.15	bonding and structure info	75I007
BaSnO3	R	Na2SnSSe2	N	1.51	1.15	bonding and structure info	75I007
BaSnO3	R	Na2SnSSe3	N	1.57	.98	bonding and structure info	75I007
BaSnO3	R	Na4SnS2Se2	N	1.33		bonding and structure info	75I007
BaSnO3	R	Na4SnS4	N	1.32		bonding and structure info	75I007
BaSnO3	R	Na4SnS4	N	1.20		bonding and structure info	75I007
BaSnO3	R	Na4SnS4-14H2O	N	1.17		bonding and structure info	75I007
BaSnO3	R	Na4SnSSe4	N	1.42		bonding and structure info	75I007
BaSnO3	R	Na6Sn2S3Se4	N	1.43	.71	bonding and structure info	75I007
BaSnO3	R	Na6Sn2S7	N	1.34	.64	bonding and structure info	75I007
BaSnO3	R	PbSnS2Se					75K010
BaSnO3	R	PbSnS3		1.10			75K010
xx(IS/BaSnO3)	R	Sn-TaS2	78	3.192	1.145	lattice dynamics data	75H022
xx(IS/BaSnO3)	R	Sn-TaS2	295	3.149	1.24	lattice dynamics data	75H022
xx(IS/beta-Sn)		Sn.33TaS2		1.23		structure information	75K029
xx(IS/beta-Sn)		Sn.56NbS2	R	1.06	--	structure information	75K029
xx(IS/BaSnO3)	R	SnS	78	3.46	.93	lattice dynamics data	75H022
xx(IS/BaSnO3)	R	SnS	295	3.53	1.05	lattice dynamics data	75H022
CaSnO3		SnS	v	--	--	77 K<T<300 K, fa vs temp	750004
Mg2Sn		SnS2	80	1.30	.40	electronic structure info	75S033
BaSnO3	R	SnS2	v	--	--	study of thermal decomposition 298<T<963	75K010
BaSnO3	R	SnSSe	v			298<T<963 K, study of thermal decomposition	75K010
xx(IS/BaSnO3)	R	Sn3TaS2	78	3.941		lattice dynamics data	75H022
xx(IS/BaSnO3)	R	Sn3TaS2	295	3.851		lattice dynamics data	75H022
BaSnO3	R	SrSnS3-nH2O	N	1.15	.63	bonding and structure info	75I007
BaSnO3	R	Sr2SnS3	N	1.18	.62	bonding and structure info	75I007
BaSnO3	R	Sr2SnS4	N	1.21		bonding and structure info	75I007
BaSnO3	R	Sr2SnS4-nH2O	N	1.17		bonding and structure info	75I007
Mg2Sn		TlAsS2	80	1.36	.40	electronic structure info	75S033
Mg2Sn		TlAsS2(glass)	80	1.58	.40	electronic structure info	75S033

TERRESTRIAL AND EXTRATERRESTRIAL MINERALS

SOURCE	S-TEMP	ABSORBER	A-TEMP	SHIFT	QS	REMARKS	REF
xx		Sn ore	88			SnO2 analysis, discussion of a systematic error	75P019

CODE	TOPIC	REFERENCE

73A046   SN-119   S Ambe and F Ambe, Radiochim Acta 20,141(1973), A Mossbauer Study of the Oxidation State of 119Sn after the Successive EC Decays of 119mTe in Telluric Acid

73F033   SN-119   D S Faleev, Mossbauer Effect on Tin-119 Nuclei in Some Magnesium Alloys, in "Mater Nauchn Konf Molodykh Uch Kafedr Inst - Khabar Inst Inzh Zheleznodorozhn Transp," Vol 3, pp 3-9 (In Russian)

74C039   SN-119   J Chappert, R H Herber, and M Winterberger, Electric and Magnetic Interaction in Cu_2FeSnS_4, Cu_2MnSnS_4, and Related Solids, in "Hyperfine Interactions Studied in Nuclear Reactions and Decay: Contributed Papers" (Conference, Uppsala, Sweden, 1974), edited by E Karlsson and R Wappling (Upplands Grafiska, Uppsala, 1974), pp 198-9

74G065   SN-119   P L Gruzin, Yu F Bychkov, I A Evstyukhina, and L A Alekseev, Prikl Yad Spektrosk 4,12-6(1974), State of Tin-119 Nuclei in Superconducting Niobium Compounds Studied by Nuclear Gamma Resonance (In Russian)

74G067   SN-119   V I Gudov, V I Stepanenko, V L Fedorin, and V S Shkalikov, Tr Metrol Inst SSSR 151,106-7(1974), Resonance Mossbauer Converter for Measuring Vibrational Parameters (In Russian)

74H053   SN-119   L Haggstrom, Determination of Hyperfine Parameters from Mossbauer Spectra Involving $1/2 \to 3/2$ Transitions, in "Hyperfine Interactions Studied in Nuclear Reactions and Decay: Contributed Papers" (Conference, Uppsala, Sweden, 1974), edited by E Karlsson and R Wappling (Upplands Grafiska, Uppsala, 1974), pp 240-1

74H055   SN-119   R H Herber, Failure of the Point-charge Model for Quadrupole Hyperfine Interactions in Organo-tin Compounds, in "Hyperfine Interactions Studied in Nuclear Reactions and Decay: Contributed Papers" (Conference, Uppsala, Sweden, 1974), edited by E Karlsson and R Wappling (Upplands Grafiska, Uppsala, 1974), pp 234-5

74K064   SN-119   D K Kaipov, T A Orazbaev, E F Galyutina, and L S Sergeeva, Prib Tekh Eksp (6),75-6(1974)/Instrum Exp Tech (USSR) 17, 1633-5(1974), Properties of a Scintillation Resonance Detector for Mossbauer Investigations

74K068   SN-119   M M Kadykenov, Izv Akad Nauk Kaz SSR, Ser Fiz-Mat (4), 75-6(1974), Effect of the Chemical Environment of an Atom on the Total Density of Its Electrons in the Region of the Nucleus (In Russian)

74L049   SN-119   I S Lyubutin and Yu S Vishnyakov, Reorientation of the Magnetic Field on the Nucleus of a Diamagnetic Atom in Rare Earth Orthoferrites, in "Proceedings of the International Conference of Magnetism ICM-73" (Moscow, 1973) ("Nauka," Moscow, 1974), Vol VI, pp 278-82 (In Russian)

74M060   SN-119   B Mehliss and H H Duncken, Wiss Z Friedrich-Schiller-Univ, Jena, Math-Naturwiss Reihe 23,755-9(1974), Adsorption von Ferrocen und Phthalocyanin an dispersem SiO_2

74M061   SN-119   M Mishima, M Idogaki, and H Negita, Shimane Daigaku Bunrigakubu Kiyo, Rigakka Hen (7),73-7(1974), Mossbauer Study of the Complex Compounds of Tin(II) Halides

74M062   SN-119   M Mishima, M Nakamura, M Izawa, and M Idogaki, Shimane Daigaku Bunrigakubu Kiyo, Rigakka Hen (7),79-84(1974), Mossbauer Effect of 119Sn in Complexes of Tin(IV) Chloride and Dialkyltin Dichlorides with Some Nitrogen Donors

74P057   SN-119   M V Plotnikova, K P Mitrofanov, and Yu N Venevtsev, Anharmonic Atom Vibration in Ferroelectric Type Perovskites According to Mossbauer Data, in "Probl issled svoistv segnetoelektrikov" (1974), Chapter 2, pp 15-6 (In Russian)

74R043   SN-119   P Roggwiller and W Kundig, The Isomer Shift and the Influ-
ence of the Chemical Environment on the Lifetime of the
Mossbauer Nucleus 119Sn, in "Hyperfine Interactions
Studied in Nuclear Reactions and Decay: Contributed
Papers" (Conference, Uppsala, Sweden, 1974), edited by E
Karlsson and R Wappling (Upplands Grafiska, Uppsala, 1974),
pp 240-1

74W027   SN-119   G Weyer, B I Deutch, A Nylandsted-Larsen, and J U
Andersen, Mossbauer and Channeling Experiments on Implants
of 119Sn in Group IV Elements in "Hyperfine Interactions
Studied in Nuclear Reactions and Decay: Contributed
Papers" (Conference, Uppsala, Sweden, 1974), edited by E
Karlsson and R Wappling (Upplands Grafiska, Uppsala, 1974),
pp 38-9

75A005   SN-119   S Ambe and F Ambe, Inorg Nucl Chem Lett 11,139-43(1975),
Mossbauer Emission Spectrum of 119Sn in 119Sb(OH)(C2O4)

75A010   SN-119   S Akselrod, M Pasternak, and S Bukshpan, Phys Rev B 11,
1040-4(1975), Mossbauer-effect Studies of the Lattice Dynamics of Granular Tin

75A021   SN-119   N E Ablesimov and S I Bondarevskii, Khim Vys Energ 9,174-5
(1975)/High Energ Chem (USSR) 9,147-8(1975), Consequences
of a Converted Isomeric Transition at 298 K in Surface
Atoms of Tin-119m

75A029   SN-119   I A Avenarius and R N Kuz'min, Temperature Dependence of
Mossbauer Effect Probability for Impurities of 119Sn in
Monocrystal of Antimony, in "Int Conf Mossbauer Spectrosc,
Proc," Vol 1, p 21 (See 75H027) (In Russian)

75B001   SN-119   G M Bancroft and K D Butler, Can J Phys 53,307-10(1975),
119Sn Mossbauer Spectra of Sn-Ni Compounds

75B004   SN-119   J Bolz and F Pobell, Z Phys B 20,95-103(1975), Mossbauer
Effect of 119Sn in Amorphous Superconducting Metals

75B006   SN-119   A Bos and A T Howe, J Chem Soc, Faraday Trans 2 71,28-40
(1975), Mossbauer and Infra-red Studies of the Diffusion
and Reactivity of (SnO)n Species (n>1) Initially Isolated
in Solid Nitrogen

75B009   SN-119   R Barbieri, N Bertazzi, C Tomarchio, and R H Herber, J
Organomet Chem 84,39-46(1975), Organotin(IV) Azido and
Mixed Azidothiocyanato Complex Anions; A Mossbauer and
Vibrational Spectroscopic Study

75B015   SN-119   D Barb, E Burzo, S Constantinescu, L Diamandescu, A Marian,
and D Tarina, Rev Roum Phys 20,103-6(1975), Mossbauer
Effect on 119Sn in Zircaloy 2

75B016   SN-119   T Birchall and A R Pereira, J Chem Soc, Dalton Trans
1087-92(1975), Nuclear Magnetic Resonance and Mossbauer
Spectra of Some Organotin Anions

75B017   SN-119   T Birchall and A W Sleight, J Solid State Chem 13,118-30
(1975), Nonstoichiometric Phases in the Sb-Nb-O and Sn-Ta-O
Systems Having Pyrochlore-related Structures

75B032   SN-119   A A Bekker, E N Efremov, and A N Nesmeyanov, Complex Investigation of the Chemical Bond Character and the Local
Surroundings Symmetry of Sn and Te in Solid Solutions on
the Base of Lead and Tin Tellurides, in "5th Int Conf Mossbauer Spec, Proc," Part 1, pp 267-9 (See 75H017)
(In Russian)

75B036   SN-119   N Bertazzi, G Alonzo, F Di Bianca, and G C Stocco, Inorg
Chim Acta 12,123-6(1975), Infrared and Mossbauer Studies
on Adducts R3SnOH-R3MX (M=Sn,Pb; X = Pseudohalide)

75B039   SN-119   N B Brandt and V G Snigirev, Fiz Tverd Tela (Leningrad)
17,910-3(1975)/Sov Phys-Solid State 17,578-9(1975),
Weakening of Force Constants in SnCd and SnBi Alloys

CODE	TOPIC	REFERENCE

75B043 SN-119   G M Bancroft and A T Rake, Inorg Chim Acta 13,175-9(1975),
                 57Fe and 119Sn Mossbauer Spectra of Phosphine and Phos-
                 phite Derivatives of Ph(3-n)Cl(n)SnFe(CO)2(pi-C5H5)
                 Compounds

75B044 SN-119   G M Bancroft, K D Butler, and T K Sham, J Chem Soc, Dalton
                 Trans 1483-6(1975), Room Temperature Tin-119 Mossbauer
                 Spectra of Unassociated Tin Compounds

75B049 SN-119   A Bos and A T Howe, J Chem Soc, Faraday Trans 2 71,28-40
                 (1975), Mossbauer and Infra-red Studies of the Diffusion
                 and Reactivity of (SnO)n Species (n > 1) Initially Iso-
                 lated in Solid Nitrogen

75B052 SN-119   Cv Bontschev, A Minkova, and C K Tkhiep, Applications
                 of Mossbauer Effect to the Investigation of the Copper-Tin
                 Intermetallic System, in "5th Int Conf Mossbauer Spec,
                 Proc," Part 2, pp 458-62 (See 75H017) (In Russian)

75B065 SN-119   B Busch, J Pebler, and K Dehnicke, Z Anorg Allg Chem 416,
                 203-10(1975), Darstellung, Mossbauer- und Schwingungsspek-
                 tren der Azidokomplexe (SnCl4(N3)2)(-2) und (SnCl4N3)2(-2)

75B067 SN-119   R Bacaud, P Bussiere, F Figueras, and J P Mathieu, C R Acad
                 Sci (Paris), Ser C 281,159-61(1975), Caracterisation par
                 Spectrometrie Mossbauer de Catalyseyrs Platine-etain
                 Deposes sur Alumine

75B068 SN-119   A Sh Bakhtyarov and L N Vasil'ev, Izv Akad Nauk SSSR, Neorg
                 Mater 11,741-2(1975)/Inorg Mater (USSR) 11,636-7(1975),
                 Investigation of the System As-Se-Ge-Sn by the Mossbauer
                 Method

75B089 SN-119   C I Balcombe, E C Macmullin, and M E Peach, J Inorg Nucl
                 Chem 37,1353-7(1975), Some Reactions and Spectroscopic
                 Studies on the Organotin Thiolates Bu3SnSR and
                 Bu2Sn(SR)2

75B090 SN-119   J G Ballard and T Birchall, Can J Chem 53,3371-3(1975),
                 119Sn Mossbauer Spectra of Some Tin(II)-Oxygen Compounds

75B097 SN-119   G N Belozerskii, Sh Z Bashaikin, and E Yu Bessonova,
                 Mossbauer Study of the Kinetics of Glass-crystal Transfor-
                 mation in Semiconductive Crystals, in "Int Conf Mossbauer
                 Spectrosc, Proc," Vol 1, pp 351-2 (See 75H027)

75B106 SN-119   R Bacaud, P Bussiere, R Dutartre, F Figueras, G A Martin,
                 and J P Mathieu, Uses of Mossbauer Spectroscopy in Hetero-
                 geneous Catalysis Research, in "Int Conf Mossbauer Spec-
                 trosc, Proc," Vol 1, pp 527-8 (See 75H027)

75B110 SN-119   K Burin, Cv Bontschev, M Grozdanov, and K K'nchev, Depen-
                 dence of the Mossbauer Line Shape of a Suspension on the
                 Size Distribution of Floating Particles, in "5th Int Conf
                 Mossbauer Spectrosc, Proc," Vol 3, pp 640-2 (See 75H017)
                 (In Russian)

75B111 SN-119   Cv Bontschev, I Vasil'ev, A Yanev, and Z Sapundshiev,
                 Investigation of the Reaction of a Group of Ants by the
                 Mossbauer Effect, in "5th Int Conf Mossbauer Spectrosc,
                 Proc," Vol 3, pp 646-8 (See 75H017) (In Russian)

75B114 SN-119   R Barbieri, F Di Bianca, G Alonzo, A Silvestri, L
                 Pellerito, N Bertazzi, and G C Stocco, Z Anorg Allg Chem
                 411,173-81(1975), Infrared and Mossbauer Spectroscopic
                 Studies on Complexes of Hal2Sn(IV) Moieties with Tri-
                 dentate Ligands

75B115 SN-119   R Barbieri, L Pellerito, N Bertazzi, G Alonzo, and J G
                 Noltes, Inorg Chim Acta 15,201-4(1975), Mossbauer Spectro-
                 scopic Studies on Compounds Containing Tin-Cadmium and
                 Tin-Zinc Bonds

CODE	TOPIC	REFERENCE

75B129   SN-119   V N Bogomolov, A I Zadorozhnii, and N A Klushin, Fiz Tverd Tela 17,2452-3(1975)/Sov Phys-Solid State 17,1627(1976), The Mossbauer Effect of Sn119 in Holes in NaX Zeolite and the Filtration of Sn119 Atoms from a Ga Melt by Zeolite

75B130   SN-119   J Bolz and F Pobell, Mossbauer Effect of 119Sn in Amorphous Tin, in "Proceedings of the 14th International Conference on Low Temperature Physics: Volume 2 Superconductivity" (Otaniemi, Finland, 1975), edited by M Krusius and M Vuorio (North-Holland Publishing Co, Amsterdam/American Elsevier Publishing Co, New York, 1975), pp 425-8

75B133   SN-119   J S Brooks and J M Williams, Phys Status Solidi A 32,413-7 (1975), Magnetic Hyperfine Interactions in Ferromagnetic Co2TiSn

75B137   SN-119   G M Bancroft, I Adams, H Lampe, and T K Sham, Chem Phys Lett 32,173-7(1975), Linewidths and Line Shapes in Solid State ESCA Studies: Electric Field Gradient Broadening of Sn 3d Lines

75B138   SN-119   G M Bancroft, B W Davies, N C Payne, and T K Sham, J Chem Soc, Dalton Trans 973-8(1975), Preparation and Spectroscopic Studies of Five-co-ordinate beta-Diketonato-tri-(organo)tin Compounds. Crystal Structure of (1,3-Diphenylpropane-1,3-dionato)triphenyltin(IV)

75B145   SN-119   K D Bos, E J Bulten, and J G Noltes, J Organomet Chem 99, 397-405(1975), Oxidative Addition Reactions of Dicyclopentadienyltin(II) and of Tin(II) Bis(acetylacetonate) with Organic Halides. The Preparation of Compounds of the Type RSn(acac)2X

75B148   SN-119   S Bukshpan, T Sonnino, and J G Dash, Surf Sci 52,466-72 (1975), Debye-Waller Factor of Molecules Adsorbed on Graphite

75C006   SN-119   D Cunningham, I Douek, M J Frazer, M McPartlin, and J D Matthews, J Organomet Chem 90,C23-4(1975), The X-ray Structure and Mossbauer Parameters of the Dichlorodimethyltin(IV) 1/1 Salicylaldehyde Adduct

75C013   SN-119   W R Cullen, R K Pomeroy, J R Sams, and T B Tsin, J Chem Soc, Dalton Trans 1216-21(1975), Tin-119 Mossbauer Study of Complexes with Chlorine-bridged Tin-Molybdenum and Int-Tungsten Bonds

75C022   SN-119   B Csakvari, E Csakvari, P Gomory, and A Vertes, J Radioanal Chem 25,275-82(1975), Mossbauer Study of the Donor Character of Oxygen in Si-O Bonds

75C036   SN-119   C C M Campbell, J Phys F 5,1931-45(1975), Hyperfine Field Systematics in Heusler Alloys

75C043   SN-119   W R Cullen and R K Pomeroy, Inorg Chem 14,939-41(1975), Preparation of (diphosMo(CO)4SnCl3)+(SnCl5OH2)- and Related Derivatives

75C046   SN-119   A B Cornwell and P G Harrison, J Chem Soc, Dalton Trans 1486-90(1975), Derivatives of Bivalent Germanium, Tin, and Lead. Part VII. Chromium, Molybdenum, and Tungsten Pentacarbonyl Complexes of Tin(II) Bis(beta-ketoenolates)

75C047   SN-119   A B Cornwell and P G Harrison, J Chem Soc, Dalton Trans 1722-6(1975), Derivatives of Divalent Germanium, Tin, and Lead. Part IX. Tin(II) Derivatives of Alkyl Acetoacetates, 4,Phenylbutane-2,4-dione, 1,3-diphenylpropane-1,3-dione, Cyclohexane-1,2- and -1,3-diones, and 2-Hydroxycyclohepta-2,4,6-trien-1-one

75C048   SN-119   A B Cornwell and P G Harrison, J Chem Soc, Dalton Trans 2017-22(1975), Derivatives of Bivalent Germanium, Tin, and Lead. Part XI. The Interaction of Tin(II) Halides and Bis(beta-ketoenolates) with Di-iron Enneacarbonyl

CODE	TOPIC	REFERENCE

75C059   SN-119   D Christov, Cv Bontschev, and N Nenov, Kautsch Gummi, Kunstst 4,201-4(1975), Untersuchung der Beschleunigung der Harz-Vulkanisation von Butylkautschuk mit Hilfe des Mossbauer-Effektes. 9, Beschleunigende Wirkung von wasserfreien Metallhalogeniden

75C060   SN-119   D Christov, Cv Bontschev, and N Nenov, Kautsch Gummi, Kunstst 4,260-3(1975), Untersuchung der Beschleunigung der Harz-Vulkanisation von Butylkautschuk mit Hilfe des Mossbauer-Effektes. 9, Beschleunigende Wirkung von wasserfreien Metallhalogeniden (continuation of 75C059)

75C064   SN-119   D Christov and Cv Bontschev, Kaut Gummi, Kunstst 12,724-5 (1975), Untersuchung der Beschleunigung der Harz-Vulkanisation von Butylkautschuk mit Hilfe des Mossbauer-Effektes 10. Beschleunigende Wirkung von Stannochlorid in Abhangigkeit von der Art der Einfuhrung in die Kautschuk-Mischung

75D003   SN-119   M A Delmas, J C Maire, W McFarlane, and Y Richard, J Organomet Chem 87,285-93(1975), Etude Comparative de la Structure d'Oxa- et Thia-cyclostannanes

75D007   SN-119   J D Donaldson, D C Puxley, and M J Tricker, J Inorg Nucl Chem 37,655-9(1975), The Interpretation of the 119mSn Mossbauer Data for Some Complexes of Tin(II)

75D030   SN-119   R J Dickinson, R V Parish, P J Rowbotham, A R Manning, and P Hackett, J Chem Soc, Dalton Trans 424-8(1975), Studies in Mossbauer Spectroscopy. Part VII. Tin-119 and Iron-57 Spectra of Compounds Involving Tin Bonded to Chromium, Molybdenum, Tungsten, Manganese, Iron, or Cobalt

75D047   SN-119   J L K F De Vries, J M Trooster, and P Ros, J Chem Phys 63, 5256-62(1975), Numerical Relativistic Self-consistent Field Calculation of Electron Density and $<r-3>$ for a Number of Electron Configurations of Iron and Tin. Application to Tin Mossbauer Spectroscopy

75D049   SN-119   J D Donaldson, J Silver, S Hadjiminolis, and S D Ross, J Chem Soc, Dalton Trans 1500-6(1975), Effects of the Presence of Valence-shell Non-bonding Electron Pairs on the Properties and Structures of Caesium Tin(II) Bromides and of Related Antimony and Tellurium Compounds

75D056   SN-119   J D Donaldson, S D Ross, J Silver, and P J Watkiss, J Chem Soc, Dalton Trans 1980-3(1975), Solid-state Effects and the Vibrational Spectra of Hexahalogeno-stannates(IV) and -tellurates(IV)

75D071   SN-119   N N Delyagin, Yu D Zonnenberg, and V I Nesterov, Zh Eksp Teor Fiz 69,1372-81(1975)/Sov Phys-JETP, Isomer Shift and Magnetic Hyperfine Interaction for Sn in Ni in the Critical Temperature Range

75D072   SN-119   N N Delyagin, Yu D Zonnenberg, and V I Nesterov, Fiz Tverd Tela 17,3036-8(1975)/Sov Phys-Solid State 17,2013-4(1976), The Magnetic Hyperfine Interaction of 119Sn in Co2TiSn

75E002   SN-119   E N Efremov and A A Bekker, Effective Magnetic Fields at the 119Sn Nuclei in Nickel and Isomer Shifts in the Ni-Sn System, in "5th Int Conf Mossbauer Spec, Proc," Part 1, pp 26-8 (See 75H017) (In Russian)

75E017   SN-119   P F R Ewings, P G Harrison, and D E Fenton, J Chem Soc, Dalton Trans 821-6(1975), Derivatives of Divalent Germanium, Tin, and Lead. Part V. Bis-(pentane-2,4-dionato)-, Bis(1,1,1-trifluoropentane-2,4-dionato)-, and Bis(1,1,1, 5,5,5,-hexafluoropentane-2,4-dionato)-tin(II)

75E018   SN-119   P F R Ewings and P G Harrison, J Chem Soc, Dalton Trans 1717-21(1975), Derivatives of Divalent Germanium, Tin, and Lead. Part VIII. Tin(II) Aryl-carboxylates, -sulphonates, and Halide Methoxides

CODE	TOPIC	REFERENCE

75E019   SN-119   P F R Ewings and P G Harrison, J Chem Soc, Dalton Trans 2015-7(1975), Derivatives of Divalent Germanium, Tin, and Lead. Part X. Tin(II) Bis(phenoxides), Bis(O-methyl dithiocarbonate), and Bis(diethyldithio-carbamate)

75E024   SN-119   J Ensling, P Gutlich, and L Rosch, Z Naturforsch, Teil B 30,850-3(1975), 57Fe- und 119Sn-Mossbauer-Untersuchungen an Tetracarbonyl(organoelement IV a-phosphin)eisen(0)-Komplexen

75E025   SN-119   E M Eremenko, V E Listovnichy, V M Sergyenkova, and A V Murzin, Dopov Akad Nauk Ukr RSR, Ser A 37,560-4(1975), Mass Transfer of Tin from Its Melt to Glass Mass in Hydrogen Medium (In Russian)

75E026   SN-119   E M Eremenko, V E Listovnichy, V M Sergyenkova, and A V Murzin, Dopov Akad Nauk Ukr RSR, Ser A 37,655-9(1975), Mass Transfer of Tin from Its Melt to Glass Mass in the Argon Medium at P(O2) =10(-16)-10(-18) atm (In Russian)

75F006   SN-119   J A Feiccabrino and E J Kupchik, J Organomet Chem 73,319-25 (1975), Preparation of Some Triethylammonium (Organo-cyanoamino)chlorotriphenylstannates

75G013   SN-119   N N Greenwood and B Youll, J Chem Soc, Dalton Trans 158-62 (1975), Reactions of Some Tin(II) and Tin(IV) Compounds with the Dodecahydro-nido-decaborate(2-) Ion, (B10H12)2-

75G025   SN-119   M Goldstein and P Tiwari, J Inorg Nucl Chem 37,1550-1 (1975), Barium Tetrachlorostannate(II)

75G028   SN-119   E A Gorlich, R Kmiec, K Latka, T Matlak, K Ruebenbauer, A Szytula, and K Tomala, Phys Status Solidi A 30,331-6 (1975), Magnetic Hyperfine Field Distribution at the Tin Site in the C1 Structure Alloys IrMnSn and PtMnSn

75G029   SN-119   E A Gorlich, R Kmiec, K Latka, T Matlak, K Ruebenbauer, A Szytula, and K Tomala, Phys Status Solidi A 30,765-70 (1975), Transferred Hyperfine Fields at the Tin Site in the Heusler-type Alloys Co2YSn (Y = Ti,Zr,Hf,V)

75G031   SN-119   R Gsell and M Zeldin, J Inorg Nucl Chem 37,1133-7(1975), Synthesis and Spectroscopic Properties of Tin(II) Alkoxides

75G035   SN-119   U Gonser, B Schmitt, and H D Pfannes, Sn119 - Polarimetry, in "5th Int Conf Mossbauer Spec, Proc," Part 3, pp 594-6 (See 75H017)

75G044   SN-119   E A Gorlich, R Kmiec, K Latka, T Matlak, K Ruebenbauer, A Szytula, and K Tomala, Transferred Hyperfine Fields at the Tin Site in the Heusler Type Alloys Co2YSn /Y=Ti,Zr,Hf,V/, in "Int Conf Mossbauer Spectrosc, Proc," Vol 1, pp 85-6 (See 75H027)

75G072   SN-119   P L Gruzin, Yu V Petrikin, and A M Rodin, At Energ/Sov At Energy 38,207(1975), Simulation of Resonance Absorption of Gamma Rays by the Monte Carlo Method

75H005   SN-119   C H Huang, O Knop, D A Othen, F W D Woodhams, and R A Howie, Can J Chem 53,79-91(1975), Pyrophosphates of Tetra-valent Elements and a Mossbauer Study of SnP2O7

75H007   SN-119   L Haggstrom, T Ericsson, R Wappling, and K Chandra, Phys Scr 11,47-54(1975), Studies of the Magnetic Structure of FeSn Using the Mossbauer Effect

75H009   SN-119   L Haggstrom, T Ericsson, and R Wappling, Phys Scr 11,94-6 (1975), An Investigation of CoSn Using Mossbauer Spectroscopy

75H011   SN-119   L Haggstrom, J Gullman, T Ericsson, and R Wappling, J Solid State Chem 13,204-7(1975), Mossbauer Study of Tin Phosphides

CODE	TOPIC	REFERENCE

75H018   SN-119   B Y K Ho and J J Zuckerman, J Organomet Chem 96,41-7(1975), Solid-state Association in Organotin Compounds Containing Bulky Organic Groups. Tricyclohexyltin Hydroxide

75H022   SN-119   R H Herber and R F Davis, J Chem Phys 63,3668-9(1975), Lattice Dynamics and Hyperfine Interactions of Tin in Tantalum Sulfide Layer Compounds

75H040   SN-119   P G Harrison and R C Phillips, J Organometal Chem 99,79-91 (1975), Structural Studies in Main Group Chemistry XI. Tin-119m Mossbauer Investigations of Triorganotin Derivatives of Substituted Pyridines

75I007   SN-119   S Ichiba, M Katada, and H Negita, J Inorg Nucl Chem 37, 2249-51(1975), Mossbauer Effect of 119Sn in Thiostannates, Selenostannates and Selenothiostannates

75I011   SN-119   P A Ioffe, A A Baklagin, and V A Kozlova, Zh Neorg Khim 20, 1712(1975)/Russ J Inorg Chem 20,960(1975), Mossbauer Spectra and State of Tin in the Compound Zn2SnO4

75I015   SN-119   Y Ishida and T Ozawa, Scr Metall 9,1103-6(1975), Grain Boundary Segregation of Tin and the Electronic and Vibrational State in Zn-Al Eutectoid

75K010   SN-119   M Katada, J Sci Hiroshima Univ, Ser A 39,45-72(1975), Mossbauer Effect of 119Sn in Tin Sulfides and Their Related Compounds

75K017   SN-119   H J Kroth, H Schumann, H G Kuivila, C D Schaeffer, Jr, and J J Zuckerman, J Am Chem Soc 97,1754-60(1975), Tin-119 Chemical Shifts of Ortho, Meta, Para, 2,6- and Polysubstituted Aryltrimethyltin Derivatives and Related Organotin Compounds

75K027   SN-119   E J Kupchik and J A Feiccabrino, J Organomet Chem 93,325-9 (1975), Preparation of Some N-substituted N-(triphenylstannyl)-cyanamides

75K029   SN-119   N Karnezos, L B Welsh, and M W Shafer, Phys Rev B 11,1808-17(1975), Structural and NMR Properties of Niobium Dichalcogenides Intercalated with Post Transition Metals

75K035   SN-119   I Ya Kuramshin, Sh Sh Bashkirov, A A Muratova, R A Manapov, A S Khramov, and A N Pudovik, Zh Obshch Khim 45, 701-2(1975)/J Gen Chem (USSR) 45,684-5(1975), Localization Centers of Donor-acceptor Bond in Complexes of Tin Halides with Phosphorous Acid Amides

75K064   SN-119   P Kamenov, E Vapirev, B Slavov, K Bourin, and Cv Bontschev, Time Measurements of Resonance Scattered Gammarays, in "Int Conf Mossbauer Spectrosc, Proc," Vol 1, pp 503-4   (See 75H027)

75K075   SN-119   H Kohler, L Neef, L Korecz, and K Burger, J Organomet Chem 90,159-71(1975), Pseudochalkogenvergindungen VII. Synthese und Struktur von Pseudochalkogenooxoacyl-organozinnverbindungen

75K079   SN-119   V M Koshkin, E E Ovechkina, and V P Romanov, Zh Eksp Teor Fiz 69,2218-21(1975)/Sov Phys-JETP, Nuclear Gamma Resonance in Neutral Tin Atoms in the In2Te3 Crystal Lattice

75K081   SN-119   H G Kurvila, J E Dixon, P L Maxfield, N M Scarpa, T M Topka, K H Tsai, and K R Wursthorn, J Organomet Chem 86, 89-107(1975), Preparation of Some Ketoorganostannanes and Ketoorganochlorostannanes. Intramolecular Coordination in Ketoorganochlorostannanes

75K083   SN-119   T S Khodashova, V A Varnek, E N Yurchenko, and M A Porai-Koshits, Dokl Akad Nauk SSSR 224,1323-6(1975)/ Dokl Chem 224,617-20(1975), The Crystal Structure and Mossbauer Parameters of a Complex of Rhodium with Tin(II) Fluoride, Cs4(Rh(SnF2(H2O)2)2(SnF15))-4H2O

CODE	TOPIC	REFERENCE

75L007   SN-119   I S Lyubutin and T V Dmitrieva, Pis'ma Zh Eksp Teor Fiz 21, 132-5(1975)/JETP Lett 21,59-60(1975), Induction of Strong Magnetic Fields at the Nuclei of Diamagnetic Tin Atoms in Chalcogenide Spinels

75L013   SN-119   I S Lyubutin and T V Dmitrieva, Strong Magnetic Fields at Nuclei of Diamagnetic Tin Atoms in Chromium Chalcogenide Spinels, in "Int Conf Mossbauer Spectrosc, Proc," Vol 1, pp 147-8 (See 75H027)

75L031   SN-119   I S Lyubutin, A P Dodokin, and E N Ageeva, Zh Eksp Teor Fiz 68,1363-7(1975)/Sov Phys-JETP 41,678-80(1976), Anomalies in the Mossbauer Spectra of Sn119 Nuclei in the Region of the Magnetic Phase Transition in Yttrium Iron Garnets

75M003   SN-119   J P Motte and N N Greenwood, J Solid State Chem 13,41-8 (1975), Etude par Effect Mossbauer de la Structure et des Proprietes de Diffusion de la Phase Antifluorine Nonstoe-chimetrique: Li8SnP4

75M010   SN-119   K Matsui, R R Hasiguti, T Shoji, and A Ohkawa, Tin-vacancy Interaction in Silicon Monitored by 119mSn Mossbauer Probe, in "Lattice Defects in Semiconductors-1974: Institute of Physics Conference Series-No 23" (Freiburg, 1975), edited by F A Huntley (The Institute of Physics, Bristol, 1975), pp 572-8

75M037   SN-119   A Minkova, B Slavov, and Cv Bontschev, On the Application of the Diffusion Theory in the Depth-selective Mossbauer Spectroscopy, in "Int Conf Mossbauer Spectrosc, Proc," Vol 1, pp 45-6 (See 75H027)

75M046   SN-119   F P Mullins and C Curran, Can J Chem 53,3200-5(1975), Moss-bauer and Related Studies of Complexes of R2Sn(NCS)2 with Neutral Ligands Containing Oxygen Donor Atoms

75M075   SN-119   R A Manapov, I Ya Kuramshin, A A Muratova, and A N Pudovik, Zh Obshch Khim 45,1975-9(1975)/J Gen Chem (USSR) 45,1940-3(1975), Mossbauer Spectra of Complexes of Organo-phosphorus Compounds with Tin Halides

75M088   SN-119   K P Mitrofanov, L P Benderskaya, S I Reiman, and V I Kongauz, Vestn Mosk Univ, Fiz 30,487-9(1975)/Moscow Univ Phys Bull 30,82-4(1975), The Mossbauer Effect on Sn119 Nuclei in a System of mMgSnO4-(1-m)-Mg3(BO3)2 Solid Solutions

75N004   SN-119   D L Nagy, G J Zimmer, T Lohner, J P Senateur, and I Bibicu, Mossbauer Study of the Magnetic Phase Transforma-tions in SnMn3N, in "5th Int Conf Mossbauer Spec, Proc," Part 1, pp 75-8 (See 75H017)

75N007   SN-119   F S Nasredinov, B T Melekh, L N Vasil'ev, and L N Seregina, Fiz Tverd Tela (Leningrad) 17,633-5(1975)/ Sov Phys-Solid State 17,413(1975), Crystal-glass Transition in Ge(0.15)Te(0.85) and Its Influence on the Local Environment of Germanium Atoms

75N008   SN-119   I N Nikolaev, A P Shotov, A F Volkov, and V P Mar'in, Pis'ma Zh Eksp Teor Fiz 21,144-7(1975)/JETP Lett 21,65-6 (1975), "Softening" of Phonon Spectrum in Semiconductors of the Pb(1-x)Sn(x)Te System on Going to the Gapless State

75N009   SN-119   V I Nikolaev and V S Rusakov, Fiz Tverd Tela (Leningrad) 17,326-7(1975)/Sov Phys-Solid State 17,200-1(1975), "Magnetic Anomalies" of the Parameters of the Mossbauer Spectra of the Nuclei Fe57 and Sn119 in Antiferromagnet FeSn2

75N013   SN-119   I V Nistiryuk and P P Seregin, Fiz Tverd Tela (Leningrad) 17,1192-4(1975)/Sov Phys-Solid State 17,768-9(1975), State of Tin Impurity Atoms in Silicon

75N026   SN-119   I Nowik, Magnetic Structure and Transferred Hyperfine In-teractions, in "Int Conf Mossbauer Spectrosc, Proc," Vol 2, pp 83-98 (See 75H038)

CODE	TOPIC	REFERENCE

75N027   SN-119   E N Nikitin, P P Seregin, V I Tarasov, and E Yu Turaev, Fiz Tverd Tela 17,2176-8(1975)/Sov Phys-Solid State 17, 1441(1976), Electronic Structure of MnSi1.7 from Mossbauer Spectroscopy Data

750001   SN-119   S Onaka and H Sano, Bull Chem Soc Jpn 48,258-61(1975), The Syntheses of R3Sn-Mn(CO)(5-n)L(n)(n=0 or 1) Compounds and Their 119Sn-Mossbauer and 1H-NMR Studies

750003   SN-119   N S Ovanesyan and V A Trukhtanov, Angular Dependence of the Superexchange Interactions Cr3+-O2-Fe3+ in the Ortho-chromites, in "5th Int Conf Mossbauer Spec, Proc," Part 1, pp 157-61 (See 75H017) (In Russian)

750004   SN-119   I Ortalli and V Fano, The Mossbauer Effect in Binary Tin Chalcogenides of Tin-129, in "5th Int Conf Mossbauer Spec, Proc," Part 1, pp 263-6 (See 75H017)

750014   SN-119   H J Odenthal, T Kruck, and K Ehlert, Z Naturforsch, Teil B 30,696-8(1975), Metallkomplexe mit anionischen Liganden von Elementen der 4. Hauptgruppe, X(1). Diskussion der Bindungsverhaltnisse in Trihalogenstannidometallat(0)-Komplexen anhand der Ergebnisse von 119mSn-Mossbauer-Untersuchungen

75P005   SN-119   L Pellerito, N Bertazzi, G C Stocco, A Silvestri, and R Barbieri, Spectrochim Acta, Part A 31,303-8(1975), Infrared and Mossbauer Spectroscopic Studies on N, N'ethylenebis (salicylideneiminato) Sn(VI)hal2

75P008   SN-119   J Pebler, Study of Phase Transitions in Cd2Nb(2-2x)Sn(2x)O(7-2x)F(2x), in "5th Int Conf Mossbauer Spec, Proc," Part 1, pp 145-51 (See 75H017)

75P011   SN-119   F Petillon and J E Guerchais, J Inorg Nucl Chem 37,1863-70 (1975), Complexes Soufres (Partie VII) du Titane(III), de l'Antimoine(V) et (III), du Bismuth(III) et de l'Etain(IV). Etude Mossbauer

75P017   SN-119   R K Puri and D A O'Connor, A Mossbauer Study of the Lattice Dynamics of Iron and Tin Impurities in Titanium and Palladium, in "5th Int Conf Mossbauer Spec, Proc," Part 2, pp 448-50 (See 75H017)

75P019   SN-119   P A Pella and J R DeVoe, J Radioanal Chem 25,185-8(1975), Systematic Error in Tin Ore Assay by Mossbauer Spectrometry

75P023   SN-119   R K Puri and L R Gupta, Phys Status Solidi B 70,785-92 (1975), An Estimation of Force Constant Change of Mossbauer 119Sn Impurity Nuclei in Palladium Host

75P026   SN-119   V N Panyushkin, L Bogner, and G Wortmann, Quadrupole Splitting of the 23,8 keV Mossbauer Line of 119Sn in Beta-Tin, in "Int Conf Mossbauer Spectrosc, Proc," Vol 1, pp 99 (See 75H027)

75P054   SN-119   O Kh Poleshchuk, Yu K Maksyutin, and I G Orlov, Koord Khim 1,666-9(1975), Charge Transfer in Tin Chloride Complexes (In Russian)

75R010   SN-119   P Roggwiller and W Kundig, Phys Rev B 11,4179-83(1975), Isomer Shift and Influence of the Chemical Environment on the Lifetime of the Mossbauer Nucleus 119Sn

75R011   SN-119   E Realo and A Lijn, Mossbauer Investigation of alpha-Fe2O3-Sn, in "5th Int Conf Mossbauer Spec, Proc," Part 1, pp 151-6 (See 75H017) (In Russian)

75R016   SN-119   J N R Ruddick and J R Sams, J Inorg Nucl Chem 37,564-6 (1975), A Mossbauer Spectroscopic Study of Some Schiff Base Complexes of Tin(IV) Halides

75R018   SN-119   A J Rein and R H Herber, J Chem Phys 63,1021-9(1975), Molecular Spectroscopy of Organometallic Compounds: Organotin(IV) Tropolonates

162

CODE	TOPIC	REFERENCE

75R021   SN-119   E Realo, Mossbauer Spectroscopy of Optical Sn-centers in Crystals of NaCl and CaS, in "Int Conf Mossbauer Spectrosc, Proc," Vol 1, pp 199-200   (See 75H027)   (In Russian)

75R025   SN-119   E Realo and S I Reiman, The Influence of the Electron Arrangement in Characteristic Admixture of Ionic Tin in Fe3O4, in "Int Conf Mossbauer Spectrosc, Proc," Vol 1, pp 373-4   (See 75H027)   (In Russian)

75R032   SN-119   V O Reikhsfel'd, V A Ivanov, and I E Saratov, Zh Obshch Khim 45,2243-5(1975)/J Gen Chem (USSR) 45,2202-4(1975), Infrared- and Mossbauer-spectral Study of Hydride Organo-Silanes, -Germanes, and -Stannanes

75R034   SN-119   P F Rodesiler, T Auel, and E L Amma, J Am Chem Soc 97, 7405-10(1975), Metal Ion-aromatic Complexes. XXII. The Preparation, Structure, and Stereochemistry of Tin(II) in pi-C6H6Sn(AlCl4)2-C6H6

75R036   SN-119   N M Rubinina, V B Shagdarov, V K Yanovskii, and R N Kuz'min, Kvan Elektron (Moscow) 2,1024-9(1975)/Sov J Quantum Electron, Study of Impurity Centers in Iron-doped Lithium Metaniobate by Mossbauer Spectroscopy

75S001   SN-119   H Sano and Y Mekata, Chem Lett 155-60(1975), Mossbauer Spectroscopic Studies of Bis(tri-n-butyltin) Sulfate, Selenate, and Chromate

75S005   SN-119   J P Schunck, J M Friedt, and Y Llabador, Rev Phys Appl 10, 121-6(1975), Spectroscopie Mossbauer de 57Fe et 119Sn par Detection des Electrons de Conversion et Auger Application a des Etudes de Surface

75S019   SN-119   S Solacolu, E Barbulescu, D Barb, and M Morariu, Rev Roum Chim 20,69-73(1975), Polymorphic Transformations in the BaTiO3-BaSnO3 System Revealed by Mossbauer Effect

75S030   SN-119   T K Sham and G M Bancroft, Inorg Chem 14,2281(1975), Tin-119 Mossbauer Quadrupole Splittings for Distorted Me2Sn(IV) Structures

75S031   SN-119   J Sitek, Mossbauer Study of the Lattice Dynamics of Tin Atoms in Antimony, in "5th Int Conf Mossbauer Spec, Proc," Part 2, pp 451-3   (See 75H017)

75S032   SN-119   Yu A Samarskii, N E Alekseevskii, and A P Kiryanov, Zh Eksp Teor Fiz 68,2330-4(1975)/Sov Phys-JETP, "Magnetic Anomaly" of the Probability of the Mossbauer Effect in Dilute Pd-Co Alloys

75S033   SN-119   P P Seregin, M A Sagatov, T F Mazets, and L N Vasil'ev, Phys Status Solidi A 28,127-32(1975), The Influence of the Crystal-glass Transition on the State of Impurity Tin Atoms in Chalcogenide Semiconductors

75S036   SN-119   V I Shtanov, V P Zlomanov, and A V Novoselova, Izv Akad Nauk SSSR, Neorg Mater 11,358-60(1975)/Inorg Mater (USSR) 11,301-3(1975), Physicochemical Investigation of the System PbSe-SnSe2

75S066   SN-119   A Schichl, F J Litterst, H Micklitz, J P Devort, and J M Friedt, 119Sn Resonance in Matrix Isolated Sn(II) and Sn(IV) Halides, in "Int Conf Mossbauer Spectrosc, Proc," Vol 1, pp 239-40   (See 75H027)

75S093   SN-119   P P Seregin, I V Nistiryuk, and F S Nasredinov, Fiz Tverd Tela 17,2330-4(1975)/Sov Phys-Solid State 17,1540-2(1976), Tin as an Isotopic Impurity in Silicon and Germanium

75S097   SN-119   F E Smith and B V Liengme, J Organomet Chem 91,C31-2 (1975), A Novel Series of Triorganotin Compounds

CODE	TOPIC	REFERENCE

75S098　SN-119　V K Sokolova, V A Varnek, N F Yudanov, and I I Tychinskaya, Izv Sib Otd Akad Nauk SSSR, Ser Khim Nauk (6),13-7(1975), Mossbauer Effect in Iron Hexafluoro-stannates　(In Russian).

75S102　SN-119　G C Stocco, G Alonzo, N Bertazzi, and F Di Bianca, Gazz Chim Ital 105,355-60(1975),36/Mossbauer, Infrared and Other Studies on Tin(II) Halide Derivatives of Multidentate Ligands

75S104　SN-119　T N Sumarokova, R A Slavinskaya, and T A Tember, Zh Obshch Khim 45,2687-92(1975)/J Gen Chem (USSR) 45,2651-4(1975), Coordination Compounds of Sn(IV) with N,N'-Diacyldicar-boxamides

75T013　SN-119　E W Thornton and P G Harrison, J Chem Soc, Faraday Trans 1 71,461-72(1975), Tin Oxide Surfaces.　Part 1, Surface Hy-droxyl Groups and the Chemisorption of Carbon Dioxide and Carbon Monoxide on Tin(IV) Oixde

75T016　SN-119　M Takano, Y Takeda, M Shimada, T Matsuzawa, T Shinjo, and T Takada, J Phys Soc Jpn 39,656-60(1975), Mossbauer Study of Supertransferred Hyperfine Field of 119Sn (Sn4+) in Ca(1-x)Sn(x)MnO3

75T023　SN-119　M Takeda and N N Greenwood, J Chem Soc, Dalton Trans 2207-12(1975), Tellurium-125 Mossbauer Spectra of Some Mixed Oxides of Tellurium(IV) and Some Mixed-valence Oxides of Tellurium(IV, VI)

75U001　SN-119　D L Uhrich, V O Aimiuwu, P I Ktorides, and W J LaPrice, Phys Rev A 12,211-8(1975), Smectic B Liquid-crystalline Glass (at 77 K) as Seen by the Mossbauer Effect of Sn-bearing Solute Molecules

75V003　SN-119　A Vertes, S Nagy, I Czako-Nagy, and E Csakvari, J Phys Chem 79,149-51(1975), Mossbauer Study of Equilibrium Con-stants of Solvates.　I. Determination of Equilibrium Con-stants of Tetraiodotin-trimethylisopropoxysilane and Tetra-bromotin-acetic Anhydride Solvates

75V009　SN-119　V A Varnek, L I Strugova, and E G Avvakumov, Fiz Tverd Tela (Leningrad) 17,561-4(1975)/Sov Phys-Solid State 17,355-6 (1975), Magnetic Structure of Particles Formed as a Result of a Solid State Reaction Between Tin and Manganese

75V015　SN-119　W Vogl and G Vogl, Solid State Commun 17,1029-33(1975), Defect Cascades and Point Defects in Low-temperature Neutron Irradiated alpha-Tin Monitored by Mossbauer Spec-troscopy

75V019　SN-119　Kh Kh Valiev, K A Duldina, and R N Kuz'min, Mossbauer Effect in Heusler Alloys (Co,Ni)2(Zr,Hf)Sn, in "Int Conf Mossbauer Spectrosc, Proc," Vol 1, pp 87-8　(See 75H027)

75V023　SN-119　M Van Rossum, G Langouche, P Boolchand, M Rots, F Namavar, and R Coussement, Electric Field Gradient and Lattice Lo-cation of Tin in Tellurium, in "Int Conf Mossbauer Spec-trosc, Proc," Vol 1, pp 205-6　(See 75H027)

75V039　SN-119　M B Varfolomeev, A S Mironova, F Kh Chibirova, and V E Plyushchev, Izv Akad Nauk SSSR, Neorg Mater 11,2242-4 (1975/Inorg Mater (USSR), Interaction of Indium Sesqui-oxide with Stannic Oxide

75V040　SN-119　V A Varnek, E N Yurchenko, V A Kogan, L N Mazalov, Yu K Maksyutin, O Kh Poleshchuk, A S Egorov, and O A Osipov, Zh Strukt Khim 16,359-66(1975)/J Struct Chem (USSR) 16, 337-43(1975), Temperature Dependence of the Resonance Ab-sorption of Gamma Quanta in Complexes of Tin(IV) Chloride with Organic Ligands

CODE	TOPIC	REFERENCE
75V047	SN-119	V A Varnek, E N Yurchenko, G L Elizarova, A I Shan'ko, L G Matvienko, P G Antonov, and Yu N Kukushkin, Koord Khim 1 161-4(1975), Mossbauer Effect in Pentacoordinated Complexes of the Platinum Metals with Tin-containing Ligands (In Russian)
75V048	SN-119	L N Vasil'ev and A Sh Bakhtyarov, Izv Akad Nauk SSSR, Neorg Mater 11,2074-6(1975)/Inorg Mater (USSR) 11,1780-1 (1975), Mossbauer Spectra of Alloys Ti2Se-As2Se3-SnSe
75W001	SN-119	W Wilder, H D Pfannes, and U Gonser, Z Metallkd 66,161-4 (1975), Mossbauer Spektroskopie am Schmelzpunkt von Zinn
75W002	SN-119	G Weyer, J U Andersen, B I Deutch, J A Golovchenko, and A Nylandsted-Larsen, Radiat Eff 24,117-21(1975), Direct Comparison of Mossbauer and Channeling Studies of Implanted 119Sn in Silicon Single Crystals
75W007	SN-119	E Wenschuh, W D Riedmann, L Korecz, and K Burger, Z Anorg Allg Chem 413,143-9(1975), N-Triorganostannyl-sulfinamide
75W022	SN-119	J L Wardell, J Chem Soc, Dalton Trans 1786-93(1975), Sulphur-substituted Organometallic Compounds. Part II. Reaction of Vinyltin Compounds with Arenesulphenyl Halides and Thiocyanates and Some Properties and Reactions of the (2-Arylthio-1-halogenoethyl)-triphenyltin Addition Products
75W026	SN-119	G Weyer, A Nylandsted-Larsen, B I Deutch, J U Andersen, and E Antoncik, Hyperfine Interac 1,93-112(1975), Covalency Effects on Implanted 119Sn in Group IV Semiconductors Studied by Mossbauer and Channeling Experiments
75Y006	SN-119	E N Yurchenko, V A Varnek, G L Elizarova, and L G Matvienko, Koord Khim 1,1406-14(1975), Study of Tin-containing Ligands in Complexes of Platinum Metals by Nuclear Gamma Resonance, X-ray Photoelectron and IR Spectroscopic, and MO LCAO Methods (In Russian)
75Z022	SN-119	V S Zavgorodnii, E T Bogoradovskii, V L Maksimov, V B Lebedev, B I Rogozev, and A A Petrov, Zh Obshch Khim 45, 2466-71(1975)/J Gen Chem (USSR) 45,2421-5(1975), Unsaturated Stanna Hydrocarbons. Application of 19F NMR Spectra in the Investigation of the Electron Structure of Pentafluorophenylethynl Derivatives of Group IV B Elements

$^{181}_{73}$Ta (6.2 keV, 136.2 keV)

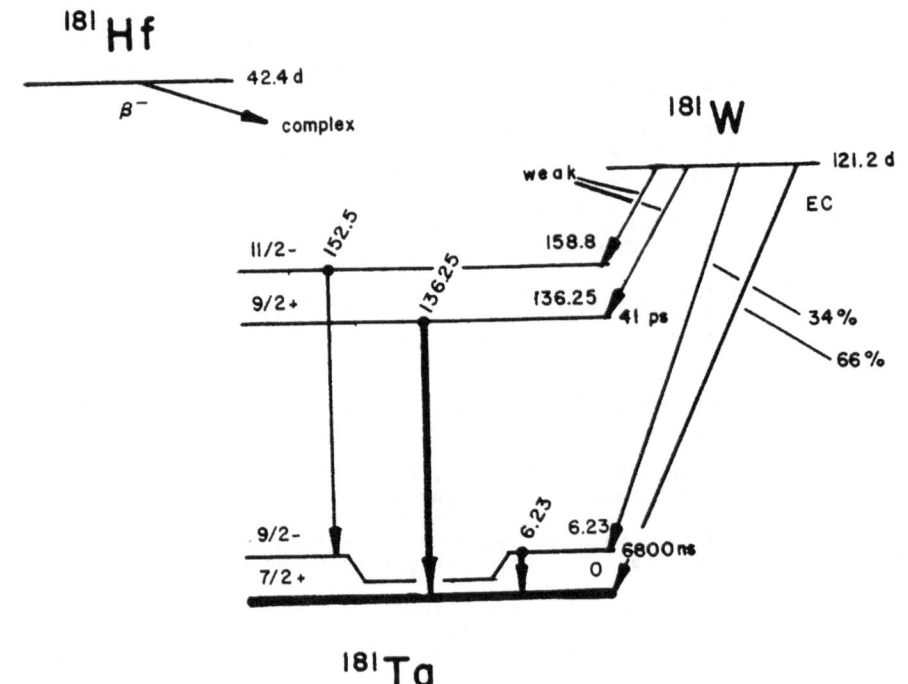

Measured Properties ($E_\gamma = 6.23$ keV)

$E_\gamma = 6.238 \pm 0.017$ keV (1),(2)

$t_{\frac{1}{2}} = 6800$ ns

$\alpha_T = 46 \pm 8$ (3)

IA $= 99.988\%$

$\mu = +2.356 \pm 0.007$ nm (4),(5),(6)

$R_\mu = +2.23 \pm 0.03$ (7),(8)

$Q = +3.9 \pm 0.4$ b (9)

$R_Q = +1.133 \pm 0.010$ (10)

Derived Parameters ($E_\gamma = 6.23$ keV)

$\sigma_0 = 1.68\,(28) \times 10^{-18}$ cm^2

$\Gamma = 6.7\,(5) \times 10^{-11}$ eV

$W_0 = 0.0064\,(5)$ mm/s

$E_r = 1.154\,(4) \times 10^{-4}$ eV

Measured Properties ($E_\gamma = 136.2$ keV)

$E_\gamma = 136.25 \pm 0.02$ keV (11)

$t_{\frac{1}{2}} = 40.0 \pm 1.5$ ps (12),(13),(14)

$\alpha_T = 1.76 \pm 0.02$ (14)

Derived Parameters ($E_\gamma = 136.2$ keV)

$\sigma_0 = 5.97\,(4) \times 10^{-20}$ cm^2

$\Gamma = 1.14\,(4) \times 10^{-5}$ eV

$W_0 = 50.2\,(19)$ mm/s

$E_r = 5.5054\,(11) \times 10^{-2}$ eV

Energy Conversions (E$_\gamma$ = 6.2 keV)

1 mm/s = 5.031 (14) MHz

1 mm/s = 2.081 (6) x 10^{-8} eV

Energy Conversion (E$_\gamma$ = 136.2 keV)

1 mm/s = 109.892 (16) MHz

1 mm/s = 4.5448 (7) x 10^{-7} eV

Energies and Intensities

	E$_\gamma$	I$_{Hf}$ (15)
γ M1 :	6.23 keV	
γ M2 :	136.2 keV	58
γ_1 :	133., keV	460
γ_2 :	137. keV	8.3
K$_{\alpha 1}$:	57.6 keV	

(1)   H. Blumberg, R. S. Hager, and E. C. Seltzer, Nucl. Phys. A136, 624 (1969).
(2)   P. Alexander, H. Ryde, and E. C. Seltzer, Nucl. Phys. 76, 1967 (1966).
(3)   A. H. Muir, Jr., Nucl. Phys. 68, 305 (1965).
(4)   L. C. Erich, A. C. Gossard, and R. L. Hartless, J. Chem. Phys. 59, 3911 (1973).
(5)   L. H. Bennett and J. I. Budnick, Phys. Rev. 120, 1812 (1960).
(6)   J. Sugar and V. Kaufman, Phys. Rev. C 12, 1336 (1975).
(7)   68S010.
(8)   70K030.
(9)   K. Murakawa and T. Kamei, Phys. Rev. 105, 671 (1957).
(10) 73K011.
(11) F. Boehm and P. Marnier, Phys. Rev. 103, 342 (1956).
(12) 69S15.
(13) A. E. Blaugrund, Y. Dar, and G. Goldring, Phys. Rev. 120, 1328 (1960).
(14) R. Armbuster, Y. Dar, J. Gerber, A. Macher, and J. P. Vivien, Nucl. Phys. A143, 315 (1970).
(15) F. T. Avignone, III and J. H. Trueblood, Nucl. Phys. A167, 129 (1971).

[Other data from Y. A. Ellis, Nucl. Data Sheets 9, 319 (1973)]

SOURCE	S-TEMP	ABSORBER	A-TEMP	SHIFT	QS	REMARKS	REF
W		Ta				study of the interference effect between photo- and conversion electrons	75W019
W		Ta	v	--		300<T<2300 K, IS vs T	75S059
xx		alpha-Ta-H	v	--		230<T<400 K, line broadening as function of H conc & temp	75H035
W		TaHx	R			IS & W vs lattice constant	74H054

CODE   TOPIC   REFERENCE

74H054 TA-181 A Heidemann, G Kaindl, D Salomon, and G Wortmann, High-resolution Mossbauer Study of Hydrided Tantalum, in "Hyperfine Interactions Studied in Nuclear Reactions and Decay: Contributed Papers" (Conference, Uppsala, Sweden, 1974), edited by E Karlsson and R Wappling (Upplands Grafiska, Uppsala, 1974), pp 238-9

75H035 TA-181 A Heidemann, G Kaindl, D Salomon, H Wipf, and G Wortmann, A Study of the Dynamical Properties of Hydrogen in the alpha-Phase of Ta-H, in "Int Conf Mossbauer Spectrosc, Proc," Vol 1, pp 411-2 (See 75H027)

75S059 TA-181 D Salomon, W Wallner, and P J West, High Temperature Behavior of the 6.2 keV Mossbauer Transition of 181Ta in Tantalum Metal, in "Int Conf Mossbauer Spectrosc, Proc," Vol 1, pp 105-6 (See 75H027)

75W019 TA-181 P J West, E Matthias, D Salomon, W Wallner, and G Weyer, Mossbauer Conversion Electron Studies of the Interference Between Photoeffect and Internal Conversion in the Absorption of 6.2 keV -E1 Gamma Radiation of 181Ta, in "Int Conf Mossbauer Spectrosc, Proc," Vol 1, pp 457-8 (See 75H027)

$^{159}_{65}$Tb (58.0 keV)

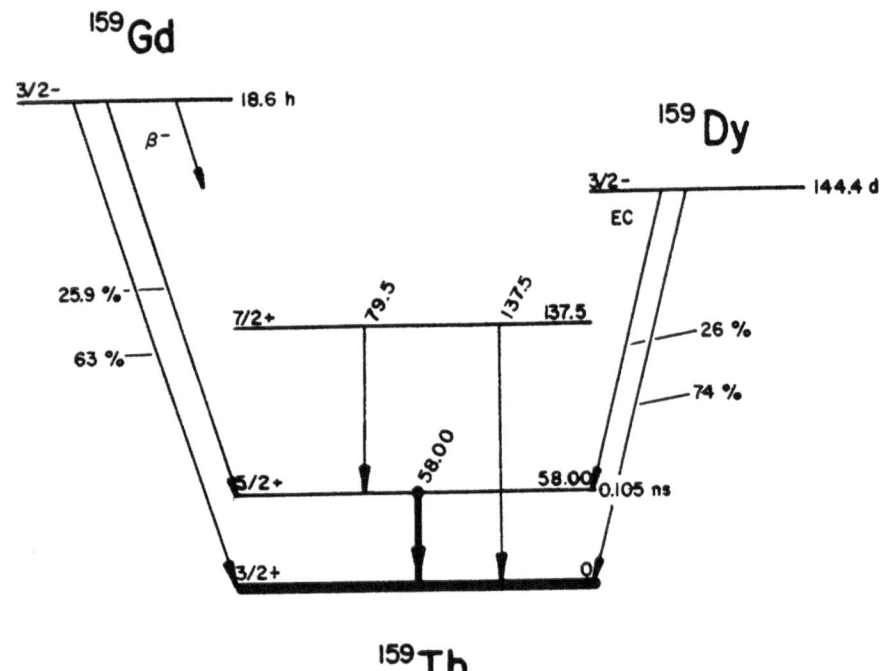

Measured Properties

E_γ = 57.995 ± 0.007 keV (1),(2)

$t_{\frac{1}{2}}$ = 0.105 ± 0.015 ns (3)

α_T = 9.36 ± 0.06 (4)

IA = 100%

μ = +2.008 ± 0.004 nm (5)

R_μ = 0.80 ± 0.05 or 1.15 ± 0.05 (6)

Q = +1.34 ± 0.11 b (7)

Energies and Intensities

	E_γ	I_{Gd}(1)
γ_M :	58.00 keV	18
γ_1 :	79.5 keV	0.4
γ_2 :	137.5 keV	0.04
K_{α_1} :	45.5 keV	

Derived Parameters

σ_o = 1.053 (6) × 10^{-19} cm^2

Γ = 4.3 (6) × 10^{-6} eV

W_o = 45 (6) mm/s

E_r = 1.13548 (19) × 10^{-2} eV

Energy Conversions

1 mm/s = 46.776 (6) MHz

1 mm/s = 1.93450 (23) × 10^{-7} eV

(1) J. C. Hill and M. L. Wiedenbeck, Nucl. Phys. A111, 457 (1968).
(2) E. L. Chupp, J. W. M. DuMond, F. J. Gordon, R. C. Jopson, and H. Mark, Phys. Rev. 112, 518 (1958).
(3) 66A009.
(4) D. Ashery, A. E. Blaugrund, and R. Kalish, in Internal Conversion Processes, edited by J. H. Hamilton (Academic Press, Inc., New York, 1966), p. 263.
(5) J. M. Baker, J. R. Chadwick, G. Garton, and J. P. Hurrell, Proc. Roy. Soc. (London) 286A, 352 (1965).
(6) M. Atac, P. Debrunner, and H. Frauenfelder, Phys. Lett. 21, 699 (1966).

[Other data from J. K. Tuli, Nucl. Data Sheets 9, 435 (1973)]

SOURCE	S-TEMP	ABSORBER	A-TEMP	SHIFT	QS	REMARKS	REF
Tb(CE)		xx				fs experiment	75C024

CODE	TOPIC		REFERENCE
75C024	TB-159		C L Chien and J C Walker, Studies of Temperature Spikes in Solids Using the Mossbauer Effect Following Coulomb Excitation, in "5th Int Conf Mossbauer Spec, Proc," Part 3, pp 560-3 (See 75H017)

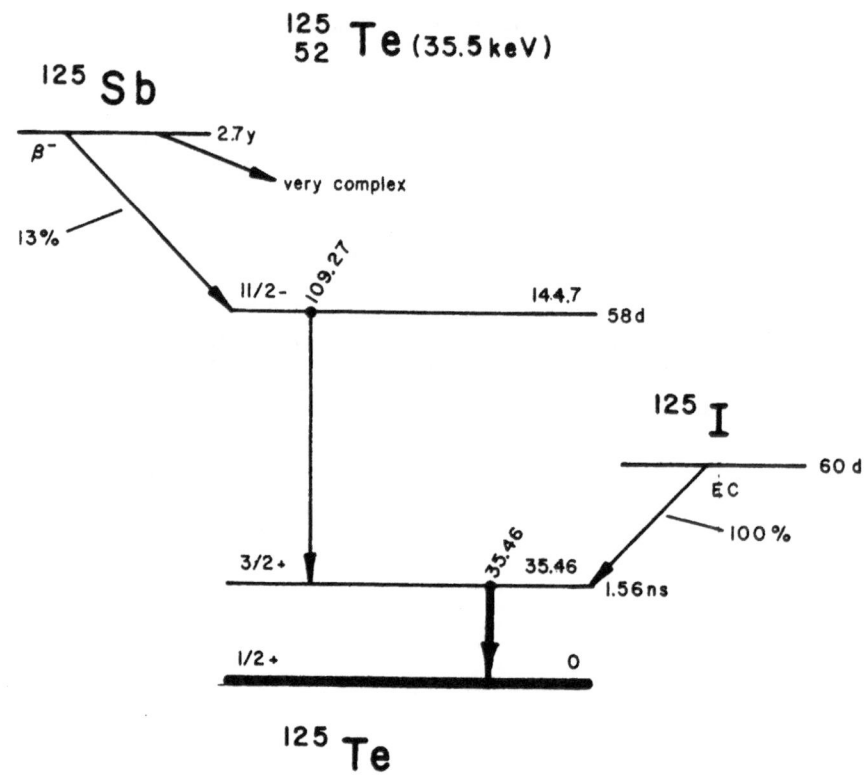

$^{125}_{52}$ Te (35.5 keV)

Measured Properties

E_γ = 35.46 ± 0.03 keV (1)

$t_{\frac{1}{2}}$ = 1.481 ± 0.004 ns (2),(3),(4)

α_T = 13.65 ± 0.55 (5)

IA = 6.99%

μ = −0.88716 ± 0.00025 nm (6)

R_μ = −0.681 ± 0.004 (7)

Q_* = −0.200 ± 0.023 b (8),(9),(10)

Energy Conversions

1 mm/s = 28,600 (24) MHz

1 mm/s = 1.1828 (10) × 10^{-7} eV

Derived Parameters

σ_o = 2.66 (10) × 10^{-19} cm^2

Γ = 3.081 (8) × 10^{-7} eV

W_o = 5.209 (15) mm/s

E_r = 5.400 (6) × 10^{-3} eV

Energies and Intensities

	E_γ	I_{Sb} (11),(12)
γ_M :	35.46 keV	20
γ_1 :	109.4 keV	0.39
γ_2 :	116.9 keV	1.10
K_{α_1} :	27.47 keV	

(1)   E. P. Mazets and Y. V. Sergeenkov, Izv. Akad. Nauk. SSSR, Ser. Fiz. 30, 1185 (1966).
(2)   C. Hohenemser and R. Rosner, Nucl. Phys. A109, 364 (1968).
(3)   E. E. Berlovich, V. V. Lukashevich, A. V. Popov, and V. M. Romanov, Yad. Fiz. 12, 217 (1970).
(4)   T. Badica, C. Ciortea, S. Dima, A. Gelberg, I. Popescu, and I. Vata, Z. Phys. 255, 390 (1972).
(5)   E. Karttunen, H. U. Freund, and R. W. Fink, Nucl. Phys. A131, 343 (1969).
(6)   H. E. Weaver, Phys. Rev. 89, 923 (1953).
(7)   75B061.
(8)   63S07.
(9)   63V02.
(10) 67P005.
(11) T. S. Nagpal and R. E. Gaucher, Can. J. Phys. 48, 2978 (1970).
(12) J. B. Gupta, N. C. Singhal, and J. H. Hamilton, Z. Phys. 261, 137 (1973).

[Other data from R. L. Auble, Nucl. Data Sheets B7, 465 (1972)]

SOURCE	S-TEMP	ABSORBER	A-TEMP	SHIFT	QS	REMARKS	REF
xx		AgAlTe2			4.2		75M065
ZnTe		AgAlTe2		+.07	4.2		75D063
xx		AgGaTe2			3.4		75M065
ZnTe		AgGaTe2		.0	3.4		75D063
xx		AgInTe2			3.3		75M065
ZnTe		AgInTe2		-.3	3.3		75D063
ZnTe		Ag2SnTe3		+.4	2.5		75D063
ZnTe		Al2Te		+.60	6.8		75D063
ZnTe(IS/Cu(I))	80	Al2Te309	4.2	+.37	6.54	bonding and structure info	75T023
Cu(I)	80	Al2Te309	80	+.71	6.70	bonding and structure info	75T023
ZnTe		BeTe	77	-.4		fa=.25, TM=145 K, bonding info	74Z009
ZnTe		CaTe	77	-.2		fa=.17, TM=128 K, bonding info	74Z009
xx		CdAl2Te4			5.6		75M065
ZnTe		CdAl2Te4					75D063
xx		CdGa2Te4			3.4		75M065
ZnTe		CdGa2Te4					75D063
xx		CdIn2Te4			3.2		75M065
ZnTe		CdIn2Te4					75D063
xx		CdTe			0		75M065
ZnTe		CdTe	77	+.6		fa=.20, TM=135 K, bonding info	74Z009
Cu(II)		CrTe	85	-.59	--	EQ=3.5 mm/s	75G026
Cu(II)		Cr2Te	85	-.22	--	EQ=5.5 mm/s, HI data	75G026
ZnTe(IS/Cu(I))	80	Cr2Te309	4.2	+.38	6.31	bonding and structure info	75T023
Cu(I)	80	Cr2Te309	80	+.31	6.25	bonding and structure info	75T023
xx		CuAlTe2			4.4		75M065
ZnTe		CuAlTe2		+.10	4.4		75D063
xx		CuGaTe2			3.7		75M065
ZnTe		CuGaTe2		+.08	3.7		75D063
xx		CuInTe2			3.4		75M065
ZnTe		CuInTe2		-.01	3.4		75D063
ZnTe		Cu2GeTe3		+.11	5.75		75D063
ZnTe		Cu2SiTe3		+.18	4.10		75D063
ZnTe		Cu2SnTe3		+.24	4.47		75D063
ZnTe		DyTe	77	0		fa=.18, TM=131 K, bonding info	74Z009
PbTe(IS/Cu(I))	4.2	p-EtOC6H4TeBr3	4.2	1.0	8.0	bonding and structure info	75B046
PbTe(IS/Cu(I))	4.2	p-EtOC6H4TeCl3	4.2	1.1	9.2	bonding and structure info	75B046
PbTe(IS/Cu(I))	4.2	p-EtOC6H4TeI3	4.2	1.0	5.2	bonding and structure info	75B046
ZnTe		Et2Te3	77	-.6	5.4	fa=.12, TM=118 K, bonding info	74Z009
ZnTe(IS/Cu(I))	80	Fe2Te309	4.2	+.47	6.42	bonding and structure info	75T023
Cu(I)	80	Fe2Te309	80	+.51	6.69	bonding and structure info	75T023
ZnTe		GaTe		+.49	6.5		75D063
xx		Ga2Te3			4.5		75M065
ZnTe		Ga2Te3		+.37	4.5		75D063
ZnTe		Ge.15Te.85				study of crystal-glass trans	75N007
ZnTe		GeTe	77	0		fa=.26, TM=147 K, bonding info	74Z009
ZnTe(IS/Cu(I))	80	H2TeO4	4.2	-1.18	0	bonding and structure info	75T023
Cu(I)	80	H2TeO4	80	-1.22	0	bonding and structure info	75T023
beta-TeO3		H6TeO6		-.03	--		75L024
Cu(I)	80	HfTe308	80	+.89	7.10	bonding and structure info	75T023
xx		In2Te3			3.5		75M065
ZnTe		In2Te3		+.20	3.5		75D063
beta-Te		La-Te alloys	77				75G062
Cu(I)	80	LiVTeO5	80	+.63	7.06	bonding and structure info	75T023
PbTe(IS/Cu(I))	4.2	p-MeOC6H4TeCl3	4.2	.9	9.2	bonding and structure info	75B046
Cu(Sb)	4.2	Me2TeBr2	4.2	+.241	8.763	bonding information	75S035
Cu(Sb)	4.2	Me2TeCl2	4.2	+.272	9.970	bonding information	75S035
Cu(Sb)	4.2	Me2TeI2	4.2	+.248	7.052	bonding information	75S035
Cu(Sb)	4.2	Me2TeI4	4.2	+.209	7.078	bonding information	75S035
ZnTe		MnTe2	77	0		fa=.22, TM=110 K, bonding info	74Z009
ZnTe		MnTe2	293	0		bonding information	74Z009
Cu(I)	80	NaVTeO5	80	+.63	6.70	bonding and structure info	75T023
beta-TeO3		Na2TeO3		-1.6	--		75L024
beta-TeO3		Na2TeO3-nH2O		-1.6	--		75L024
Cu(I)	80	Nb2Te209	80	+.85	6.55	bonding and structure info	75T023
beta-Te		Nd-Te alloys	77				75G062
ZnTe		PbTe	77	-.2		fa=.23, TM=141 K, bonding info	74Z009
Cu(I)	4.2	PhTeI	4.2	.6	9.5	bonding and structure info	75B046
Cu(I)	4.2	PhTeI3	4.2	.9	3.9	bonding and structure info	75B046
PbTe(IS/Cu(I))	4.2	Ph2TeBr2	4.2	.5	8.1	bonding and structure info	75B046
PbTe(IS/Cu(I))	4.2	Ph2TeCl2	4.2	.5	9.2	bonding and structure info	75B046
Cu(I)	4.2	Ph2TeI2	4.2	.7	5.9	bonding and structure info	75B046
beta-Te		Pr-Te alloys	77				75G062

SOURCE	S-TEMP	ABSORBER	A-TEMP	SHIFT	QS	REMARKS	REF
ZnTe		RhTe2	77	0		fa=.40, TM=110 K, bonding info	74Z009
ZnTe		RhTe2	293	0		fa=.10, TM=140 K, bonding info	74Z009
Cu(I)	80	ScTaTe6O16	80	+.78	7.32	bonding and structure info	75T023
ZnTe		ScTe	77	+.4	6.7	fa=.09, TM=110 K, bonding info	74Z009
beta-TeO3		SnTe		1.20		bonding and structure info	75B032
ZnTe		SnTe	77	+.2		fa=.22, TM=140 K, bonding info	74Z009
Cu(I)	80	SnTe3O8	80	+.69	7.34	bonding and structure info	75T023
beta-TeO3	80	Te				single crystal study, fs=.58	75O015
xx		Te				discussion of previous results	75K076
xx		Te				Mossbauer diffraction exp	75Z020
Cu(Sb)	4.2	Te	4.2	+.228	7.489		75S035
beta-TeO3	R	Te	90			nuclear diffraction from single crystals	75Z001
xx		Te	v			fa calculations	75P031
Cu(Sb)	4.2	TeCl4	4.2	.74	3.45		75B047
Cu(Sb)	4.2	TeCl4 in Ar		+1.0	3.5		75M048
Cu(Sb)	4.2	TeCl4 in Ar	4.2	1.0	3.5	nuclear radius experiment	75B047
Cu(Sb)	4.2	TeF6	4.2	-1.64			75B047
Cu(Sb)	4.2	TeF6 in Ar		-1.54			75M048
Cu(Sb)	4.2	TeF6 in Ar	4.2	-1.54		nuclear radius experiment	75B047
beta-TeO3	80	alpha-TeO2				single crystal study, fs=.58	75O015
beta-TeO3		TeO2		-1.92	7.12		75V034
beta-TeO3		TeO2		-2.1	--		75L024
xx		TeO2					75K076
beta-TeO3		beta-TeO3		.0	--		75L024
beta-Ag2Te		beta-TeO3		-1.1		structure information	75E003
Ag7Te4		beta-TeO3		-1.2		structure information	75E003
AuTe2		beta-TeO3		-1.5		structure information	75E003
Cu(Sb)		beta-TeO3		-.9	0	structure information	75E003
Cu(2-x)Te		beta-TeO3		-1.1	1	structure information	75E003
Cu(4-x)Te2		beta-TeO3		-1.1	3.0	structure information	75E003
CuTe		beta-TeO3		-1.3	6.2	structure information	75E003
NaSbO3		beta-TeO3		.0	--	.	75L024
Sb2O3-nH2O		beta-TeO3		+1.7	--		75L024
Sb2O3		beta-TeO3		+1.7	--		75L024
Sb2O4		beta-TeO3		--	---		75L024
Sb2O5		beta-TeO3		-.2	--		75L024
Sb2O5-nH2O		beta-TeO3		-.1	--		75L024
Te		beta-TeO3		-1.7	7.4	structure information	75E003
Cu(I)	80	TeU	4.2	+.16		HI=113.5 kOe	75S040
Cu(I)	80	TeU	78	+.49		HI=72.5 kOe	75S040
Cu(I)	80	TeU	138	+.56			75S040
ZnTe(IS/Cu(I))	80	alpha-TeVO4	4.2	--	--	bonding and structure info	75T023
Cu(I)	80	alpha-TeVO4	80	+.73	6.04	bonding and structure info	75T023
ZnTe(IS/Cu(I))	80	beta-TeVO4	4.2	+.60	6.59	bonding and structure info	75T023
Cu(I)	80	beta-TeVO4	80	+.57	7.00	bonding and structure info	75T023
Cu(Sb)	4.2	Te2 in Ar		+.38	9.64		75M048
Cu(Sb)	4.2	Te2 in Ar	4.2	+.38	9.64	nuclear radius experiment	75B047
Cu(Sb)	4.2	Te2 in Kr		+.31	9.55		75M048
Cu(Sb)	4.2	Te2 in Kr	4.2	+.31	9.55	nuclear radius experiment	75B047
xx(IS/Cu(I))	80	Te2O5	v	--	--	bonding and structure info	75T023
xx(IS/Cu(I))	80	Te4O9	v	--	--	bonding and structure info	75T023
Cu(I)	80	Te4U3	4.2	+.15		HI=80.0 kOe	75S040
Cu(I)	80	Te4U3	78	+.48			75S040
Cu(I)	80	Te12U7	4.2	+.16		HI=84.5 kOe	75S040
Cu(I)	80	Te12U7	78	+.59			75S040
Cu(I)	80	TiTe3O8	80	+.73	6.95	bonding and structure info	75T023
Cu(I)	80	UOTe	4.2	+.25		HI=127.5 kOe	75S040
Cu(I)	80	UOTe	78	+.31		HI=99.5 kOe	75S040
Cu(I)	80	UOTe	180	+.47			75S040
Cu(I)	80	UTeO5	80	+.63	6.55	bonding and structure info	75T023
ZnTe(IS/Cu(I))	80	UTe3O9	4.2	+.67	7.08	bonding and structure info	75T023
Cu(I)	80	UTe3O9	80	+.74	7.48	bonding and structure info	75T023
Cu(I)	80	V2Te2O9	80	+.66	6.88	bonding and structure info	75T023
ZnTe		YTe	77	+.5	4.8	fa=.11, TM=115 K, bonding info	74Z009
ZnTe		YbTe	77	+1.3		fa=.26, TM=147 K, bonding info	74Z009
xx		ZnAl2Te4			6.4		75M065
ZnTe		ZnAl2Te4		+.73	6.4		75D063
xx		ZnGa2Te4			3.5		75M065
ZnTe		ZnGa2Te4					75D063
xx		ZnIn2Te4			3.7		75M065
ZnTe		ZnIn2Te4					75D063

174

SOURCE	S-TEMP	ABSORBER	A-TEMP	SHIFT	QS	REMARKS	REF
xx		ZnTe			0		75M065
MnI2	v	ZnTe			--	T=1.4 & 77 K, study of the charge state	74P055
Pd2MnSb	4.2	ZnTe				HI=+857 kOe, Rg=-.2270	75B061
Pd2MnSb	4.2	ZnTe				HI=848 kOe	75M029
Te	N	ZnTe		-.66	7.25		75L011
Te	He	ZnTe	He	-.66	7.74		75L011
Te	He	ZnTe	He		--	single crystal experiment	75L011
Pd2MnSb	4.2	ZnTe	4.2			HI=848 kOe	75M038
ZnTe		ZnTe	77	0		fa=.26, TM=147 K, bonding info	74Z009
Cu(Sb)		ZnTe	v			discussion of previous data, fa vs temperature	75P002
Cu(I)	80	ZrTe3O8	80	+.68	7.13	bonding and structure info	75T023
PbTe(IS/Cu(I))	4.2	(p-EtOC6H4)2Te	4.2	.7	9.1	=Cl2, bonding & structure info	75B046
Cu(I)	4.2	(p-MeC6H4)2Te	4.2	.7	10.1	bonding and structure info	75B046
Cu(I)	4.2	(p-MeC6H4)2Te=	4.2	1.6	8.1	=Cl2, bonding and struc info	75B046
PbTe(IS/Cu(I))	4.2	(p-MeC6H4)2TeI2	4.2	.6	6.3	bonding and structure info	75B046
Cu(I)	4.2	(p-MeC6H4)2Te2	4.2	.6	9.9	bonding and structure info	75B046
PbTe(IS/Cu(I))	4.2	(p-MeOC6H4)2Te	4.2	.3	11.3	bonding and structure info	75B046
PbTe(IS/Cu(I))	4.2	(p-MeOC6H4)2Te2	4.2	.3	10.3	bonding and structure info	75B046
PbTe(IS/Cu(I))	4.2	(p-PhOC6H4)2Te2	4.2	.3	10.3	bonding and structure info	75B046

CODE	TOPIC	REFERENCE

74P055   TE-125   M Pasternak, The Ultimate Charge State of Te in Insulator MnI2, in "Hyperfine Interactions Studied in Nuclear Reactions and Decay: Contributed Papers" (Conference, Uppsala, Sweden, 1974), edited by E Karlsson and R Wappling (Upplands Grafiska, Uppsala, 1974), pp 82-3

74Z009   TE-125   V S Zasimov, R N Kuz'min, and A I Firov, Mossbauer Effect on Tellurium-125 Nuclei in Tellurium Compounds, in "Str Svoistva Primen Metallid, (Mater Simp), 2nd" (1972), edited by I I Kornilov and N M Matveeva ("Nauka," Moscow, 1974), pp 93-6 (In Russian)

75B032   TE-125   A A Bekker, E N Efremov, and A N Nesmeyanov, Complex Investigation of the Chemical Bond Character and the Local Surroundings Symmetry of Sn and Te in Solid Solutions on the Base of Lead and Tin Tellurides, in "5th Int Conf Mossbauer Spec, Proc," Part 1, pp 267-9 (See 75H017) (In Russian)

75B046   TE-125   F J Berry, E H Kustan, and B C Smith, J Chem Soc, Dalton Trans 1323-4(1975), Tellurium-125 Mossbauer Spectra of Some Aryltellurium-(II) and -(IV) Compounds

75B047   TE-125   P H Barrett, P A Montano, H Micklitz, and J B Mann, Phys Rev B 12,1676-80(1975), Mossbauer Study of Rare-gas-matrix Isolated 125Te Dimers, 125TeF6, and 125TeCl4 Molecules

75B061   TE-125   P Boolchand, M Tenhover, S Jha, G Langouche, B W Triplett, S S Hanna, and P Jena, Phys Lett 54A,293-4(1975), Magnetic Hyperfine Structure of 125Te in Ferromagnetic Pd2MnSb

75C015   TE-125   R M Cheyne, C H W Jones, and S Husebye, Can J Chem 53, 1855-60(1975), The 125Te Mossbauer Absorption Spectra of Tellurium Complexes with Sulfur-containing Ligands

75D063   TE-125   A K Dragunas and K V Makaryunas, Nuclear Gamma-resonance Spectra of Polycomponent and Diamond-like Semiconductors, in "5th Int Conf Mossbauer Spectrosc, Proc," Vol 3, pp 517-21 (See 75H017) (In Russian)

75E003   TE-125   E N Efremov, A A Bekker, and R V Baranova, Structure Peculiarities of the Copper, Silver and Gold Telluries, in "5th Int Conf Mossbauer Spec, Proc," Part 1, pp 270-3 (See 75H017) (In Russian)

CODE	TOPIC	REFERENCE

75G026   TE-125   J Granot and S Bukshpan, J Phys C 8,1435-42(1975), Mossbauer Effect Measurements in Ferromagnetic Cr(x)Te(y) Compounds

75G062   TE-125   V V Gorbachev, L A Linskii, S P Ionov, O A Sadovskaya, and A Yu Aleksandrov, Mossbauer Effect in Some Tellurides of Rare-earth Elements of the Cerium Subgroup, in "5th Int Conf Mossbauer Spectrosc, Proc," Vol 3, pp 522-4 (See 75H017) (In Russian)

75K076   TE-125   A V Kolpakov, E N Ovchinnikova, and R N Kuz'min, Kristallografiya 20,221-5(1975)/Sov Phys-Crystallogr 20,135-8 (1975), Symmetry of Electric Field Gradients in Crystals

75L011   TE-125   G Langouche, M Van Rossum, K P Schmidt, and R Coussement, The Quadrupole Interaction of 125Te and 129I in Polycrystalline Te and in Te Single Crystals, in "5th Int Conf Mossbauer Spec, Proc," Part 3, pp 531-6 (See 75H017)

75L024   TE-125   R A Lebedev, Yu D Perfil'ev, L A Kulikov, M I Afanasov, A M Babeshkin, and A N Nesmeyanov, Gamma-resonance Spectroscopy Investigation of the Chemical Subsequences of the Isomeric Transition, Electron Capture, Beta-minus-decay in Solid Substances, in "5th Int Conf Mossbauer Spectrosc, Proc," Vol 3, pp 537-45 (See 75H017) (In Russian)

75M029   TE-125   G R Mackay, C Blaauw, and W Leiper, J Phys F 5,L166-70 (1975), The Hyperfine Field at Tellurium Impurity Sites in the Heusler Alloy Pd2MnSb

75M038   TE-125   G R Mackay, C Blaauw, and W Leiper, The Hyperfine Field at Tellurium Impurity Sites in the Heusler Alloys Pd2MnSb125, in "Int Conf Mossbauer Spectrosc, Proc," Vol 1, pp 89-90 (See 75H027)

75M048   TE-125   H Micklitz, P H Barrett, and P A Montano, Mossbauer Studies of Rare-gas Matrix Isolated 125Te Compounds, in "Int Conf Mossbauer Spectrosc, Proc," Vol 1, pp 241-2 (See 75H027)

75M065   TE-125   K V Makaryunas, E K Makaryuene, A K Dragunas, and M L Bal'chyuene, Electric Field Gradients at the Nuclei of Tellurium and Impurity Iodine in the A(1)B(2)Te2 and A(2)B(3)Te4 Crystals, in "5th Int Conf Mossbauer Spectrosc, Proc," Vol 3, pp 529-30 (See 75H017) (In Russian)

75N007   TE-125   F S Nasredinov, B T Melekh, L N Vasil'ev, and L N Seregina, Fiz Tverd Tela (Leningrad) 17,633-5(1975)/ Sov Phys-Solid State 17,413(1975), Crystal-glass Transition in Ge(0.15)Te(0.85) and Its Influence on the Local Environment of Germanium Atoms

75O015   TE-125   A A Opalenko, I A Avenarius, R P Vardapetyan, and R N Kuz'min, Phys Status Solidi B 72,K125-30(1975), The Anisotropy of Atomic Vibrations in Te and TeO2 Crystals

75P002   TE-125   A K Prabhakaran, S B Raju, and R G Mendiratta, Solid State Commun 16,407-8(1975), Mossbauer f Factor for 35.5 keV-Te125 Transition in Zinc-Blende Type Crystals

75P031   TE-125   B M Powell and P Martel, J Phys Chem Solids 36,1287-98 (1975), The Lattice Dynamics of Tellurium

75S035   TE-125   K V Smith, J S Thayer, and B J Zabransky, Inorg Nucl Chem Lett 11,441-6(1975), Mossbauer Spectra of Some Alkyltellurium(IV) Compounds

75S040   TE-125   J Suwalski, L Dabrowski, J Leciejewicz, J Piekoszewski, and W Suski, The Mossbauer Effect of 125Te in Uranium-Tellurium Compounds, in "5th Int Conf Mossbauer Spec, Proc," Part 3, pp 514-6 (See 75H017)

75T023   TE-125   M Takeda and N N Greenwood, J Chem Soc, Dalton Trans 2207-12(1975), Tellurium-125 Mossbauer Spectra of Some Mixed Oxides of Tellurium(IV) and Some Mixed-valence Oxides of Tellurium(IV, VI)

CODE	TOPIC	REFERENCE

75V034   TE-125   R P Vardapetyan, R N Kuz'min, A A Opalenko, and W Fischer, Mossbauer Effect in the Polycrystals of Paratellurite TeO2, in "5th Int Conf Mossbauer Spectrosc, Proc," Vol 3, pp 525-8 (See 75H017) (In Russian)

75Z001   TE-125   V S Zasimov and R N Kuz'min, Phys Status Solidi B 70,K55-7 (1975), Diffraction of Resonant Gamma-quanta in Tellurium Single Crystals for Three Orders of Reflection

75Z020   TE-125   V S Zasimov, R N Kuz'min, and A Yu Aleksandrov, Fiz Tverd Tela 17,3083-6(1975)/Sov Phys-Solid State 17,2044-5(1976), Diffraction of 35.6 keV Resonance Gamma Rays in a Tellurium Single Crystal

$^{169}_{69}$ Tm (8.40 keV)

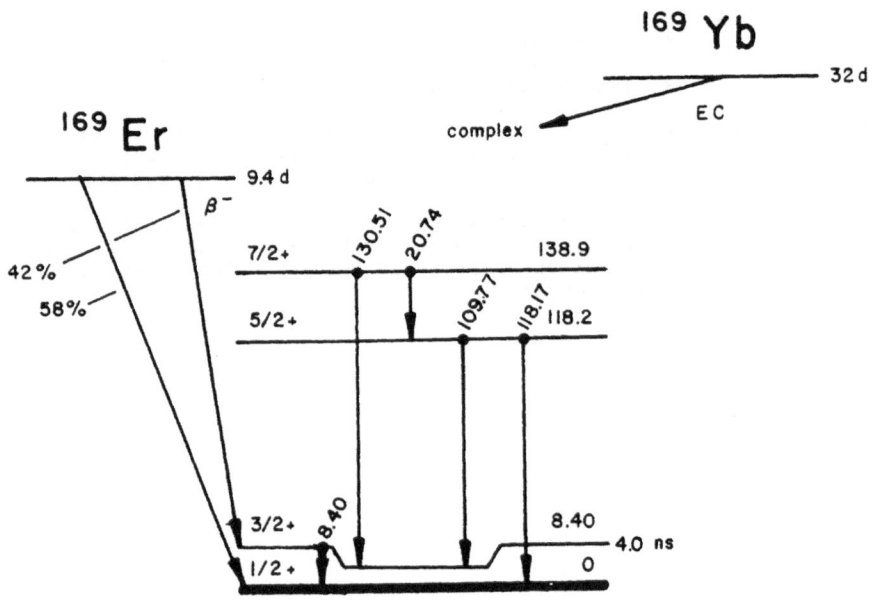

Measured Properties

E_γ = 8.401 ± 0.008 keV (1)

$t_{\frac{1}{2}}$ = 4.0 ± 0.1 ns

α_T = 268

IA = 100%

μ = −0.2310 ± 0.0015 nm (2)

R_μ = −2.31 ± 0.05 (3), (4), (5), (6)

Q^* = −1.20 ± 0.07 b (4), (5) ⊥

Derived Parameters

σ_0 = 2.58 (5) × 10^{-19} cm^2

Γ = 1.14 (3) × 10^{-7} eV

W_0 = 8.14 (20) mm/s

E_r = 2.242 (3) × 10^{-4} eV

Energy Conversions

1 mm/s = 6.776 (6) MHz

1 mm/s = 2.802 (3) × 10^{-8} eV

Energies and Intensities

	E_γ	I_{Yb} (7)
γ_M :	8.41 keV	
γ_1 :	20.7 keV	0.66
γ_2 :	63.1 keV	124
γ_3 :	93.6 keV	7.2
γ_4 :	109.8 keV	50
γ_5 :	118.2 keV	5.4
γ_6 :	130.5 keV	34
K_{α_1} :	50.7 keV	

⊥ Sternheimer corrected

(1) T. A. Carlson, P. Erman, and K. Fransson, Nucl. Phys. A111, 371 (1968).
(2) D. Giglberger and S. Penselin, Z. Phys. 199, 244 (1967).
(3) 63K09.
(4) 64C01.
(5) 64K18C.
(6) 68C018.
(7) S. K. Sen, D. L. Salie, and E. Tomchuk, Can. J. Phys. 50, 2348 (1972).

[Other data from B. Harmatz, Nucl. Data Sheets 10, 359 (1973)]

SOURCE	S-TEMP	ABSORBER	A-TEMP	SHIFT	QS	REMARKS	REF
Fe(Tm)	v	xx			--	ion implantation study	75N021
Ni(Dy)	5	xx			--	ion implantation study	75N021
Al-Er alloy		AuTm	v			4.2<T<295 K	75K011

CODE	TOPIC	REFERENCE
75K011	TM-169	C W Kimball, A E Dwight, G M Kalvius, B D Dunlap, and M V Nevitt, Phys Rev B 12,819-23(1975), Low-temperature Phase Transition and Isomer-shift Systematics in Intermediate Phases of Rare-earth-gold Compounds
75N021	TM-169	L Niesen, H P Wit, P J Kikkert, and H De Waard, Lattice Location of Rare-earth Impurities Implanted in Ferromagnetic Metals Derived from Mossbauer Experiments, in "Int Conf Mossbauer Spectrosc, Proc," Vol 1, pp 207-8 (See 75H027)
75S095	TM-169	J Sivardiere, M Blume, and M J Clauser, Hyperfine Interac 1,227-50(1975), Magnetic Relaxation and Paramagentic Mossbauer Spectra: Influence of the Off-diagonal Hyperfine Coupling

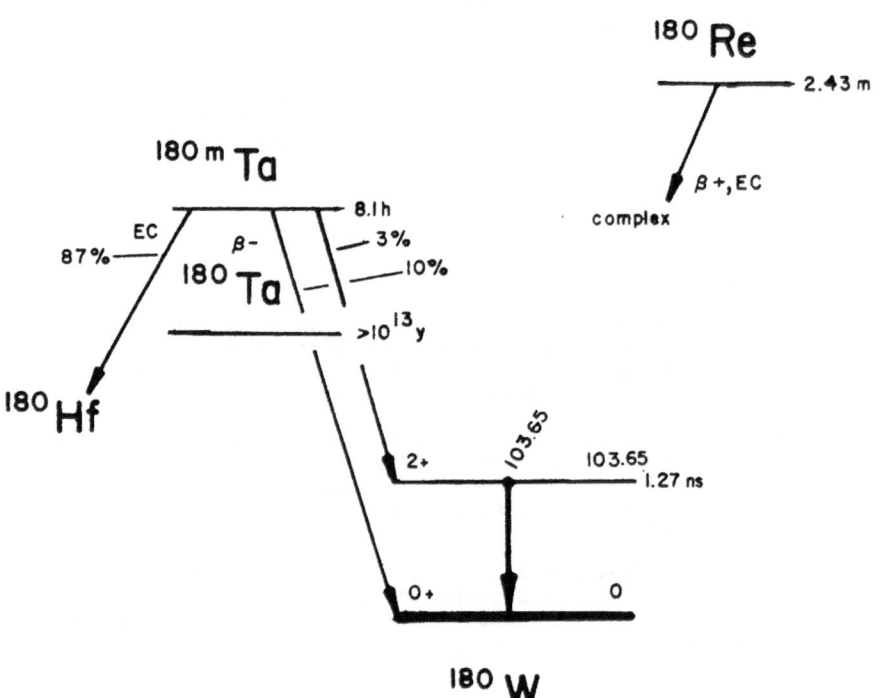

$$^{180}_{74}W\ (103\,keV)$$

Measured Properties

E_γ = 103.7 ± 0.2 keV (1),(2)

$t_{\frac{1}{2}}$ = 1.27 ± 0.06 ns (3),(4),(5)

α_T = 3.44 (theory)

IA = 0.14%

μ^* = 0.520 ± 0.034 nm (6)

Q^* = –1.82 ± 0.04 b (6),(7)

Energy Conversions

1 mm/s = 83.56 (24) MHz

1 mm/s = 3.456 (10) × 10^{-7} eV

Derived Parameters

σ_o = 2.57 (3) × 10^{-19} cm^2

Γ = 3.59 (17) × 10^{-7} eV

W_o = 2.08 (10) mm/s

E_r = 3.2007 (13) × 10^{-2} eV

Energies and Intensities

	E_γ	I_{Re}(7)
γ_M :	103. keV	23 (2)
γ_1 :	76.5 keV	0.7
γ_2 :	234.8 keV	0.5 (1)
K_{α_1} :	59.3 keV	87 (15)

(1) P. F. A. Goudsmit, J. Konijn, and F. W. N. de Boer, Nucl. Phys. A104, 497 (1967).
(2) K. J. Hofstetter and P. J. Daly, Phys. Rev. 159, 1000 (1967).
(3) W. M. Currie, Nucl. Phys. 47, 551 (1963).
(4) T. J. de Boer, E. W. ten Napel, and J. Blok, Physica 29, 1013 (1963).
(5) A. Hübner, Z. Phys. 183, 25 (1965).
(6) 73Z002.
(7) 72H001.

[Other data from L. R. Greenwood, Nucl. Data Sheets 15, 559 (1975)]

SOURCE	S-TEMP	ABSORBER	A-TEMP	SHIFT	QS	REMARKS	REF
Ta(IS/W)		WO3		.00		quadrupole moment ratio data	75B104

CODE	TOPIC	REFERENCE
75B104	W-180	H Bokemeyer, K Wohlfahrt, E Kankeleit, and D Eckardt, Z Phys A 274,305-18(1975), Mossbauer Conversion Spectroscopy: Measurements on the First Excited States of 180,182W and 145Pm

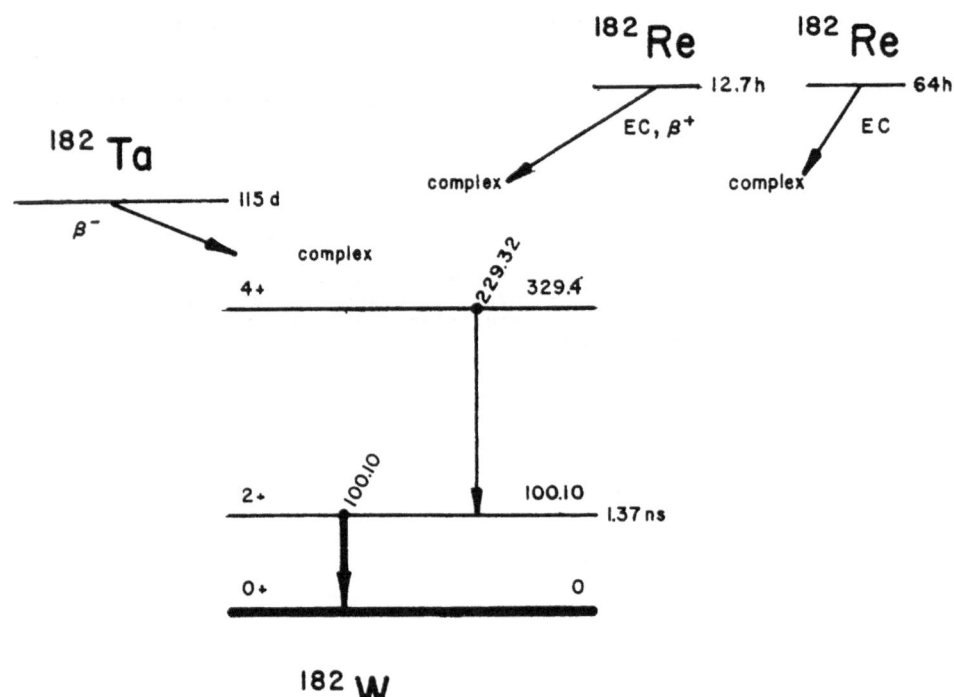

$^{182}_{74}$W (100.1 keV)

^{182}W

Measured Properties

E_γ = 100.10399 ± 0.00006 keV (1)

$t_{\frac{1}{2}}$ = 1.31 ± 0.06 ns (2), (3)

α_T = 3.85 ± 0.14 (4)

IA = 26.41 %

μ^* = + 0.512 ± 0.025 nm (5), (6), (7)

Derived Parameters

σ_o = 2.52 (7) × 10^{-19} cm^2

Γ = 3.48 (16) × 10^{-7} cm^2

W_o = 2.09 (10) mm/s

E_r = 2.95548 (3) × 10^{-2} eV

Energy Conversions

1 mm/s = 80.7386 (6) MHz

1 mm/s = 3.339110 (17) × 10^{-7} eV

Energies and Intensities

	E_γ	I_{Ta} (8)
γ_M :	100.10 keV	41
γ_1 :	65.7 keV	8.1
γ_2 :	67.6 keV	119
γ_3 :	84.7 keV	7.6
γ_4 :	113.7 keV	5.5
γ_5 :	116.4 keV	1.3
γ_6 :	152.4 keV	21
K_{α_1} :	59.3 keV	

(1) G. L. Borchert, W. Scheck, and O. W. B. Schult, Nucl. Instrum. Methods 124, 107 (1975).
(2) A. Höglund, S. G. Malmskog, A. Marelius, K. G. Välivaara, and J. Kozyczkowski, Nucl. Phys. A169, 49 (1971).
(3) D. Bloess, A. Krusche, and F. Münnich, Z. Phys. 192, 358 (1966).
(4) O. Nilsson, S. Högberg, S. E. Karlsson, and G. M. El-Sayad, Nucl. Phys. A100, 351 (1967).
(5) 65C05.
(6) 68P024.
(7) 69P15.
(8) D. H. White, R. E. Birkett, and T. Thomson, Nucl. Instrum. Methods 77, 261 (1970).

[Other data from M. R. Schmorak, Nucl. Data Sheets 14, 559 (1975).]

SOURCE	S-TEMP	ABSORBER	A-TEMP	SHIFT	QS	REMARKS	REF
Ta		Ag3W(CN)8				bonding and structure info	75C004
Ta		Cd2W(CN)8-8H2O			--	EQ=+12.8, eta=.8, struc info	75C004
Ta		Co(NH3)6W(CN)8				bonding and structure info	75C004
Ta	10	Cs2WS4	10	-.030		EQ=1.04 mm/s, eta=0	75B104
Ta		(EtNC)4W(CN)4			--	EQ=-8.6, eta=.7, struc info	75C004
Ta	10	FeWO4	10			EQ=-9.57 mm/s, eta=0	75B104
Ta		H4W(CN)8 in H2O			--	EQ=+14.8 mm/s, bond/struc info	75C004
Ta		H4W(CN)8-6H2O			--	EQ=+12.6 mm/s, bond/struc info	75C004
Ta	10	K2WS4	10	-.009		EQ=2.79 mm/s, eta=0	75B104
Ta		K3W(CN)8-H2O				bonding and structure info	75C004
Ta		K4W(CN)8 in H2O			--	EQ=-16.5 mm/s, bond/struc info	75C004
Ta		K4W(CN)8-2H2O	He		--	EQ=-16.0 mm/s, bond/struc info	75C004
Ta		Li4W(CN)8-nN2O			--	EQ=-15.9 mm/s, bond/struc info	75C004
Ta	10	(NH4)2WO2S2	10	-.041		EQ=-9.31 mm/s, eta=0	75B104
Ta	10	(NH4)2WS4	10	+.011		EQ=-3.30 mm/s, eta=0	75B104
Ta	10	Na2WO4	10	-.022			75B104
Ta		Na3W(CN)8-4H2O				bonding and structure info	75C004
Ta	10	Rb2WS4	10	-.045		EQ=1.72 mm/s, eta=0	75B104
Ta		W				W=2.1 mm/s	75K043
Ta	10	W	10	0			75B104
Ta	10	WC	10	+.010		EQ=+6.27 mm/s, eta=.34	75B104
Ta	10	WO2	10	+.028		EQ=5.38 mm/s, eta=0	75B104
Ta	10	WO3	10	-.031		EQ=-8.34 mm/s, eta=.614	75B104
Ta	10	WS2	10	+.054		EQ=10.16 mm/s, eta=0	75B104

CODE	TOPIC	REFERENCE

75B104   W-182   H Bokemeyer, K Wohlfahrt, E Kankeleit, and D Eckardt,
Z Phys A 274,305-18(1975), Mossbauer Conversion
Spectroscopy: Measurements on the First Excited States of
180,182W and 145Pm

75C004   W-182   M G Clark, J R Gancedo, A G Maddock, and A F Williams, J
Chem Soc, Dalton Trans 120-4(1975), Mossbauer Study of
Octacyanotungstate Anions

75K043   W-182   D K Kaipov, U M Makhanov, A V Kuz'minov, D N Smirin, and
Zh I Adymov, The Study of Mossbauer Effect for the
100.1 keV State of 182W with the Current Registration
Method, in "Int Conf Mossbauer Spectrosc, Proc," Vol 1,
pp 35-6 (See 75H027) (In Russian)

$^{129}_{54}$Xe (39.6 keV)

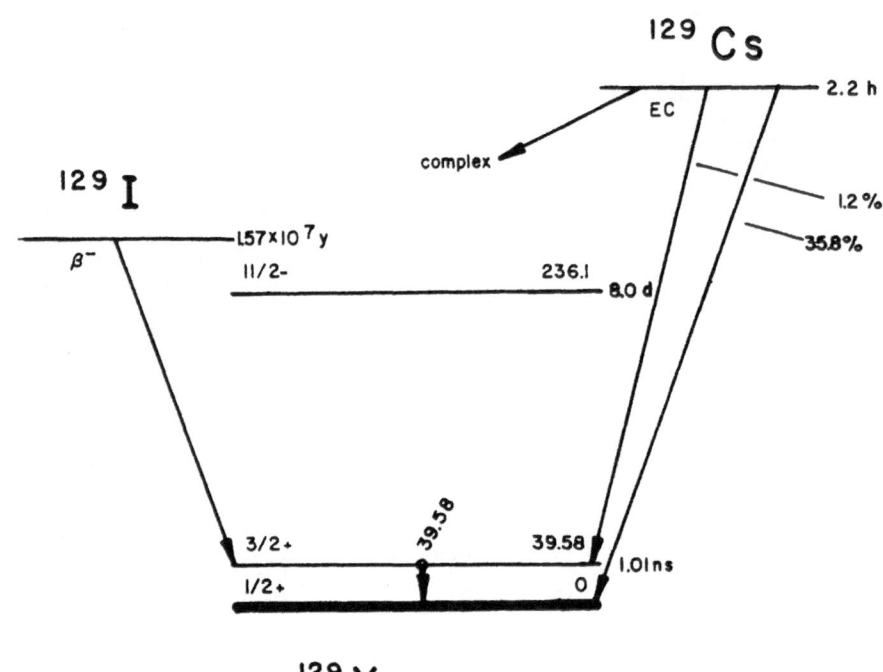

Measured Properties

E_γ = 39.58 ± 0.04 keV (1)

$t_{\frac{1}{2}}$ = 1.01 ± 0.04 ns (2)

α_T = 12.3 (2)

IA = 26.44%

μ = −0.77682 ± 0.00023 nm (3)

R_μ = −0.75 ± 0.12 (4)

Q^* = −0.41 ± 0.04 b (5)

Derived Parameters

σ_o = 2.35 (9) × 10^{-19} cm^2

Γ = 4.52 (18) × 10^{-7} eV

W_o = 6.8 (3) mm/s

E_r = 6.519 (9) × 10^{-3} eV

Energy Conversions

1 mm/s = 31.92 (3) MHz

1 mm/s = 1.3202 (13) × 10^{-7} eV

Energies and Intensities

	E_γ	I_{Cs}
γ_M :	39.58 keV	10.5
γ_1 :	93.2 keV	2.4
K_{α_1} :	29.78 keV	300

(1) H. W. Taylor and B. Singh, J. Phys. Soc. Jap. 32, 1472 (1972).
(2) J. S. Geiger, R. L. Graham, I. Bergström, and F. Brown, Nucl. Phys. 68, 352 (1965).
(3) D. Brinkman, Helv. Phys. Acta 41, 367 (1968).
(4) 74V024.
(5) 64P01.

[Other data from D. J. Horen, Nucl. Data Sheets B8, 123 (1972)]

SOURCE	S-TEMP	ABSORBER	A-TEMP	SHIFT	QS	`	REMARKS	REF
Fe(Xe)	4.2	Na4XeO6-2H2O	4.2				ion implantation experiment	75V036

CODE	TOPIC	REFERENCE
75V036	XE-129	M Van Rossum, G Langouche, H Pattyn, G Dumont, J Odeurs, A Meykens, R Coussement, and P Boolchand, Lattice Location of Xe in Iron, in "AIP Conference Proceedings-No 24, Magnetism and Magnetic Materials-1974" (20th Annual Conference, San Francisco), edited by C D Graham, Jr, G H Lander, and J J Rhyne (American Institute of Physics, New York, 1975), pp 460-1

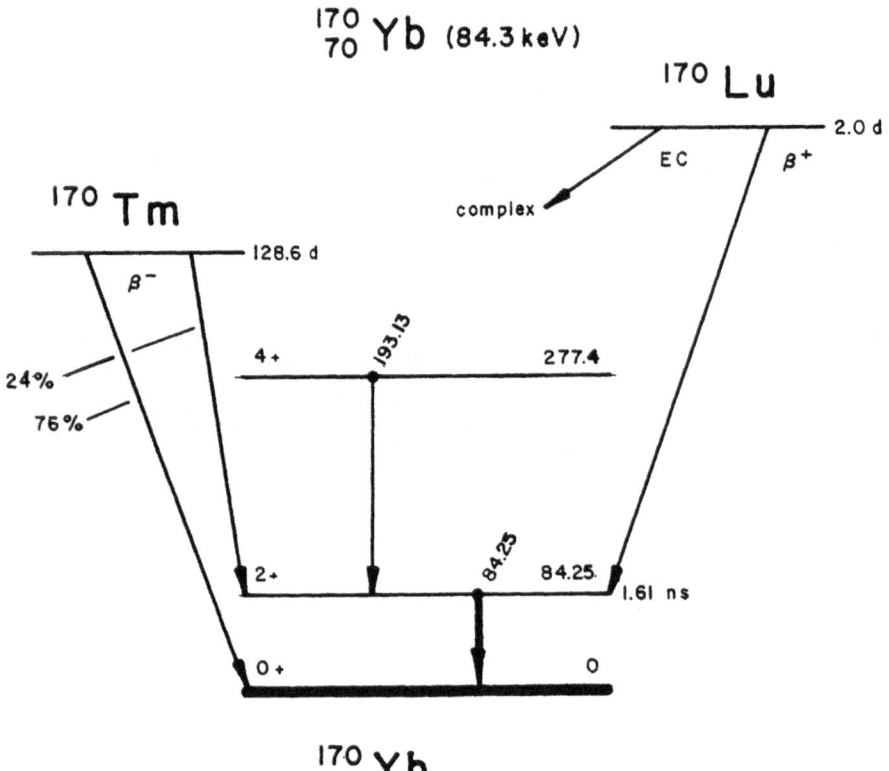

$^{170}_{70}$ Yb (84.3 keV)

^{170} Yb

Measured Properties

E_γ = 84.2529 ± 0.0012 keV (1),(2),(3)

$t_{\frac{1}{2}}$ = 1.608 ± 0.017 ns (4),(5)

α_T = 8.05 ± 0.20 (4)

IA = 3.03%

μ^* = 0.669 ± 0.008 nm (7),(8),(9)

Q^* = -2.11 ± 0.11 b (10)

Derived Parameters

σ_0 = 1.90 (4) × 10^{-19} cm^2

Γ = 2.84 (3) × 10^{-7} eV

W_0 = 2.019 (21) mm/s

E_r = 2.241323 (19) × 10^{-2} eV

Energies and Intensities

	E_γ	I_{Lu}(1)
γ_M :	84.26 keV	195
γ_1 :	152.6 keV	6
γ_2 :	194. keV	46
K_{α_1} :	54.4 keV	

Energy Conversions

1 mm/s = 67.9529 (5) MHz

1 mm/s = 2.81033 (1) × 10^{-7} eV

(1) W. Beer and J. Kern, Nucl. Imstrum. Methods 117, 183 (1974).
(2) D. E. Raeside, Nucl. Instrum. Methods 87, 7 (1970).
(3) G. L. Borchert, W. Scheck, and K. P. Wieder, Z. Naturforsch, Teil A 30, 274 (1975).
(4) D. K. Gupta and G. N. Rao, Nucl. Phys. A182, 669 (1972).
(5) A. Bäcklin, S. G. Malmskog, and H. Solhod, Ark. Fys. 34, 495 (1967).
(6) J. Plch, J. Zderadicka, and L. Kokta, Czech. J. Phys. B 23, 1181 (1973).
(7) 65H05.
(8) 68M011.
(9) 75B038.
(10) H. Ryde, G. D. Symons, and Z. Szymunski, Nucl. Phys. 80, 249 (1966).

[Other data from M. R. Schmorak and R. L. Auble, Nucl. Data Sheets 15, 371 (1975)]

SOURCE	S-TEMP	ABSORBER	A-TEMP	SHIFT	QS	REMARKS	REF
Au(Tm)	v					study of Kondo anomaly	74G059
Au(Tm)	v	xx				.6 K<T<26 K, relaxation	75G016
Fe2Tm	4.2	Al3Tb		--	--	study of electronic rearrangement effects	74H052
Au(Tm)		Al3Yb	4.2	--	--	4.2 K<T<40 K, relaxation	75S022
xx		Au				HA=400 G	74S131
TmB12	4.2	CaF2	4.2	--		g1=.334 nm	75B038
xx		Cs2Yb(1-x)Gdx=				=Cl6, spin relaxation	75A037
Fe2Ho(1-x)Tmx	v	Yb	4.1		--	4.1<T<80 K, study of crystalline fields, exchange interactions & spin reorientations	75Y005
CaF2(Tm)	4.2	YbB6	4.2	--		source experiment, g1=.334 nm	75B038
TmB12		YbEtSO4	v	--	--	1.3<T<30 relaxation	75B018
Al2Tm		YbFeO3	v			obsvn of the spin reorientation transition at Yb sites, 4.2<T<10 K	75D069
Al2Tm		YbFeO3	v		--	spin reorientation, 4.2K<T<10K	75D019

CODE	TOPIC	REFERENCE
74G059	YB-170	F Gonzalez-Jimenez and P Imbert, Mossbauer Study of a Kondo Anomaly in Au-Yb, in "Proceedings of the International Conference of Magnetism ICM-73" (Moscow, 1973) ("Nauka," Moscow, 1974), Vol V, pp 73-8
74H052	YB-170	L L Hirst, J Stohr, E Zech, G K Shenoy, and G M Kalvius, Mossbauer Studies of Magnetic Rare Earth Intermetallics: Electronic Re-arrangement Effects in Source Experiments, in "Magnetic Resonance and Related Phenomena, Proceedings of AMPERE Congress, 18th" (Nottingham, 1974), edited by P S Allen, E R Andrew, and C A Bates (North-Holland Publishing Co, Amsterdam, 1974), pp 67-8
74S130	YB-170	J Stohr, G M Kalvius, G K Shenoy, and L L Hirst, Electronic Rearrangement and Relaxation Effects Following the Nuclear Transformation in Mossbauer Source Experiments, in "Hyperfine Interactions Studied in Nuclear Reactions and Decay: Contributed Papers" (Conference, Uppsala, Sweden, 1974), edited by E Karlsson and R Wappling (Upplands Grafiska, Uppsala, 1974), pp 80-1
74S131	YB-170	J Stohr, W Wagner, and G M Kalvius, Mossbauer Spectroscopy of Dilute Au:Yb Alloys in Small External Magnetic Fields, in "Hyperfine Interactions Studied in Nuclear Reactions and Decay: Contributed Papers" (Conference, Uppsala, Sweden, 1974), edited by E Karlsson and R Wappling (Upplands Grafiska, Uppsala, 1974), pp 188-9
75A037	YB-170	L Asch, G K Shenoy, B D Dunlap, and G M Kalvius, Spin Relaxation Studies in Cubic Cs2Na(Yb(1-x)Gdx)Cl6, in "Int Conf Mossbauer Spectrosc, Proc," Vol 1, pp 403-4 (See 75H027)
75B018	YB-170	C Borely, F Gonzalez-Jimenez, P Imbert, and F Varret, J Phys Chem Solids 36,605-9(1975), Mossbauer Study of Ytterbium Ethylsulphate
75B038	YB-170	C Borely, F Gonzalez-Jimenez, P Imbert, and F Varret, J Phys Chem Solids 36,683-8(1975), Mossbauer Study of 170Yb in CaF2
75C008	YB-170	J D Cashion, M A Coulthard, and D B Prowse, J Phys C 8, 1267-75(1975), Mossbauer Isomer Shifts in Rare Earth Compounds: II. Nuclear Charge Radius and Electron Density Studies
75D019	YB-170	G R Davidson, B D Dunlap, M Eibschutz, and L G Van Uitert, Phys Rev B 12,1681-8(1975), Mossbauer Study of Yb Spin Reorientation and Low-temperature Magnetic Configuration in YbFeO3

CODE	TOPIC	REFERENCE

75D069   YB-170   G R Davidson, B D Dunlap, M Eibschutz, and L G Van Uitert, Direct Observation of the Spin-reorientation Transition at Yb Sites in YbFeO3, in "AIP Conference Proceedings-No 24, Magnetism and Magnetic Materials-1974" (20th Annual Conference, San Francisco), edited by C D Graham, Jr, G H Lander, and J J Rhyne (American Institute of Physics, New York 1975), pp 63-4

75G016   YB-170   F Gonzalez-Jimenez, B Cornut, and B Coqblin, Phys Rev B 11,4674-82(1975), Influence of the Crystalline Field and Kondo Effects on the Relaxation Rate: Application to Mossbauer Experiments of Ytterbium Diluted in Gold

75S022   YB-170   J Stohr, Phys Rev B 11,3559-72(1975), Mossbauer Relaxation Studies of 170Yb in Dilute Au: 170Tm Sources in External Magnetic Fields up to 55 kG

75Y005   YB-170   R Yanovsky, E R Bauminger, D Levron, I Nowik, and S Ofer, Solid State Commun 17,1511-4(1975), Crystalline Fields, Exchange Interactions and Spin Reorientations in TmxHo(1-x)Fe2 Systems, Studied by a Yb170 Mossbauer Probe

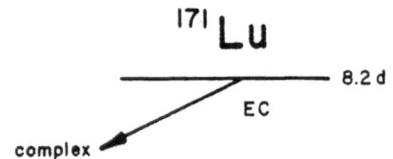

$^{171}_{70}$ **Yb** (66.7 keV, 75.9 keV)

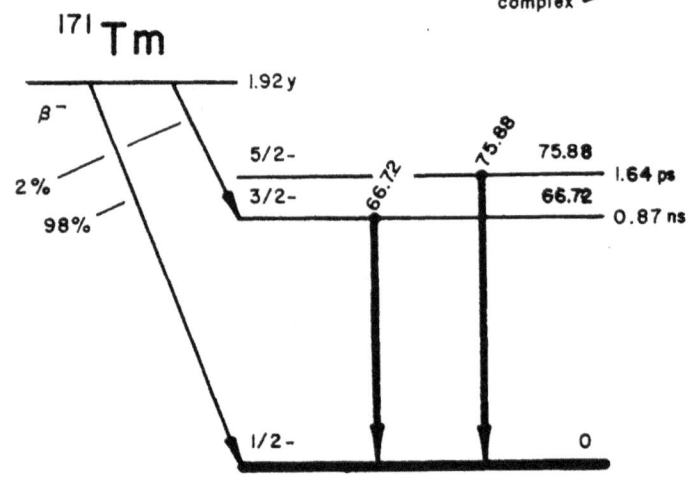

^{171} **Tm**

^{171} **Yb**

Measured Properties (E_γ = 66.7 keV)

E_γ = 66.719 ± 0.006 keV (1),(2)

$t_{\frac{1}{2}}$ = 870 ± 100 ps (3)

α_T = 13.0 (theory)

IA = 14.31%

μ = + 0.49188 ± 0.00020 nm (4)

R_μ = 0.710 ± 0.004 (3),(5)

Q^* = − 1.59 ± 0.03 b (5),(6),(7)

Derived Parameters (E_γ = 66.7 keV)

σ_o = 7.8 (3) × 10^{-20} cm^2

Γ = 5.2 (6) × 10^{-7} eV

W_o = 4.7 (5) mm/s

E_r = 1.39733 (21) × 10^{-2} eV

Energy Conversions (E_γ = 66.7 keV)

1 mm/s = 53.812 (6) MHz

1 mm/s = 2.2251 (23) × 10^{-7} eV

Measured Properties (E_γ = 75.9 keV)

E_γ = 75.875 ± 0.006 keV (1),(2)

$t_{\frac{1}{2}}$ = 1.64 ± 0.16 ps (3)

α_T = 7.83 (theory)

R_μ = 2.055 ± 0.010 nm (8)

Q^* = − 2.21 ± 0.04 b (8),(9)

Derived Parameters E_γ = 75.9 keV

σ_o = 1.443(8) × 10^{-19} cm^2

Γ = 2.8 (3) × 10^{-7} eV

W_o = 2.20 (21) mm/s

E_r = 1.80726 (24) × 10^{-2} eV

Energy Conversions (E_γ = 75.9 keV)

1 mm/s = 61.198 (6) MHz

1 mm/s = 2.53098 (23) × 10^{-7} eV

Energies and Intensities

	E_γ	I_{Lu} (2)
γ_{M1} :	66.73 keV	35(3)
γ_{M2} :	75.88 keV	108(10)
γ_1 :	55.68 keV	18(2)
γ_2 :	72.36 keV	43(4)
K_{α_1} :	53.39 keV	

(1)   E. L. Chupp, J. W. M. DuMond, F. J. Gordon, R. C. Jopson, and H. Mark, Phys. Rev. 112, 518 (1958).
(2)   N. A. Bonch-Osmolovskaya, Ts. Vylov, K. Ya. Gromov, and A. Sh. Khaminov, Izv. Akad. Nauk. SSSR, Ser. Fiz. 38, 2516 (1974).
(3)   66H002.
(4)   L. Olschewski, Z. Phys. 249, 205 (1972).
(5)   66G014.
(6)   65K01.
(7)   71P009.
(8)   70H032.
(9)   66K015.

[Other data from D. J. Horen and B. Hamatz, Nucl. Data Sheets 11, 549 (1974)]

SOURCE	S-TEMP	ABSORBER	A-TEMP	SHIFT	QS	REMARKS	REF
Al2Tm		AuYb	4.2				75K011

CODE	TOPIC	REFERENCE
75K011	YB-171	C W Kimball, A E Dwight, G M Kalvius, B D Dunlap, and M V Nevitt, Phys Rev B 12,819-23(1975), Low-temperature Phase Transition and Isomer-shift Systematics in Intermediate Phases of Rare-earth-gold Compounds

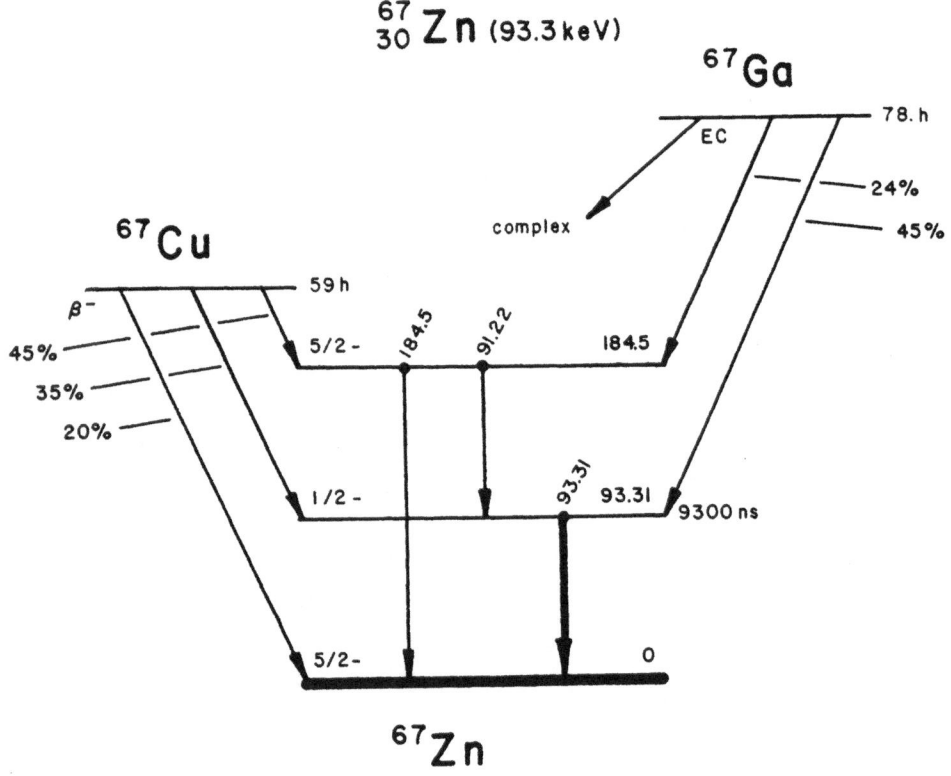

$^{67}_{30}Zn$ (93.3 keV)

Measured Properties

E_γ = 93.317 ± 0.020 keV (1)

$t_{\frac{1}{2}}$ = 9.15 ± 0.15 μs (2)

α_T = 0.89 (1)

IA = 4.11 %

μ = + 0.8755 ± 0.0003 nm (3),(4)

R_μ = + 0.66 ± 0.06 (5)

Derived Parameters

σ_o = 4.96 (13) x 10^{-20} cm^2

Γ = 4.99 (8) x 10^{-11} eV

W_o = 0.000320 (5) mm/s

E_r = 6.9766 (21) x 10^{-2} eV

Energy Conversions

1 mm/s = 75.264 (16) MHz

1 mm/s = 3.1127 (7) x 10^{-7} eV

Energies and Intensities

	E_γ	I_{Cu}(6)	I_{Ga}
γ_M :	93. keV	360	70
γ_1 :	91. keV	155	2.9
γ_2 :	184. keV	1000	20.7
K_{α_1} :	8.64 keV		

(1)   M. S. Freedman, F. T. Porter, and F. Wagner, Phys. Rev. 151, 886 (1966).
(2)   V. E. Lewis, M. J. Wood, and I. W. Goodier, Int. J. Appl. Radiat. Isotopes 23, 279 (1972).
(3)   P. W. Spence, M. N. Mermott, Phys. Lett. 24A , 430 (1967).
(4)   H. E. Weaver, Jr., Phys. Rev. 89, 923 (1953).
(5)   73P012.
(6)   S. Raman and J. J. Pinajian, Nucl. Phys. A131, 393 (1969).

[Other data from S. C. Pancholi and W. B. Ewbank, Nucl. Data Sheets B2 (6), 79 (1968)]

SOURCE	S-TEMP	ABSORBER	A-TEMP	SHIFT	QS	REMARKS	REF
ZnO(Ga)		ZnO			--	study of hyperfine interaction	74P056

CODE	TOPIC	REFERENCE
74P056	ZN-67	W Potzel, G J Perlow, and H De Waard, Hyperfine Interaction in 67ZnO by Frequency Modulation Mossbauer Spectroscopy, in "Hyperfine Interactions Studied in Nuclear Reactions and Decay: Contributed Papers" (Conference, Uppsala, Sweden, 1974), edited by E Karlsson and R Wappling (Upplands Grafiska, Uppsala, 1974), pp 266-7

1975 Topical Reference Lists

CODE	TOPIC	REFERENCE

75B102   ANALYS   J Bainbridge, Nucl Instrum Methods 128,531-5(1975), Quantitative Analysis of Mossbauer Backscatter Spectra from Multi-layer Films

75B140   ANALYS   D Barb, D Tarina, and D P Lazar, Rev Roum Phys 20,673-90 (1975), Polarization Patterns for Mossbauer Hyperfine Spectra

75D004   ANALYS   W A Dollase, Am Mineral 60,257-64(1975), Statistical Limitations of Mossbauer Spectral Fitting

75D042   ANALYS   J M Daniels, Nucl Instrum Methods 128,473-81(1975), An Approximation to the Mossbauer Lineshape Integral in Terms of Rational Functions

75D043   ANALYS   J M Daniels, Nucl Instrum Methods 128,483-93(1975), The Interpretation of Mossbauer Spectra Obtained with Polarized Gamma-rays and Single Crystal Absorbers

75K042   ANALYS   J Kansy, J E Frackowiak, B M Jankowski, and T J Panek, The Recurrent Formula for Transmission Integral Evaluation, in "Int Conf Mossbauer Spectrosc, Proc," Vol 1, pp 27-8 (See 75H027)

75M009   ANALYS   S Morup and E Both, Nucl Instrum Methods 124,445-8(1975), Interpretation of Mossbauer Spectra with Broadened Lines

75M016   ANALYS   T Mukoyama, Nucl Instrum Methods 126,153-4(1975), Fitting of Lorentzian to Mossbauer Spectra by Non-iterative Method

75N014   ANALYS   C I Nistor, Several Remarks Concerning the Fitting of Mossbauer Spectra, in "5th Int Conf Mossbauer Spec, Proc," Part 3, pp 637-9 (See 75H017)

75N019   ANALYS   D L Nagy, I Dezsi, and K Kulcsar, Can We Prepare Texture-free Mossbauer Samples? in "Int Conf Mossbauer Spectrosc, Proc," Vol 1, pp 25-6 (See 75H027)

75N025   ANALYS   S Nagy, B Levay, and A Vertes, Acta Chim Acad Sci Hung 85, 273-88(1975), Investigations on the Correlation Between the Optimum Thickness of Mossbauer Absorbents and Certain Experimental Parameters

75N034   ANALYS   Y Nishihara, Denshi Gijutsu Sogo Kenkyujo Iho 12,13-28 (1975), Least-squares Fitting of Mossbauer Spectrum (In Japanese)

75P050   ANALYS   J Piekoszewski, K Kisynska, and L Dabrowski, Nukleonika 20, 947-52(1975), Determination of the Optimum Absorber Thickness in Mossbauer Effect Experiments (In Russian)

75R005   ANALYS   K Ruebenbauer, Acta Phys Pol A 47,11-5(1975), Some Remarks on the Fourier Analysis of Mossbauer Spectra

75V004   ANALYS   E D Von Meerwall, Comput Phys Commun 9,117-28(1975), A Least-squares Spectral Curve Fitting Routine for Strongly Overlapping Lorentzians or Gaussians

75V025   ANALYS   J Van Dongen Torman, R Jagannathan, and J M Trooster, Hyperfine Interac 1,135-44(1975), Analysis of 57Fe Mossbauer Hyperfine Spectra

75V050   ANALYS   E D Von Meerwall, Comput Phys Commun 9,351-9(1975), A Fortran Code for Automatic Spectrum Analysis on Medium Scale Computers

75W028   ANALYS   J M Williams and J S Brooks, Nucl Instrum Methods 128, 363-72(1975), The Thickness Dependence of Mossbauer Absorption Line Areas in Unpolarized and Polarized Absorbers

CODE	TOPIC	REFERENCE
65S034	GENRAL	H Sano, Genshiryoku Kogyo 11(8),49-55(1965), Mossbauer Effect and Its Application in Chemistry (In Japanese)
72K076	GENRAL	R N Kuz'min, Appar Metody Rentgenovskogo Anal 10,85-8 (1972), Selection of X-ray Diagram Lines for Excitation of Mossbauer Gamma Transitions (In Russian)
72M075	GENRAL	W Meisel, Urania 48(2),28-31(1972), Werkzeug der Wissenschaft: Gammastrahlen
74C037	GENRAL	D Christov, Cv Bontschev, and Z Bursova, Die Mossbauer-Spektroskopie als Forschungs-methode in angewandten organischen Chemie. Stand und Perspektive (Bulgarian Academy of Science, Sofia, 1974), 86 pages
75F012	GENRAL	H Frauenfelder, Summary of the 5th International Conference on Mossbauer Spectroscopy, in "5th Int Conf Mossbauer Spec, Proc," Part 3, pp 763-5 (See 75H017)
75G006	GENRAL	U Gonser, From Strange Effect to Mossbauer Effect, in "Topics in Applied Physics, Volume 5 - Mossbauer Spectroscopy," edited by U Gonser (Springer-Verlag, Berlin, 1975), pp 1-51
75G015	GENRAL	U Gonser, Editor, "Topics in Applied Physics: Volume 5, Mossbauer Spectroscopy" (Springer-Verlag, Berlin, 1975)
75G057	GENRAL	V I Gol'danskii, Opening Talk, in "Int Conf Mossbauer Spectrosc, Proc," Vol 2, pp 9-16 (See 75H038)
75G061	GENRAL	U Gonser, Concluding Remarks, in "Int Conf Mossbauer Spectrosc, Proc," Vol 2, pp 475-8 (See 75H038)
75H017	GENRAL	M Hucl and T Zemcik, Editors, "The 5th International Conference on Mossbauer Spectroscopy, Proceedings" (Bratislava, Czechoslovakia, September, 1973) (Czechoslovak Atomic Energy Commision, Nuclear Information Centre, Praha 5, Zbraslav, 1975), 769 pages
75H027	GENRAL	A Z Hrynkiewicz and J A Sawicki, Editors, International Conference on Mossbauer Spectroscopy, Proceedings (Cracow, Poland, 1975) (Wykonano w Powielarni Akademii Gorniczo-Hutniczej im S Staszica, Cracow, 1975), Vol 1, 541 pages
75H038	GENRAL	A Z Hrynkiewicz and J A Sawicki, Editors, International Conference on Mossbauer Spectroscopy, Proceedings (Cracow, Poland, 1975) (Wykonano w Powielarni Akademii Gorniczo-Hutniczej im S Staszica, Cracow, 1975), Vol 2, 481 pages
75M074	GENRAL	T V Malysheva, Mossbauer Effect in Geochemistry and Cosmo-Chemistry ("Nauka," Moscow, 1975), 166 pages (In Russian)
75S006	GENRAL	N J Seeley, Nature 254,479(1975), Mossbauer Spectroscopy in Archaeology
75S023	GENRAL	I P Suzdalev, Vestn Akad Nauk SSSR (3),93-5(1975), Conference on Applications of Mossbauer Spectroscopy (In Russian)

CODE	TOPIC	REFERENCE

73D047   INSTRM   G I Danilov, S V Mamikoyan, and G N Shlokov, Tr VNII
Radiats Tekhn (9),289-96(1973), Apparatus and Method for
Quantitative Phase Analysis of Ferromagnetic Substances
Based on the Mossbauer Effect (In Russian)

74K064   INSTRM   D K Kaipov, T A Orazbaev, E F Galyutina, and L S Sergeeva,
Prib Tekh Eksp (6),75-6(1974)/Instrum Exp Tech (USSR) 17,
1633-5(1974), Properties of a Scintillation Resonance Detector for Mossbauer Investigations

74V037   INSTRM   A M Voronin, Vestn Akad Nauk Kaz (9),71-4(1974), Digital
Storing Device for Nuclear Gamma Resonance Spectroscopy
(In Russian)

75B010   INSTRM   J Becker, L Eriksson, L C Moberg, and Z H Cho, Nucl
Instrum Methods 123,199-201(1975), On the Use of Tin-loaded
Plastic Scintillators in Mossbauer Spectroscopy

75B112   INSTRM   Yu F Babikova, M R Gryaznov, L M Isakov, N S Kolpakov, and
M N Uspenskii, Prib Tekh Eksp (3),152-5(1975)/Instrum Exp
Tech (USSR) 18,850-53(1975), Scattered Gamma-ray Registra-
tion Channel for Nuclear Gamma-resonance Spectrometer

75B149   INSTRM   S I Bondarevskii and V V Eremin, Prib Tekh Eksp (6),57-8
(1975)/Instrum Exp Tech (USSR), Scintillation Resonance
Detector for Mossbauer Studies

75D067   INSTRM   D Dora, E Kisch, and L Varhaemi, Design and Measurement
Possibilities of the NP-255 Type Spectrometer for the
Mossbauer Effect Measurements, in "5th Int Conf Mossbauer
Spectrosc, Proc," Vol 3, pp 656-60 (See 75H017)
(In Russian)

75G043   INSTRM   P L Gruzin, Yu V Petrikin, and R A Stukan, Prib Tekh Eksp
(3),48-9(1975)/Instrum Exp Tech (USSR) 18,716-8(1975),
Installation for the Observation of Nuclear Gamma-resonance
According to Conversion Electrons

75G063   INSTRM   I Ya Garzanov, V I Karlashuk, V A Kotelnikov, V K
Labushkin, and E F Makarov, Mossbauer Resonance Analyzer
of Cassiterite MAKR-4, in "5th Int Conf Mossbauer Spectrosc, Proc," Vol 3, pp 660-4 (See 75H017) (In Russian)

75G066   INSTRM   S K Godovikov and V G Snigirev, Prib Tekh Eksp (3),243-4
(1975)/Instrum Exp Tech (USSR) 18,970-1(1975), Magnesium
Windows with a Miniature Indium Seal for Mossbauer Helium
Cryostats

75I001   INSTRM   Y Isozumi and M Takafuchi, Bull Inst Chem Res, Kyoto Univ
53,63-7(1975), Mossbauer Spectroscopy by Scattered Elec-
trons at 77 K

75K007   INSTRM   M N Kamthe, K V Bankar, and S K Date, J Phys Educ 2,15-20
(1975), Design and Construction of Versatile Mossbauer
Spectrometer

75K066   INSTRM   E Kankeleit, Some Technical Developments in Mossbauer Spectroscopy, in "Int Conf Mossbauer Spectrosc, Proc,"
Vol 2, pp 43-58 (See 75H038)

75L028   INSTRM   R A Levy, P A Flinn, and R A Hartzell, Nucl Instrum
Methods 131,559-62(1975), A Proportional Counter for
Efficient Mossbauer Scattering Experiments

75S041   INSTRM   K P Schmidt, J De Raedt, G Langouche, M Van Rossum, and
R Coussement, A Computerized Piezoelectric Mossbauer Spectrometer, in "5th Int Conf Mossbauer Spec, Proc," Part 3,
pp 649-52 (See 75H017)

75S056   INSTRM   A Kh Sherif, A A Opalenko, and R N Kuz'min, Unit for
Measuring of the Mossbauer Effect under High Pressure, in
"Int Conf Mossbauer Spectrosc, Proc," Vol 1, pp 33-4
(See 75H027)

CODE	TOPIC	REFERENCE

75S092 INSTRM M Seberini and J Cirak, Jad Energ 21,263-4(1975), Determination of Zero Velocity of a Mossbauer Spectrometer Using the Method of Electronic Magnifying Lens (In Slovak)

75T030 INSTRM T Tomov, T Ruskov, S Georgiev, and N Pavlov, Design of an Electro-mechanical Vibrator and Its Temperature Stabilization, in "5th Int Conf Mossbauer Spectrosc, Proc," Vol 3, pp 653-6 (See 75H017) (In Russian)

75W004 INSTRM H P Wit, Rev Sci Instrum 46,927-8(1975), Simple Moire Calibrator for Velocity Transducers Used in Mossbauer Effect Measurements

75W012 INSTRM M Wroblewski, J Mirkowski, and J Wroblewska, Camac Mossbauer Spectrometer System with Small On-line Computer for Multi-spectra Recording and Processing, in "Int Conf Mossbauer Spectrosc, Proc," Vol 1, pp 31-2 (See 75H027)

75Y001 INSTRM T Yoshimura, M Fujiwara, and N Wakabayashi, Jpn J Appl Phys 14,691-5(1975), A Derivative Mossbauer Spectrometer Stabilized by Digital Method

CODE	TOPIC	REFERENCE
73B101	PROPSL	A P Buev, U Kh Kopvillem, and L N Shakhmuratova, Uch Zap Kazan Gos Pedagog Inst 125,60-5(1973), Gamma Echo, Bragg Echo, and the Phenomenon of Echo Signals in the Correlation Properties of Nuclear Radiations (In Russian)
73M059	PROPSL	A V Mitin and G P Chugunova, Uch Zap Kazan Gos Pedagog Inst 125,73-8(1973), Optical Principles of Detecting Double Gamma A-nuclear Resonances (In Russian)
74S129	PROPSL	P I Smolyanskii, Zap Vses Mineral Obshch (2),272-8(1974), On a Possible Application of Mossbauer Effect of Optical Analogue in Mineralogy (On Example of Fluorite) (In Russian)
75A042	PROPSL	V G Alpatov, A G Beda, G E Bizina, A V Davydov, and M M Korotkov, State of the Experiments on the Gamma-resonance Excitation of the Long-lived Isomeric State of 107Ag, in "5th Int Conf Mossbauer Spectrosc, Proc," Vol 3, pp 499-504 (See 75H017) (In Russian)
75B008	PROPSL	G C Baldwin and R V Khokhlov, Phys Today February, 32-9 (1975), Prospects for a Gamma-ray Laser
75B100	PROPSL	D Barb and D Tarina, Use of Polarized Gamma Ray for Effective Hyperfine Parameter Determinations, in "Int Conf Mossbauer Spectrosc, Proc," Vol 1, pp 467-8 (See 75H027)
75B113	PROPSL	D Barb and M Rogalski, J Chim Phys Phys-Chim Biol 72,470-6 (1975), Conditions pour Obtenir.l'Effet Faraday Classique avec un Rayonnement Gamma Emis sans Recul
75B118	PROPSL	Sh Sh Bashkirov and E K Sadykov, Fiz Tverd Tela 17,1864-6 (1975)/Sov Phys-Solid State 17,1226(1975), A Mechanism of Double Gamma Resonance
75G034	PROPSL	V I Gol'danskii, On the General Principles and Problems of Realization of Gasers (Gamma-lasers), in "5th Int Conf Mossbauer Spec, Proc," Part 3, pp 584-93 (See 75H017)
75H025	PROPSL	A Z Hrynkiewicz, E Popiel, and J A Sawicki, Angular Correlations in a 122.0 keV - 14.4 keV 57Fe Mossbauer Coincidence Experiment, in "5th Int Conf Mossbauer Spec, Proc," Part 3, pp 591-3 (See 75H017)
75H028	PROPSL	J P Hannon and G T Trammell, Opt Commun 15,330-4(1975), Anomalous Emission Effect and the Possibility of Gamma-ray Lasers
75K061	PROPSL	E Kankeleit, Z Phys A 275,119-21(1975), The Effect of Decaying Atomic States on Integral and Time Differential Mossbauer Spectra
75K087	PROPSL	P Kamenov and Cv Bontschev, Dokl Bolg Akad Nauk 28,1175-7 (1975), On the Possibility of Realizing a Gamma Laser with Long-living Isomer Nuclei
75P025	PROPSL	E Popiel and J A Sawicki, Evaluations of ME-PAC Experiments in 57Fe, in "Int Conf Mossbauer Spectrosc, Proc," Vol 1, pp 13-4 (See 75H027)
75T024	PROPSL	G T Trammell and J P Hannon, Opt Commun 15,325-9(1975), Threshold Conditions for Pulsed Gamma-ray Lasers
75Z023	PROPSL	E V Zolotoyabko and E M Iolin, Latv PSR Zinat Akad Vestis, Fiz Teh Zinat Ser (4),46-50(1975), Study of Structural Phase Transitions by Using the Coherent Scattering of Mossbauer Radiation (In Russian)

CODE	TOPIC	REFERENCE

65S035   REVIEW   H Sano, Bussei 6,315-21(19765), Chemical Application of the Nuclear Gamma Ray Resonance - Mossbauer Effect (In Japanese)

68H050   REVIEW   G Harris, Theor Chim Acta 10,155-80(1968), Spin-mixing and the Different Spin States of Ferric Ion in Tetragonal Symmetry. II. Localized Properties: Zero Field Splittings, Effective Magnetic Moments, Magnetic Field Energies and Electric Field Gradients of Ferric Ion

72B096   REVIEW   K Burger, Kem Ujabb Eredmenyei 9,89-199(1972), Application of Mossbauer Spectroscopy to Complex Chemistry (In Hungarian)

73S088   REVIEW   I P Suzdalev, V I Gol'danskii, G M Zabrova, and B M Kadenaci, Untersuchung topochemischer Reaktionen mit Hilfe der Mossbauer-Spektroskopie und der Derivatographie, in "Festkorperchemie," edited by V Boldyrev and K Meyer (Verlag Grundstoffind, Leipzig, 1973), pp 160-78

74B101   REVIEW   V N Bykov, S S Budagovskii, M I Gavrilyuk, V S Golovkin, V A Levdik, I I Rudnev, and V A Solov'ev, Electronic Structure and Anomalies of Macroscopic Physical and Technological Properties of Transition Metals, in "Tr Fiz-Energ Inst," edited by V A Kuznetsov (Atomizdat, Moscow, 1974), pp 469-81 (In Russian)

74D055   REVIEW   D W Dale, Mossbauer Spectroscopy, in "Modern Physical Techniques in Materials Technology," edited by T Mulvey and R K Webster (Oxford University Press, London, 1974), pp 281-90

74F035   REVIEW   J Flouquet, Ann Phys (Paris) 8,5-51(1974), L'Etude des Alliages Dilues par Orientation Nucleaire

74G058   REVIEW   V I Gol'danskii, Usp Khim 43,2113-45(1974)/Russ Chem Rev 43,1021-42(1974), The Periodic Law and Progress in the Study of the Structure of Matter

74J018   REVIEW   C Janot and P Delcroix, Ethnol Fr 3,179-88(1974), Caracterisation de Materiaux Archeologiques par Spectrometrie Mossbauer

74M058   REVIEW   B N Maimur and V F Moroz, Use of Gamma Resonance Spectroscopy for Studying Phase Transformations during Heating and Reduction of Iron Oxides, in "Nov Metody Issled Protsessov Vosstanov Chern Met Dokl Simp" (1971), edited by S T Rostovtsev ("Nauka," Moscow, 1974), pp 56-62 (In Russian)

74S119   REVIEW   C M Scala, Aust Gemmol 12,119-24(1974), A Survey of Some Research Work on Impurities in Gems

74S127   REVIEW   H Sano, Boshoku Gijutsu 23,311-6(1974), Mossbauer Spectroscopy and Its Applications to Corrosion Studies (In Japanese)

74S128   REVIEW   D A Shirley, Effects of Extranuclear Fields on Nuclear Radiations, in "Nuclear Spectroscopy and Reactions" (Volume 40-C in "Pure and Applied Physics"), edited by J Cerny (Academic Press, New York, 1974), pp 513-49

74W026   REVIEW   R E Watson and L H Bennett, Charge Transfer in Alloys: The Blind Men and the Elephant, in "Charge Transfer/Electronic Structure of Alloys" (Twin Symposium, Philadelphia, 1973), edited by L H Bennett and R H Willens (Metallurgical Society of the American Institute of Mining, Metallurgical, and Petroleum Engineers, 1974), pp 1-21

75B002   REVIEW   C M P Barton and N N Greenwood, Europium-151 Mossbauer Spectroscopy, in "Mossbauer Effect Data Index, Covering the 1973 Literature," edited by J G Stevens and V E Stevens (Plenum Publishing Corp, New York, 1975), pp 395-445

CODE	TOPIC	REFERENCE
75B056	REVIEW	V A Belyakov, Usp Fiz Nauk 115,553-601(1975)/Adv Phys Sci (USSR), Crystal Diffraction of Mossbauer Gamma-radiation
75B062	REVIEW	D Barb, Polarized Mossbauer Transitions, in "5th Int Conf Mossbauer Spec, Proc," Part 3, pp 665-704 (See 75H017)
75B107	REVIEW	D Barb, The Role of Polarization State of Gamma Rays in Mossbauer Spectroscopy, in "Int Conf Mossbauer Spectrosc, Proc," Vol 2, pp 379-97 (See 75H038)
75B108	REVIEW	G C Baldwin, J W Pettit, and H R Schwenn, A Critical Review of Current Proposals for Mossbauer Gasers, in "Int Conf Mossbauer Spectrosc, Proc," Vol 2, pp 413-28 (See 75H038)
75B147	REVIEW	O Brummer, K Berndt, G Dlubek, U Berg, G Drager, W Beier, and G Dworzak, Wiss Z Martin-Luther-Univ, Halle-Wittengerg, Math-Naturwiss Reihe 24(6),5-28(1975), Some Solid State Spectroscopic Studies by the Physics Section
75C010	REVIEW	R Cammack, The Mossbauer Effect and Its Applications in Biology, in "New Techniques in Biophysics and Cell Biology," edited by R H Pain and B Smith (John Wiley & Sons, Inc, New York, 1975), Vol 2, pp 341-84
75C021	REVIEW	M G Clark, Hyperfine Interactions and Molecular Structure, in "MTP International Review of Science, Physical Chemistry Series 2, Volume 2," edited by A D Buckingham (Butterworths, London, 1975), Chap 7, pp 239-97
75C056	REVIEW	J M D Coey, The Clay Minerals; Use of the Mossbauer Effect to Characterize Them and Study Their Transformations, in "Int Conf Mossbauer Spectrosc, Proc," Vol 2, pp 333-54 (See 75H038)
75D001	REVIEW	H De Waard, 129I Mossbauer Spectroscopy, in "Mossbauer Effect Data Index, Covering the 1973 Literature," edited by J G Stevens and V E Stevens (Plenum Publishing Corp, New York, 1975), pp 447-94
75D021	REVIEW	B W Dale, Contemp Phys 16,127-46(1975), Mossbauer Spectroscopy
75D023	REVIEW	H De Waard, Phys Scr 11,157-66(1975), The Investigation of Radiation Damage and Lattice Location by Nuclear Hyperfine Interaction Techniques
75D060	REVIEW	L M Dautov, M M Kadykenov, and D K Kaipov, The Ratio of Mossbauer Nuclei delta(R2) Parameters, in "Int Conf Mossbauer Spectrosc, Proc," Vol 1, pp 497-8 (See 75H027)
75D062	REVIEW	I Dezsi, Mossbauer Effect Studies on Phase Transitions, in "Int Conf Mossbauer Spectrosc, Proc," Vol 2, pp 221-38 (See 75H038)
75D064	REVIEW	L M Dautov, M M Kadykenov, and D K Kaipov, Systematics of the Radius Changes of Nuclei during the Excitations on Mossbauer Levels, in "5th Int Conf Mossbauer Spectrosc, Proc," Vol 3, pp 619-32 (See 75 H017) (In Russian)
75F007	REVIEW	P A Flinn, Metallurgical Applications of the Mossbauer Effect, in "5th Int Conf Mossbauer Spec, Proc," Part 2, pp 275-88
75F009	REVIEW	H Frauenfelder, Biological Molecules, in "5th Int Conf Mossbauer Spec, Proc," Part 2, pp 401-7 (See 75H017)
75F014	REVIEW	F E Fujita, Mossbauer Spectroscopy in Physical Metallurgy, in "Topics in Applied Spectroscopy, Volume 5 - Mossbauer Spectroscopy," edited by U Gonser (Springer-Verlag, Berlin, 1975), pp 201-36
75F025	REVIEW	H Frauenfelder, Dynamics of Biomolecules, in "Int Conf Mossbauer Spectrosc, Proc," Vol 2, pp 305-17 (See 75H038)

CODE	TOPIC	REFERENCE

75F026   REVIEW   E N Frolov, G I Likhtenshtein, and V I Gol'danskii, The Dynamic Structure of Proteins and Water-protein Interactions, in "Int Conf Mossbauer Spectrosc, Proc," Vol 2, pp 319-31   (See 75H038)

75F027   REVIEW   A J Freeman, Status of Theoretical Charge and Spin Determinations, in "Int Conf Mossbauer Spectrosc, Proc," Vol 2, pp 435-65   (See 75H038)

75G010   REVIEW   H M Gager and M C Hobson, Jr, Catal Rev 11,1-40(1975), Mossbauer Spectroscopy

75G040   REVIEW   R W Grant, Mossbauer Spectroscopy in Magnetism: Characterization of Magnetically-ordered Compounds, in "Topics in Applied Physics, Volume 5 Mossbauer Spectroscopy," edited by U Gonser (Springer-Verlag, Berlin, 1975), pp 97-137

75G042   REVIEW   P Gutlich, Mossbauer Spectroscopy in Chemistry, in "Topics in Applied Physics, Volume 5 - Mossbauer Spectroscopy," edited by U Gonser (Springer-Verlag, Berlin, 1975), pp 53-96

75G058   REVIEW   U Gonser, Defects, in "Int Conf Mossbauer Spectrosc, Proc," Vol 2, pp 113-31   (See 75H038)

75G059   REVIEW   V I Gol'danskii, Yu V Maksimov, and I P Suzdalev, Application of Mossbauer Spectroscopy in Catalysis, in "Int Conf Mossbauer Spectrosc, Proc," Vol 2, pp 163-88 (See 75H038)

75G060   REVIEW   V I Gol'danskii and V A Naimot, Note on the Non-Mossbauer Gamma Ray Laser, in "Int Conf Mossbauer Spectrosc, Proc," Vol 2, pp 429-33   (See 75H038)

75G078   REVIEW   R Greatrex, Mossbauer Spectroscopy, in "Spectroscopic Properties of Inorganic and Organometallic Compounds, Volume 8," edited by N N Greenwood (The Chemical Society, Burlington House, London, 1975), pp 415-509

75H004   REVIEW   W B Holzapfel, CRC Crit Rev - Solid State Sci 5,89-123 (1975), Mossbauer Studies at High Pressure

75H006   REVIEW   S S Hafner, Mossbauer Spectroscopy in Lunar Geology and Mineralogy, in "Topics in Applied Physics, Volume 5-Mossbauer Spectroscopy," edited by U Gonser (Springer-Verlag, Berlin, 1975), pp 167-99

75H024   REVIEW   A Z Hrynkiewicz, Mossbauer Spectroscopy Compared with Other Spectroscopic Techniques, in "5th Int Conf Mossbauer Spec, Proc," Part 3, pp 573-83   (See 75H017)

75H036   REVIEW   S S Hafner, Mossbauer Results from the Lunar Samples Returned by the Apollo Missions. A Summary, in "Int Conf Mossbauer Spectrosc, Proc," Vol 1, pp 471-2   (See 75H027)

75H037   REVIEW   C Hohenemser, A Review of Mossbauer Effect Studies of Magnetic Critical Phenomena, in "Int Conf Mossbauer Spectrosc, Proc," Vol 2, pp 239-56   (See 75H038)

75H046   REVIEW   F Hartmann-Boutron, Ann Phys (Paris) 9,285-356(1975), Theories de la Relaxation des Impuretes Radioactives dans les Solides

75I013   REVIEW   S P Ionov, G V Ionova, V S Lubimov, and E F Makarov, Phys Status Solidi B 71,11-57(1975), Instability of Crystal Lattices with Respect to Electron Density Redistributions

75J005   REVIEW   C E Johnson, Mossbauer Spectroscopy in Biology, in "Topics in Applied Physics, Volume 5 - Mossbauer Spectroscopy," edited by U Gonser (Springer-Verlag, Berlin, 1975), pp 139-66

CODE	TOPIC	REFERENCE

75J011 REVIEW C E Johnson, Mossbauer Spectroscopy, in "Amino-acids, Peptides, and Proteins, Volume 6" (A Specialist Periodical Report: A Review of the Literature Published during 1973), edited by R C Sheppard (The Chemical Society, Burlington House, London, 1974), pp 256-61

75K001 REVIEW B Keisch, Mossbauer Effect Spectroscopy Without Sampling: Application to Art and Archaeology, in "Advances in Chemistry Series, Number 138: Archaeological Chemistry" (American Chemical Society, New York, 1975), pp 186-206

75K036 REVIEW G M Kalvius, Electronic Structure of the Light Actinides from Mossbauer Spectroscopy, in "5th Int Conf Mossbauer Spec, Proc," Part 3, pp 485-98 (See 75H017)

75K039 REVIEW W Keune, U Gonser, and H Vollmar, Mossbauer-Spektroskopie als zerstorungsfreie Materialprufmethode, in "Handbuch der zerstorungsfreien Materialprufung," edited by E A W Muller (R Oldenbourg Verlag, Munich, 1975), U 3, 31, 32, 321, 322

75K065 REVIEW Yu M Kagan, Coherent Phenomenon on Interaction of Mossbauer Radiation with Crystals (Progress in Theory), in "Int Conf Mossbauer Spectrosc, Proc," Vol 2, pp 17-41 (See 75H038) (In Russian)

75K066 REVIEW E Kankeleit, Some Technical Developments in Mossbauer Spectroscopy, in "Int Conf Mossbauer Spectrosc, Proc," Vol 2, pp 43-58 (See 75H038)

75K067 REVIEW W Kundig, Calibration of the Isomer Shift, in "Int Conf Mossbauer Spectrosc, Proc," Vol 2, pp 355-67 (See 75H038)

75L005 REVIEW G Longworth, Nature 256,367(1975), Mossbauer Effect and Order-disorder Transitions in Alloys

75L023 REVIEW F J Litterst and G M Kalvius, Investigations of Non-crystalline Materials and Liquid Crystals by the Mossbauer Effect, in "Int Conf Mossbauer Spectrosc, Proc," Vol 2, pp 189-220 (See 75H038)

75M021 REVIEW W Meisel, Some Analytical Aspects of the Mossbauer Spectrometry, in "5th Int Conf Mossbauer Spec, Proc," Part 1, pp 200-13 (See 75H017)

75M064 REVIEW E F Makarov and R A Stukan, Emission Gamma-resonance Spectroscopy, in "Int Conf Mossbauer Spectrosc, Proc," Vol 2, pp 133-61 (See 75H038)

75M068 REVIEW B Manuschev, Technological Applications of the Mossbauer Effect, in "5th Int Conf Mossbauer Spectrosc, Proc," Vol 3, pp 687-98 (See 75H017) (In Russian)

75M086 REVIEW E Munck and P M Champion, Ann N Y Acad Sci 244,142-62 (1975), Heme Proteins and Model Compounds: Mossbauer Absorption and Emission Spectroscopy

75N018 REVIEW S Nasu and U Gonser, Zairyo 24,1083-91(1975), Mossbauer Effect and Its Applications in Material Science (In Japanese)

75N026 REVIEW I Nowik, Magnetic Structure and Transferred Hyperfine Interactions, in "Int Conf Mossbauer Spectrosc, Proc," Vol 2, pp 83-93 (See 75H038)

75O013 REVIEW D A O'Connor, Crystallography Using the Rayleigh Scattering of Gamma Rays, in "Int Conf Mossbauer Spectrosc, Proc," Vol 2, pp 369-77 (See 75H038)

75P022 REVIEW H D Pfannes and U Gonser, Texture Problems in Mossbauer Spectroscopy, in "5th Int Conf Mossbauer Spec, Proc," Part 3, pp 596-9 (See 75H017)

CODE	TOPIC	REFERENCE

75P048   REVIEW   F Parak, Spatial Arrangement and Electronic Structure of the Active Center of Iron Containing Biomolecules, in "Int Conf Mossbauer Spectrosc, Proc," Vol 2, pp 285-303 (See 75H038)

75R004   REVIEW   L D Roberts, The Mossbauer Effect for 197Au, in "Mossbauer Effect Data Index, Covering the 1973 Literature," edited by J G Stevens and V E Stevens (Plenum Publishing Corp, New York, 1975), pp 349-91

75R013   REVIEW   K Raclavsky, Mossbauer Effect Technique in Mineral Science, in "5th Int Conf Mossbauer Spec, Proc," Part 2, pp 347-55 (See 75H017)

75S086   REVIEW   G K Shenoy and B D Dunlap, Mossbauer Relaxation Lineshapes in the Presence of Hyperfine Interactions Containing Off-diagonal Terms, in "Int Conf Mossbauer Spectrosc, Proc," Vol 2, pp 275-84 (See 75H038)

75S087   REVIEW   J G Stevens and V E Stevens, Mossbauer Effect Data Center, in "Int Conf Mossbauer Spectrosc, Proc," Vol 2, pp 467-73 (See 75H038)

75S090   REVIEW   R H Sands and W R Dunham, Quart Rev Biophys 7,443-504 (1975), Spectroscopic Studies of Two-iron Ferredoxins

75T006   REVIEW   T Tominaga, Gendai Kagaku (46),32-9(1975), Mossbauer Spectroscopy (In Japanese)

75V008   REVIEW   A Vertes, Mossbauer Effect Studies of Chemical Reactions, in "5th Int Conf Mossbauer Spec, Proc," Part 1, pp 179-99 (See 75H017)

75W011   REVIEW   J M Williams, Cryogenics 15,307-22(1975), The Mossbauer Effect and Its Applications at Very Low Temperatures

75W024   REVIEW   H H F Wegener, The Study of Relaxation Mechanisms by Means of the Mossbauer Spectroscopy, in "Int Conf Mossbauer Spectrosc, Proc," Vol 2, pp 257-74 (See 75H038)

75W025   REVIEW   W K Wojtowiecki and S B Sazonov, The Amplitude and Phase Modulation of Mossbauer Gamma-quanta, in "Int Conf Mossbauer Spectrosc, Proc," Vol 2, pp 399-411 (See 75H038)

75Z014   REVIEW   T Zemcik, Cesk Cas Fyz 25,464-75(1975), Information on the Electronic Structure of Metals and Alloys from the Hyperfine Structure of Mossbauer Spectra (In Czech)

75Z018   REVIEW   T Zemcik, Progress in Mossbauer Spectroscopy of Metallic Systems, in "Int Conf Mossbauer Spectrosc, Proc," Vol 2, pp 59-81 (See 75H038)

CODE	TOPIC	REFERENCE

73B101 THEORY A P Buev, U Kh Kopvillem, and L N Shakhmuratova, Uch Zap Kazan Gos Pedagog Inst 125,60-5(1973), Gamma Echo, Bragg Echo, and the Phenomenon of Echo Signals in the Correlation Properties of Nuclear Radiations (In Russian)

74B097 THEORY V A Belyakov and Yu M Aivazyan, Tr Metrol Inst SSSR 119, 38-43(1974), Mossbauer Diffraction on Crystals with a Complex Structure of Electric Field Gradients (In Russian)

74B098 THEORY V A Belyakov and Yu M Aivazyan, Tr Metrol Inst SSSR 119, 43-9(1974), Interference Effects in the Resonance Scattering of Gamma Radiation from Nuclei with Two Mossbauer Transitions on Crystals (In Russian)

74D057 THEORY A Ya Dzyublik, Zh Eksp Teor Fiz 67,1534-8(1974)/Sov Phys-JETP 40,763-5(1975), Mossbauer Effect in Ellipsoidal Brownian Particles

74G067 THEORY V I Gudov, V I Stepanenko, V L Fedorin, and V S Shkalikov, Tr Metrol Inst SSSR 151,106-7(1974), Resonance Mossbauer Converter for Measuring Vibrational Parameters (In Russian)

74I017 THEORY Ya A Iosilevskii, Kristallografiya 19,677-83(1974)/Sov Phys-Crystallogr 19,421-4(1975), Anisotropy of the Thermal Fluctuations of Atoms in a Crystal

74S121 THEORY A Singh and K N Shrivastava, Crystal Field Contribution to Mossbauer Isomer Shift, in "Proceedings of the Nuclear Physics and Solid State Physics Symposium" (Bombay, 1974) (Department of Atomic Energy, Bombay, 1974), Vol 17C, pp 357-60

74S130 THEORY J Stohr, G M Kalvius, G K Shenoy, and L L Hirst, Electronic Rearrangement and Relaxation Effects Following the Nuclear Transformation in Mossbauer Source Experiments, in "Hyperfine Interactions Studied in Nuclear Reactions and Decay: Contributed Papers" (Conference, Uppsala, Sweden, 1974), edited by E Karlsson and R Wappling (Upplands Grafiska, Uppsala, 1974), pp 80-1

74V038 THEORY V I Vorontsov and V I Vysotskii, Kvan Elektron (Kiev) (8), 63-9(1974), On Induced Mossbauer Gamma Emission (In Russian)

74V039 THEORY V I Vorontsov and V I Vysotskii, Kvan Elektron (Kiev) (8), 69-79(1974), On the Kinetics of Forced Gamma Emission (In Russian)

75A006 THEORY S K Arif, D St P Bunbury, G J Bowden, and R K Day, J Phys F 5,1037-47(1975), Nuclear Energy Levels: Applications to Complex 57Fe Mossbauer Spectra

75A044 THEORY T M Aivazyan, Yu M Aivazyan, and A R Mkrtchyan, On the Possibility of the Acoustic Modulation of the Mossbauer Radiation Passing Through an Absorber, in "5th Int Conf Mossbauer Spectrosc, Proc," Vol 3, pp 607-8 (See 75H017) (In Russian)

75A046 THEORY A V Andreev and Yu A Il'inskii, Zh Eksp Teor Fiz 68,811-6 (1975)/Sov Phys-JETP 41,403-5(1976), Amplification in a Gamma Laser when the Bragg Condition is Satisfied

75A047 THEORY M A Andreeva, R N Kuz'min, and S F Oparina, Phys Status Solidi B 71,K143-6(1975), Variation of the Resonantly Scattered Wave by Means of Ultrasonic Vibrations for the Phase Determination

75A048 THEORY M A Andreeva and R N Kuz'min, Phys Status Solidi B 71, K201-4(1975), Evolution of the Mossbauer Spectrum of the Bragg Reflection with Increasing Crystal Thickness (Kinematic Theory of Diffraction)

CODE	TOPIC	REFERENCE

75B030   THEORY   C Boekema, F Van der Woude, and G A Sawatzky, Spin and Charge Density Oscillations in Doped Magnetite: A Classical Selfconsistent Charge Density Oscillation Model, in "5th Int Conf Mossbauer Spec, Proc," Part 1, pp 96-8 (See 75H017)

75B033   THEORY   E V Baklanov and V P Chebotaev, Pis'ma Zh Eksp Teor Fiz 21, 286-9(1975)/JETP Lett 21,131-2(1975), Concerning One Possibility of Lasing in the Gamma Band

75B063   THEORY   D Barb and D Tarina, Polarized Mossbauer Transitions in Mixed Hyperfine Interactions, in "5th Int Conf Mossbauer Spec, Proc," Part 3, pp 699-704 (See 75H017)

75B071   THEORY   V A Belyakov and R Ch Bokun, Acta Crystallogr, Sec A 31, 737-41(1975), Kinematical Theory of Mossbauer Diffraction by Magnetically Ordered Crystals

75B098   THEORY   K Byrin, Cv Bontschev, and K K'nchev, Mossbauer Effect in Suspension and Observation of the Velocity Spectrum of Brownian Particles, in "Int Conf Mossbauer Spectrosc, Proc," Vol 1, pp 431-2 (See 75H027)

75B101   THEORY   D Barb and M Rogalski, Birefringence of Arbitrary Polarized Mossbauer Radiation, in "Int Conf Mossbauer Spectrosc, Proc," Vol 1, pp 469-70 (See 75H027)

75B126   THEORY   V A Belyakov and E V Smirnov, Zh Eksp Teor Fiz 68,608-22 (1975)/Sov Phys-JETP 41,301-7(1975), On the Polarization Characteristics of Mossbauer Coherent Scattering by Perfect Magnetic-ordered Crystals

75B135   THEORY   D Barb and L Diamandescu, Rev Roum Phys 20,259-71(1975), Simulation of Mossbauer Relaxation Spectra Case of Effective Spin S = 1/2

75B139   THEORY   D Barb and D Tarina, Rev Roum Phys 20,399-402(1975), Polarized Mossbauer Spectra - Tool for Effective Hyperfine Parameters Determinations

75C016   THEORY   R W Cochrane, R Harris, M Plischke, D Zobin, and M J Zuckermann, Phys Rev B 12,1969-70(1975), Mossbauer Absorption Spectra in Amorphous Metallic Alloys

75D011   THEORY   S Dattagupta, Phys Rev B 12,47-57(1975), Effect of Nuclear Motion on Mossbauer Relaxation Spectra

75D017   THEORY   P G De Gennes, J Phys (Paris) 36,603-6(1975), Sur une Eventuelle Application de l'Effet Mossbauer ou des Neutrons a l'Etude des Interfaces Fluides et des Smectiques

75D025   THEORY   S Dattagupta, Phys Rev B 12,3584-95(1975), Effects of Off-diagonal Hyperfine Interaction on Mossbauer Relaxation Spectra

75D059   THEORY   L M Dautov, M M Kadykenov, and D K Kaipov, The Influence of the Chemical Bond on the Coefficient of the Electron Internal Conversion, in "Int Conf Mossbauer Spectrosc, Proc," Vol 1, pp 493-4 (See 75H027)

75D065   THEORY   L M Dautov, Isomer Shifts Calibration of the Mossbauer Spectra, in "5th Int Conf Mossbauer Spectrosc, Proc," Vol 3, pp 629-32 (See 75H017) (In Russian)

75D066   THEORY   L Dabrowski, J Piekoszewski, and J Suwalski, Analytical Determination of Hyperfine Interaction Parameters in the General Case on the Base of the NGR Spectra for $I3/2 \rightarrow I1/2$, in "5th Int Conf Mossbauer Spectrosc, Proc," Vol 3, pp 635-7 (See 75H017) (In Russian)

75F024   THEORY   H Fischer, U Gonser, H D Pfannes, and T Shinjo, Texture Determination by Polarized Recoil-free Gamma Rays, in "Int Conf Mossbauer Spectrosc, Proc," Vol 1, pp 463-4 (See 75H027)

CODE	TOPIC	REFERENCE

75G008 THEORY L Gunther and J Zitkova-Wilcox, J Stat Phys 12,205-15 (1975), The Mossbauer Effect: A Potentially Ideal Probe into Brownian Motion

75G019 THEORY V I Gol'danskii and S V Karyagin, Influence of Anisotropic Atomic Motion on the Mossbauer Spectrum of Polycrystalline Systems, in "5th Int Conf Mossbauer Spec, Proc," Part 2, pp 439-47 (See 75H017) (In Russian)

75G024 THEORY V I Gol'danskii and S V Karyagin, Phys Status Solidi B 68, 693-702(1975), Effect of the Anisotropy of Atomic Motions on Mossbauer Spectra of Polycrystals

75G081 THEORY R K Gupta, Phys Rev B 12,4452-9(1975), Mean-square Amplitudes of Vibration for Ionic Crystals

75H001 THEORY F Hartmann-Boutron and D Spanjaard, J Phys (Paris) 36, 307-14(1975), On the Use of Liouville Relaxation Supermatrices in Mossbauer Studies

75I012 THEORY E M Iolin and E V Zolotoyabko, Zh Eksp Teor Fiz 68,1331-6 (1975)/Sov Phys-JETP 41,662-4(1976), Inelastic Scattering of Mossbauer Gamma Rays by Crystals near the Temperature of a Structural Phase Transiton

75I019 THEORY E M Iolin and V V Martyshchenko, Latv PSR Zinat Akad Vestis, Fiz Teh Ser (3),29-34(1975), Mossbauer Irradiation from Thick Crystal near Kossel Cone Intersection (In Russian)

75I020 THEORY L N Ivanov and V S Letokhov, Zh Eksp Teor Fiz 68,1748-56 (1975)/Sov Phys-JETP 41,877-81(1976), Spectrum of Electron-nuclear Gamma Transitions in Atomic Nuclei

75K004 THEORY O Knop, E M Palmer, and R W Robinson, Acta Crystallogr, Sec A 31,19-31(1975), Arrangements of Point Charges Having Zero Electric-field Gradient (Errata: personal communication)

75K016 THEORY B Krishnamurthy and K P Sinha, J Magn Res 17,189-92(1975), Mossbauer Lineshape in the Presence of NMR Transitions

75K019 THEORY L Korecz, On Comparison of Molecular Orbital and Isomeric Shift Data of Some Low-spin Iron Compounds, in "5th Int Conf Mossbauer Spec, Proc," Part 1, pp 215-7 (See 75H017)

75K034 THEORY S Kumar, Phys Status Solidi B 69,K145-7(1975), On the Usability of Lattice Dynamical Theory in Mossbauer Spectroscopic Studies

75K065 THEORY Yu M Kagan, Coherent Phenomenon on Interaction of Mossbauer Radiation with Crystals (Progress in Theory), in "Int Conf Mossbauer Spectrosc, Proc," Vol 2, pp 17-41 (See 75H038) (In Russian)

75K072 THEORY S V Karyagin, Fiz Tverd Tela 17,1856-8(1975)/Sov Phys-Solid State 17,1220-1(1975), Influence of Dimensions of Jump Diffusion on the Broadening of a Mossbauer Line of a Polycrystalline Sample

75K073 THEORY V N Kashcheev, Fiz Tverd Tela 17,3552-8(1975)/Sov Phys-Solid State 17,2194-8(1976), Critical Anomalies of the Mossbauer Line of a Ferromagnet

75M007 THEORY A V Mitin and G P Chugunova, Phys Status Solidi A 28,39-48 (1975), Polarization Detection Methods of Double Gamma-ray Nuclear Resonances

75M030 THEORY G Meissner and K Binder, Phys Rev B 12,3948-55(1975), Debye-Waller Factor, Compressibility Sum Rule, and Central Peak at Structural Phase Transitions

CODE	TOPIC	REFERENCE

75M037   THEORY   A Minkova, B Slavov, and Cv Bontschev, On the Application of the Diffusion Theory in the Depth-selective Mossbauer Spectroscopy, in "Int Conf Mossbauer Spectrosc, Proc," Vol 1, pp 45-6 (See 75H027)

75M039   THEORY   A S Moskvin, N S Ovanesyan, and V A Trukhtanov, Angular Dependence of Supertransferred Hyperfine Interaction, in "Int Conf Mossbauer Spectrosc, Proc," Vol 1, pp 145-6 (See 75H027)

75M063   THEORY   K V Makaryunas, The Problem of Scale Calibration in Mossbauer Isomer Shift and Chemical Influence on Electron Capture, in "Int Conf Mossbauer Spectrosc, Proc," Vol 1, pp 495-6 (See 75H027) (In Russian)

75M067   THEORY   A V Mitin, Influence of the Alternating Magnetic Field on the Polarization of the Gamma-radiation, in "5th Int Conf Mossbauer Spectrosc, Proc," Vol 3 pp 615-8 (See 75H017) (In Russian)

75N006   THEORY   C I Nistor and T Beica, The Multiplicity of the Quadrupole Interaction in the Mixed Spinelic Ferrites $Fe(Me(x)Fe(2-x))O4$, in "5th Int Conf Mossbauer Spec, Proc," Part 1, pp 104-6 (See 75H017)

75O006   THEORY   L G Onoprienko, Fiz Tverd Tela (Leningrad) 17,609-11 (1975)/Sov Phys-Solid State 17,394(1975), Mossbauer Effect in Uniaxial Ferromagnets with Domain Structure

75O016   THEORY   L G Onoprienko, Fiz Met Metalloved 39,751-6(1975)/Phys Met Metallogr (USSR), Mossbauer Effect in Magnetically Uniaxial Ferromagnets

75P049   THEORY   M Piecuch and C Janot, J Phys Chem Solids 36,1135-45 (1975), Contribution Electronique au Gradient de Champ Electrique sur une Impurete de Transition dans un Environnement Metallique Anisotrope

75R012   THEORY   S N Ray, T Lee, and T P Das, Phys Rev B 12,58-63(1975), Effect of Many-body Interactions on Isomer Shift in Iron Compounds

75R035   THEORY   S K Roy, M Singh, and B P Srivastava, Indian J Pure Appl Phys 13,217-20(1975), Study of Mossbauer Effect of a Lattice Containing a Pair of Point Defects at High Temperature Limit

75S017   THEORY   R R Sharma and P Moutsos, Phys Rev B 11,1840-6(1975), Mossbauer Quadrupole Splitting in Ferric Hemin

75S038   THEORY   A Szczepanski, Bull Acad Pol Sci, Ser Sci Tech 23,47-52 (1975), Coincidence Mossbauer Spectroscopy and the Dynamics of Point Defects. I. The Classical Theory

75S051   THEORY   A Singh and K N Shrivastava, Solid State Commun 17,1123-4 (1975), Crystal Field Contribution to Mossbauer Isomer Shift

75S054   THEORY   E V Smirnov and V A Belyakov, On the Dynamic Theory of the Diffraction of Resonant Gamma Rays on Magnetically Ordered Crystals, in "Int Conf Mossbauer Spectrosc, Proc," Vol 1, pp 11-2 (See 75H027) (In Russian)

75S055   THEORY   S O Svensson, S Morup, and G Trumpy, Anisotropic Diffusional Line Broadening in Single Crystals of fcc Metals in "Int Conf Mossbauer Spectrosc, Proc," Vol 1, pp 19-20 (See 75H027)

75S071   THEORY   G Ya Selyutin, Mossbauer Spectrum near the Critical Point in Ferromagnet, in "Int Conf Mossbauer Spectrosc, Proc," Vol 1, pp 361-2 (See 75H027) (In Russian)

CODE	TOPIC	REFERENCE

75T012   THEORY   A Trautwein, F E Harris, A J Freeman, and J P Desclaux, Phys Rev B 11,4101-5(1975), Relativistic Electron Densities and Isomer Shifts of 57Fe in Iron-Oxygen and Iron-Fluorine Clusters and of Iron in Solid Noble Gases

75V037   THEORY   R P Vardapetyan, Fiz Tverd Tela 17,1850-2(1975)/Sov Phys-Solid State 17,1215(1975), Inhomogeneous Broadening of the Components of a Mossbauer Doublet of Single-crystal and Polysrcstalline Samples

75V046   THEORY   V I Vorontsov and V I Vysotskii, Fiz Tverd Tela 17,2944-52 (1975)/Sov Phys-Solid State 17,1959-64(1976), Bragg Diffraction of Mossbauer Radiation in the Case of Hyperfine Splitting

75W009   THEORY   C Wissel, Solid State Commun 17,1011-2(1975), Anomalies in the Mossbauer Spectrum due to Soft Modes

75Z009   THEORY   R Zimmermann, Chem Phys Lett 34,416-8(1975), Description of the Angular Dependence of Dipole Transitions by Intensity Tensors

Addendum Reference List

CODE	TOPIC	REFERENCE

65S034   GENRAL   H Sano, Genshiryoku Kogyo 11(8),49-55(1965), Mossbauer Effect and Its Application in Chemistry (In Japanese)

65S035   REVIEW   H Sano, Bussei 6,315-21(19765), Chemical Application of the Nuclear Gamma Ray Resonance - Mossbauer Effect (In Japanese)

68H050   REVIEW   G Harris, Theor Chim Acta 10,155-80(1968), Spin-mixing and the Different Spin States of Ferric Ion in Tetragonal Symmetry. II. Localized Properties: Zero Field Splittings, Effective Magnetic Moments, Magnetic Field Energies and Electric Field Gradients of Ferric Ion

69E022   FE-57   Y Endoh, Y Ishikawa, and T Shinjo, Phys Lett 29A,310-1 (1969), Iron Impurities in the Antiferromagnetic Manganese Copper (Mn0.95Cu0.05) Alloy

72B096   REVIEW   K Burger, Kem Ujabb Eredmenyei 9,89-199(1972), Application of Mossbauer Spectroscopy to Complex Chemistry (In Hungarian)

72K076   GENRAL   R N Kuz'min, Appar Metody Rentgenovskogo Anal 10,85-8 (1972), Selection of X-ray Diagram Lines for Excitation of Mossbauer Gamma Transitions (In Russian)

72M075   GENRAL   W Meisel, Urania 48(2),28-31(1972), Werkzeug der Wissenschaft: Gammastrahlen

72M076   FE-57   Y Murakami and Y Ikai, Shindo Gijutsu Kenkyukai-Shi 11, 182-9(1972), Environment Sensitive Mechanical Properties of Copper and Its Alloys - Correlation with Stress-corrosion Cracking (In Japanese)

72S098   FE-57   G N Shlokov and G I Danilov, Radiats Tekh 7,96-103(1972), Using Nuclear Gamma Resonance to Determine Degree of Ferrization (In Russian)

73A046   SN-119   S Ambe and F Ambe, Radiochim Acta 20,141(1973), A Mossbauer Study of the Oxidation State of 119Sn after the Successive EC Decays of 119mTe in Telluric Acid

73B100   FE-57   V S Belyaev and L I Koshkin, Uch Zap Kuibyshev Gos Pedagog Inst 125,66-71(1973), Deposition of Magnetite Films on a Single-crystal Substrate by Mossbauer Effect (In Russian)

217

CODE	TOPIC	REFERENCE

73B101 PROPSL A P Buev, U Kh Kopvillem, and L N Shakhmuratova, Uch Zap
Kazan Gos Pedagog Inst 125,60-5(1973), Gamma Echo, Bragg
Echo, and the Phenomenon of Echo Signals in the Correlation
Properties of Nuclear Radiations   (In Russian)

73B101 THEORY A P Buev, U Kh Kopvillem, and L N Shakhmuratova, Uch Zap
Kazan Gos Pedagog Inst 125,60-5(1973), Gamma Echo, Bragg
Echo, and the Phenomenon of Echo Signals in the Correlation
Properties of Nuclear Radiations   (In Russian)

73D047 INSTRM G I Danilov, S V Mamikoyan, and G N Shlokov, Tr VNII
Radiats Tekhn (9),289-96(1973), Apparatus and Method for
Quantitative Phase Analysis of Ferromagnetic Substances
Based on the Mossbauer Effect   (In Russian)

73F033 SN-119 D S Faleev, Mossbauer Effect on Tin-119 Nuclei in Some
Magnesium Alloys, in "Mater Nauchn Konf Molodykh Uch Kafedr
Inst - Khabar Inst Inzh Zheleznodorozhn Transp," Vol 3,
pp 3-9   (In Russian)

73K053 FE-57 K K Khristoforov, L P Nikitina, L M Krizhanskii, and S P
Eknmov, Distribution of Iron(II) Ion and Structural Features of Orthorhombic Pyroxenes under Various Thermodynamic Conditions in "Probl Izuch Osvoeniya Prir Resur Sev,"
(1973), pp 64-5   (In Russian)

73M059 PROPSL A V Mitin and G P Chugunova, Uch Zap Kazan Gos Pedagog Inst
125,73-8(1973), Optical Principles of Detecting Double
Gamma A-nuclear Resonances   (In Russian)

73N027 FE-57 M Nakamura, N Maeda, and K Morinaga, Kyushu Daigaku Kogaku
Shuho 46,538-45(1973), Mossbauer Spectra Analysis by Non-
linear Filters   (In Japanese)

73S088 REVIEW I P Suzdalev, V I Gol'danskii, G M Zabrova, and B M
Kadenaci, Untersuchung topochemischer Reaktionen mit Hilfe
der Mossbauer-Spektroskopie und der Derivatographie, in
"Festkorperchemie," edited by V Boldyrev and K Meyer
(Verlag Grundstoffind, Leipzig, 1973), pp 160-78

74A049 I-129 N E Ablesimov, S I Bondarevskii, I S Kirin, and V A
Tarasov, Radiokhimiya 16,919-22(1974)/Sov Radiochem 16,
895-7(1975), Post Effects of beta-Decay in Frozen
Tellurium-containing Solutions

74B092   FE-57   C Bansal, J Ray, and G Chandra, Effect of Environment on
                     the Hyperfine and Exchange Fields of Fe57 Atoms in Disor-
                     dered Ni0.48Fe0.52 Alloy by Mossbauer Effect, in "Pro-
                     ceedings of the Nuclear Physics and Solid State Physics
                     Symposium" (Bombay, 1974) (Department of Atomic Energy,
                     Bombay, 1974), Vol 17C, pp 369-71

74B093   FE-57   D Barb, M Morariu, L Diamandescu, and I Bibicu, Rev Roum
                     Phys 19,425-9(1974), Internal Magnetic Fields in Fe-Si
                     Electrotechnical Steels

74B094   FE-57   E Burzo, M Bodea, and D Barb, Mossbauer Effect Investiga-
                     tion in Some Pseudobinary Iron Laves Phases, in "Magnetic
                     Resonance and Related Phenomena, Proceedings of AMPERE
                     Congress, 18th" (Nottingham, 1974), edited by P S Allen,
                     E R Andrew, and C A Bates (North-Holland Publishing Co,
                     Amsterdam, 1974), pp 71-2

74B095   FE-57   Sh Sh Bashkirov, A B Liberman, V I Sinyavskii, N N
                     Yefimova, and Yu A Mamaluy, Ukr Fiz Zh 19,1949-54(1975)/
                     Ukr Phys J, Magnetic Structure of Indium-substituted
                     Hexaferrites

74B096   FE-57   Sh Sh Bashkirov, N G Ivoilov, I K Kosterina, and V A
                     Chistyakov, Local Fields on Iron Nuclei in an Yttrium Gar-
                     net Ferrite in the Phase Transition Region, in "Strukt
                     Svoistva Ferritov," edited by N N Sirota ("Nauka i Tekh-
                     nika," Minsk, 1974), pp 181-5   (In Russian)

74B097   THEORY   V A Belyakov and Yu M Aivazyan, Tr Metrol Inst SSSR 119,
                     38-43(1974), Mossbauer Diffraction on Crystals with a Com-
                     plex Structure of Electric Field Gradients   (In Russian)

74B098   THEORY   V A Belyakov and Yu M Aivazyan, Tr Metrol Inst SSSR 119,
                     43-9(1974), Interference Effects in the Resonance Scat-
                     tering of Gamma Radiation from Nuclei with Two Mossbauer
                     Transitions on Crystals   (In Russian)

74B099   FE-57   W J S Blackburn and B P Tilley, J Mater Sci 9,1265-9(1975),
                     The Magnetic Properties of Glass-ceramics in the
                     CoO-Fe2O3-B2O3 System

74B100   FE-57   C Blaauw, C Boekema, F Van der Woude, and G A Sawatzky,
                     Investigation of Some Transition Metal Oxides Exhibiting
                     Metal - Non-metal Transitions, in "Proccedings of the In-
                     ternational Conference on Physics of Semiconductors, 12th,"
                     edited by M H Pilkuhn (Teubner, Stuttgart, 1974), pp 583-7

74B101   REVIEW   V N Bykov, S S Budagovskii, M I Gavrilyuk, V S Golovkin, V
                     A Levdik, I I Rudnev, and V A Solov'ev, Electronic Struc-
                     ture and Anomalies of Macroscopic Physical and Technologi-
                     cal Properties of Transition Metals, in "Tr Fiz-Energ
                     Inst," edited by V A Kuznetsov (Atomizdat, Moscow, 1974),
                     pp 469-81   (In Russian)

74C037   GENRAL   D Christov, Cv Bontschev, and Z Bursova, Die Mossbauer-
                     Spektroskopie als Forschungs-methode in angewandten organ-
                     ischen Chemie. Stand und Perspektive (Bulgarian Academy
                     of Science, Sofia, 1974), 86 pages

74C038   FE-57   J M D Coey, J C Bruyere, H Roux-Buisson, and R Brusetti,
                     Magnetic Properties of NiS at the Metal Non-metal Transi-
                     tion, in "Proceedings of the International Conference of
                     Magnetism ICM-73" (Moscow, 1973) ("Nauka," Moscow, 1974),
                     Vol V, pp 166-71

CODE	TOPIC	REFERENCE

74C039   SN-119   J Chappert, R H Herber, and M Winterberger, Electric and Magnetic Interaction in Cu2FeSnS4, Cu2MnSnS4, and Related Solids, in "Hyperfine Interactions Studied in Nuclear Reactions and Decay: Contributed Papers" (Conference, Uppsala, Sweden, 1974), edited by E Karlsson and R Wappling (Upplands Grafiska, Uppsala, 1974), pp 198-9

74C040   EU-151   R L Cohen, G Beyer, and B I Deutch, Origins of hf Field and Exchange Coupling of Eu Substitutional Impurities in Iron, in "Hyperfine Interactions Studied in Nuclear Reactions and Decay: Contributed Papers" (Conference, Uppsala, Sweden, 1974), edited by E Karlsson and R Wappling (Upplands Grafiska, Uppsala, 1974), pp 158-9

74C041   FE-57   J M D Coey, Can J Earth Sci 11,1489-93(1974), Iron Compounds in Lake Sediments

74D055   REVIEW   D W Dale, Mossbauer Spectroscopy, in "Modern Physical Techniques in Materials Technology," edited by T Mulvey and R K Webster (Oxford University Press, London, 1974), pp 281-90

74D056   FE-57   I Ya Dekhtyar, M M Nishchenko, and V P Romashko, Positron Annihilation and the Mossbauer Effect in Dilute Solid Solutions of Group V and VII Metals with Iron, in "Elektron struktura perekhod met, ikh splavov soedin" (1974), pp 299-302 (In Russian)

74D057   THEORY   A Ya Dzyublik, Zh Eksp Teor Fiz 67,1534-8(1974)/Sov Phys-JETP 40,763-5(1975), Mossbauer Effect in Ellipsoidal Brownian Particles

74D058   FE-57   B Ts Dudreva and R K Pirinchieva, Bulg J Phys 11,126-4 (1974), Mossbauer Spectra of Fe57 in Phthalocyanine

74E021   FE-57   H Ebiko, H Yamamoto, W Suetaka, and S Shimodaira, Spectroscopic Studies of Passive Films on Iron, in "Fifth International Congress on Metallic Corrosion, Proceedings" (Tokyo, 1972) (National Association of Corrosion Engineers, Houston, Texas, 1974), pp 285-9

74E022   FE-57   M D Evdokimov and S B Tomilov, Crystal Chemistry and Properties of Aegirite-diopside Series Pyroxenes, in "Miner Parageneзisy Miner Magmat Metasomaticheskikh Gorn Porod," edited by P M Tatarinov ("Nauka," Leningrad Otd, Leningrad, 1974), pp 81-6 (In Russian)

74F035   REVIEW   J Flouquet, Ann Phys (Paris) 8,5-51(1974), L'Etude des Alliages Dilues par Orientation Nucleaire

74G058   REVIEW   V I Gol'danskii, Usp Khim 43,2113-45(1974)/Russ Chem Rev 43,1021-42(1974), The Periodic Law and Progress in the Study of the Structure of Matter

74G059   YB-170   F Gonzalez-Jimenez and P Imbert, Mossbauer Study of a Kondo Anomaly in Au-Yb, in "Proceedings of the International Conference of Magnetism ICM-73" (Moscow, 1973) ("Nauka," Moscow, 1974), Vol V, pp 73-8

CODE	TOPIC	REFERENCE

74G060 FE-57 P L Gruzin, O P Elyutin, V S Mkrtchyan, Yu L Rodionov, and M Kh Khachatryan, Izv Akad Nauk Arm SSR, Fiz 9,397-401 (1974), Nuclear Gamma Resonance Study of the Ordering of Iron-(Aluminum, Germanium) (Fe3(Al,Ge)) Alloys (In Russian)

74G061 FE-57 P C M Gubbens and K H J Buschow, Mossbauer and Magnetization Studies of Some Rare Earth-iron Compounds of the Type R2Fe17, in "Proccedings of the International Conference of Magnetism ICM-73" (Moscow, 1973) ("Nauka," Moscow, 1974), Vol V, pp 60-6

74G062 FE-57 T S Gendler and R N Kuz'min, Possibilities of Gamma Resonance Spectroscopy for Phase Analysis, in "Nov Metody Issled Protsessov Vosstanov Chern Met, Dokl Simp" (1971), edited by S T Rostovtsev ("Nauka," Moscow, 1974), pp 62-8 (In Russian)

74G063 FE-57 A I Gorbanev, A M Babeshkin, N N Savvateev, Yu D Perfil'ev, V B Margulis, and B E Dzevitskii, Izv Akad Nauk SSSR, Met (6),37-40(1974)/Russ Metall (6),29-32(1974), Study of the Valency State of Iron in Iron Ores by Nuclear Gamma Resonance (NGR) Spectroscopy

74G064 FE-57 N S Greschnykova, S M Irkaev, R N Kuz'min, and B V Mill, Izv Akad Nauk Tadzh SSR, Otd Fiz-Mat Geol-Khim Nauk 54(4), 29-34(1974), The Cation Distribution in the Sb-substituted Yttrium Ferrite Garnets (In Russian)

74G065 SN-119 P L Gruzin, Yu F Bychkov, I A Evstyukhina, and L A Alekseev, Prikl Yad Spektrosk 4,12-6(1974), State of Tin-119 Nuclei in Superconducting Niobium Compounds Studied by Nuclear Gamma Resonance (In Russian)

74G066 FE-57 P L Gruzin, Yu L Rodionov, Yu A Li, V E Kaluzhskii, L K Mikhailova, V Yu Kolontsov, Metallofizika, Kiev 56,54-9 (1974), The Effect of Concentration Heterogeneities in the Sub-microstructure on Martensite Transformations (In Russian)

74G067 SN-119 V I Gudov, V I Stepanenko, V L Fedorin, and V S Shkalikov, Tr Metrol Inst SSSR 151,106-7(1974), Resonance Mossbauer Converter for Measuring Vibrational Parameters (In Russian)

74G067 THEORY V I Gudov, V I Stepanenko, V L Fedorin, and V S Shkalikov, Tr Metrol Inst SSSR 151,106-7(1974), Resonance Mossbauer Converter for Measuring Vibrational Parameters (In Russian)

74G068 FE-57 F Grandjean and A Gerard, Analysis of Mossbauer Spectra in Presence of Electronic Hopping, in "Hyperfine Interactions Studied in Nuclear Reactions and Decay: Contributed Papers" (Conference, Uppsala, Sweden, 1974), edited by E Karlsson and R Wappling (Upplands Grafiska, Uppsala, 1974), pp 276-7

74H051 FE-57 K Haneda, C Miyakawa, and H Kojima, J Am Ceram Soc 57,354-7 Preparation of High-coercivity BaFe12O19

74H052 YB-170 L L Hirst, J Stohr, E Zech, G K Shenoy, and G M Kalvius, Mossbauer Studies of Magnetic Rare Earth Intermetallics: Electronic Re-arrangement Effects in Source Experiments, in "Magnetic Resonance and Related Phenomena, Proceedings of AMPERE Congress, 18th" (Nottingham, 1974), edited by P S Allen, E R Andrew, and C A Bates (North-Holland Publishing Co, Amsterdam, 1974), pp 67-8

CODE	TOPIC	REFERENCE

74H053   FE-57   L Haggstrom, Determination of Hyperfine Parameters from Mossbauer Spectra Involving 1/2→3/2 Transitions, in "Hyperfine Interactions Studied in Nuclear Reactions and Decay: Contributed Papers" (Conference, Uppsala, Sweden, 1974), edited by E Karlsson and R Wappling (Upplands Grafiska, Uppsala, 1974), pp 240-1

74H053   SN-119   L Haggstrom, Determination of Hyperfine Parameters from Mossbauer Spectra Involving 1/2→3/2 Transitions, in "Hyperfine Interactions Studied in Nuclear Reactions and Decay: Contributed Papers" (Conference, Uppsala, Sweden, 1974), edited by E Karlsson and R Wappling (Upplands Grafiska, Uppsala, 1974), pp 240-1

74H054   TA-181   A Heidemann, G Kaindl, D Salomon, and G Wortmann, High-resolution Mossbauer Study of Hydrided Tantalum, in "Hyperfine Interactions Studied in Nuclear Reactions and Decay: Contributed Papers" (Conference, Uppsala, Sweden, 1974), edited by E Karlsson and R Wappling (Upplands Grafiska, Uppsala, 1974), pp 238-9

74H055   I-129   R H Herber, Failure of the Point-charge Model for Quadrupole Hyperfine Interactions in Organo-tin Compounds, in "Hyperfine Interactions Studied in Nuclear Reactions and Decay: Contributed Papers" (Conference, Uppsala, Sweden, 1974), edited by E Karlsson and R Wappling (Upplands Grafiska, Uppsala, 1974), pp 234-5

74H055   SN-119   R H Herber, Failure of the Point-charge Model for Quadrupole Hyperfine Interactions in Organo-tin Compounds, in "Hyperfine Interactions Studied in Nuclear Reactions and Decay: Contributed Papers" (Conference, Uppsala, Sweden, 1974), edited by E Karlsson and R Wappling (Upplands Grafiska, Uppsala, 1974), pp 234-5

74I016   FE-57   H Inoue, M Izumi, and E Imoto, Bull Chem Soc Jpn 47,1712-6 (1974), Oxidation by Iron(III) Complexes. VII. The Reactions of Iron(III) Chloride with Toluene and Substituted Toluenes

74I017   THEORY   Ya A Iosilevskii, Kristallografiya 19,677-83(1974)/Sov Phys-Crystallogr 19,421-4(1975), Anisotropy of the Thermal Fluctuations of Atoms in a Crystal

74J016   FE-57   J Jach, J Nonmetals 2,89-93(1974), A Mossbauer Study of the Devitrification of an Iron-containing Borate Glass

74J017   EU-151   V G Jadhao, G N Rao, R M Singru, D Bahadur, and C N R Rao, Mossbauer Spectroscopic Studies of EuCoO3 Employing 57Fe and 151Eu Gamma Rays, in "Proceedings of the Nuclear Physics and Solid State Physics Symposium" (Bombay, 1974) (Department of Atomic Energy, Bombay, 1974), Vol 17C, pp 361-4

74J017   FE-57   V G Jadhao, G N Rao, R M Singru, D Bahadur, and C N R Rao, Mossbauer Spectroscopic Studies of EuCoO3 Employing 57Fe and 151Eu Gamma Rays, in "Proceedings of the Nuclear Physics and Solid State Physics Symposium" (Bombay, 1974) (Department of Atomic Energy, Bombay, 1974), Vol 17C, pp 361-4

74J018   REVIEW   C Janot and P Delcroix, Ethnol Fr 3,179-88(1974), Caracterisation de Materiaux Archeologiques par Spectrometrie Mossbauer

CODE	TOPIC	REFERENCE

74K064 INSTRM D K Kaipov, T A Orazbaev, E F Galyutina, and L S Sergeeva, Prib Tekh Eksp (6),75-6(1974)/Instrum Exp Tech (USSR) 17, 1633-5(1974), Properties of a Scintillation Resonance Detector for Mossbauer Investigations

74K064 SN-119 D K Kaipov, T A Orazbaev, E F Galyutina, and L S Sergeeva, Prib Tekh Eksp (6),75-6(1974)/Instrum Exp Tech (USSR) 17, 1633-5(1974), Properties of a Scintillation Resonance Detector for Mossbauer Investigations

74K065 DY-161 C Kellershohn, J N Rimbert, F Soubirou, and C Hubert, Bone Uptake of Rare Earths Observed by Mossbauer Effect, in "Magnetic Resonance and Related Phenomena, Proceedings of AMPERE Congress, 18th" (Nottingham, 1974), edited by P S Allen, E R Andrew, and C A Bates (North-Holland Publishing Co, Amsterdam, 1974), pp 289-90

74K066 FE-57 V Kothekar, Proc Ind Nat Sci Acad, Part A 40,112-23(1974), Molecular Orbital Approach to the Chemical Interpretation of Mossbauer Effect

74K067 FE-57 M M L Kwan and R W Hoffman, Jpn J Appl Phys, Suppl 2, Part 1 (Proceedings of the Sixth International Vacuum Congress, Kyoto, 1974), Electrical and Magnetic Properties of Pure Amorphous Cobalt Films, pp 729-32

74K068 SN-119 M M Kadykenov, Izv Akad Nauk Kaz SSR, Ser Fiz-Mat (4), 75-6(1974), Effect of the Chemical Environment of an Atom on the Total Density of Its Electrons in the Region of the Nucleus (In Russian)

74K069 FE-57 Yu F Krupanskii and I P Suzdalev, Magnetic Properties of the Iron Oxide Ultrafine Particles, in "Proceedings of the International Conference of Magnetism ICM-73" (Moscow, 1973) ("Nauka," Moscow, 1974), Vol VI, pp 170-5 (In Russian)

74L045 NP-237 D J Lam, B D Dunlap, A R Harvey, M H Mueller, A T Aldred, I Nowik, and G H Lander, The Magnetic Properties of Monosulphides of Neptunium and Plutonium, in "Proceedings of the International Conference on Magnetism ICM-73" (Moscow, 1973) ("Nauka," Moscow, 1974), Vol VI, pp 74-8

74L046 FE-57 S Larsson, E K Viinikka, M L De Siqueira, and J W D Connolly, The Electronic Structure of Octahedral Transition Metal Halides as Calculated by the Multiple Scattering Method, in "International Journal of Quantum Chemistry, Symposium No 8" (Proceedings of the International Symposium on Atomic, Molecular, and Solid-state Theory and Quantum Statistics, Sanibel Island, Florida, 1974), edited by P-O Lowdin (Interscience Publication/John Wiley & Sons, New York, 1974), pp 145-60

74L047 FE-57 G V Loseva, N V Murashko, and A V Polosin, Izv Akad Nauk SSSR, Neorg Mater 13,473-6(1974)/Inorg Mater (USSR) 10, 405-7(1974), Dehydration Mechanism of Lepidocrocite

74L048 FE-57 S A Losievskaya and I M Puzei, Mossbauer Spectra of Ordered Iron-Aluminum Alloys after Various Heat Treatments, in "Dokl IV Vses soveshch po uporyadocheniyu atomov i ego vliyaniyu na svoistva splavov" (1974), Chapter I, pp 169-74 (In Russian)

CODE	TOPIC	REFERENCE

74L049   FE-57   I S Lyubutin and Yu S Vishnyakov, Reorientation of the Magnetic Field on the Nucleus of a Diamagnetic Atom in Rare Earth Orthoferrites, in "Proceedings of the International Conference of Magnetism ICM-73" (Moscow, 1973) ("Nauka," Moscow, 1974), Vol VI, pp 278-82 (In Russian)

74L049   SN-119   I S Lyubutin and Yu S Vishnyakov, Reorientation of the Magnetic Field on the Nucleus of a Diamagnetic Atom in Rare Earth Orthoferrites, in "Proceedings of the International Conference of Magnetism ICM-73" (Moscow, 1973) ("Nauka," Moscow, 1974), Vol VI, pp 278-82 (In Russian)

74M058   REVIEW   B N Maimur and V F Moroz, Use of Gamma Resonance Spectroscopy for Studying Phase Transformations during Heating and Reduction of Iron Oxides, in "Nov Metody Issled Protsessov Vosstanov Chern Met Dokl Simp" (1971), edited by S T Rostovtsev ("Nauka," Moscow, 1974), pp 56-62 (In Russian)

74M059   GE-73   K Matsui, O Konno, and S Ishino, Kakuriken Kenkyu Hokoku 7,147-55(1974), Preparation of Ga73 Mossbauer Source in Germanium Through Photonuclear Reactions

74M060   FE-57   B Mehliss and H H Duncken, Wiss Z Friedrich-Schiller-Univ, Jena, Math-Naturwiss Reihe 23,755-9(1974), Adsorption von Ferrocen und Phthalocyanin an dispersem SiO2

74M060   SN-119   B Mehliss and H H Duncken, Wiss Z Friedrich-Schiller-Univ, Jena, Math-Naturwiss Reihe 23,755-9(1974), Adsorption von Ferrocen und Phthalocyanin an dispersem SiO2

74M061   SN-119   M Mishima, M Idogaki, and H Negita, Shimane Daigaku Bunrigakubu Kiyo, Rigakka Hen (7),73-7(1974), Mossbauer Study of the Complex Compounds of Tin(II) Halides

74M062   SN-119   M Mishima, M Nakamura, M Izawa, and M Idogaki, Shimane Daigaku Bunrigakubu Kiyo, Rigakka Hen (7),79-84(1974), Mossbauer Effect of 119Sn in Complexes of Tin(IV) Chloride and Dialkyltin Dichlorides with Some Nitrogen Donors

74M063   FE-57   V A Makarov and I M Puzei, Mossbauer Study of Concentration Inhomogeneities in Fe-Ni Invar Alloys, in "Dokl VI Vses soveshch po uporyadocheniyu atomov i ego vliyaniyu na svoistva splavov" (1974), Chapter 1, pp 175-80 (In Russian)

74M064   FE-57   V A Makarov, Critical Superparamagnetism of Iron-Nickel Invar Alloys, in "Proceedings of the International Conference of Magnetism ICM-73" (Moscow, 1973) ("Nauka," Moscow, 1974), Vol III, pp 154-9 (In Russian)

74M065   FE-57   R I Mints, V A Semenkin, and Yu A Shevchenko, Izv Akad Nauk SSSR, Met (6),153-7(1974)/Russ Metall (6),124-7(1974), The Influence of Plastic Deformation on the Mossbauer Effect in Fe-Mn Solid Solutions

74N020   ER-166   L Niesen and P J Kikkert, Mossbauer Effect of Er Impurities Implanted in Iron, in "Hyperfine Interactions Studied in Nuclear Reactions and Decay: Contributed Papers" (Conference, Uppsala, Sweden, 1974), edited by E Karlsson and R Wappling (Upplands Grafiska, Uppsala, 1974), pp 160-1

74N021   FE-57   L P Nikitina, S P Ekimov, L M Krizhanskii, R G Grebenshchikov, and V V Motsartov, Dokl Akad Nauk SSSR 219,148-51(1974)/Dokl Phys Chem 219,1033-6(1974), The Mossbauer Spectroscopy of Mg(1-x)FexGeO3 Solid Solutions

CODE	TOPIC	REFERENCE

740014   FE-57   H Ohashi, M Koizumi, and T Morozumi, Hokkaido Daigaku Kogakubu Kenkyu Hokoku 72,145-53(1974), Study on Oxidation of Iron by Mossbauer Spectroscopy (In Japanese)

740015   NP-237   A Ohkawa, T Shoji, and K Matsui, Kakurikin Kenkyu Hokoku 7, 140-6(1974), Np237 Mossbauer Probe Embedded in Uranium Dioxide by Means of U238 (gamma/n)U237

74P053   FE-57   Yu D Perfil'ev, M I Afanasov, L A Kulikov, and A M Babeshkin, Vestn Mosk Univ, Khim 29,742-3(1974)/Moscow Univ, Chem Bull 29(6),76-7(1974), Mossbauer Emission Spectra of 57Co in Quadrivalent Compounds of Cobalt and Iron

74P054   EU-151   E Polaczkowa and A Polaczek, Pol Akad Nauk, Oddzial Krakowie, Pr Kom Ceram, Ceram 21,29-44(1974), Recent Studies on the Cubic Tungsten Bronzes of Sodium and Rare Earth Metals: Structural Problems and Magnetic Properties

74P055   TE-125   M Pasternak, The Ultimate Charge State of Te in Insulator MnI2, in "Hyperfine Interactions Studied in Nuclear Reactions and Decay: Contributed Papers" (Conference, Uppsala, Sweden, 1974), edited by E Karlsson and R Wappling (Upplands Grafiska, Uppsala, 1974), pp 82-3

74P056   ZN-67   W Potzel, G J Perlow, and H De Waard, Hyperfine Interaction in 67ZnO by Frequency Modulation Mossbauer Spectroscopy, in "Hyperfine Interactions Studied in Nuclear Reactions and Decay: Contributed Papers" (Conference, Uppsala, Sweden, 1974), edited by E Karlsson and R Wappling (Upplands Grafiska, Uppsala, 1974), pp 266-7

74P057   SN-119   M V Plotnikova, K P Mitrofanov, and Yu N Venevtsev, Anharmonic Atom Vibration in Ferroelectric Type Perovskites According to Mossbauer Data, in "Probl issled svoistv segnetoelektrikov" (1974), Chapter 2, pp 15-6 (In Russian)

74R034   FE-57   K R P M Rao and P K Iyengar, Mossbauer Spectroscopy Study of Short-range Magnetic Ordering in Co-Ga(57Fe) Intermetallic Compounds, in "Proceedings of the Nuclear Physics and Solid State Physics Symposium" (Bombay, 1974) (Department of Atomic Energy, Bombay, 1974), Vol 17C, pp 365-8

74R035   FE-57   J Ray, C Bansal, and G Chandra, Distribution of 57Fe Hyperfine Fields in NiMn: 57Fe Systems by Mossbauer Effect Studies, in "Proceedings of the Nuclear Physics and Solid State Physics Symposium" (Bombay, 1974) (Department of Atomic Energy, Bombay, 1974), Vol 17C, pp 373-6

74R036   FE-57   K V Reddy, S C Chetty, and A S Bommanavar, Mossbauer Studies on FeTex, in "Proceedings of the Nuclear Physics and Solid State Physics Symposium" (Bombay, 1974) (Department of Atomic Energy, Bombay, 1974), Vol 17C, pp 381-2

74R037   FE-57   C N R Rao, J Indian Chem Soc 51,979-87(1974), Localized Versus Collective Behaviour of d-Electrons in Transition Metal Oxide Systems of Perovskite Structure

74R038   FE-57   V P Romashko, Yader Priborostr (24),157-63(1974), Mossbauer Studies of Invar-type Iron-Nickel Alloys (In Russian)

CODE	TOPIC	REFERENCE

74R039   FE-57   B V Ryzhenko, V Ya El'ner, and F A Sidorenko, Isomeric Shift of the Mossbauer Spectra of Iron-57 in Iron-Cobalt-Aluminum (Fe(1-x)CoxAl) and Iron-Nickel-Aluminum (Fe(1-x)NixAl) Solid Solutions in "Fiz met i ikh soedin" (1974), Volume 2, 18-22   (In Russian)

74R040   GE-73   R S Raghavan and L N Pfeiffer, Ge73: A New High-resolution Mossbauer Nuclide, in "Hyperfine Interactions Studied in Nuclear Reactions and Decay: Contributed Papers" (Conference, Uppsala, Sweden, 1974), edited by E Karlsson and R Wappling (Upplands Grafiska, Uppsala, 1974), pp 282-3

74R041   FE-57   T Raman, V K Nagesh, D Chakravorty, and G N Rao, Mossbauer Studied in the Glass System Na2OB2O3-Fe2O3, in "Hyperfine Interactions Studied in Nuclear Reactions and Decay: Contributed Papers" (Conference, Uppsala, Sweden, 1974), edited by E Karlsson and R Wappling (Upplands Grafiska, Uppsala, 1974), pp 280-1

74R042   CS-133   S R Reintsema, S A Drentje, and H De Waard, New Results on the Location of Xenon Impurities Implanted in Iron Derived from Mossbauer Spectra, in "Hyperfine Interactions Studied in Nuclear Reactions and Decay: Contributed Papers" (Conference, Uppsala, Sweden, 1974), edited by E Karlsson and R Wappling (Upplands Grafiska, Uppsala, 1974), pp 74-5

74R043   SN-119   P Roggwiller and W Kundig, The Isomer Shift and the Influence of the Chemical Environment on the Lifetime of the Mossbauer Nucleus 119Sn, in "Hyperfine Interactions Studied in Nuclear Reactions and Decay: Contributed Papers" (Conference, Uppsala, Sweden, 1974), edited by E Karlsson and R Wappling (Upplands Grafiska, Uppsala, 1974), pp 240-1

74S119   REVIEW   C M Scala, Aust Gemmol 12,119-24(1974), A Survey of Some Research Work on Impurities in Gems

74S120   FE-57   T Shigematsu and Y Sasaki, Bull Inst Chem Res, Kyoto Univ 52,658-63(1974), Preparation and Characterization of Bis(2-(2-pyridyl)benzimidazole)iron(II) Complexes

74S121   THEORY   A Singh and K N Shrivastava, Crystal Field Contribution to Mossbauer Isomer Shift, in "Proceedings of the Nuclear Physics and Solid State Physics Symposium" (Bombay, 1974) (Department of Atomic Energy, Bombay, 1974), Vol 17C, pp 357-60

74S122   FE-57   K Spartalian, W T Oosterhuis, N S VanderVen, and J Ashkin, Mossbauer and ESR Spectroscopy of Fe3+ in Some Biological Iron Transport Compounds, in "Magnetic Resonance and Related Phenomena, Proceedings of AMPERE Congress 18th" (Nottingham, 1974), edited by P S Allen, E R Andrew, and C A Bates (North-Holland Publishing Co, Amsterdam, 1974), pp 267-8

74S123   FE-57   V B Spiridonov, V S Fridman, Yu L Rodionov, and P L Gruzin, Metalloved Term Obrab Met (10),28-32(1974)/ Met Sci Heat Treatment 844-8(1974), Structural Changes during Aging of Maraging Steel 03Kh11N10M2T

74S124   FE-57   R G Srivastava, S N Shringi, and C M Srivastava, Mossbauer Study of Ferrous Zinc Ferrite System, in "Proceedings of the Nuclear Physics and Solid State Physics Symposium" (Bombay, 1974) (Department of Atomic Energy, Bombay, 1974), Vol 17C, pp 377-80

CODE	TOPIC	REFERENCE

74S125   FE-57   P Steiner, W Von Zdrojewski, D Gumprecht, and S Hufner, Mossbauer Effect Investigations of the Kondo System Fe:Cu, in "Proceedings of the International Conference of Magnetism ICM-73" (Moscow, 1973) ("Nauka," Moscow, 1974), Vol V, pp 79-2

74S126   FE-57   W Steiner and H Ortbauer, Phys Status Solidi A 26,451-7 (1974), Magnetische Messungen an Y(FexCo(1-x))2 im Bereich kleiner Fe-Konzentrationen

74S127   REVIEW   H Sano, Boshoku Gijutsu 23,311-6(1974), Mossbauer Spectroscopy and Its Applications to Corrosion Studies (In Japanese)

74S128   REVIEW   D A Shirley, Effects of Extranuclear Fields on Nuclear Radiations, in "Nuclear Spectroscopy and Reactions" (Volume 40-C in "Pure and Applied Physics"), edited by J Cerny (Academic Press, New York, 1974), pp 513-49

74S129   PROPSL   P I Smolyanskii, Zap Vses Mineral Obshch (2),272-8(1974), On a Possible Application of Mossbauer Effect of Optical Analogue in Mineralogy (On Example of Fluorite) (In Russian)

74S130   THEORY   J Stohr, G M Kalvius, G K Shenoy, and L L Hirst, Electronic Rearrangement and Relaxation Effects Following the Nuclear Transformation in Mossbauer Source Experiments, in "Hyperfine Interactions Studied in Nuclear Reactions and Decay: Contributed Papers" (Conference, Uppsala, Sweden, 1974), edited by E Karlsson and R Wappling (Upplands Grafiska, Uppsala, 1974), pp 80-1

74S130   YB-170   J Stohr, G M Kalvius, G K Shenoy, and L L Hirst, Electronic Rearrangement and Relaxation Effects Following the Nuclear Transformation in Mossbauer Source Experiments, in "Hyperfine Interactions Studied in Nuclear Reactions and Decay: Contributed Papers" (Conference, Uppsala, Sweden, 1974), edited by E Karlsson and R Wappling (Upplands Grafiska, Uppsala, 1974), pp 80-1

74S131   YB-170   J Stohr, W Wagner, and G M Kalvius, Mossbauer Spectroscopy of Dilute Au:Yb Alloys in Small External Magnetic Fields, in "Hyperfine Interactions Studied in Nuclear Reactions and Decay: Contributed Papers" (Conference, Uppsala, Sweden, 1974), edited by E Karlsson and R Wappling (Upplands Grafiska, Uppsala, 1974), pp 188-9

74T033   IR-193   G Tanner, F E Wagner, G M Kalvius, G K Shenoy, and K H J Buschow, Mossbauer Studies of (RE)Ir2 Intermetallic Compounds, in "Magnetic Resonance and Related Phenomena, Proceedings of AMPERE Congress, 18th" (Nottingham, 1974), edited by P S Allen, E R Andrew, and C A Bates (North-Holland Publishing Co, Amsterdam, 1974), pp 87-8

74T034   SB-121   S T Tamaev, R N Kuz'min, Kh Kh Valiev, and S M Irkaev, The Effective Magnetic Fields on the 121Sb Nuclei in the Heusler-type Alloys, in "Proceedings of the International Conference of Magnetism ICM-73" (Moscow, 1973) ("Nauka," Moscow, 1974), Vol V, pp 83-7 (In Russian)

74V036   FE-57   D J Vaughan and J R Craig, Am Mineral 59,926-33(1974), The Crystal Chemistry and Magnetic Properties of Iron in the Monosulfide Solid Solution of the Fe-Ni-S System

74V037   FE-57   A M Voronin, Vestn Akad Nauk Kaz (9),71-4(1974), Digital Storing Device for Nuclear Gamma Resonance Spectroscopy (In Russian)

CODE	TOPIC	REFERENCE

74V037 INSTRM   A M Voronin, Vestn Akad Nauk Kaz (9),71-4(1974), Digital Storing Device for Nuclear Gamma Resonance Spectroscopy (In Russian)

74V038 THEORY   V I Vorontsov and V I Vysotskii, Kvan Elektron (Kiev) (8), 63-9(1974), On Induced Mossbauer Gamma Emission (In Russian)

74V039 THEORY   V I Vorontsov and V I Vysotskii, Kvan Elektron (Kiev) (8), 69-79(1974), On the Kinetics of Forced Gamma Emission (In Russian)

74V040 FE-57   P O Voznyuk, V N Dubinin, V V Kuz'movich, and N A Ivkina, Fiz-Khim Mekh Liofil'nost Dispersnykh Sist 6,72-5(1974), Metallopolymers of Iron Based on Silicomolybdenum Blue Studied by the Mossbauer Effect (In Russian)

74V041 FE-57   P O Voznyuk and V N Dubinin, Magnetic Hyperfine Structure of Mossbauer Spectra of the Ultra-fine Particles of Iron Hydroxides, in "Proceedings of the International Conference of Magnetism ICM-73" (Moscow, 1973) ("Nauka," Moscow, 1974), Vol VI, pp 165-9 (In Russian)

74W026 REVIEW   R E Watson and L H Bennett, Charge Transfer in Alloys: The Blind Men and the Elephant, in "Charge Transfer/Electronic Structure of Alloys" (Twin Symposium, Philadelphia, 1973), edited by L H Bennett and R H Willens (Metallurgical Society of the American Institute of Mining, Metallurgical, and Petroleum Engineers, 1974), pp 1-21

74W027 SN-119   G Weyer, B I Deutch, A Nylandsted-Larsen, and J U Andersen, Mossbauer and Channeling Experiments on Implants of 119Sn in Group IV Elements in "Hyperfine Interactions Studied in Nuclear Reactions and Decay: Contributed Papers" (Conference, Uppsala, Sweden, 1974), edited by E Karlsson and R Wappling (Upplands Grafiska, Uppsala, 1974), pp 38-9

74W028 DY-161   H P Wit and L Niesen, Spin Relaxation Phenomena in Mossbauer Spectra of Dysprosium Impurities Implanted in Iron, in "Hyperfine Interactions Studied in Nuclear Reactions and Decay: Contributed Papers" (Conference, Uppsala, Sweden, 1974), edited by E Karlsson and R Wappling (Upplands Grafiska, Uppsala, 1974), pp 246-7

74W029 FE-57   J M Wilson and D L Uhrich, Mol Cryst Liq Cryst 25,113-21 (1974), Reinterpretation of the Fe-57 Mossbauer Effect of 1-1'-Diacetylferrocene in 4-4'-Bis(heptyloxy)azoxybenzene

74Z009 TE-125   V S Zasimov, R N Kuz'min, and A I Firov, Mossbauer Effect on Tellurium-125 Nuclei in Tellurium Compounds, in "Str Svoistva Primen Metallid, (Mater Simp), 2nd" (1972), edited by I I Kornilov and N M Matveeva ("Nauka," Moscow, 1974), pp 93-6 (In Russian)

1975 Master Reference List

CODE	TOPIC	REFERENCE
75A001	FE-57	C R Abeledo, R B Frankel, and A A Misetich, Chem Phys Lett 31,108-10(1975), Hyperfine Interactions in MgF2:Fe2+ and ZnF2:Fe2+ by Mossbauer Spectroscopy
75A002	FE-57	M Ableiter and U Gonser, Z Metallkd 66,86-92(1975), Untersuchungen am System Niob-Wasserstoff
75A003	NP-237	A T Aldred, D J Lam, A R Harvey, and B D Dunlap, Phys Rev B 11,1169-75(1975), Magnetic Properties of Neptunium Laves Phases: NpOs(2-x)Ru(x) Pseudobinary System
75A004	NP-237	A T Aldred, B D Dunlap, D J Lam, G H Lander, M H Mueller, and I Nowik, Phys Rev B 11,530-44(1975), Magnetic Properties of Neptunium Laves Phases: NpMn2, NpFe2, NpCo2, and NpNi2
75A005	SN-119	S Ambe and F Ambe, Inorg Nucl Chem Lett 11,139-43(1975), Mossbauer Emission Spectrum of 119Sn in 119Sb(OH)(C2O4)
75A006	THEORY	S K Arif, D St P Bunbury, G J Bowden, and R K Day, J Phys F 5,1037-47(1975), Nuclear Energy Levels: Applications to Complex 57Fe Mossbauer Spectra
75A007	FE-57	S K Arif, D St P Bunbury, G J Bowden, and R K Day, J Phys F 5,1048-63(1975), 57Fe Mossbauer Studies of the (RE)Fe3 Intermetallic Compounds
75A008	FE-57	L Asch, G K Shenoy, J M Friedt, J P Adloff, and R Kleinberger, J Chem Phys 62,2335-42(1975), Mossbauer Effect and X-ray Studies of the Phase Transition in Iron Hexammine Salts
75A009	FE-57	E J Ansaldo, Nature 254,501(1975), Mossbauer Absorption in a Metamict Mineral
75A010	SN-119	S Akselrod, M Pasternak, and S Bukshpan, Phys Rev B 11, 1040-4(1975), Mossbauer-effect Studies of the Lattice Dynamics of Granular Tin
75A011	FE-57	F Aramu, V Maxia, and C Muntoni, Lett Nuovo Cimento Soc Ital Fis 12,225-7(1975), Resonant-gamma-ray Absorption in Iron Oxalate
75A012	NP-237	A T Aldred, B D Dunlap, D J Lam, G H Lander, M H Mueller, and I Nowik, Magnetic Properties of the Neptunium Laves Phases: NpMn2, NpFe2, NpCo2, NpNi2, in "AIP Conference Proceedings—No 24, Magnetism and Magnetic Materials-1974" (20th Annual Conference, San Francisco), edited by C D Graham, Jr, G H Lander, and J J Rhyne (American Institute of Physics, New York, 1975), pp 347-8
75A013	FE-57	L Asch, G K Shenoy, J M Friedt, and J P Adloff, J Chem Soc, Dalton Trans 1235-8(1975), Mossbauer Spectroscopy of Hexaammineiron(II) Nitrate, Thiocyanate, and Sulphate
75A014	FE-57	S K Arif, D St P Bunbury, and G J Bowden, J Phys F 5,1785-91(1975), Temperature Dependent Hyperfine Interactions in the Intermetallic Compounds YFe3, TbFe3 and ErFe3
75A015	FE-57	S K Arif, D St P Bunbury, and G J Bowden, J Phys F 5,1792-1800(1975), Crystallographic and Mossbauer Investigations of Dy(Fe(1-x)Co(x))3 and Y(Fe(1-x)Co(x))3 Intermetallic Compounds
75A016	FE-57	G Albanese, M Carbucicchio, and A Deriu, Temperature Dependence of the Sublattice Magnetization in Al and Ga Substituted M-type Hexagonal Ferrites, in "5th Int Conf Mossbauer Spec, Proc," Part 1, pp 136-9 (See 75H017)
75A017	FE-57	H Annersten, Neues Jahrb Mineral Monatsh 8,378-84(1975), A Mossbauer Characteristic of Ordered Glauconite

CODE	TOPIC	REFERENCE

75A018   SB-121   S Ambe and F Ambe, J Chem Phys 63,4077-8(1975), Mossbauer Emission Spectrum of 121Sb after the Beta- Decay of 121mSn in SnS2: Nuclear Decay Synthesis of Antimony(V) Sulfide

75A019   FE-57   R A Arents, Yu V Maksimov, and I P Suzdalev, Mossbauer Investigation of Local Magnetic Properties of the Iron Epsilon-carbide and Intermediary Carbides Formed at the epsilon to-chi-to-theta Phase Transformations, in "5th Int Conf Mossbauer Spec, Proc," Part 2, pp 328-31   (See 75H017) (In Russian)

75A020   FE-57   A Antonov, A Z Hrynkiewicz, D S Kulgawczuk, S Lasocki, and B Jezowska-Trzebiatowska, Mossbauer Spectroscopy in Adenine, Guanine, and Glutathione Iron Complexes, in "5th Int Conf Mossbauer Spec, Proc," Part 2, pp 408-12 (See 75H017)

75A021   SN-119   N E Ablesimov and S I Bondarevskii, Khim Vys Energ 9,174-5 (1975)/High Energ Chem (USSR) 9,147-8(1975), Consequences of a Converted Isomeric Transition at 298 K in Surface Atoms of Tin-119m

75A022   FE-57   A M Afanas'ev, S S Yakimov, and V N Zarubin, J Phys C 8, L368-70(1975), Zeeman Electronic Splitting in the Absorption Spectra of Mossbauer Gamma Rays in Single-crystal Al(NO3)3-9H2O:Fe3+

75A023   FE-57   M I Afanasov, Yu D Perfil'ev, and A M Babeshkin, J Radioanal Chem 27,125-8(1975), Mossbauer Study of Post-effects in Substituted Cobalt beta-Diketonates Caused by 57Co Decay

75A024   FE-57   I P Alenchikova, B E Dzevitskii, V S Neporezov, N N Savvateev, and V F Sukhoverkhov, Zh Neorg Khim 20,2156-61 (1975)/Russ J Inorg Chem, Nuclear Gamma-resonance Study of Iron Fluoride-based Phases

75A025   SB-121   S Ambe, J Inorg Nucl Chem 37,2023(1975), Chemical Properties of Sb(III)(C2O4)OH

75A026   FE-57   E B Andersen, J Fenger, and J Rose-Hansen, Lithos 8, 237-46(1975), Determination of (Fe2+)/(Fe3+)-ratios in Arfvedsonite by Mossbauer Spectroscopy

75A027   FE-57   G Albanese and C Ghezzi, Determination of Elastic and Inelastic Scattering of Gamma-rays in Vitreous Silica by Means of the Mossbauer Effect, in "5th Int Conf Mossbauer Spec, Proc," Part 3, pp 704-9   (See 75H017)

75A028   FE-57   A N Artem'ev, V V Sklyarevskii, G V Smirnov, and E P Stepanov, The Energy Analysis of Resonant Gamma-rays Diffracted by alpha-Fe2O3 Single Crystal, in "5th Int Conf Mossbauer Spec, Proc," Part 3, pp 707-9   (See 75H017)

75A029   SN-119   I A Avenarius and R N Kuz'min, Temperature Dependence of Mossbauer Effect Probability for Impurities of 119Sn in Monocrystal of Antimony, in "Int Conf Mossbauer Spectrosc, Proc," Vol 1, p 21   (See 75H027)   (In Russian)

75A030   FE-57   A Abras and R A Mansur, Application of the Mossbauer Effect to the Study of Brazilian Ilmenites, in "Int Conf Mossbauer Spectrosc, Proc," Vol 1, pp 135-6   (See 75H027)

75A031   FE-57   R A Abramovitch, J L Atwood, M L Good, and B A Lampert, Inorg Chem 14,3085-9(1975), Crystal Structure and Mossbauer Spectrum of (2)Ferrocenophanethiazine 1,1-Dioxide

75A032   FE-57   C C Addison, P G Harrison, N Logan, L Blackwell, and J H Jones, J Chem Soc, Dalton Trans 830-3(1975), Mossbauer Study of the Thermal Decomposition of Dinitrogen Tetroxide Solvates of Iron(III) Nitrate

CODE	TOPIC	REFERENCE

75A033   FE-57   M I Afanasov, S Nagy, Yu D Perfil'ev, A Vertes, and A M Babeshkin, Radiochem Radioanal Lett 23,181-6(1975), Mossbauer Study of Cobalt(II) Chloride Solutions in Non-aqueous Solvents

75A034   FE-57   M I Afanasov, Yu D Perfil'ev, and A M Babeshkin, Khim Vys Energ 9,283-5(1975)/High Energ Chem (USSR) 9,250-2(1975), Mossbauer Emission Spectra of 57Co in Cobalt Dipyridyl Complexes

75A035   FE-57   H Annersten, Fortschr Mineral 52,583-90(1975), Mossbauer Study of Iron in Natural and Synthetic Biotites

75A036   FE-57   R A Arents, Yu V Maksimov, I P Suzdalev, and E F Makarov, Study of the Influence of Defects in Doped Fe Catalizator under Nitration of the Crystal and by Formation of Catalytically Active Structure (Using the NGR Method), in "Int Conf Mossbauer Spectrosc, Proc," Vol 1, pp 303-4   (See 75H027)   (In Russian)

75A037   YB-170   L Asch, G K Shenoy, B D Dunlap, and G M Kalvius, Spin Relaxation Studies in Cubic Cs2Na(Yb(1-x)Gdx)Cl6, in "Int Conf Mossbauer Spectrosc, Proc," Vol 1, pp 403-4 (See 75H027)

75A038   FE-57   A M Afanas'ev, S S Yakimov, and V N Zarubin, Manifestation of the Zeeman Electronic Splitting in the Absorption Spectra of the Mossbauer Gamma-rays in the Al(NO3)3-9H2O:Fe3+ Single Crystal, in "Int Conf Mossbauer Spectrosc, Proc," Vol 1, pp 413-5 (See 75H027)

75A039   FE-57   G Albanese, M Carbucicchio, A Deriu, G Asti, and S Rinaldi, Appl Phys 7,227-38(1975), Influence of the Cation Distribution on the Magnetization of Y-type Hexagonal Ferrites

75A040   FE-57   J Arai, J Phys Soc Jpn 39,1409-10(1975), Anomalous Mossbauer Spectra Induced by Inhomogeneity in Fe-Ni Invar Alloy

75A041   FE-57   F Aramu, V Maxia, D De Filippo, and E F Trogu, Lett Nuovo Cimento Soc Ital Fis 14,517-9(1975), Resonant Gamma Ray Absorption in Iron(III) Dithio Chelates

75A042   PROPSL   V G Alpatov, A G Beda, G E Bizina, A V Davydov, and M M Korotkov, State of the Experiments on the Gamma-resonance Excitation of the Long-lived Isomeric State of 107Ag, in "5th Int Conf Mossbauer Spectrosc, Proc," Vol 3, pp 499-504 (See 75H017)   (In Russian)

75A043   FE-57   T M Aivazyan, Yu M Aivazyan, L A Kocharyan, and A R Mkrtchyan, To the Determination of the Mossbauer Absorption Spectra Parameters at the Ultra-sound Excitations, in "5th Int Conf Mossbauer Spectrosc, Proc," Vol 3, pp 600-6 (See 75H017)   (In Russian)

75A044   THEORY   T M Aivazyan, Yu M Aivazyan, and A R Mkrtchyan, On the Possibility of the Acoustic Modulation of the Mossbauer Radiation Passing Through an Absorber, in "5th Int Conf Mossbauer Spectrosc, Proc," Vol 3, pp 607-8   (See 75H017) (In Russian)

75A045   FE-57   R E Anderson, W R Dunham, B H Sands, A J Bearden, and H L Crespi, Biochim Biophys Acta 408,306-18(1975), On the Nature of the Iron Sulfur Cluster in a Deuterated Algal Ferredoxin

75A046   THEORY   A V Andreev and Yu A Il'inskii, Zh Eksp Teor Fiz 68,811-6 (1975)/Sov Phys-JETP 41,403-5(1976), Amplification in a Gamma Laser when the Bragg Condition is Satisfied

CODE	TOPIC	REFERENCE

75A047   THEORY   M A Andreeva, R N Kuz'min, and S F Oparina, Phys Status Solidi B 71,K143-6(1975), Variation of the Resonantly Scattered Wave by Means of Ultrasonic Vibrations for the Phase Determination

75A048   THEORY   M A Andreeva and R N Kuz'min, Phys Status Solidi B 71, K201-4(1975), Evolution of the Mossbauer Spectrum of the Bragg Reflection with Increasing Crystal Thickness (Kinematic Theory of Diffraction)

75A049   FE-57   J Arai, J Phys Soc Jpn 39,692-7(1975), Anomalous Decrease of Thermal Expansion Due to Inhomogeneity in Fe-Ni Invar Alloy

75A050   FE-57   B M Arakelyan, L A Alekseev, A A Vasil'ev, and P L Gruzin, Zavod Lab 41,1115-7(1975)/Ind Lab (USSR) 41,1382-3(1975), A Mossbauer Spectroscopic Study of the Phase Transitions in Gamma-irradiated Iron Oxide

75A051   FE-57   A V Astakhov, Yu B Voitkovskii, O N Generalov, and S V Sidorov, Kristallografiya 20,769-74(1975)/Sov Phys-Crystallogr 20,471-4(1976), NGR Investigation of Some Lamellar and Boron-containing Silicates

75A052   FE-57   U Atzmony and M P Dariel, Spin Reorientation Studies in Cubic Laves Rare-earth Iron Compounds, in "AIP Conference Proceedings—No 24, Magnetism and Magnetic Materials-1974" (20th Annual Conference, San Francisco), edited by C D Graham, Jr, G H Lander, and J J Rhyne (American Institute of Physics, New York, 1975), pp 662-3

75A053   FE-57   A A Ashe, III, E Meyers, P Shu, T Von Lehmann, and J Bastide, J Am Chem Soc 97,6865-6(1975), Bis(1-substituted-borabenzene)iron Complexes

75A054   FE-57   Kh I Amirkhanov, L K Anokhina, L K Palivoda, R I Chalabov, and R U Gabitova, Dokl Akad Nauk SSSR 223,1218-9(1975)/ Dokl Chem Technol, Use of the Mossbauer Effect for Dating Ores and for Evaluating the Thermodynamic Conditions of Their Formation (Illustrated by the Kizil-Dere Chalcopyrite Deposit in Dagestan)

75A055   FE-57   Kh I Amirkhanov and L K Anokhina, Dokl Akad Nauk SSSR 225, 659-60(1975)/Dokl Acad Sci, Earth Sci Sect, Age-dependent Variations of Iron-57 Nuclear Levels in Minerals

75B001   SN-119   G M Bancroft and K D Butler, Can J Phys 53,307-10(1975), 119Sn Mossbauer Spectra of Sn-Ni Compounds

75B002   EU-151   C M P Barton and N N Greenwood, Europium-151 Mossbauer Spectroscopy, in "Mossbauer Effect Data Index, Covering the 1973 Literature," edited by J G Stevens and V E Stevens (Plenum Publishing Corp, New York, 1975), pp 395-445

75B002   REVIEW   C M P Barton and N N Greenwood, Europium-151 Mossbauer Spectroscopy, in "Mossbauer Effect Data Index, Covering the 1973 Literature," edited by J G Stevens and V E Stevens (Plenum Publishing Corp, New York, 1975), pp 395-445

75B003   FE-57   T Birchall and A F Reid, J Solid State Chem 13,351-9(1975), An 57Fe Mossbauer Effect Study of Magnetic Ordering in the Fe2O3-Cr2O3 System

75B004   SN-119   J Bolz and F Pobell, Z Phys B 20,95-103(1975), Mossbauer Effect of 119Sn in Amorphous Superconducting Metals

75B005   FE-57   R J Borg and G J Dienes, J Appl Phys 46,99-104(1975), Short-range Order in Au-Fe Radiation-enhanced Diffusion and the Effectiveness of 14-MeV Neutrons

CODE	TOPIC	REFERENCE

75B006   SN-119   A Bos and A T Howe, J Chem Soc, Faraday Trans 2 71,28-40 (1975), Mossbauer and Infra-red Studies of the Diffusion and Reactivity of (SnO)n Species (n>1) Initially Isolated in Solid Nitrogen

75B007   FE-57   Yu V Baldokhin, V A Makarov, E F Makarov, and V A Povitskii, Phys Status Solidi A 27,265-71(1975), Mossbauer Investigation of Radio Frequency Striction Vibrations in Magnetic Alloys

75B008   PROPSL   G C Baldwin and R V Khokhlov, Phys Today February, 32-9 (1975), Prospects for a Gamma-ray Laser

75B009   SN-119   R Barbieri, N Bertazzi, C Tomarchio, and R H Herber, J Organomet Chem 84,39-46(1975), Organotin(IV) Azido and Mixed Azidothiocyanato Complex Anions; A Mossbauer and Vibrational Spectroscopic Study

75B010   INSTRM   J Becker, L Eriksson, L C Moberg, and Z H Cho, Nucl Instrum Methods 123,199-201(1975), On the Use of Tin-loaded Plastic Scintillators in Mossbauer Spectroscopy

75B011   FE-57   J Bosse, H Gabriel, and W Vollmann, Phys Status Solidi B 68,81-91(1975), Nuclear Spin Dynamics in the Presence of Electronic Relaxation

75B012   FE-57   B Brunot, Chem Phys Lett 32,187-9(1975), Mossbauer Study of Deuteration Effects on Some Ferrous Hexammine Salts

75B013   I-129   S Bukshpan, C Goldstein, T Sonnino, L May, and M Pasternak, J Chem Phys 62,2606-9(1975), Molecular Complexes of I2: A Mossbauer Effect Study

75B014   I-129   S Bukshpan, M Pasternak, and T Sonnino, J Chem Phys 62, 2916-7(1975), Mossbauer Effect Results for I2, IBr, and ICl in Different Chemical States

75B015   SN-119   D Barb, E Burzo, S Constantinescu, L Diamandescu, A Marian, and D Tarina, Rev Roum Phys 20,103-6(1975), Mossbauer Effect on 119Sn in Zircaloy 2

75B016   SN-119   T Birchall and A R Pereira, J Chem Soc, Dalton Trans 1087-92(1975), Nuclear Magnetic Resonance and Mossbauer Spectra of Some Organotin Anions

75B017   SN-119   T Birchall and A W Sleight, J Solid State Chem 13,118-30 (1975), Nonstoichiometric Phases in the Sb-Nb-O and Sn-Ta-O Systems Having Pyrochlore-related Structures

75B018   YB-170   C Borely, F Gonzalez-Jimenez, P Imbert, and F Varret, J Phys Chem Solids 36,605-9(1975), Mossbauer Study of Ytterbium Ethylsulphate

75B019   FE-57   A Bristoti, P J Viccaro, J I Kunrath, and D E Brandao, Inorg Nucl Chem Lett 11,253-8(1975), Mossbauer Analysis and Thermal Decomposition Studies of Fe2(SO4)3-nH2O

75B020   FE-57   A Bristoti, J I Kunrath, P J Viccaro, and L Bergter, J Inorg Nucl Chem 37,1149-51(1975), Mossbauer and Thermogravimetric Analysis of the Oxidation Pathway in the Thermal Decomposition of FeSO4-7H2O

75B021   SB-121   L Brattas, J D Donaldson, A Kjekshus, D G Nicholson, and J T Southern, Acta Chem Scand, Ser A 29,217-9(1975), 121Sb Mossbauer Studies on NbSb2 and TaSb2

75B022   GD-155   E R Bauminger, A Diamant, I Felner, I Nowik, and S Ofer, Phys Rev Lett 34,962-5(1975), Anisotropic Hyperfine Interactions in Gadolinium Metal

75B023   FE-57   L Billard and A Chamberod, Solid State Commun 17,113-8 (1975), On the Dissymmetry of Mossbauer Spectra in Iron-Nickel Alloys

CODE	TOPIC	REFERENCE
75B024	FE-57	C Boekema, F Van der Woude, and G A Sawatzky, Phys Rev B 11,2705-6(1975), Interpretation of the High-pressure Effects in the Rare-earth Orthoferrites
75B025	FE-57	C Blaauw, H Hanson, F Van der Woude, and G A Sawatzky, The Effect of Si on the Electronic Structure of Iron in FeSi Alloys, in "5th Int Conf Mossbauer Spec, Proc," Part 1, pp 10-4 (See 75H017)
75B026	FE-57	C Blaauw, H Hanson, and F Van der Woude, Mossbauer Effect in Pure and Impurity Doped FeSi2, in "5th Int Conf Mossbauer Spec, Proc," Part 1, pp 28-32 (See 75H017)
75B027	FE-57	D Barb, E Burzo, and M Morariu, The Temperature Dependence of the Mossbauer Parameters and the Magnetic Properties of the RFe2 (R=Y or Rare-earth) Compounds, in "5th Int Conf Mossbauer Spec, Proc," Part 1, pp 37-40 (See 75H017)
75B028	FE-57	V G Bhide and D S Rajoria, 57Fe Mossbauer Effect Studies of Magnetic Ordering in La(1-x)Sr(x)CoO3, in "5th Int Conf Mossbauer Spec, Proc," Part 1, pp 55-70 (See 75H017)
75B029	FE-57	J Baumann, D Seyboth, and F Sontheimer, The Determination of Material Constants in Antiferromagnetic Fe-Ge Alloys by Means of the Mossbauer Effect, in "5th Int Conf Mossbauer Spec, Proc," Part 1, pp 71-4 (See 75H017)
75B030	THEORY	C Boekema, F Van der Woude, and G A Sawatzky, Spin and Charge Density Oscillations in Doped Magnetite: A Classical Selfconsistent Charge Density Oscillation Model, in "5th Int Conf Mossbauer Spec, Proc," Part 1, pp 96-8 (See 75H017)
75B031	FE-57	I Bibicu, I Dezsi, T Lohner, and D L Nagy, Ionic Species and Frozen Solutions, in "5th Int Conf Mossbauer Spec, Proc," Part 1, pp 247-50 (See 75H017)
75B032	SN-119	A A Bekker, E N Efremov, and A N Nesmeyanov, Complex Investigation of the Chemical Bond Character and the Local Surroundings Symmetry of Sn and Te in Solid Solutions on the Base of Lead and Tin Tellurides, in "5th Int Conf Mossbauer Spec, Proc," Part 1, pp 267-9 (See 75H017) (In Russian)
75B032	TE-125	A A Bekker, E N Efremov, and A N Nesmeyanov, Complex Investigation of the Chemical Bond Character and the Local Surroundings Symmetry of Sn and Te in Solid Solutions on the Base of Lead and Tin Tellurides, in "5th Int Conf Mossbauer Spec, Proc," Part 1, pp 267-9 (See 75H017) (In Russian)
75B033	THEORY	E V Baklanov and V P Chebotaev, Pis'ma Zh Eksp Teor Fiz 21, 286-9(1975)/JETP Lett 21,131-2(1975), Concerning One Possibility of Lasing in the Gamma Band
75B034	FE-57	O A Bayukov, G I Gashimov, A I Drokin, V P Klonnikov, M I Petrov, A G Rustamov, and E A Eivazov, Izv Akad Nauk SSSR, Ser Fiz 39,219-21(1975)/Bull Acad Sci USSR, Phys Ser 39, 196-8(1975), Magnetic, Mossbauer, and Electrical Investigations of Ferrite-spinels in the NiFe2O(4-x)S(x) System
75B035	FE-57	V F Belov, V G Pyl'nev, I S Zheludev, V V Korovushkin, E V Korneev, and Yu N Yarmukhamedov, Kristallografiya 20,167-8 (1975)/Sov Phys-Crystallogr 20,96-7(1975), Study of Ferroelectric Boracite Mn3B7O13Cl by the Nuclear Gamma Resonance Method
75B036	SN-119	N Bertazzi, G Alonzo, F Di Bianca, and G C Stocco, Inorg Chim Acta 12,123-6(1975), Infrared and Mossbauer Studies on Adducts R3SnOH-R3MX (M=Sn,Pb; X = Pseudohalide)
75B037	FE-57	S S Bhandari and J Varma, Phys Status Solidi A 29,K59-64 (1975), Mossbauer Studies of Ilmenite Single Crystal

CODE	TOPIC	REFERENCE
75B038	YB-170	C Borely, F Gonzalez-Jimenez, P Imbert, and F Varret, J Phys Chem Solids 36,683-8(1975), Mossbauer Study of 170Yb in CaF2
75B039	SN-119	N B Brandt and V G Snigirev, Fiz Tverd Tela (Leningrad) 17,910-3(1975)/Sov Phys-Solid State 17,578-9(1975), Weakening of Force Constants in SnCd and SnBi Alloys
75B040	EU-151	K H J Buschow, W J Huiskamp, H T Le Fever, F J Steenwijk, and R C Thiel, J Phys F 5,1625-36(1975), Mossbauer Effect and Magnetic Properties of Some Eu-Zn Compounds
75B041	FE-57	P Bussiere, R Dutartre, G A Martin, and J P Mathieu, C R Acad Sci, Ser C 280,1133-6(1975), Etude par Spectrometrie Mossbauer d'un Catalyseur au Fer Depose sur l'Oxyde de Magnesium et de Son Interaction avec l'Hydrogene
75B042	FE-57	G M Bancroft and J R Brown, Am Mineral 60,265-72(1975), A Mossbauer Study of Coexisting Hornblendes and Biotites: Quantitative Fe3+/Fe2+ Ratios
75B043	FE-57	G M Bancroft and A T Rake, Inorg Chim Acta 13,175-9(1975), 57Fe and 119Sn Mossbauer Spectra of Phosphine and Phosphite Derivatives of Ph(3-n)Cl(n)SnFe(CO)2(pi-C5H5) Compounds
75B043	SN-119	G M Bancroft and A T Rake, Inorg Chim Acta 13,175-9(1975), 57Fe and 119Sn Mossbauer Spectra of Phosphine and Phosphite Derivatives of Ph(3-n)Cl(n)SnFe(CO)2(pi-C5H5) Compounds
75B044	SN-119	G M Bancroft, K D Butler, and T K Sham, J Chem Soc, Dalton Trans 1483-6(1975), Room Temperature Tin-119 Mossbauer Spectra of Unassociated Tin Compounds
75B045	FE-57	G M Bancroft and P L Sears, Inorg Chem 14,2716-20(1975), Iron-57 Mossbauer Study of Iron(II)-carbene Compounds. Bonding Characteristics of Carbenes
75B046	TE-125	F J Berry, E H Kustan, and B C Smith, J Chem Soc, Dalton Trans 1323-4(1975), Tellurium-125 Mossbauer Spectra of Some Aryltellurium-(II) and -(IV) Compounds
75B047	TE-125	P H Barrett, P A Montano, H Micklitz, and J B Mann, Phys Rev B 12,1676-80(1975), Mossbauer Study of Rare-gas-matrix Isolated 125Te Dimers, 125TeF6, and 125TeCl4 Molecules
75B048	FE-57	A E Berkowitz, J A Lahut, I S Jacobs, L M Levinson, and D W Forester, Phys Rev Lett 34,594-7(1975), Spin Spinning at Ferrite-organic Interfaces
75B049	SN-119	A Bos and A T Howe, J Chem Soc, Faraday Trans 2 71,28-40 (1975), Mossbauer and Infra-red Studies of the Diffusion and Reactivity of (SnO)n Species (n > 1) Initially Isolated in Solid Nitrogen
75B050	FE-57	D V Balashov, V S Vartanov, B G Zemskov, and E F Makarov, Investigation of the Properties of the Fe-Ni Alloys by the Nuclear Gamma-resonance at High Pressure, in "5th Int Conf Mossbauer Spec, Proc," Part 2, pp 292-5 (See 75H017) (In Russian)
75B051	FE-57	O Brummer, G Drager, and D Katzer, Determination of the Short/Range Order in Fe-Al Alloys of Different Thermal and Mechanical Treatment Using the Mossbauer Effect, in "5th Int Conf Mossbauer Spec, Proc," Part 2, pp 306-10 (See 75H017)
75B052	SN-119	Cv Bontschev, A Minkova, and C K Tkhiep, Applications of Mossbauer Effect to the Investigation of the Copper-Tin Intermetallic System, in "5th Int Conf Mossbauer Spec, Proc," Part 2, pp 458-62 (See 75H017) (In Russian)

CODE	TOPIC	REFERENCE

75B053   FE-57   B Balko and G R Hoy, Relaxation Studies Using Selective Excitation Double Mossbauer Techniques, in "5th Int Conf Mossbauer Spec, Proc," Part 2, pp 480-4   (See 75H017)

75B054   FE-57   C Bansal, J Ray, and G Chandra, J Phys F 5,1663-6(1975), Distribution of Hyperfine Fields in a Disordered Ni48Fe52 Alloy by Mossbauer Effect

75B055   FE-57   G N Belozerskii and Yu P Khimich, Fiz Tverd Tela (Leningrad) 17,1352-7(1975)/Sov Phys-Solid State 17,871-3(1975), Hyperfine Interactions and Thermal Motion of Iron Nuclei in Hexagonal Barium Ferrite

75B056   REVIEW   V A Belyakov, Usp Fiz Nauk 115,553-601(1975)/Adv Phys Sci (USSR), Crystal Diffraction of Mossbauer Gamma-radiation

75B057   FE-57   K N Binnatov, A A Katsnel'son, and Yu L Rodionov, Izv Vyssh Uchebn Zaved, Fiz 18(3),125-6(1975)/Sov Phys J, Reciprocal Distribution of Atoms in a Ternary Ni-Fe-Cr Solid Solution

75B058   FE-57   M Boudart, A Delbouille, J A Dumesic, S Khammouma, and H Topsoe, J Catal 37,486-502(1975), Surface, Catalytic and Magnetic Properties of Small Iron Particles

75B059   FE-57   R A Buckwald and A A Hirsch, Solid State Commun 17,621-5 (1975), Mossbauer Effect Study of Multi-stage Electronic Transition in Magnetite

75B060   FE-57   S S Budagovskii, V N Bykov, M I Gavrilyuk, E M Ivanyushkin, and I I Rudnev, Ukr Fiz Zh (Russ Ed) 20,392-6(1975), Decay Process of Solid BCC-solutions of the System Tungsten-Rhenium   (In Russian)

75B061   TE-125   P Boolchand, M Tenhover, S Jha, G Langouche, B W Triplett, S S Hanna, and P Jena, Phys Lett 54A,293-4(1975), Magnetic Hyperfine Structure of 125Te in Ferromagnetic Pd2MnSb

75B062   REVIEW   D Barb, Polarized Mossbauer Transitions, in "5th Int Conf Mossbauer Spec, Proc," Part 3, pp 665-704   (See 75H017)

75B063   THEORY   D Barb and D Tarina, Polarized Mossbauer Transitions in Mixed Hyperfine Interactions, in "5th Int Conf Mossbauer Spec, Proc," Part 3, pp 699-704   (See 75H017)

75B064   FE-57   B Brzoska, J Baumann, H Jena, D Seyboth, and F Sontheimer, Mossbauer Experiments with Coulomb Excited 57Fe after Recoil Implantation into Cu, Be and Mn, in "5th Int Conf Mossbauer Spec, Proc," Part 3, pp 736-9   (See 75H017)

75B065   SN-119   B Busch, J Pebler, and K Dehnicke, Z Anorg Allg Chem 416, 203-10(1975), Darstellung, Mossbauer- und Schwingungsspektren der Azidokomplexe (SnCl4(N3)2)(-2) und (SnCl4N3)2(-2)

75B066   FE-57   E Byrom, A J Freeman, and D E Ellis, Covalency Effects on Tetrahedral and Octahedral Fe3+ Sites in YIG: Charge and Spin Densities and Neutron Form Factors, in "AIP Conference Proceedings No 24, Magnetism and Magnetic Materials-1974" (20th Annual Conference, San Francisco), edited by C D Graham, Jr, G H Lander, and J J Rhyne (American Institute of Physics, New York, 1975), pp 209-10

75B067   SN-119   R Bacaud, P Bussiere, F Figueras, and J P Mathieu, C R Acad Sci (Paris), Ser C 281,159-61(1975), Caracterisation par Spectrometrie Mossbauer de Catalyseyrs Platine-etain Deposes sur Alumine

75B068   SN-119   A Sh Bakhtyarov and L N Vasil'ev, Izv Akad Nauk SSSR, Neorg Mater 11,741-2(1975)/Inorg Mater (USSR) 11,636-7(1975), Investigation of the System As-Se-Ge-Sn by the Mossbauer Method

CODE	TOPIC	REFERENCE

75B069   FE-57   E Barnighausen, J Appl Crystallogr 8,477-87(1975), Ein Verfahren zur Bestimmung der elastischen und inelastischen Streuanteile im Bereich der Bragg-Reflexe mit Hilfe von Mossbauer-Quelle und -Absorber

75B070   FE-57   J M Bellerby and M J Mays, J Chem Soc, Dalton Trans 1281-3 (1975), Organonitriles as Ligands in Low-spin Di(1,2-bis (diethylphosphino)-ethane)iron Complexes

75B071   THEORY   V A Belyakov and R Ch Bokun, Acta Crystallogr, Sec A 31, 737-41(1975), Kinematical Theory of Mossbauer Diffraction by Magnetically Ordered Crystals

75B072   FE-57   Yu V Baldokhin, V A Povitskii, V A Makarov, and E F Makarov, Radiofrequency Modulation in Mossbauer Spectra of Magnetic Alloys, in "Int Conf Mossbauer Spectrosc Proc," Vol 1, pp 55-6 (See 75H027)

75B073   FE-57   Sh Sh Bashkirov and E K Sadykov, Quadrupole Mechanism of the Double Gamma Resonance, in "Int Conf Mossbauer Spectrosc, Proc," Vol 1, pp 57-8 (See 75H027) (In Russian)

75B074   EU-151   E R Bauminger, I Felner, D Levron, I Nowik, and S Ofer, Interconfiguration Fluctuations in Metallic Rare Earth Compounds, in "Int Conf Mossbauer Spectrosc, Proc," Vol 1, pp 69-70 (See 75H027)

75B075   FE-57   R J Borg and R D Taylor, Magnetic Hyperfine Splitting in Dilute Au-Fe Alloys, in "Int Conf Mossbauer Spectrosc, Proc," Vol 1, pp 77-8 (See 75H027)

75B076   FE-57   M Baran and A Polaczek, Mossbauer Effect Study of Iron Vanadium Oxide Bronzes, in "Int Conf Mossbauer Spectrosc, Proc," Vol 1, pp 91-2 (See 75H027)

75B077   FE-57   Sh Sh Bashkirov and V N Lebedev, Calculation of the Quadrupole Splitting of the Mossbauer Spectrum of Pentacoordinated $Fe3+$ Ion in Ferrite $BaFe12019$, in "Int Conf Mossbauer Spectrosc, Proc," Vol 1, pp 95-6 (See 75H027) (In Russian)

75B078   FE-57   G N Belozerskii, V M Erkin, V A Malyshevskii, O G Sokolov, and Yu P Khimich, NGR Studies of the Quasibinary Alloys on the Base of alpha-Iron, in "Int Conf Mossbauer Spectrosc, Proc," Vol 1, pp 125-6 (See 75H027)

75B079   FE-57   Sh Sh Bashkirov, R K Gubaidullin, N G Ivoilov, and V A Chistyakov, Hyperfine Fields on Fe Nuclei in Ga-substituted Yttrium Orthoferrite, in "Int Conf Mossbauer Spectrosc, Proc," Vol 1, pp 151-2 (See 75H027) (In Russian)

75B080   FE-57   M F Vereshchak and A K Zhetbaev, Mossbauer Effect in Al-substituted Sodium Orthoferrites, in "Int Conf Mossbauer Spectrosc, Proc," Vol 1, pp 153-4 (See 75H027) (In Russian)

75B081   FE-57   G N Belozerskii and Yu P Khimich, Mossbauer Effect in Hexagonal Barium Ferrite $BaFe12019$, in "Int Conf Mossbauer Spectrosc, Proc," Vol 1, pp 159-60 (See 75H027)

75B082   FE-57   Sh Sh Bashkirov, A B Liberman, and V I Sinyavskii, Magnetic Transformations in Hexagonal Ferrites, in "Int Conf Mossbauer Spectrosc, Proc," Vol 1, pp 163-4 (See 75H027) (In Russian)

75B083   FE-57   Sh Sh Bashkirov, A B Liberman, and V I Sinyavskii, Local Fields and Exchange Interactions in Ferrimagnetic Compounds with the Hexagonal Structure, in "Int Conf Mossbauer Spectrosc, Proc," Vol 1, pp 165-6 (See 75H027) (In Russian)

CODE	TOPIC	REFERENCE

75B084 FE-57 R J Borg and I Y Borg, The Magnetic Structure of an Alkali Amphibole, in "Int Conf Mossbauer Spectrosc, Proc," Vol 1 pp 167-8 (See 75H027)

75B085 FE-57 E Burzo, D Barb, and M Bodea, Thermal Variation of the Hyperfine Fields in Some Ternary Iron Compounds, in "Int Conf Mossbauer Spectrosc, Proc," Vol 1, pp 169-70 (See 75H027)

75B086 FE-57 E Burzo, M Bodea, D Barb, and J Laforest, Mossbauer Effect Study of $Th(Fe(x)Co(1-x))5$ Compounds, in "Int Conf Mossbauer Spectrosc, Proc," Vol 1, pp 171-2 (See 75H027)

75B087 FE-57 W Bergholz and W Schroter, Mossbauer Spectroscopy of Cobalt in Silicon, in "Int Conf Mossbauer Spectrosc, Proc," Vol 1, pp 215-6 (See 75H027)

75B088 AU-197 E Baggio-Saitovitch, U Wagner, F E Wagner, and J Danon, Mossbauer Studies of the Reduction of Complex Gold Cyanides by Electron Irradiation, in "Int Conf Mossbauer Spectrosc, Proc," Vol 1, pp 223-4 (See 75H027)

75B089 SN-119 C I Balcombe, E C Macmullin, and M E Peach, J Inorg Nucl Chem 37,1353-7(1975), Some Reactions and Spectroscopic Studies on the Organotin Thiolates $Bu3SnSR$ and $Bu2Sn(SR)2$

75B090 SN-119 J G Ballard and T Birchall, Can J Chem 53,3371-3(1975), 119Sn Mossbauer Spectra of Some Tin(II)-Oxygen Compounds

75B091 FE-57 R Bau, B Don, R Greatrex, R J Haines, R A Love, and R D Wilson, Inorg Chem 14,3021-5(1975), Mossbauer and X-ray Diffraction Studies on $HFe3(CO)9(SR)$ ($R=i-C3H7$ and $t-C4H9$). Example of Noncorrelation of M-M Distance with M-H-M Bonding

75B092 DY-161 M Belakhovsky, Solid State Commun 17,349-52(1975), Contact Charge Density in Trivalent Dysprosium Intermetallics Through APW Calculations

75B093 FE-57 N Burriesci and G Parravano, Diluted Iron Atoms in Solid Substrates: Poly(4 Vinylpyridine)-$FeCl3-6H2O$ Complexes, in "Int Conf Mossbauer Spectrosc, Proc," Vol 1, pp 247-8 (See 75H027)

75B094 FE-57 O A Bayukov, V M Buznik, V P Ikonnikov, and M I Petrov, Mossbauer Effect Determination of the Magnetic Structure and Electric Field Gradient (EFG) Tensor in $Fe3B06$ Single Crystals, in "Int Conf Mossbauer Spectrosc, Proc," Vol 1, pp 263-4 (See 75H027)

75B095 FE-57 E Baggio-Saitovitch and M A De Paoli, Mossbauer Spectroscopy of Iron(II) Trisdiimine Complexes, in "Int Conf Mossbauer Spectrosc, Proc," Vol 1, pp 277-8 (See 75H027)

75B096 FE-57 J J Bara, A Gutsze, A T Pedziwiatr, and Z M Stadnik, Investigation of Iron Doped Sodium A-type Zeolite with the Mossbauer Effect Method, in "Int Conf Mossbauer Spectrosc, Proc," Vol 1, pp 311-2 (See 75H027)

75B097 SN-119 G N Belozerskii, Sh Z Bashaikin, and E Yu Bessonova, Mossbauer Study of the Kinetics of Glass-crystal Transformation in Semiconductive Crystals, in "Int Conf Mossbauer Spectrosc, Proc," Vol 1, pp 351-2 (See 75H027)

75B098 THEORY K Byrin, Cv Bontschev, and K K'nchev, Mossbauer Effect in Suspension and Observation of the Velocity Spectrum of Brownian Particles, in "Int Conf Mossbauer Spectrosc, Proc," Vol 1, pp 431-2 (See 75H027)

CODE	TOPIC	REFERENCE

75B099   FE-57   D Bade, F Parak, U F Thomanek, H Eicher, and G M Kalvius, Mossbauer Effect and Magnetic Susceptibility Investigations of Human Hemoglobin, in "Int Conf Mossbauer Spectrosc, Proc," Vol 1, pp 439-40 (See 75H027)

75B100   PROPSL   D Barb and D Tarina, Use of Polarized Gamma Ray for Effective Hyperfine Parameter Determinations, in "Int Conf Mossbauer Spectrosc, Proc," Vol 1, pp 467-8 (See 75H027)

75B101   THEORY   D Barb and M Rogalski, Birefringence of Arbitrary Polarized Mossbauer Radiation, in "Int Conf Mossbauer Spectrosc, Proc," Vol 1, pp 469-70 (See 75H027)

75B102   ANALYS   J Bainbridge, Nucl Instrum Methods 128,531-5(1975), Quantitative Analysis of Mossbauer Backscatter Spectra from Multi-layer Films

75B103   FE-57   L Bogner and E R Seidel, Cryogenics 15,680-2(1975), Thermal Contact of Mossbauer Sources Far below 4.2 K

75B104   PM-145   H Bokemeyer, K Wohlfahrt, E Kankeleit, and D Eckardt, Z Phys A 274,305-18(1975), Mossbauer Conversion Spectroscopy: Measurements on the First Excited States of 180,182W and 145Pm

75B104   W-180   H Bokemeyer, K Wohlfahrt, E Kankeleit, and D Eckardt, Z Phys A 274,305-18(1975), Mossbauer Conversion Spectroscopy: Measurements on the First Excited States of 180,182W and 145Pm

75B104   W-182   H Bokemeyer, K Wohlfahrt, E Kankeleit, and D Eckardt, Z Phys A 274,305-18(1975), Mossbauer Conversion Spectroscopy: Measurements on the First Excited States of 180,182W and 145Pm

75B105   FE-57   V G Bhide, D S Rajoria, C N R Rao, G Rama Rao, and V G Jadhao, Phys Rev B 12,2832-43(1975), Itinerant-electron Ferromagnetism in La(1-x) SrxCoO3: A Mossbauer Study

75B106   FE-57   R Bacaud, P Bussiere, R Dutartre, F Figueras, G A Martin, and J P Mathieu, Uses of Mossbauer Spectroscopy in Heterogeneous Catalysis Research, in "Int Conf Mossbauer Spectrosc, Proc," Vol 1, pp 527-8 (See 75H027)

75B106   SN-119   R Bacaud, P Bussiere, R Dutartre, F Figueras, G A Martin, and J P Mathieu, Uses of Mossbauer Spectroscopy in Heterogeneous Catalysis Research, in "Int Conf Mossbauer Spectrosc, Proc," Vol 1, pp 527-8 (See 75H027)

75B107   REVIEW   D Barb, The Role of Polarization State of Gamma Rays in Mossbauer Spectroscopy, in "Int Conf Mossbauer Spectrosc, Proc," Vol 2, pp 379-97 (See 75H038)

75B108   REVIEW   G C Baldwin, J W Pettit, and H R Schwenn, A Critical Review of Current Proposals for Mossbauer Gasers, in "Int Conf Mossbauer Spectrosc, Proc," Vol 2, pp 413-28 (See 75H038)

75B109   FE-57   Yu V Baldokhin, E F Makarov, A V Mitin, and V A Povitskii, Change of the Mossbauer Spectra in a Strong Radio-frequency Field, in "5th Int Conf Mossbauer Spectrosc, Proc," Vol 3, pp 609-14 (See 75H017) (In Russian)

75B110   SN-119   K Burin, Cv Bontschev, M Grozdanov, and K K'nchev, Dependence of the Mossbauer Line Shape of a Suspension on the Size Distribution of Floating Particles, in "5th Int Conf Mossbauer Spectrosc, Proc," Vol 3, pp 640-2 (See 75H017) (In Russian)

75B111   SN-119   Cv Bontschev, I Vasil'ev, A Yanev, and Z Sapundshiev, Investigation of the Reaction of a Group of Ants by the Mossbauer Effect, in "5th Int Conf Mossbauer Spectrosc, Proc," Vol 3, pp 646-8 (See 75H017) (In Russian)

CODE	TOPIC	REFERENCE

75B112   FE-57   Yu F Babikova, M R Gryaznov, L M Isakov, N S Kolpakov, and M N Uspenskii, Prib Tekh Eksp (3),152-5(1975)/Instrum Exp Tech (USSR) 18,850-53(1975), Scattered Gamma-ray Registration Channel for Nuclear Gamma-resonance Spectrometer

75B112   INSTRM   Yu F Babikova, M R Gryaznov, L M Isakov, N S Kolpakov, and M N Uspenskii, Prib Tekh Eksp (3),152-5(1975)/Instrum Exp Tech (USSR) 18,850-53(1975), Scattered Gamma-ray Registration Channel for Nuclear Gamma-resonance Spectrometer

75B113   PROPSL   D Barb and M Rogalski, J Chim Phys Phys-Chim Biol 72,470-6 (1975), Conditions pour Obtenir l'Effet Faraday Classique avec un Rayonnement Gamma Emis sans Recul

75B114   SN-119   R Barbieri, F Di Bianca, G Alonzo, A Silvestri, L Pellerito, N Bertazzi, and G C Stocco, Z Anorg Allg Chem 411,173-81(1975), Infrared and Mossbauer Spectroscopic Studies on Complexes of Hal2Sn(IV) Moieties with Tridentate Ligands

75B115   SN-119   R Barbieri, L Pellerito, N Bertazzi, G Alonzo, and J G Noltes, Inorg Chim Acta 15,201-4(1975), Mossbauer Spectroscopic Studies on Compounds Containing Tin-Cadmium and Tin-Zinc Bonds

75B116   FE-57   Sh Sh Bashkirov, G D Kurbatov, E S Makhnev, and V A Chistyakov, Dokl Akad Nauk SSSR 223,622-4(1975)/Dokl Phys Chem 223,757-9(1975), Use of the Mossbauer Effect for Measuring the Short-range Order Parameters in a Solid Solution

75B117   FE-57   Sh Sh Bashkirov, A B Liberman, and V I Sinyavskii, Fiz Tverd Tela 17,1788-90(1975)/Sov Phys-Solid State 17,1162-3 (1975), Magnetic Structure and Exchange Interaction Parameters in Complex Ferrimagnetic Compounds

75B118   PROPSL   Sh Sh Bashkirov and E K Sadykov, Fiz Tverd Tela 17,1864-6 (1975)/Sov Phys-Solid State 17,1226(1975), A Mechanism of Double Gamma Resonance

75B119   FE-57   Sh Sh Bashkirov, A B Liberman, and V I Sinyavskii, Fiz Tverd Tela 17,1876-7(1975)/Sov Phys-Solid State 17,1236 (1975), "Anomalous" Behavior of the Debye-Waller Factor of Ferrimagnets with a Weak Exchange Coupling Between Magnetic Sublattices

75B120   FE-57   Sh Sh Bashkirov and V N Lebedev, Fiz Tverd Tela 17,2450-1 (1975)/Sov Phys-Solid State 17,1625-6(1976), An Unusually Large Quadrupole Splitting of the Mossbauer Line from the Fivefold-coordinated Fe^{3+} Ion in the Ferrite $BaFe12O19$

75B121   FE-57   Sh Sh Bashkirov, A B Liberman, and V I Sinyavskii, Zh Eksp Teor Fiz 69,1841-3(1975)/Sov Phys-JETP, Magnetic Transformations in Indium-substituted Hexaferrites

75B122   FE-57   D G Batyr, I I Bulgak, K I Turta, and R A Stukan, Koord Khim 1,655-8(1975), Mossbauer Spectra of Iron(II) Dioximines with Physiologically Active Ligands (In Russian)

75B123   FE-57   A F Belov, E V Korneev, T A Khimich, V F Belov, V V Korovushkin, and I M Kolesnikov, Zh Fiz Khim 49,1679-82 (1975)/Russ J Phys Chem 49,990-2(1975), Crystal Chemistry of Cobalt Aluminates

75B124   FE-57   A F Belov, E V Korneev, T A Khimich, V F Belov, V V Korovushkin, I S Zheludev, and I M Kolesnikov, Zh Fiz Khim 49,1683-8(1975)/Russ J Phys Chem 49,992-6(1975),The Electronic-nuclear Interactions and Non-equivalent Positions of the Iron Ions in Tourmaline

CODE	TOPIC	REFERENCE
75B125	FE-57	L M Belyaev, T V Dmitrieva, I S Lyubutin, A P Mazhara, and V E Fedorov, Zh Eksp Teor Fiz 68,1176-82(1975)/Sov Phys-JETP 41,583-6(1976), Features of Electric and Magnetic Hyperfine Interactions of Fe57 Nuclei in Chalcogenide Spinels
75B126	THEORY	V A Belyakov and E V Smirnov, Zh Eksp Teor Fiz 68,608-22 (1975)/Sov Phys-JETP 41,301-7(1975), On the Polarization Characteristics of Mossbauer Coherent Scattering by Perfect Magnetic-ordered Crystals
75B127	FE-57	N Benczer-Koller, J M Trooster, and C Song, Direct Measurements of the Spin Contact Density in Magnetic Materials, in "AIP Conference Proceedings-No 24, Magnetism and Magnetic Materials-1974" (20th Annual Conference, San Francisco), edited by C D Graham, Jr, G H Lander, and J J Rhyne (American Institute of Physics, New York, 1975), pp 785-7
75B128	FE-57	N I Bezmen, V K Egorov, G V Novikov, G Yu Odinets, and A V Chichagov, Ocherki Fiz-Khim Petrol 5,5-8(1975), Nuclear Gamma Resonance Study of Ordering Vacancies in Hexagonal Pyrrhotines (In Russian)
75B129	SN-119	V N Bogomolov, A I Zadorozhnii, and N A Klushin, Fiz Tverd Tela 17,2452-3(1975)/Sov Phys-Solid State 17,1627(1976), The Mossbauer Effect of Sn119 in Holes in NaX Zeolite and the Filtration of Sn119 Atoms from a Ga Melt by Zeolite
75B130	SN-119	J Bolz and F Pobell, Mossbauer Effect of 119Sn in Amorphous Tin, in "Proceedings of the 14th International Conference on Low Temperature Physics: Volume 2 Superconductivity" (Otaniemi, Finland, 1975), edited by M Krusius and M Vuorio (North-Holland Publishing Co, Amsterdam/ American Elsevier Publishing Co, New York, 1975), pp 425-8
75B131	FE-57	R J Borg, F R Szofran, W L Burmester, and D J Sellmyer, Magnetic Ordering in Several Fe-chain Silicate Compounds, in "AIP Conference Proceedings-No 24, Magnetism and Magnetic Materials-1974" (20th Annual Conference, San Francisco), edited by C D Graham, Jr, G H Lander, and J J Rhyne (American Institute of Physics, New York, 1975), pp 365-6
75B132	FE-57	A Brecher, W H Menke, J B Adams, and M J Gaffey, The Effects of Heating and Subsolidus Reduction on Lunar Materials: An Analysis by Magnetic Methods, Optical, Mossbauer, and X-ray Diffraction Spectroscopy, in "Proceedings of the Sixth Lunar Science Conference" (Houston, 1975), Vol 3, pp 3091-9 (Also referenced as Geochim Cosmochim Acta, Suppl 6)
75B133	SN-119	J S Brooks and J M Williams, Phys Status Solidi A 32,413-7 (1975), Magnetic Hyperfine Interactions in Ferromagnetic Co2TiSn
75B134	FE-57	E Burzo, M Bodea, and D Barb, C R Acad Sci (Paris), Ser B 280,345-8(1975), Champs Internes au Noyau 57Fe dans les Composes R(FexNi(1-x))2 ou R Represente Dy ou Ho
75B135	THEORY	D Barb and L Diamandescu, Rev Roum Phys 20,259-71(1975), Simulation of Mossbauer Relaxation Spectra Case of Effective Spin S = 1/2
75B136	FE-57	V M Byr'ko, A I Busev, T I Tikhonova, N V Baibakova, and L I Shepel', Zh Anal Khim 30,1885-91(1975)/J Anal Chem USSR, Some Analytical Properties of Hydrozonium Hydrazinedithiocarbaminate
75B137	SN-119	G M Bancroft, I Adams, H Lampe, and T K Sham, Chem Phys Lett 32,173-7(1975), Linewidths and Line Shapes in Solid State ESCA Studies: Electric Field Gradient Broadening of Sn 3d Lines

CODE	TOPIC	REFERENCE
75B138	SN-119	G M Bancroft, B W Davies, N C Payne, and T K Sham, J Chem Soc, Dalton Trans 973-8(1975), Preparation and Spectroscopic Studies of Five-co-ordinate beta-Diketonato-tri-(organo)tin Compounds. Crystal Structure of (1,3-Diphenyl-propane-1,3-dionato)triphenyltin(IV)
75B139	THEORY	D Barb and D Tarina, Rev Roum Phys 20,399-402(1975), Polarized Mossbauer Spectra - Tool for Effective Hyperfine Parameters Determinations
75B140	ANALYS	D Barb, D Tarina, and D P Lazar, Rev Roum Phys 20,673-90 (1975), Polarization Patterns for Mossbauer Hyperfine Spectra
75B141	FE-57	Sh Sh Bashkirov, R A Iskhakov, A B Liberman, L K Manenkova, and V I Sinyavskii, Izv Vyssh Uchebn Fiz (12),89-92(1975/ Sov J Phys, Influence of Manganese and Gallium Ions on the Magnetic Structure of a Lithium Ferrite
75B142	FE-57	I Bibicu, G C Moisil, D Barb, and M Romanescu, Rev Roum Phys 20,531-5(1975), Mossbauer Technique for Investigating the Action of an Organic Inhibitor of Corrosion
75B143	FE-57	K N Binnatov, A A Katsnel'son, and Yu L Rodionov, Fiz Met Metalloved 39,1021-5(1975)/Phys Met Metallogr (USSR), Effect of Alloying with Chromium on the Atomic Ordering of Nickel-Iron Alloys
75B144	FE-57	I N Bogachev, G A Charushnikova, V V Ovchinnikov, and V Litvinov, Fiz Met Metalloved 39,1269-74(1975)/Phys Met Metallogr (USSR), Banding in Manganese Steel G8 in the Region of Irreversible Temper Brittleness
75B145	SN-119	K D Bos, E J Bulten, and J G Noltes, J Organomet Chem 99, 397-405(1975), Oxidative Addition Reactions of Dicyclopen-tadienyltin(II) and of Tin(II) Bis(acetylacetonate) with Organic Halides. The Preparation of Compounds of the Type $RSn(acac)2X$
75B146	FE-57	W D Brewer and E Wehmeier, Phys Rev B 12,4608-16(1975), Hyperfine Fields at Transition-element Impurities in Gd
75B147	REVIEW	O Brummer, K Berndt, G Dlubek, U Berg, G Drager, W Beier, and G Dworzak, Wiss Z Martin-Luther-Univ, Halle-Wittengerg, Math-Naturwiss Reihe 24(6),5-28(1975), Some Solid State Spectroscopic Studies by the Physics Section
75B148	SN-119	S Bukshpan, T Sonnino, and J G Dash, Surf Sci 52,466-72 (1975), Debye-Waller Factor of Molecules Adsorbed on Graphite
75B149	INSTRM	S I Bondarevskii and V V Eremin, Prib Tekh Eksp (6),57-8 (1975)/Instrum Exp Tech (USSR), Scintillation Resonance Detector for Mossbauer Studies
75C001	FE-57	D C Champeney, E S M Higgy, and R G Ross, J Phys C 8,507-18 (1975), High-pressure Mossbauer Study of Ferrous-ion Motion in Glycerol
75C002	NP-237	C A Clausen, III and J A Stone, J Inorg Nucl Chem 37, 261-4(1975), Mossbauer Spectra of 237Np in Ion Exchange Resins and in a Solvent Extractant
75C003	FE-57	Y W Chow and Z Mukerji, Biochem Biophys Res Commun 62, 989-96(1975), Mossbauer Studies of Anhydrous Hemoglobin and Its Subunits
75C004	W-182	M G Clark, J R Gancedo, A G Maddock, and A F Williams, J Chem Soc, Dalton Trans 120-4(1975), Mossbauer Study of Octacyanotungstate Anions

CODE	TOPIC	REFERENCE
75C005	FE-57	D C Champeney and G W Dean, J Phys C 8,1276-84(1975), Molecular Vibrations in Glassy Glycerol Measured by Mossbauer Scattering
75C006	SN-119	D Cunningham, I Douek, M J Frazer, M McPartlin, and J D Matthews, J Organomet Chem 90,C23-4(1975), The X-ray Structure and Mossbauer Parameters of the Dichlorodimethyltin(IV) 1/1 Salicylaldehyde Adduct
75C007	FE-57	G J Cain, J A Barclay, and J D Cashion, J Low Temp Phys 19, 513-30(1975), Nuclear Magnetic Resonance on Polarized Nuclei Using the Mossbauer Effect in 57CoFe
75C008	DY-161	J D Cashion, M A Coulthard, and D B Prowse, J Phys C 8, 1267-75(1975), Mossbauer Isomer Shifts in Rare Earth Compounds: II. Nuclear Charge Radius and Electron Density Studies
75C008	EU-151	J D Cashion, M A Coulthard, and D B Prowse, J Phys C 8, 1267-75(1975), Mossbauer Isomer Shifts in Rare Earth Compounds: II. Nuclear Charge Radius and Electron Density Studies
75C008	GD-155	J D Cashion, M A Coulthard, and D B Prowse, J Phys C 8, 1267-75(1975), Mossbauer Isomer Shifts in Rare Earth Compounds: II. Nuclear Charge Radius and Electron Density Studies
75C008	ND-145	J D Cashion, M A Coulthard, and D B Prowse, J Phys C 8, 1267-75(1975), Mossbauer Isomer Shifts in Rare Earth Compounds: II. Nuclear Charge Radius and Electron Density Studies
75C008	PR-141	J D Cashion, M A Coulthard, and D B Prowse, J Phys C 8, 1267-75(1975), Mossbauer Isomer Shifts in Rare Earth Compounds: II. Nuclear Charge Radius and Electron Density Studies
75C008	SM-149	J D Cashion, M A Coulthard, and D B Prowse, J Phys C 8, 1267-75(1975), Mossbauer Isomer Shifts in Rare Earth Compounds: II. Nuclear Charge Radius and Electron Density Studies
75C008	YB-170	J D Cashion, M A Coulthard, and D B Prowse, J Phys C 8, 1267-75(1975), Mossbauer Isomer Shifts in Rare Earth Compounds: II. Nuclear Charge Radius and Electron Density Studies
75C009	FE-57	J M D Coey, Geochim Cosmochim Acta 39,401-15(1975), Iron in a Post-glacial Lake Sediment Core: A Mossbauer Effect Study
75C010	REVIEW	R Cammack, The Mossbauer Effect and Its Applications in Biology, in "New Techniques in Biophysics and Cell Biology," edited by R H Pain and B Smith (John Wiley & Sons, Inc, New York, 1975), Vol 2, pp 341-84
75C011	FE-57	J Chappert and J R Regnard, Colloq Int CNRS, No 242 (Physiques sous Champs Magnetiques Intenses), 257-60(1975), High Field Mossbauer Spectroscopy Using Water-cooled Magnets
75C012	RU-99	C A Clausen, III and M L Good, J Catal 38,92-100(1975), Mossbauer Effect Studies of Supported Ruthenium Catalysts
75C013	SN-119	W R Cullen, R K Pomeroy, J R Sams, and T B Tsin, J Chem Soc, Dalton Trans 1216-21(1975), Tin-119 Mossbauer Study of Complexes with Chlorine-bridged Tin-Molybdenum and Int-Tungsten Bonds
75C014	FE-57	W R Cares and J W Hightower, J Catal 39,36-43(1975), Mossbauer Spectra of Ferrite Catalysts Used in Oxidative Dehydrogenation

CODE	TOPIC	REFERENCE

75C015   TE-125   R M Cheyne, C H W Jones, and S Husebye, Can J Chem 53, 1855-60(1975), The 125Te Mossbauer Absorption Spectra of Tellurium Complexes with Sulfur-containing Ligands

75C016   FE-57   R W Cochrane, R Harris, M Plischke, D Zobin, and M J Zuckermann, Phys Rev B 12,1969-70(1975), Mossbauer Absorption Spectra in Amorphous Metallic Alloys

75C016   THEORY   R W Cochrane, R Harris, M Plischke, D Zobin, and M J Zuckermann, Phys Rev B 12,1969-70(1975), Mossbauer Absorption Spectra in Amorphous Metallic Alloys

75C017   FE-57   J M Crowell and J C Walker, Antiferromagnetic Iron in Small Gamma-phase Fe-Ni Crystals, in "5th Int Conf Mossbauer Spec, Proc," Part 2, pp 289-91 (See 75H017)

75C018   FE-57   P M Champion, J D Lipscomb, E Munck, P Debrunner, and I C Gunsalus, Biochemistry 14,4151-8(1975), Mossbauer Investigations of High-spin Ferrous Heme Proteins, I. Cytochrome P-450

75C019   FE-57   P M Champion, R Chiang, E Munck, P Debrunner, and L P Hager, Biochemistry 14,4159-65(1975), Mossbauer Investigations of High-spin Ferrous Heme Proteins II. Chloroperoxidase, Hoseradish Peroxidase, and Hemoglobin

75C020   FE-57   B K Chaudhuri, Solid State Commun 16,767-72(1975), A New Type of Phase Transition in $M(ClO4)_2(H2O)_6$ (M = Fe,Co,Ni and Mn)

75C021   REVIEW   M G Clark, Hyperfine Interactions and Molecular Structure, in "MTP International Review of Science, Physical Chemistry Series 2, Volume 2," edited by A D Buckingham (Butterworths, London, 1975), Chap 7, pp 239-97

75C022   SN-119   B Csakvari, E Csakvari, P Gomory, and A Vertes, J Radioanal Chem 25,275-82(1975), Mossbauer Study of the Donor Character of Oxygen in Si-O Bonds

75C023   I-129   R Coussement, G Dumont, G Langouche, H Pattyn, M Rots, K P Schmidt, and M Van Rossum, Implantation of 129mTe in Fe and Ni Foils and Determination of the Magnetic Hyperfine Fields, in "5th Int Conf Mossbauer Spec, Proc," Part 3, pp 549-52 (See 75H017)

75C024   ER-167   C L Chien and J C Walker, Studies of Temperature Spikes in Solids Using the Mossbauer Effect Following Coulomb Excitation, in "5th Int Conf Mossbauer Spec, Proc," Part 3, pp 560-3 (See 75H017)

75C024   HO-165   C L Chien and J C Walker, Studies of Temperature Spikes in Solids Using the Mossbauer Effect Following Coulomb Excitation, in "5th Int Conf Mossbauer Spec, Proc," Part 3, pp 560-3 (See 75H017)

75C024   LU-175   C L Chien and J C Walker, Studies of Temperature Spikes in Solids Using the Mossbauer Effect Following Coulomb Excitation, in "5th Int Conf Mossbauer Spec, Proc," Part 3, pp 560-3 (See 75H017)

75C024   TB-159   C L Chien and J C Walker, Studies of Temperature Spikes in Solids Using the Mossbauer Effect Following Coulomb Excitation, in "5th Int Conf Mossbauer Spec, Proc," Part 3, pp 560-3 (See 75H017)

75C025   FE-57   S Caric, L Marinkov, and J Slivka, Phys Status Solidi A 13,263-8(1975), A Mossbauer Study of the Thermal Decomposition of FeC2O4-2H2O

75C026   FE-57   F A B Chaves and V K Garg, J Inorg Nucl Chem 37,2283-5 (1975), Mossbauer Frozen Solution Studies of Fe(III) Nitrate and Fe(III) Perchlorate

CODE	TOPIC	REFERENCE
75C027	SM-149	J M D Coey, S K Ghatak, and F Holtzberg, Semiconductor-metal Transition in SmS - A 149Sm Mossbauer Study, in "AIP Conference Proceedings No 24, Magnetism and Magnetic Materials-1974" (20th Annual Conference, San Francisco), edited by C D Graham, Jr, G H Lander, and J J Rhyne (American Institute of Physics, New York, 1975), pp 38-9
75C028	FE-57	C G Cooke and M J Mays, J Chem Soc, Dalton Trans 455-60 (1975), Reaction of (HFeCo3(CO)12)with Phosphorus Donor Ligands
75C029	FE-57	J Chappert, B D Sawicka, and J A Sawicki, The Positive Sign of the Internal Magnetic Field in Ni3(Fe(CN)6)2, in "Int Conf Mossbauer Spectrosc, Proc," Vol 1, pp 71-2 (See 75H027)
75C030	NI-61	G Czjzek, J Fink, H Schmidt, F Gautier, G Krill, M F Lapierre, P Panissod, and C Robert, An Investigation of Magnetic Structures and Phase Transitions in NiS(2-x)Se(x) by 61Ni-Mossbauer Spectroscopy, in "Int Conf Mossbauer Spectrosc, Proc," Vol 1, pp 81-2 (See 75H027)
75C031	FE-57	G Chandra, C Bansal, and J Ray, Mossbauer Effect Study of Equiatomic Disordered CoFe and NiFe Alloys, in "Int Conf Mossbauer Spectrosc, Proc," Vol 1, pp 129-30 (See 75H027)
75C032	FE-57	G Chandra, J Ray, and C Bansal, 57Fe Hyperfine Fields in a Disordered Ni(0.67)Mn(0.33) Alloy, in "Int Conf Mossbauer Spectrosc, Proc," Vol 1, pp 131-2 (See 75H027)
75C033	DY-161	G Crecelius and H Maletta, Electronic Structure and Transferred Hyperfine Interactions in the Series Dy(Fe(x)Ni(1-x))3, in "Int Conf Mossbauer Spectrosc, Proc," Vol 1, pp 149-50 (See 75H027)
75C034	FE-57	J Chappert, G Jehanno, and F Varret, Asymmetry Parameter in Ferrous Fluosilicate, in "Int Conf Mossbauer Spectrosc, Proc," Vol 1, pp 161-2 (See 75H027)
75C035	FE-57	C L Chein and J C Walker, Origin of rf Sidebands in Mossbauer Spectra, in "Int Conf Mossbauer Spectrosc, Proc," Vol 1, pp 47-8 (See 75H027)
75C036	SB-121	C C M Campbell, J Phys F 5,1931-45(1975), Hyperfine Field Systematics in Heusler Alloys
75C036	SN-119	C C M Campbell, J Phys F 5,1931-45(1975), Hyperfine Field Systematics in Heusler Alloys
75C037	FE-57	J Chappert, B D Sawicka, and J A Sawicki, Phys Status Solidi B 72,K139-41(1975), The Positive Sign of the Internal Magnetic Field in Ni3(Fe(CN)6)2
75C038	FE-57	L S Chia, W R Cullen, J R Sams, and J C Scott, Can J Chem 53,2232-9(1975), Monosubstituted Derivatives of (L-L)Fe2(CO)6 (L-L = Fluorocarbon Bridged Ligand)
75C039	FE-57	S S Cohen, R H Nussbaum, and D G Howard, Phys Rev B 12, 4095-101(1975), Determination of an Unambiguous Parameter for the Impurity-lattice Interaction
75C040	FE-57	C L Chien and R Hasegawa, Mossbauer Studies of the Amorphous Alloys Fe-Pd-Si, in "Int Conf Mossbauer Spectrosc, Proc," Vol 1, pp 343-4 (See 75H027)
75C041	DY-161	J M D Coey, J Chappert, J P Rebouillat, and T S Wang, 161Dy Mossbauer Spectra of an Amorphous Dy-Co Alloy, in "Int Conf Mossbauer Spectrosc, Proc," Vol 1, pp 347-8 (See 75H027)
75C042	FE-57	M C Cadee, Acta Crystallogr, Sec B 31,2012-5(1975), The Crystal Structure of Hexagonal BaSrFe4O8

CODE	TOPIC	REFERENCE
75C043	SN-119	W R Cullen and R K Pomeroy, Inorg Chem 14,939-41(1975), Preparation of (diphosMo(CO)4SnCl3)+(SnCl5OH2)- and Related Derivatives
75C044	FE-57	J P Collman, R R Gagne, C A Reed, T R Halbert, G Lang, and W T Robinson, J Am Chem Soc 97,1427-39(1975), "Picket Fence Porphyrins." Synthetic Models for Oxygen Binding Hemoproteins
75C045	FE-57	J P Collman, J L Hoard, N Kim, G Lang, and C A Reed, J Am Chem Soc 97,2676-81(1975), Synthesis, Stereochemistry, and Structure-related Properties of alpha, beta, gamma, delta-Tetraphenylporphinatoiron(II)
75C046	SN-119	A B Cornwell and P G Harrison, J Chem Soc, Dalton Trans 1486-90(1975), Derivatives of Bivalent Germanium, Tin, and Lead. Part VII. Chromium, Molybdenum, and Tungsten Pentacarbonyl Complexes of Tin(II) Bis(beta-ketoenolates)
75C047	SN-119	A B Cornwell and P G Harrison, J Chem Soc, Dalton Trans 1722-6(1975), Derivatives of Divalent Germanium, Tin, and Lead. Part IX. Tin(II) Derivatives of Alkyl Acetoacetates, 4,Phenylbutane-2,4-dione, 1,3-diphenylpropane-1,3-dione, Cyclohexane-1,2- and -1,3-diones, and 2-Hydroxycyclohepta-2,4,6-trien-1-one
75C048	FE-57	A B Cornwell and P G Harrison, J Chem Soc, Dalton Trans 2017-22(1975), Derivatives of Bivalent Germanium, Tin, and Lead. Part XI. The Interaction of Tin(II) Halides and Bis(beta-ketoenolates) with Di-iron Enneacarbonyl
75C048	SN-119	A B Cornwell and P G Harrison, J Chem Soc, Dalton Trans 2017-22(1975), Derivatives of Bivalent Germanium, Tin, and Lead. Part XI. The Interaction of Tin(II) Halides and Bis(beta-ketoenolates) with Di-iron Enneacarbonyl
75C049	SM-149	J M D Coey and S K Ghatak, Electronic Configuration Fluctuations of Samarium in (Sm(1-x)Y(x))S, in "Int Conf Mossbauer Spectrosc, Proc," Vol 1, pp 369-70 (See 75H027)
75C050	FE-57	I D Cherkes and V P Shumeyko, Study of Phase Transformations in FeS2 during Thermal Treatment, in "Int Conf Mossbauer Spectrosc, Proc," Vol 1, pp 389-90 (See 75H027) (In Russian)
75C051	FE-57	J M D Coey, Freshwater Ferromanganese Nodules, in "Int Conf Mossbauer Spectrosc, Proc," Vol 1, pp 481-2 (See 75H027)
75C052	FE-57	L F Checherckaya, V P Romanov, and A E Borovik, Hyperfine Structure of NGR Spectra and Electric Properties of Wustite Between 80 and 300 K, in "Int Conf Mossbauer Spectrosc, Proc," Vol 1, pp 513-4 (See 75H027) (In Russian)
75C053	FE-57	S Caric, L Marinkov, and J Slivka, A Mossbauer Study of the Thermal Decomposition of FeC2O4-2H2O, in "Int Conf Mossbauer Spectrosc, Proc," Vol 1, pp 515-6 (See 75H027)
75C054	FE-57	D Cvjeticanin, I Savic, L Marinkov, and S Caric, Mossbauer Spectroscopic Studies of Gels of Colloidal Trivalent Iron, in "Int Conf Mossbauer Spectrosc, Proc," Vol 1, pp 517-8 (See 75H027)
75C055	FE-57	T E Cranshaw and R C Mercader, Mossbauer Spectra and Quadrupole Couplings on FeSi Crystals under Tetragonal and Trigonal Distortions, in "Int Conf Mossbauer Spectrosc, Proc," Vol 1, pp 523-4 (See 75H027)
75C056	REVIEW	J M D Coey, The Clay Minerals; Use of the Mossbauer Effect to Characterize Them and Study Their Transformations, in "Int Conf Mossbauer Spectrosc, Proc," Vol 2, pp 333-54 (See 75H038)

CODE	TOPIC	REFERENCE
75C057	FE-57	G Chandra, J Ray, and C Bansal, A Mossbauer Study of Very Dilute Fe Impurities in Ni-Mn Systems, in "Proceedings of the 14th International Conference on Low Temperature Physics: Volume 3 Low Temperature Properties of Solids" (Otaniemi, Finland, 1975), edited by M Krusius and M Vuorio (North-Holland Publishing Co, Amsterdam/American Elsevier Publishing Co, New York, 1975), pp 358-61
75C058	FE-57	C L Chien and J C Walker, Observation of rf-induced Side-band Effects in an Amorphous Magnetic Material, in "AIP Conference Proceedings-No 24, Magnetism and Magnetic Materials-1974" (20th Annual Conference, San Francisco), edited by C D Graham, Jr, G H Lander, and J J Rhyne (American Institute of Physics, New York, 1975), pp 127-8
75C059	SN-119	D Christov, Cv Bontschev, and N Nenov, Kautsch Gummi, Kunstst 4,201-4(1975), Untersuchung der Beschleunigung der Harz-Vulkanisation von Butylkautschuk mit Hilfe des Moss-bauer-Effektes. 9, Beschleunigende Wirkung von wasser-freien Metallhalogeniden
75C060	SN-119	D Christov, Cv Bontschev, and N Nenov, Kautsch Gummi, Kunstst 4,260-3(1975), Untersuchung der Beschleunigung der Harz-Vulkanisation von Butylkautschuk mit Hilfe des Moss-bauer-Effektes. 9, Beschleunigende Wirkung von wasser-freien Metallhalogeniden (continuation of 75C059)
75C061	FE-57	S J Campbell and T J Hicks, J Phys F 5,27-35(1975), Atomic Short Range Order in CuFe Alloys
75C062	NP-237	A A Chaikhorskii, Radiokhimiya 17,910-8(1975)/Sov Phys-Radiochem, Results of Studying Neptunium Compounds by a Mossbauer Spectroscopic Method
75C063	FE-57	R Courrier and G Le Caer, C R Acad Sci (Paris), Ser C 280,637-40(1975), Mise en Evidence d'un Stade de Prepreci-pitation au Cours du Viellissement d'un Alliage de Type Maragin Fe-19 at% Ni-3 at% Mo.
75C064	SN-119	D Christov and Cv Bontschev, Kaut Gummi, Kunstst 12,724-5 (1975), Untersuchung der Beschleunigung der Harz-Vulkani-sation von Butylkautschuk mit Hilfe des Mossbauer-Effektes 10. Beschleunigende Wirkung von Stannochlorid in Abhangig-keit von der Art der Einfuhrung in die Kautschuk-Mischung
75D001	I-129	H De Waard, 129I Mossbauer Spectroscopy, in "Mossbauer Effect Data Index, Covering the 1973 Literature," edited by J G Stevens and V E Stevens (Plenum Publishing Corp, New York, 1975), pp 447-94
75D001	REVIEW	H De Waard, 129I Mossbauer Spectroscopy, in "Mossbauer Effect Data Index, Covering the 1973 Literature," edited by J G Stevens and V E Stevens (Plenum Publishing Corp, New York, 1975), pp 447-94
75D002	FE-57	D B DeYoung and R G Barnes, J Chem Phys 62,1726-38(1975), A Mossbauer Effect Study of 57Fe in Transition Metal Monoborides
75D003	SN-119	M A Delmas, J C Maire, W McFarlane, and Y Richard, J Organomet Chem 87,285-93(1975), Etude Comparative de la Structure d'Oxa- et Thia-cyclostannanes
75D004	ANALYS	W A Dollase, Am Mineral 60,257-64(1975), Statistical Limitations of Mossbauer Spectral Fitting
75D005	FE-57	A E Dwight, C W Kimball, R S Preston, S P Taneja, and L Weber, J Less-Common Met 40,285-91(1975), Crystallo-graphic and Mossbauer Study of (Sc,Y,Ln)(Fe,Al)2 Intermetallic Compounds

CODE	TOPIC	REFERENCE
75D006	FE-57	G Dehe and B Seidel, Phys Status Solidi A 29,K47-50(1975), A Mossbauer Effect Study of Al-substituted Magnetite
75D007	SN-119	J D Donaldson, D C Puxley, and M J Tricker, J Inorg Nucl Chem 37,655-9(1975), The Interpretation of the 119mSn Mossbauer Data for Some Complexes of Tin(II)
75D008	SB-121	J D Donaldson, A Kjekshus, D G Nicholson, and J T Southern, Acta Chem Scand, Ser A 29,220-4(1975),121Sb Mossbauer Studies on Antimony(III) Chalcogenohalides
75D009	FE-57	J W Drijver, F Van der Woude, and S Radelaar, Phys Rev Lett 34,1026-9(1975), Order-disorder Transition in Ni3Fe Studied by Mossbauer Spectroscopy
75D010	NP-237	B D Dunlap, A T Aldred, D J Lam, and G R Davidson, High-field Susceptibility in Ferromagnetic NpOs2, in "AIP Conference Proceedings-No 24, Magnetism and Magnetic Materials-1974" (20th Annual Conference, San Francisco), edited by C D Graham, Jr, G H Lander, and J J Rhyne (American Institute of Physics, New York, 1975), pp 351-2
75D011	THEORY	S Dattagupta, Phys Rev B 12,47-57(1975), Effect of Nuclear Motion on Mossbauer Relaxation Spectra
75D012	FE-57	H DeGraaf and J M Trooster, Solid State Commun 16,1387-91 (1975), Mossbauer Investigation of Magnetic Phase Transitions in FeI2
75D013	SB-121	J D Donaldson, A Kjekshus, D G Nicholson, and T Rakke, J Less-Common Met 41,255-63(1975), Properties of TiSb2 and VSb2
75D014	FE-57	J W Drijver and F Van der Woude, Mossbauer Effect Measurements on the Intermetallic Compounds Ni3Al and Ni3Ga, in "5th Int Conf Mossbauer Spec, Proc," Part 1, pp 33-6 (See 75H017)
75D015	FE-57	G Dehe, B Seidel, C Michalk, and K Melzer, Mossbauer Effect Study of Fe(3-x)Al(x)O4, in "5th Int Conf Mossbauer Spec, Proc," Part 1, pp 106-13 (See 75H017)
75D016	FE-57	L Dabrowski, J Piekoszewski, J Suwalski, and S Makolagwa, Fiz Tverd Tela (Leningrad) 17,765-9(1975)/Sov Phys-Solid State 17,489-91(1975), Effect of V5+ Ions on Superexchange Interaction in Yttrium Iron Garnet
75D017	THEORY	P G De Gennes, J Phys (Paris) 36,603-6(1975), Sur une Eventuelle Application de l'Effet Mossbauer ou des Neutrons a l'Etude des Interfaces Fluides et des Smectiques
75D018	IR-193	G J Davies, A G Maddock and A F Williams, J Chem Soc, Chem Commun 264(1975), A Single-line Source for 193Iridium Mossbauer Spectroscopy
75D019	YB-170	G R Davidson, B D Dunlap, M Eibschutz, and L G Van Uitert, Phys Rev B 12,1681-8(1975), Mossbauer Study of Yb Spin Reorientation and Low-temperature Magnetic Configuration in YbFeO3
75D020	FE-57	B Dzienis and M Kopcewicz, Geophysical Aspects of the Mossbauer Study of Iron in Atmospheric Air, in "5th Int Conf Mossbauer Spec, Proc," Part 2, pp 395-400 (See 75H017)
75D021	REVIEW	B W Dale, Contemp Phys 16,127-46(1975), Mossbauer Spectroscopy
75D022	FE-57	N De Cristofaro and R Kaplow, Scr Met 9,781-5(1975), Loss of Lattice Rigidity in Iron-Nitrogen Austenite
75D023	REVIEW	H De Waard, Phys Scr 11,157-66(1975), The Investigation of Radiation Damage and Lattice Location by Nuclear Hyperfine Interaction Techniques

250

CODE	TOPIC	REFERENCE
75D024	SB-121	J P Devort and J M Friedt, Chem Phys Lett 35,423-5(1975), 121Sb Mossbauer Spectroscopy in Alkali Antimony(V) Hexafluorides
75D025	FE-57	S Dattagupta, Phys Rev B 12,3584-95(1975), Effects of Off-diagonal Hyperfine Interaction on Mossbauer Relaxation Spectra
75D025	THEORY	S Dattagupta, Phys Rev B 12,3584-95(1975), Effects of Off-diagonal Hyperfine Interaction on Mossbauer Relaxation Spectra
75D026	FE-57	G W Durbin, C E Johnson, and M F Thomas, J Phys C 8, 3051-7(1975), Direct Observation of Field-induced Spin Reorientation in YFeO3 by the Mossbauer Effect
75D027	FE-57	H Drost, H Von Lojewski, K Palow, R Wallenstein, and G Weyer, Time Dependence of Mossbauer Resonance Scattering, in "5th Int Conf Mossbauer Spec, Proc," Part 3, pp 713-5 (See 75H017)
75D028	FE-57	B L Dickson, Am Mineral 60,98-104(1975), The Iron Distribution in Rhodonite
75D029	FE-57	V N Dubinin, P O Voznyuk, V V Kuz'movich, and N A Ivkina, Teor Eksp Khim 11,417-21(1975)/Theor Exp Chem (USSR), Nuclear Gamma-resonance Spectroscopic Study of Silicotungstic Acid-base Metal Polymers
75D030	FE-57	R J Dickinson, R V Parish, P J Rowbotham, A R Manning, and P Hackett, J Chem Soc, Dalton Trans 424-8(1975), Studies in Mossbauer Spectroscopy. Part VII. Tin-119 and Iron-57 Spectra of Compounds Involving Tin Bonded to Chromium, Molybdenum, Tungsten, Manganese, Iron, or Cobalt
75D030	SN-119	R J Dickinson, R V Parish, P J Rowbotham, A R Manning, and P Hackett, J Chem Soc, Dalton Trans 424-8(1975), Studies in Mossbauer Spectroscopy. Part VII. Tin-119 and Iron-57 Spectra of Compounds Involving Tin Bonded to Chromium, Molybdenum, Tungsten, Manganese, Iron, or Cobalt
75D031	FE-57	S M Dubiel, J Zukrowski, J Korecki, and K Krop, Acta Phys Pol A 47,199-205(1975), The Effect of Neighbouring Chromium Atoms on Hyperfine Magnetic Field at 57Fe Nuclei in Fe-Cr Alloys
75D032	FE-57	S M Dubiel and J Zukrowski, Acta Phys Pol A 48,315-8(1975), Magnetic Hyperfine Field Distributions from Mossbauer Spectra of Iron-Chromium Alloys
75D033	FE-57	S M Dubiel, J Zukrowski, and L Kozlowski, Mossbauer Effect Study of the Magnetic Hyperfine Fields, in "Int Conf Mossbauer Spectrosc, Proc," Vol 1, pp 73-4 (See 75H027)
75D034	FE-57	G Dehe, B Seidel, K Melzer, and C Michalk, A Mossbauer Effect Study of Al-, Cu-, Sn- and Ti-substituted Magnetite, in "Int Conf Mossbauer Spectrosc, Proc," Vol 1, pp 93-4 (See 75H027)
75D035	FE-57	S M Dubiel, J Zukrowski, and I Vincze, Isomer Shifts in Iron-Chromium Alloys, in "Int Conf Mossbauer Spectrosc, Proc," Vol 1, pp 115-6 (See 75H027)
75D036	FE-57	H De Graaf and J M Trooster, Magnetic Phase Transitions in FeI2 Studied with Mossbauer Spectroscopy, in "Int Conf Mossbauer Spectrosc, Proc," Vol 1, pp 183-4 (See 75H027)
75D037	FE-57	T Dosmaganbetov, S A Bashanov, A K Zhetbaev, L S Sergeeva, Angular Distributions of Resonant Scattered Gamma-quanta on 57Fe Nuclei, in "Int Conf Mossbauer Spectrosc, Proc," Vol 1, pp 187-8 (See 75H027) (In Russian)

CODE	TOPIC	REFERENCE
75D038	FE-57	V Drago, E Baggio-Saitovitch, and J Danon, Mossbauer Spectroscopy of Electron Irradiated Natural Layered Silicates, in "Int Conf Mossbauer Spectrosc, Proc," Vol 1, pp 225-6 (See 75H027)
75D039	FE-57	I Ya Dekhtyar and M M Nishchenko, Influence of the Defects Originated in alpha-Fe during Laser Irradiation on the Mossbauer Effect and the Spectra of Electron-positron Annihilation, in "Int Conf Mossbauer Spectrosc, Proc," Vol 1, pp 231-2 (See 75H027) (In Russian)
75D040	FE-57	I Ya Dekhtyar, L I Ivanov, N A Litvinova, and M M Nishchenko, Mossbauer Spectra of 57Fe Affected by Powerful Laser Pulses, in "Int Conf Mossbauer Spectrosc, Proc," Vol 1, pp 233-4 (See 75H027) (In Russian)
75D041	FE-57	G Dehe, B Seidel, and W Meisel, An Iterative Method for Absorber Thickness Corrections in Mossbauer Experiments, in "Int Conf Mossbauer Spectrosc, Proc," Vol 1, pp 29-30 (See 75H027)
75D042	ANALYS	J M Daniels, Nucl Instrum Methods 128,473-81(1975), An Approximation to the Mossbauer Lineshape Integral in Terms of Rational Functions
75D043	ANALYS	J M Daniels, Nucl Instrum Methods 128,483-93(1975), The Interpretation of Mossbauer Spectra Obtained with Polarized Gamma-rays and Single Crystal Absorbers
75D043	FE-57	J M Daniels, Nucl Instrum Methods 128,483-93(1975), The Interpretation of Mossbauer Spectra Obtained with Polarized Gamma-rays and Single Crystal Absorbers
75D044	FE-57	J M Daniels, F E Moore, and S K Panda, Can J Phys 53,2428-37(1975), A Mossbauer Study of the Beta Phase of Iron-Germanium
75D045	FE-57	C N W Darlington, W J Fitzgerald, and D A O'Connor, Phys Lett 54A,35-6(1975), On the Energy Width of the Central Mode in the Critical Scattering of X-rays by SrTiO3
75D046	FE-57	M L De Siqueira, S Larsson, and J W D Connolly, J Phys Chem Solids 36,1419-22(1975), Mossbauer Isomer Shifts and the Multiple-scattering Method
75D047	FE-57	J L K F De Vries, J M Trooster, and P Ros, J Chem Phys 63, 5256-62(1975), Numerical Relativistic Self-consistent Field Calculation of Electron Density and <r-3> for a Number of Electron Configurations of Iron and Tin. Application to Tin Mossbauer Spectroscopy
75D047	SN-119	J L K F De Vries, J M Trooster, and P Ros, J Chem Phys 63, 5256-62(1975), Numerical Relativistic Self-consistent Field Calculation of Electron Density and <r-3> for a Number of Electron Configurations of Iron and Tin. Application to Tin Mossbauer Spectroscopy
75D048	FE-57	L Di Sipio, S Calogero, G Albertin, and A A Orio, J Organometal Chem 97,257-60(1975), Mossbauer Studies of Some Hexa-coordinate Iron(II) Complexes
75D049	SN-119	J D Donaldson, J Silver, S Hadjiminolis, and S D Ross, J Chem Soc, Dalton Trans 1500-6(1975), Effects of the Presence of Valence-shell Non-bonding Electron Pairs on the Properties and Structures of Caesium Tin(II) Bromides and of Related Antimony and Tellurium Compounds
75D050	FE-57	I Dezsi, T Lohner, D L Nagy, G Ritter, and H Spiering, Mossbauer Spectroscopy of Fe-HEDTA Frozen Solutions, in "Int Conf Mossbauer Spectrosc, Proc," Vol 1, pp 257-8 (See 75H027)

252

CODE	TOPIC	REFERENCE

75D051   FE-57   J A Dumesic, H Topsoe, and M Boudart, Mossbauer Effect Evidence for Surface Sensitive Magnetic Anisotropy, in "Int Conf Mossbauer Spectrosc, Proc," Vol 1, pp 319-20 (See 75H027)

75D052   FE-57   I Dezsi, T Lohner, and D L Nagy, Ferroelectric Phase Transformation in $CH_3NH_3Fe/SO4/2-12H_2O$, in "Int Conf Mossbauer Spectrosc, Proc," Vol 1, pp 371-2 (See 75H027)

75D053   FE-57   I Dezsi, T Lohner, B Molnar, D L Nagy, and G Ritter, Morin Transition in alpha-Fe_2O_3 Diluted with Trivalent Metallic Ions, in "Int Conf Mossbauer Spectrosc, Proc," Vol 1, pp 379-80 (See 75H027)

75D054   FE-57   D J Dwight and E A Lorch, Iron-57 Mossbauer Sources for Special Applications, in "Int Conf Mossbauer Spectrosc, Proc," Vol 1, pp 459-60 (See 75H027)

75D055   FE-57   I Demeter, L Keszthelyi, and W Meisel, Origin of the Line Width of Stainless Steel in Mossbauer Spectroscopy, in "Int Conf Mossbauer Spectrosc, Proc," Vol 1, pp 461-2 (See 75H027)

75D056   SN-119   J D Donaldson, S D Ross, J Silver, and P J Watkiss, J Chem Soc, Dalton Trans 1980-3(1975), Solid-state Effects and the Vibrational Spectra of Hexahalogeno-stannates(IV) and -tellurates(IV)

75D057   FE-57   J A Dumesic, H Topsoe, and M Boudart, J Catal 37,513-22 (1975), Surface, Catalytic and Magnetic Properties of Small Iron Particles

75D058   FE-57   B Dzienis and M Kopcewicz, Investigation of the Seasonal Variations of Concentration of Iron in Atmospheric Air Using the Mossbauer Technique, in "Int Conf Mossbauer Spectrosc, Proc," Vol 1, pp 479-80 (See 75H027)

75D059   THEORY   L M Dautov, M M Kadykenov, and D K Kaipov, The Influence of the Chemical Bond on the Coefficient of the Electron Internal Conversion, in "Int Conf Mossbauer Spectrosc, Proc," Vol 1, pp 493-4 (See 75H027)

75D060   REVIEW   L M Dautov, M M Kadykenov, and D K Kaipov, The Ratio of Mossbauer Nuclei delta(R2) Parameters, in "Int Conf Mossbauer Spectrosc, Proc," Vol 1, pp 497-8 (See 75H027)

75D061   FE-57   E De Grave, C Dauwe, A Govaert, and J De Sitter, On the Magnetic Hyperfine Fields in Quenched Magnesium Ferrites, in "Int Conf Mossbauer Spectrosc, Proc," Vol 1, pp 519-20 (See 75H027)

75D062   REVIEW   I Dezsi, Mossbauer Effect Studies on Phase Transitions, in "Int Conf Mossbauer Spectrosc, Proc," Vol 2, pp 221-38 (See 75H038)

75D063   TE-125   A K Dragunas and K V Makaryunas, Nuclear Gamma-resonance Spectra of Polycomponent and Diamond-like Semiconductors, in "5th Int Conf Mossbauer Spectrosc, Proc," Vol 3, pp 517-21 (See 75H017) (In Russian)

75D064   REVIEW   L M Dautov, M M Kadykenov, and D K Kaipov, Systematics of the Radius Changes of Nuclei during the Excitations on Mossbauer Levels, in "5th Int Conf Mossbauer Spectrosc, Proc," Vol 3, pp 619-32 (See 75 H017) (In Russian)

75D065   THEORY   L M Dautov, Isomer Shifts Calibration of the Mossbauer Spectra, in "5th Int Conf Mossbauer Spectrosc, Proc," Vol 3, pp 629-32 (See 75H017) (In Russian)

CODE	TOPIC	REFERENCE

75D066   THEORY   L Dabrowski, J Piekoszewski, and J Suwalski, Analytical Determination of Hyperfine Interaction Parameters in the General Case on the Base of the NGR Spectra for $I3/2 \rightarrow I1/2$, in "5th Int Conf Mossbauer Spectrosc, Proc," Vol 3, pp 635-7 (See 75H017) (In Russian)

75D067   INSTRM   D Dora, E Kisch, and L Varhaemi, Design and Measurement Possibilities of the NP-255 Type Spectrometer for the Mossbauer Effect Measurements, in "5th Int Conf Mossbauer Spectrosc, Proc," Vol 3, pp 656-60 (See 75H017) (In Russian)

75D068   RU-99   F M DaCosta, T C Gibb, R Greatrex, and N N Greenwood, Chem Phys Lett 36,655-7(1975), The 99Ru Mossbauer Spectrum of beta-RuCl3

75D069   YB-170   G R Davidson, B D Dunlap, M Eibschutz, and L G Van Uitert, Direct Observation of the Spin-reorientation Transition at Yb Sites in YbFeO3, in "AIP Conference Proceedings-No 24, Magnetism and Magnetic Materials-1974" (20th Annual Conference, San Francisco), edited by C D Graham, Jr, G H Lander, and J J Rhyne (American Institute of Physics, New York 1975), pp 63-4

75D070   FE-57   G Dehe, B Seidel, K Melzer, and C Michalk, Phys Status Solidi A 31,439-47(1975), Determination of a Cation Distribution Model of the Spinel System Fe(3-x)AlxO4

75D071   SN-119   N N Delyagin, Yu D Zonnenberg, and V I Nesterov, Zh Eksp Teor Fiz 69,1372-81(1975)/Sov Phys-JETP, Isomer Shift and Magnetic Hyperfine Interaction for Sn in Ni in the Critical Temperature Range

75D072   SN-119   N N Delyagin, Yu D Zonnenberg, and V I Nesterov, Fiz Tverd Tela 17,3036-8(1975)/Sov Phys-Solid State 17,2013-4(1976), The Magnetic Hyperfine Interaction of 119Sn in Co2TiSn

75D073   FE-57   K J Duff, J Chem Phys 63,2259-60(1975), Comment on the Calculation of the Isomer Shift in Ferric Hemin

75D074   SB-121   J D Donaldson, A Kjekshus, D G Nicholson, and T Rakke, Acta Chem Scand, Ser A 29,803-9(1975), Properties of Sb-compounds with Rutile-like Structures

75D075   FE-57   G Dublon, U Atzmony, M P Dariel, and H Shaked, Phys Rev B 12,4628-33(1975), Spin Rotation in Ternary No0.6Tb0.4Fe2

75D076   FE-57   N D Devyatkov, V V Khrapov, R E Garibov, V A Kudryashova, V I Gaiduk, G F Bakaushina, A M Khrapko, A A Levina, and A P Andreeva, Dokl Akad Nauk SSSR 225,962-5(1975)/Dokl Biophys, Effect of Low Intensity Millimeter Radiation on Gamma Resonance Spectra of Hemoglobin

75D077   FE-57   J Dudas and T Zemcik, Cesk Cas Fys 25,476-81(1975), Mossbauer Study of the Magnetic Transformation of the Ni3Al-(57Co + 57Fe) Surface Layer (In Slovakian)

75E001   FE-57   M Eibschutz, M E Lines, and R C Sherwood, Phys Rev B 11, 4595-4605(1975), Magnetism in Orbitally Unquenched Chainar Compounds. II. The Ferromagnetic Case: RbFeCl3

75E002   SN-119   E N Efremov and A A Bekker, Effective Magnetic Fields at the 119Sn Nuclei in Nickel and Isomer Shifts in the Ni-Sn System, in "5th Int Conf Mossbauer Spec, Proc," Part 1, pp 26-8 (See 75H017) (In Russian)

75E003   TE-125   E N Efremov, A A Bekker, and R V Baranova, Structure Peculiarities of the Copper, Silver and Gold Telluries, in "5th Int Conf Mossbauer Spec, Proc," Part 1, pp 270-3 (See 75H017) (In Russian)

CODE	TOPIC	REFERENCE
75E004	FE-57	E P Elsukov, I E Startseva, L G Onoprienko, and E E Yurchikov, Fiz Tverd Tela (Leningrad) 17,35-43(1975)/ Sov Phys-Solid State 17,19-23(1975), Use of the Mossbauer Effect in Studies of the Domain Structure of Ferromagnets
75E005	FE-57	H H Ettwig and W Pepperhoff, Arch Eisenhuttenwes 46,667-8 (1975), Zur Neeltemperatur des Gamma-Eisens
75E006	FE-57	N A Eissa, H A Sallam, and Z Miligy, Observation of a Low Spin State of Iron in Olivine by Mossbauer Effect, in "5th Int Conf Mossbauer Spec, Proc," Part 2, pp 361-3 (See 75H017)
75E007	FE-57	D Eliezer, S Nadiv, and M Ron, Appl Phys Lett 26,340-1 (1975), Mossbauer Study of Eta Phase in Fe-Ge Binary System
75E008	FE-57	N A Eissa, H A Sallam, and L Keszthelyi, Mossbauer Effect Study of Ancient Egyptian Pottery, in "5th Int Conf Mossbauer Spec, Proc," Part 3, pp 749-51 (See 75H017)
75E009	FE-57	N A Eissa and H A Sallam, Mossbauer Effect of the Origin of the Colour in the Ancient Egyptian Black Ware, in "5th Int Conf Mossbauer Spec, Proc," Part 3, pp 752-8 (See 75H017)
75E010	FE-57	M Eibschutz, F J DiSalvo, G W Hull, Jr, and S Mahajan, Appl Phys Lett 27,464-6(1975), Ferromagnetism in Metallic Fe(x)TaS2 (x=0.28)
75E011	FE-57	T Ericsson, W Karner, L Haggstrom, K Chandra, and R Wappling, Magnetic Properties of Cubic FeGe, in "Int Conf Mossbauer Spectrosc, Proc," Vol 1, pp 83-4 (See 75H027)
75E012	FE-57	N A Eissa, A A Bahgat, and H A Saleh, Study of the Variation of the Debye Temperature in Ni-Cd Ferrite System, in "Int Conf Mossbauer Spectrosc, Proc," Vol 1, pp 191-3 (See 75H027)
75E013	FE-57	S P Ekimov, L P Nikitina, R G Grebenshchikov, L M Krizhanskii, O G Chigareva, and V V Motsartov, Dokl Akad Nauk SSSR 222,1126-9(1975)/Dokl Phys Chem 222,578-8(1975), Identification of Subsites and Manifestation of an Ordered Distribution of Cations in Ca(1-x)Fe(1+x)Ge2O6 Germanates According to Data of Mossbauer Spectroscopy
75E014	FE-57	N A Eissa, A L Hussein, and A G Mostafa, Study of the Effect of Al2O3 on the Structure of Iron in Lithium Borate Glass Using Mossbauer Effect (ME) Spectroscopy, in "Int Conf Mossbauer Spectrosc, Proc," Vol 1, pp 357-8 (See 75H027)
75E015	FE-57	J Ensling, P Gutlich, and M Sorai, Chemical Influences on the High Spin-Low Spin Transition in Tris(2-picolylamine) iron(II) Chloride, in "Int Conf Mossbauer Spectrosc, Proc," Vol 1, pp 383-4 (See 75H027)
75E016	FE-57	B J Evans, Experimental Studies of the Electrical Conductivity and Phase Transition in Fe3O4, in "AIP Conference Proceedings-No 24, Magnetism and Magnetic Materials-1974" (20th Annual Conference, San Francisco), edited by C D Graham, Jr, G H Lander, and J J Rhyne (American Institute of Physics, New York, 1975), pp 73-8
75E017	SN-119	P F R Ewings, P G Harrison, and D E Fenton, J Chem Soc, Dalton Trans 821-6(1975), Derivatives of Divalent Germanium, Tin, and Lead. Part V. Bis-(pentane-2,4-dionato)-, Bis(1,1,1-trifluoropentane-2,4-dionato)-, and Bis(1,1,1, 5,5,5,-hexafluoropentane-2,4-dionato)-tin(II)

CODE	TOPIC	REFERENCE
75E018	SN-119	P F R Ewings and P G Harrison, J Chem Soc, Dalton Trans 1717-21(1975), Derivatives of Divalent Germanium, Tin, and Lead. Part VIII. Tin(II) Aryl-carboxylates, -sulphonates, and Halide Methoxides
75E019	SN-119	P F R Ewings and P G Harrison, J Chem Soc, Dalton Trans 2015-7(1975), Derivatives of Divalent Germanium, Tin, and Lead. Part X. Tin(II) Bis(phenoxides), Bis(O-methyl dithiocarbonate), and Bis(diethyldithio-carbamate)
75E020	FE-57	H Eicher, Oxygenation of Hemoglobin Inferred from Mossbauer Spectroscopy, in "Int Conf Mossbauer Spectrosc, Proc," Vol 1, pp 437-8 (See 75H027)
75E021	FE-57	N A Eissa, H A Sallam, and B A Ashi, Mossbauer Effect Study of Plant and Animal Fossils, in "Int Conf Mossbauer Spectrosc, Proc," Vol 1, pp 483-5 (See 75H027)
75E022	FE-57	J C Eisenstein, M F Taragin, and D D Thornton, Antiferromagnetic Order in Amphibole Asbestos, in "AIP Conference Proceedings-No 24, Magnetism and Magnetic Materials-1974" (20th Annual Conference, San Francisco), edited by C D Graham, Jr, G H Lander, and J J Rhyne (American Institute of Physics, New York, 1975), pp 357-8
75E023	FE-57	N A Eissa, H A Sallam, and Z Miligy, Indian J Pure Appl Phys 13,23-5(1975), Observation of a Low-spin State of Iron in Olivine by Mossbauer Effect
75E024	FE-57	J Ensling, P Gutlich, and L Rosch, Z Naturforsch, Teil B 30,850-3(1975), 57Fe- und 119Sn-Mossbauer-Untersuchungen an Tetracarbonyl(organoelement IV a-phosphin)eisen(0)-Komplexen
75E024	SN-119	J Ensling, P Gutlich, and L Rosch, Z Naturforsch, Teil B 30,850-3(1975), 57Fe- und 119Sn-Mossbauer-Untersuchungen an Tetracarbonyl(organoelement IV a-phosphin)eisen(0)-Komplexen
75E025	SN-119	E M Eremenko, V E Listovnichy, V M Sergyenkova, and A V Murzin, Dopov Akad Nauk Ukr RSR, Ser A 37,560-4(1975), Mass Transfer of Tin from Its Melt to Glass Mass in Hydrogen Medium (In Russian)
75E026	SN-119	E M Eremenko, V E Listovnichy, V M Sergyenkova, and A V Murzin, Dopov Akad Nauk Ukr RSR, Ser A 37,655-9(1975), Mass Transfer of Tin from Its Melt to Glass Mass in the Argon Medium at P(O2) =10(-16)-10(-18) atm (In Russian)
75E027	FE-57	B J Evans and L J Swartzendruber, Magnetic Hyperfine Field Structure of Iron Urushibara Type Catalysts, in "AIP Conference Proceedings-No 24, Magnetism and Magnetic Materials-1974" (20th Annual Conference, San Francisco), edited by C D Graham, Jr, G H Lander, and J J Rhyne (American Institute of Physics, New York, 1975), pp 391-3
75E028	FE-57	B J Evans and E W Sergent, Jr, Contrib Mineral Petrol 53, 183-94(1975), 57Fe NGR of Fe Phases in "Magnetic Cassiterites"
75F001	SB-121	V Fraknoy-Koros, P Gelencser, B Levay, and A Vertes, J Lumin 9,467-74(1975), Application of Mossbauer Spectroscopy to Follow the Incorporation of Antimony into Calcium-halophosphate Phosphors. I
75F002	RU-99	D C Foyt, M L Good, J G Cosgrove, and R L Collins, J Inorg Nucl Chem 37,1913-6(1975), Mossbauer Data Reduction: The M1-E2 Transition of 99Ru

CODE	TOPIC	REFERENCE

75F003   FE-57   W Fischer, B N Kuz'min, and R P Vardapetyan, Mossbauer Study of the YFe3 Compound, in "5th Int Conf Mossbauer Spec, Proc," Part 1, pp 41-5 (See 75H017)

75F004   FE-57   G Filoti, A Gelberg, V Spanu, P Telnic, and H Niculescu-Majewska, Quadrupole Interaction in R3Al(4.9)Fe(0.1)O12, in "5th Int Conf Mossbauer Spec, Proc," Part 1, pp 90-3 (See 75H017)

75F005   FE-57   K Frohlich, Mossbauer Effect and Thermal Analysis Studies of Frozen Aqueous Solutions of Iron(II) Salts, in "5th Int Conf Mossbauer Spec, Proc," Part 1, pp 251-9 (See 75H017)

75F006   SN-119   J A Feiccabrino and E J Kupchik, J Organomet Chem 73,319-25 (1975), Preparation of Some Triethylammonium (Organo-cyanoamino)chlorotriphenylstannates

75F007   REVIEW   P A Flinn, Metallurgical Applications of the Mossbauer Effect, in "5th Int Conf Mossbauer Spec, Proc," Part 2, pp 275-88

75F008   FE-57   J E Frackowiak, B M Jankowski, and T Panek, Mossbauer Effect in the Fe-Ni (38%) Invar Alloy, in "5th Int Conf Mossbauer Spec, Proc," Part 2, pp 295-8 (See 75H017)

75F009   REVIEW   H Frauenfelder, Biological Molecules, in "5th Int Conf Mossbauer Spec, Proc," Part 2, pp 401-7 (See 75H017)

75F010   FE-57   E N Frolov, A D Mokrushin, O V Belonogova, G I Likhtenshtein, V A Trukhtanov, and V I Gol'danskii, Investigation of the Lattice Dynamics of Albumin by the Nuclear Gamma Resonance Method, in "5th Int Conf Mossbauer Spec, Proc," Part 2, pp 426-9 (See 75H017) (In Russian)

75F011   FE-57   P Fischer, W Halg, P Roggwiller, and E R Czerlinsky, Solid State Commun 16,987-92(1975), Cation Distribution of Y-Fe-Al Garnets

75F012   GENERAL   H Frauenfelder, Summary of the 5th International Conference on Mossbauer Spectroscopy, in "5th Int Conf Mossbauer Spec, Proc," Part 3, pp 763-5 (See 75H017)

75F013   FE-57   B W Fitzsimmons, S E Al-Mukhtar, L F Larkworthy, and R R Patel, J Chem Soc, Dalton Trans 1969-73(1975), Magnetic and Mossbauer Investigations of NN-disubstituted Bis(dithiocarbamato)iron(II) Complexes

75F014   REVIEW   F E Fujita, Mossbauer Spectroscopy in Physical Metallurgy, in "Topics in Applied Spectroscopy, Volume 5 - Mossbauer Spectroscopy," edited by U Gonser (Springer-Verlag, Berlin, 1975), pp 201-36

75F015   FE-57   W Fischer and B N Kuz'min, Spin Polarization in the Conduction Band of RFe3 Type Intermetallics, in "Int Conf Mossbauer Spectrosc, Proc," Vol 1, pp 65-7 (See 75H027)

75F016   NI-61   J Fink, G Czjzek, and H Schmidt, Magnetic Hyperfine Interactions at 61Ni in Ni-Mn Alloys, in "Int Conf Mossbauer Spectrosc, Proc," Vol 1, pp 79-80 (See 75H027)

75F017   FE-57   E Fritzsch, M Schneider, C Pietzsch, and P Deus, Stress-induced Time-dependent Mossbauer-line Broadening of Armco-Fe-foils, in "Int Conf Mossbauer Spectrosc, Proc," Vol 1, pp 119-20 (See 75H027)

75F018   FE-57   V Florescu, G Filoti, and D Barb, Mossbauer Studies of Substituted M-type Hexagonal Ferrites, in "Int Conf Mossbauer Spectrosc, Proc," Vol 1, pp 195-6 (See 75H027)

CODE	TOPIC	REFERENCE

75F019   FE-57   J Fleisch, P Gutlich, E Mohs, and G K Wolf, Mossbauer Emission Spectroscopy of 57Co Implanted into Iron Compounds, in "Int Conf Mossbauer Spectrosc, Proc," Vol 1, pp 217-8 (See 75H027)

75F020   FE-57   I Felner, I Mayer, A Grill, and M Schieber, Solid State Commun 16,1005-9(1975), Magnetic Ordering in Rare Earth Iron Silicides and Germanides of the RFe2X2 Type

75F021   FE-57   G Filoti and C N Turcanu, Iron Complexes with 8-Quinolioil and Homologs, Studied by Mossbauer Spectrometry, in "Int Conf Mossbauer Spectrosc, Proc," Vol 1, pp 281-2 (See 75H027)

75F022   FE-57   G Filoti, C N Turcanu, and C Oproiu, Mossbauer Investigation of the Thermal and Radiolytical Treatment in Iron 8-Quinolinol Complexes, in "Int Conf Mossbauer Spectrosc, Proc," Vol 1, pp 283-4 (See 75H027)

75F023   FE-57   J E Frackowiak, H Morawiec, and T J Panek, Study of Disorder-order Transformations in Ni3Fe Alloy, in "Int Conf Mossbauer Spectrosc, Proc," Vol 1, pp 385-6 (See 75H027)

75F024   THEORY   H Fischer, U Gonser, H D Pfannes, and T Shinjo, Texture Determination by Polarized Recoil-free Gamma Rays, in "Int Conf Mossbauer Spectrosc, Proc," Vol 1, pp 463-4 (See 75H027)

75F025   REVIEW   H Frauenfelder, Dynamics of Biomolecules, in "Int Conf Mossbauer Spectrosc, Proc," Vol 2, pp 305-17 (See 75H038)

75F026   REVIEW   E N Frolov, G I Likhtenshtein, and V I Gol'danskii, The Dynamic Structure of Proteins and Water-protein Interactions, in "Int Conf Mossbauer Spectrosc, Proc," Vol 2, pp 319-31 (See 75H038)

75F027   REVIEW   A J Freeman, Status of Theoretical Charge and Spin Determinations, in "Int Conf Mossbauer Spectrosc, Proc," Vol 2, pp 435-65 (See 75H038)

75F028   NP-237   V M Filin, V I Gol'danskii, and V S Nefedov, Mossbauer Scattering at 237Np, in "5th Int Conf Mossbauer Spectrosc, Proc," Vol 3, pp 563-8 (In Russian)

75F029   NP-237   V M Filin, V I Gol'danskii, and V S Nefedov, Mossbauer Effect Investigation of the State of the 237Np Atom Formed after the Alpha-decay of 241Am, in "5th Int Conf Mossbauer Spectrosc, Proc," Vol 3, pp 568-72 (See 75H017) (In Russian)

75F030   FE-57   M K Fayek, A A Bahgat, and Y M Abass, Atomkernenergie 26, 285-6(1975), Neutron Diffraction and Mossbauer Studies on Cobalt Substituted Zinc Ferrites

75F031   FE-57   G Filoti, Stud Cercet Fiz 27,665-702(1975), Mossbauer Studies on Jahn-Teller Distorted Compounds (In Romanian)

75F032   DY-161   D W Forester, R Abbundi, R Segnan, and D Sweger, Magnetic Hyperfine Structure in Amorphous DyFe2, in "AIP Conference Proceedings—No 24, Magnetism and Magnetic Materials—1974" (20th Annual Conference, San Francisco), edited by C D Graham, Jr, G H Lander, and J J Rhyne (American Institute of Physics, New York, 1975), pp 115-6

75F032   FE-57   D W Forester, R Abbundi, R Segnan, and D Sweger, Magnetic Hyperfine Structure in Amorphous DyFe2, in "AIP Conference Proceedings—No 24, Magnetism and Magnetic Materials 1974" (20th Annual Conference, San Francisco), edited by C D Graham, Jr, G H Lander, and J J Rhyne (American Institute of Physics, New York, 1975), pp 115-6

CODE	TOPIC	REFERENCE
75F033	FE-57	O M Frolova, Izv Vyssh Uchebn Zaved, Geol Razved 18(1), 59-61(1975), Mossbauer Absorption Spectra of Magnetites at Different Redox Stages (In Russian)
75G001	FE-57	A N Garg, Z Naturforsch, Teil B 30,96-8(1975), Correlation of Mossbauer Quadrupole Splitting (delta Eq) and Infrared Cyanide Frequency Separation in Alkali Metal Ferricyanides
75G002	SB-121	R G Goel, J N R Ruddick, and J R Sams, J Chem Soc, Dalton Trans 67-71(1975), Antimony-121 Mossbauer Spectroscopic Study of Bis(halogenoacetato)-trimethylantimony Derivatives
75G003	FE-57	J R Gosselin, M G Townsend, R J Tremblay, and A H Webster, Mater Res Bull 10,41-50(1975), Mossbauer Investigation of Synthetic Single Crystal Monoclinic Fe7S8
75G004	FE-57	R M Golding and K J Jessop, Aust J Chem 28,179-83(1975), Temperature Dependence of delta Eq for Iron(IV) Compounds in Strong Crystal Field
75G005	FE-57	F Grandjean and A Gerard, Solid State Commun 16,553-6 (1975), Analysis by Mossbauer Spectroscopy of the Electronic Hopping Process in Ilvaite
75G006	GENRAL	U Gonser, From Strange Effect to Mossbauer Effect, in "Topics in Applied Physics, Volume 5 - Mossbauer Spectroscopy," edited by U Gonser (Springer-Verlag, Berlin, 1975), pp 1-51
75G007	FE-57	V N Gridnev, V G Gavrilyuk, I Ya Dekhtyar, Yu Ya Meshkov, P S Nisin, and V G Prokopenko, Dopov Akad Nauk Ukr RSR, Ser A (1),75-9(1975), Mossbauer Effect in Plastically Deformed Fe-C Alloy (In Russian)
75G008	THEORY	L Gunther and J Zitkova-Wilcox, J Stat Phys 12,205-15 (1975), The Mossbauer Effect: A Potentially Ideal Probe into Brownian Motion
75G009	FE-57	T C Gibb, R Greatrex, N N Greenwood, and K G Snowdon, J Solid State Chem 14,193-202(1975), A Study of the New Perovskite Solid Solution Series SrFe(x)Ru(1-x)O(3-y) by Ruthenium-99 and Iron-57 Mossbauer Spectroscopy
75G009	RU-99	T C Gibb, R Greatrex, N N Greenwood, and K G Snowdon, J Solid State Chem 14,193-202(1975), A Study of the New Perovskite Solid Solution Series SrFe(x)Ru(1-x)O(3-y) by Ruthenium-99 and Iron-57 Mossbauer Spectroscopy
75G010	REVIEW	H M Gager and M C Hobson, Jr, Catal Rev 11,1-40(1975), Mossbauer Spectroscopy
75G011	FE-57	V K Garg, P G David, T Matsuzawa, and T Shinjo, Bull Chem Soc Jpn 48,1933-6(1975), Mossbauer and Magnetic Study of Mononuclear and Oxo-bridged Binuclear Iron(III) Complexes of 1,10-Phenanthroline and 2,2'-Bipyridine
75G012	FE-57	M J Graham, D A Channing, G A Swallow, and R D Jones, J Mater Sci 13,1175-81(1975), A Mossbauer Study of the Reduction of Hematite in Hydrogen at 535 C
75G013	SN-119	N N Greenwood and B Youll, J Chem Soc, Dalton Trans 158-62 (1975), Reactions of Some Tin(II) and Tin(IV) Compounds with the Dodecahydro-nido-decaborate(2-) Ion, (B10H12)2-
75G014	FE-57	R Grossinger, W Steiner, G Wiesinger, and H Kirchmayr, Mossbauer Investigations of the Pseudobinary System Dy(Fe(x)Al(1-x))2, in "5th Int Conf Mossbauer Spec, Proc," Part 1, pp 46-8 (See 75H017)

CODE	TOPIC	REFERENCE
75G015	GENRAL	U Gonser, Editor, "Topics in Applied Physics: Volume 5, Mossbauer Spectroscopy" (Springer-Verlag, Berlin, 1975)
75G016	YB-170	F Gonzalez-Jimenez, B Cornut, and B Coqblin, Phys Rev B 11,4674-82(1975), Influence of the Crystalline Field and Kondo Effects on the Relaxation Rate: Application to Mossbauer Experiments of Ytterbium Diluted in Gold
75G017	FE-57	J M Genin, G Le Caer, and A Simon, Mossbauer Spectroscopy of the Transformation of Epsilon Carbide during the Tempering of Fe-C Martensite - The Existence of Highly Faulted Cementite, in "5th Int Conf Mossbauer Spec, Proc," Part 2, pp 318-27 (See 75H017)
75G018	FE-57	M Greguskova, J Cirak, J Novotny, and I Cernohorsky, Investigation of Iron-containing Complexes of Deoxyribonucleic Acid Nucleosides by Mossbauer Spectroscopy, in "5th Int Conf Mossbauer Spec, Proc," Part 2, pp 415-9 (See 75H017)
75G019	THEORY	V I Gol'danskii and S V Karyagin, Influence of Anisotropic Atomic Motion on the Mossbauer Spectrum of Polycrystalline Systems, in "5th Int Conf Mossbauer Spec, Proc," Part 2, pp 439-47 (See 75H017) (In Russian)
75G020	FE-57	A N Garg, Radiochem Radioanal Lett 22,157-62(1975), On the Sign of Electric Field Gradient in Monosubstituted Ferrocyanides
75G021	FE-57	V Kh Gevorkyan, Yu G Chugunnyi, V P Ivanitskii, and N N Kovalyukh, Geol Zh (Russ Ed) 35(2),94-104(1975), Radiocarbon and Mossbauer Studies on Phosphorite Concretions of the Indian Ocean (In Russian)
75G022	FE-57	T C Gibb, J Phys C 8,229-35(1975), Determination of the Electric Field Gradient Tensor in Ferrous Ammonium Sulphate Using a Polarized Mossbauer Souce
75G023	FE-57	T C Gibb, Chem Phys 7,449-56(1975), Determination of the Electric Field Gradient Tensor in Ferrous Chloride Tetrahydrate Using a Polarized Mossbauer Source
75G024	THEORY	V I Gol'danskii and S V Karyagin, Phys Status Solidi B 68, 693-702(1975), Effect of the Anisotropy of Atomic Motions on Mossbauer Spectra of Polycrystals
75G025	SN-119	M Goldstein and P Tiwari, J Inorg Nucl Chem 37,1550-1 (1975), Barium Tetrachlorostannate(II)
75G026	I-129	J Granot and S Bukshpan, J Phys C 8,1435-42(1975), Mossbauer Effect Measurements in Ferromagnetic Cr(x)Te(y) Compounds
75G026	TE-125	J Granot and S Bukshpan, J Phys C 8,1435-42(1975), Mossbauer Effect Measurements in Ferromagnetic Cr(x)Te(y) Compounds
75G027	EU-151	E A Gorlich, H U Hrynkiewicz, R Kmiec, K Latka, K Tomala, A Czopnik, and N Iliev, Phys Status Solidi A 30,K17-20 (1975), Mossbauer Effect Studies in EuIn3 and EuPb3
75G028	SN-119	E A Gorlich, R Kmiec, K Latka, T Matlak, K Ruebenbauer, A Szytula, and K Tomala, Phys Status Solidi A 30,331-6 (1975), Magnetic Hyperfine Field Distribution at the Tin Site in the C1 Structure Alloys IrMnSn and PtMnSn
75G029	SN-119	E A Gorlich, R Kmiec, K Latka, T Matlak, K Ruebenbauer, A Szytula, and K Tomala, Phys Status Solidi A 30,765-70 (1975), Transferred Hyperfine Fields at the Tin Site in the Heusler-type Alloys Co2YSn (Y = Ti,Zr,Hf,V)

CODE	TOPIC	REFERENCE
75G030	FE-57	J C Grenier, M Pouchard, and P Hagenmuller, J Solid State Chem 13,92-8(1975), Influence de la Substitution du Fer sur les Proprietes Magnetiques des Solutions Solides Ca2Fe(2-x)M(x)O5 (M = Al,Sc,Cr,Co,Ga)
75G031	SN-119	R Gsell and M Zeldin, J Inorg Nucl Chem 37,1133-7(1975), Synthesis and Spectroscopic Properties of Tin(II) Alkoxides
75G032	FE-57	M P Gupta and H B Mathur, J Phys C 8,370-6(1975), The Cation Distribution in the Ferrite FeV2O4: Mossbauer and X-ray Diffraction Studies
75G033	EU-151	E A Gorlich, H U Hrynkiewicz, K Latka, K Tomala, and R Kmiec, Mossbauer Effect Studies in EuIn3 and EuPb3, in "5th Int Conf Mossbauer Spec, Proc," Part 3, pp 555-6 (See 75H017)
75G034	PROPSL	V I Gol'danskii, On the General Principles and Problems of Realization of Gasers (Gamma-lasers), in "5th Int Conf Mossbauer Spec, Proc," Part 3, pp 584-93 (See 75H017)
75G035	SN-119	U Gonser, B Schmitt, and H D Pfannes, Sn119 - Polarimetry, in "5th Int Conf Mossbauer Spec, Proc," Part 3, pp 594-6 (See 75H017)
75G036	FE-57	C R Guarnieri, R J Semper, and J C Walker, Measurements of Hyperfine Fields in Very Thin Iron Films as a Function of Temperature, in "5th Int Conf Mossbauer Spec, Proc," Part 3, pp 716-7 (See 75H017)
75G037	EU-151	E A Gorlich, H U Hrynkiewicz, K Latka, K Tomala, and R Kmiec, The Mossbauer Effect in Eu3S4 at Low Temperatures, in "5th Int Conf Mossbauer Spec, Proc," Part 3, pp 557-60 (See 75H017)
75G038	FE-57	U Gonser, S Nasu, W Keune, and O Weis, Solid State Commun 17,233-6(1975), Fe57 Hyperfine Field Distributions in Fe-28 Per Cent Ni-3 Per Cent C Invar Alloy
75G039	FE-57	J R Gosselin, J L Horwood, M G Townsend, R J Tremblay, and A H Webster, Magnetic Susceptibility and Mossbauer Study of Single Crystal FeS, in "AIP Conference Proceedings No 24, Magnetism and Magnetic Materials-1974" (20th Annual Conference, San Francisco), edited by C D Graham, Jr, G H Lander, and J J Rhyne (American Institute of Physics, New York, 1975), pp 55-6
75G040	REVIEW	R W Grant, Mossbauer Spectroscopy in Magnetism: Characterization of Magnetically-ordered Compounds, in "Topics in Applied Physics, Volume 5 Mossbauer Spectroscopy," edited by U Gonser (Springer-Verlag, Berlin, 1975), pp 97-137
75G041	FE-57	R Grossinger and W Steiner, Phys Status Solidi A 28,K135-8 (1975), Magnetic Investigations on Dy(Fe(x)Al(1-x))2 Laves Phase Compounds
75G042	REVIEW	P Gutlich, Mossbauer Spectroscopy in Chemistry, in "Topics in Applied Physics, Volume 5 - Mossbauer Spectroscopy," edited by U Gonser (Springer-Verlag, Berlin, 1975), pp 53-96
75G043	FE-57	P L Gruzin, Yu V Petrikin, and R A Stukan, Prib Tekh Eksp (3),48-9(1975)/Instrum Exp Tech (USSR) 18,716-8(1975), Installation for the Observation of Nuclear Gamma-resonance According to Conversion Electrons
75G043	INSTRM	P L Gruzin, Yu V Petrikin, and R A Stukan, Prib Tekh Eksp (3),48-9(1975)/Instrum Exp Tech (USSR) 18,716-8(1975), Installation for the Observation of Nuclear Gamma-resonance According to Conversion Electrons

CODE	TOPIC	REFERENCE

75G044   SN-119   E A Gorlich, R Kmiec, K Latka, T Matlak, K Ruebenbauer,
A Szytula, and K Tomala, Transferred Hyperfine Fields at
the Tin Site in the Heusler Type Alloys Co2YSn
/Y=Ti,Zr,Hf,V/, in "Int Conf Mossbauer Spectrosc, Proc,"
Vol 1, pp 85-6   (See 75H027)

75G045   FE-57   P C M Gubbens and A M Van der Kraan, Mossbauer Effect Study
of ThFe5, in "Int Conf Mossbauer Spectrosc, Proc," Vol 1,
pp 177-8   (See 75H027)

75G046   FE-57   Y Geronov, T Tomov, and S Georgiev, J Appl Electrochem 5,
351-8(1975), Mossbauer Spectroscopy Investigation of the
Iron Electrode during Cycling in Alkaline Solution

75G047   FE-57   F Grandjean and A Gerard, J Magn Magn Mat 1,64-72(1975),
Study by Mossbauer Spectroscopy of Arsenides Fe(2-x)MnxAs

75G048   FE-57   F W Grevels, D Schulz, E Koerner von Gustorf, and D St P
Bunbury, J Organometal Chem 91,341-56(1975), (n-(trans-
1,2-dihalogenathylen))-tetracarbonyleisen-Komplexe: Ther-
mische Umwandlung in mu-(1-eta: 1-2-eta(trans-2-Halovinyl)-
mu-Halogen-bis(tricarbonyleisen)-(Fe-Fe)-Komplexe und
weitere Enthalogenierungsreaktionen

75G049   FE-57   V K Garg and Y S Liu, J Chem Phys 62,56-8(1975),
Study of Electric Field Gradient Parameters and Quadrupole
Splitting of Rozenit

75G050   FE-57   A Gerard and F Grandjean, J Phys Chem Solids 36,1365-70
(1975), Mossbauer Spectra in the Presence of a Fluctuating
Electric Field Gradient - Applications to the Cases of
Semi-metallic Compounds

75G051   FE-57   V I Gol'danskii, O P Kevdin, N K Kivrina, E F Makarov, V
Ya Rochev, and R A Stukan, To the Problem of Observation
of Mossbauer Effect in Nematic and Cholesteric Mezophases,
in "Int Conf Mossbauer Spectrosc, Proc," Vol 1, pp pp 339-
40 (See 75H027)   (In Russian)

75G052   FE-57   V I Gol'danskii, O P Kevdin, N K Kivrina, E F Makarov, V
Ya Rochev, and R A Stukan, Mossbauer Study of Solutions of
Fe Acetylacetonate in Smectic and Nematic Phase of Liquid
Crystals, in "Int Conf Mossbauer Spectrosc, Proc," Vol 1,
pp 341-2 (See 75H027)   (In Russian)

75G053   FE-57   T S Gendler, R N Kuz'min, and T K Urazaeva, Relaxation
Phenomenon in Goethite-water System, in "Int Conf Mossbauer
Spectrosc, Proc," Vol 1, pp 423-4   (See 75H027)

75G054   FE-57   E E Gaubman, V I Gol'danskii, and I P Suzdalev, Two Cyclo-
hexane Phases by Resonance Rayleigh Scattering, in "Int
Conf Mossbauer Spectrosc, Proc," Vol 1, pp 455-6
(See 75H027)

75G055   FE-57   R Grimm, P Gutlich, R Link, E Kankeleit, W Reichenbacher,
and D Walcher, Time Differential Mossbauer Spectroscopy on
/57Co(phen)3/(ClO4)2, in "Int Conf Mossbauer Spectrosc,
Proc," Vol 1, pp 501-2   (See 75H027)

75G056   FE-57   A Govaert, C Dauwe, E De Grave, and J De Sitter,
Mossbauer Spectroscopy on Naturally Occuring Goethite, in
"Int Conf Mossbauer Spectrosc, Proc," Vol 1, pp 521-2
(See 75H027)

75G057   GENRAL   V I Gol'danskii, Opening Talk, in "Int Conf Mossbauer
Spectrosc, Proc," Vol 2, pp 9-16 (See 75H038)

75G058   REVIEW   U Gonser, Defects, in "Int Conf Mossbauer Spectrosc,
Proc," Vol 2, pp 113-31   (See 75H038)

CODE	TOPIC	REFERENCE

75G059 REVIEW V I Gol'danskii, Yu V Maksimov, and I P Suzdalev, Application of Mossbauer Spectroscopy in Catalysis, in "Int Conf Mossbauer Spectrosc, Proc," Vol 2, pp 163-88 (See 75H038)

75G060 REVIEW V I Gol'danskii and V A Naimot, Note on the Non-Mossbauer Gamma Ray Laser, in "Int Conf Mossbauer Spectrosc, Proc," Vol 2, pp 429-33 (See 75H038)

75G061 GENRAL U Gonser, Concluding Remarks, in "Int Conf Mossbauer Spectrosc, Proc," Vol 2, pp 475-8 (See 75H038)

75G062 TE-125 V V Gorbachev, L A Linskii, S P Ionov, O A Sadovskaya, and A Yu Aleksandrov, Mossbauer Effect in Some Tellurides of Rare-earth Elements of the Cerium Subgroup, in "5th Int Conf Mossbauer Spectrosc, Proc," Vol 3, pp 522-4 (See 75H017) (In Russian)

75G063 INSTRM I Ya Garzanov, V I Karlashuk, V A Kotelnikov, V K Labushkin, and E F Makarov, Mossbauer Resonance Analyzer of Cassiterite MAKR-4, in "5th Int Conf Mossbauer Spectrosc, Proc," Vol 3, pp 660-4 (See 75H017) (In Russian)

75G064 NI-61 F Gautier, G Krill, M F Lapierre, P Panissod, C Robert, G Czjzek, J Fink, and H Schmidt, Phys Lett 53A,31-3(1975), Existence of an Antiferromagnetic Metallic Phase (AFM) in the NiS(2-x)Sex System with Pyrite Structure

75G065 FE-57 T C Gibb, Chem Phys Lett 30,137-9(1975), Determination of the Electric Field Gradient Tensor in Sodium Nitroprusside Using a Polarized Mossbauer Source

75G066 INSTRM S K Godovikov and V G Snigirev, Prib Tekh Eksp (3),243-4 (1975)/Instrum Exp Tech (USSR) 18,970-1(1975), Magnesium Windows with a Miniature Indium Seal for Mossbauer Helium Cryostats

75G067 FE-57 G N Goncharov and S B Tomilov, Zap Vses Mineral O-va 104 (2),238-40(1975), Nuclear Gamma Resonance Spectroscopic Study of Pentlandite (In Russian)

75G068 FE-57 V A Gordienko, V I Nikolaev, and S S Yakimov, Exchange Interactions in Nickel Ferrite-chromites, in "Ferrimagnetizm," edited by K P Belov and Yu D Tret'yakov (Izd Mosk Univ, Moscow, 1975), pp 147-63 (In Russian)

75G069 FE-57 V N Gridnev, V V Nemoshkalenko, Yu Ya Meshkov, V G Gavrilyuk, V G Prokopenko, and O N Razumov, Phys Status Solidi A 31,201-10(1975), Mossbauer Effect in Deformed Fe-C Alloys

75G070 FE-57 J M Grow and H H Wickman, Magnetic Hyperfine Structure and Relaxation Effects in Bromobis(morpholydithiocarbamato) Iron(III), in "AIP Conference Proceedings-No 24, Magnetism and Magnetic Materials-1974" (20th Annual Conference, San Francisco), edited by C D Graham, Jr, G H Lander, and J J Rhyne (American Institute of Physics, New York, 1975), pp 215-6

75G071 FE-57 P L Gruzin, K N Zaitsev, R A Sadykov, R A Sizov, and M N Uspenskii, Kristallografiya 20,954-9(1975)/Sov Phys-Crystallogr 20,587-90(1976), Investigation of the Cation Distribution and Magnetic Structure of the Y3InxGaxFe(5-2x)O12 Ferrite-garnet System

75G072 FE-57 P L Gruzin, Yu V Petrikin, and A M Bodin, At Energ/Sov At Energy 38,207(1975), Simulation of Resonance Absorption of Gamma Rays by the Monte Carlo Method

75G072 SN-119 P L Gruzin, Yu V Petrikin, and A M Bodin, At Energ/Sov At Energy 38,207(1975), Simulation of Resonance Absorption of Gamma Rays by the Monte Carlo Method

CODE	TOPIC	REFERENCE

75G073   FE-57   P L Gruzin, V N Gorokhov, and Yu V Petrikin, Zavod Lab 41 984-6(1975)/Ind Lab (USSR) 41,1222-4(1975), Conversion-electron Mossbauer Spectroscopy in Plasma Metal Anodizing Research

75G074   SB-121   S E Gukasyan, V P Gor'kov, L A Sadokhina, F Kh Chibirova, and V S Shpinel', Zh Strukt Khim 16,207-11(1975)/J Struct Chem (USSR) 16,191-4(1975), NGR Spectra of Certain Complex Antimony(III) Fluorides

75G075   FE-57   G K Gurtovoi, A S Lagutin, and V M Cherepanov, Zh Eksp Teor Fiz 68,743-9(1975)/Sov Phys-JETP 41,369-72(1975), Magnetic Phase Transitions in Orthoferrite with a Morin Point

75G076   SB-121   V P Gor'kov, R L Davidovich, G V Zimina, L A Sadokhina, F Kh Chibirova, and V S Shpinel', Koord Khim 1,561-6(1975), Mossbauer Study of Antimony(III) Complex Compounds (In Russian)

75G077   FE-57   I G Gusakovskaya, T I Larkina, and V I Gol'danskii, Fiz Tverd Tela 17,1808-10(1975)/Sov Phys-Solid State 17,1181-2 (1975), Characteristics of Solid-state Reactions in Crystalline and Glassy States

75G078   REVIEW   R Greatrex, Mossbauer Spectroscopy, in "Spectroscopic Properties of Inorganic and Organometallic Compounds, Volume 8," edited by N N Greenwood (The Chemical Society, Burlington House, London, 1975), pp 415-509

75G079   FE-57   P L Gruzin, Yu L Rodionov, and Yu A Li, Fiz Met Metalloved 39,1211-7(1975)/Phys Met Metallogr (USSR), Redistribution of Carbon Atoms in Submicroscopic Steel Volumes

75G080   FE-57   P L Gruzin, Yu L Rodionov, and Yu A Li, Fiz Met Metalloved 40,1046-52(1975)/Phys Met Metallogr (USSR), Investigation of the Redistribution of Atoms in Submicroscopic Volumes in the Early Stages of Ageing of Iron-Tungsten and Iron-Molybdenum Alloys by the Nuclear Gamma Resonance Method

75G081   I-129   R K Gupta, Phys Rev B 12,4452-9(1975), Mean-square Amplitudes of Vibration for Ionic Crystals

75G081   THEORY   R K Gupta, Phys Rev B 12,4452-9(1975), Mean-square Amplitudes of Vibration for Ionic Crystals

75G082   FE-57   V I Gol'danskii and R A Stukan, Koord Khim 1,137-9(1975), Structure of Iron Cyanide Complexes (In Russian)

75G083   FE-57   P L Gruzin, Yu L Rodionov, I Ya Georgieva, and I I Nikitinia, Dokl Akad Nauk SSSR 225,1058-60(1975)/Sov Phys-Dokl, Nuclear Gamma Resonance Study of Martensite of Isothermeral and Athermal Reactions

75H001   THEORY   F Hartmann-Boutron and D Spanjaard, J Phys (Paris) 36, 307-14(1975), On the Use of Liouville Relaxation Supermatrices in Mossbauer Studies

75H002   FE-57   A T Howe and K J Gallagher, J Chem Soc, Faraday Trans 1 71, 22-34(1975), Mossbauer Studies in the Colloid System beta-FeOOH-beta-Fe2O3: Structures and Dehydration Mechanism

75H003   FE-57   R E M Hedges, Nature 254,501-3(1975), Mossbauer Spectroscopy of Chinese Glazed Ceramics

75H004   REVIEW   W B Holzapfel, CRC Crit Rev - Solid State Sci 5,89-123 (1975), Mossbauer Studies at High Pressure

CODE	TOPIC	REFERENCE
75H005	SN-119	C H Huang, O Knop, D A Othen, F W D Woodhams, and R A Howie, Can J Chem 53,79-91(1975), Pyrophosphates of Tetravalent Elements and a Mossbauer Study of SnP2O7
75H006	REVIEW	S S Hafner, Mossbauer Spectroscopy in Lunar Geology and Mineralogy, in "Topics in Applied Physics, Volume 5-Mossbauer Spectroscopy," edited by U Gonser (Springer-Verlag, Berlin, 1975), pp 167-99
75H007	FE-57	L Haggstrom, T Ericsson, R Wappling, and K Chandra, Phys Scr 11,47-54(1975), Studies of the Magnetic Structure of FeSn Using the Mossbauer Effect
75H007	SN-119	L Haggstrom, T Ericsson, R Wappling, and K Chandra, Phys Scr 11,47-54(1975), Studies of the Magnetic Structure of FeSn Using the Mossbauer Effect
75H008	FE-57	L Haggstrom, T Ericsson, R Wappling, and E Karlsson, Phys Scr 11,55-9(1975), Mossbauer Study of Hexagonal FeGe
75H009	FE-57	L Haggstrom, T Ericsson, and R Wappling, Phys Scr 11,94-6 (1975), An Investigation of CoSn Using Mossbauer Spectroscopy
75H009	SN-119	L Haggstrom, T Ericsson, and R Wappling, Phys Scr 11,94-6 (1975), An Investigation of CoSn Using Mossbauer Spectroscopy
75H010	FE-57	L Haggstrom, R Wappling, T Ericsson, Y Andersson, and S Rundqvist, J Solid State Chem 13,84-91(1975), Mossbauer and X-ray Studies of Fe5PB2
75H011	SN-119	L Haggstrom, J Gullman, T Ericsson, and R Wappling, J Solid State Chem 13,204-7(1975), Mossbauer Study of Tin Phosphides
75H012	FE-57	J Hesse and A Rubartsch, Physica 80B 33-47(1975), Magnetische Eigenschaften Entmischter Fe-Cr Legierungen
75H013	FE-57	C S Hogg and R E Meads, Mineral Mag 40,79-88(1975), A Mossbauer Study of Thermal Decomposition of Biotites
75H014	FE-57	C S Hogg, P J Malden, and R E Meads, Mineral Mag 40,89-96 (1975), Identification of Iron-containing Impurities in Natural Kaolinites Using the Mossbauer Effect
75H015	FE-57	D Hanzel, Mossbauer Measurements of BaEu2Fe2O7 and SrEu2Fe2O7, in "5th Int Conf Mossbauer Spec, Proc," Part 1, pp 140-6 (See 75H017)
75H016	FE-57	H Hobert and D Arnold, Changing of the Electron Structure of Dispersed Iron Oxide during Interaction with Amines and Borofluoride, in "5th Int Conf Mossbauer Spec, Proc," Part 1, pp 227-32 (See 75H017)
75H017	GENRAL	M Hucl and T Zemcik, Editors, "The 5th International Conference on Mossbauer Spectroscopy, Proceedings" (Bratislava, Czechoslovakia, September, 1973) (Czechoslovak Atomic Energy Commision, Nuclear Information Centre, Praha 5, Zbraslav, 1975), 769 pages
75H018	SN-119	B Y K Ho and J J Zuckerman, J Organomet Chem 96,41-7(1975), Solid-state Association in Organotin Compounds Containing Bulky Organic Groups. Tricyclohexyltin Hydroxide
75H019	FE-57	S R Hong and H N Ok, Phys Rev B 11,4176-8(1975), Mossbauer Study of Antiferromagnetic FeCr2Se4
75H020	FE-57	M Hucl, Mossbauer Study of Magnetic Phases of Natural Pyrrhotites, in "5th Int Conf Mossbauer Spec, Proc," Part 2, pp 356-60 (See 75H017)

CODE	TOPIC	REFERENCE

75H021   FE-57   S S Hafner, The Effect of a One Megabar Shock on the Mossbauer Spectrum of 57Fe in a Chain Silicate, in "5th Int
Conf Mossbauer Spec, Proc," Part 2, pp 372-9 (See 75H017)

75H022   SN-119 R H Herber and R F Davis, J Chem Phys 63,3668-9(1975),
Lattice Dynamics and Hyperfine Interactions of Tin in
Tantalum Sulfide Layer Compounds

75H023   FE-57   F E Huggins, Am Mineral 60,316-9(1975), The 3d Levels of
Ferrous Ions in Silicate Garnets

75H024   REVIEW A Z Hrynkiewicz, Mossbauer Spectroscopy Compared with Other
Spectroscopic Techniques, in "5th Int Conf Mossbauer Spec,
Proc," Part 3, pp 573-83 (See 75H017)

75H025   FE-57   A Z Hrynkiewicz, E Popiel, and J A Sawicki, Angular
Correlations in a 122.0 keV - 14.4 keV 57Fe Mossbauer Coin-
cidence Experiment, in "5th Int Conf Mossbauer Spec, Proc,"
Part 3, pp 591-3 (See 75H017)

75H025   PROPSL A Z Hrynkiewicz, E Popiel, and J A Sawicki, Angular
Correlations in a 122.0 keV - 14.4 keV 57Fe Mossbauer Coin-
cidence Experiment, in "5th Int Conf Mossbauer Spec, Proc,"
Part 3, pp 591-3 (See 75H017)

75H026   FE-57   J A Helsen, M Van Deyck, G Langouche, R Coussement, M
Van Rossum, and K P Schmidt, Clays Clay Mineral 23,332-4
(1975), The Orientation of the Principal Axes System of the
Electric Field Gradient in Fe(III) Vermiculite Determined
by Mossbauer Spectroscopy

75H027   GENRAL A Z Hrynkiewicz and J A Sawicki, Editors, International
Conference on Mossbauer Spectroscopy, Proceedings (Cracow,
Poland, 1975) (Wykonano w Powielarni Akademii Gorniczo-
Hutniczej im S Staszica, Cracow, 1975), Vol 1, 541 pages

75H028   PROPSL J P Hannon and G T Trammell, Opt Commun 15,330-4(1975),
Anomalous Emission Effect and the Possibility of Gamma-ray
Lasers

75H029   IR-193   R Hoppe and K Claes, J Less-Common Metals 43,129-42(1975),
Uber Oxoiridate: Zur Kenntnis von KIrO3

75H030   FE-57   T Hori and H Sekizawa, Jpn J Appl Phys 14,1163-8(1975),
Magnetic Properties and Magnetic Annealing Effect in
Ni(0.8)Co(0.2)Fe(2-x)VxO4 Ferrite

75H031   FE-57   M A Hoselton, L J Wilson, and R S Drago, J Am Chem Soc 97,
1722-9(1975), Substituent Effects on the Spin Equilibrium
Observed with Hexadentate Ligands on Iron(II)

75H032   FE-57   G P Huffman and H H Podgurski, Acta Metall 23,1367-79
(1975), Mossbauer Study of Nitrided Fe-Mo and Fe-Ti Alloys

75H033   FE-57   M Hucl, T Gabcova, and J Cirak, Critical Coefficient beta
for Nickel, in "Int Conf Mossbauer Spectrosc, Proc," Vol 1,
pp 365-6 (See 75H027)

75H034   FE-57   C Hohenemser, T Kachnowski, and T Bergstresser, Experimen-
tal and Theoretical Study of Probe Disturbance in Hyperfine
Critical Exponents, in "Int Conf Mossbauer Spectrosc,
Proc," Vol 1, pp 367-8 (See 75H027)

75H035   TA-181 A Heidemann, G Kaindl, D Salomon, H Wipf, and G Wortmann,
A Study of the Dynamical Properties of Hydrogen in the
alpha-Phase of Ta-H, in "Int Conf Mossbauer Spectrosc,
Proc," Vol 1, pp 411-2 (See 75H027)

75H036   REVIEW S S Hafner, Mossbauer Results from the Lunar Samples Returned by the Apollo Missions. A Summary, in "Int Conf
Mossbauer Spectrosc, Proc," Vol 1, pp 471-2 (See 75H027)

CODE	TOPIC	REFERENCE

75H037　REVIEW　C Hohenemser, A Review of Mossbauer Effect Studies of Magnetic Critical Phenomena, in "Int Conf Mossbauer Spectrosc,
Proc," Vol 2, pp 239-56　(See 75H038)

75H038　GENRAL　A Z Hrynkiewicz and J A Sawicki, Editors,　International
Conference on Mossbauer Spectroscopy, Proceedings (Cracow,
Poland, 1975) (Wykonano w Powielarni Akademii Gorniczo-
Hutniczej im S Staszica, Cracow, 1975), Vol 2, 481 pages

75H039　FE-57　R J Haines, J A De Beer, and R Greatrex, J Organometal Chem
85,89-99(1975), Reactions of Metal Carbonyl Derivatives
XVII. Tri- as Well as Di-nuclear Products from the Reactions of Bis(dicarbonyl-eta-cyclopentadienyliron) with
Various Dialkyl Disulphides

75H040　SN-119　P G Harrison and R C Phillips, J Organometal Chem 99,79-91
(1975), Structural Studies in Main Group Chemistry　XI.
Tin-119m Mossbauer Investigations of Triorganotin Derivatives of Substituted Pyridines

75H041　FE-57　N Heiman and K Lee, Phys Lett 55A,297-8(1975), Mossbauer
Effect Measurement of the Internal Fields in Amorphous
Rare Earth-iron Alloys

75H042　FE-57　M T Hirvonen, A P Jauho, T E Katila, and K J Riski, On the
Spin Structure of K3Fe(CN)6 in the Ordered State, in
"Proceedings of the 14th International Conference on Low
Temperature Physics:　Volume 3　Low Temperature Properties
of Solids" (Otaniemi, Finland, 1975), edited by M Krusius
and M Vuorio (North-Holland Publishing Co, Amsterdam/
American Elsevier Publishing Co, New York 1975), pp 234-7

75H043　FE-57　R M Housley, S Geller, and G P Espinosa, Cation Deficiency
in Si-substituted YIG, in "AIP Conference Proceedings-
No 24, Magnetism and Magnetic Materials-1974" (20th Annual
Conference, San Francisco), edited by C D Graham, Jr, G H
Lander, and J J Rhyne (American Institute of Physics, New
York, 1975), pp 374-5

75H044　FE-57　G P Huffman and G R Dunmyre, Superparamagnetic Clusters
of Fe2+ Spins in Lunar Olivine:　Dissolution by High-tem-
perature Annealing, in "Proceedings of the Sixth Lunar
Science Conference" (Houston, 1975), Vol 1, pp 757-72
(Also referenced as Geochim Cosmochim Acta, Suppl 6)

75H045　FE-57　Y Y Huang and J R Anderson, J Catal 40,143-53(1975), On
the Reduction of Supported Iron Catalysts Studied by Mossbauer Spectroscopy

75H046　REVIEW　F Hartmann-Boutron, Ann Phys (Paris) 9,285-356(1975),
Theories de la Relaxation des Impuretes Radioactives dans
les Solides

75I001　INSTRM　Y Isozumi and M Takafuchi, Bull Inst Chem Res, Kyoto Univ
53,63-7(1975), Mossbauer Spectroscopy by Scattered Electrons at 77 K

75I002　FE-57　A Ito and S Morimoto, Electronic Properties of Fe2+ in
MTiO3, in "5th Int Conf Mossbauer Spec, Proc," Part 1,
pp 161-6　(See 75H017)

75I003　FE-57　A Ito and S Morimoto, Mossbauer Study of Magnetic Properties of KFeF3, in "5th Int Conf Mossbauer Spec, Proc,"
Part 1, pp 166-72　(See 75H017)

75I004　FE-57　V K Imshennik, A M Afanas'ev, V I Gol'danskii, E F Makarov,
A S Plachinda, and I P Suzdalev, Investigation of the Dy-
namic and Static Distortions of Complexes by the Gamma
Resonance Spectroscopy, in "5th Int Conf Mossbauer Spec,
Proc," Part 1, pp 259-62　(See 75H017)　(In Russian)

CODE	TOPIC	REFERENCE

75I005   FE-57   P A Ioffe, L Sh Aleksandrova, M I Kazakov, G A Mitenkov, N N Shishkin, L Sh Tsemekhman, and S E Vaisburd, Zh Fiz Khim 48,7-9(1975)/Russ J Phys Chem 49,5-6(1975), Hexagonal and Monclinic Pyrrhotite Investigated by Mossbauer Spectroscopy

75I006   FE-57   V P Ivanitskii, I V Matyash, and F I Rakovich, Geokhimiya 850-7(1975), Influence of Radioactive Radiations on the Mossbauer Spectra of Biotites (In Russian) (Possible translation in Geochem Int)

75I007   SN-119   S Ichiba, M Katada, and H Negita, J Inorg Nucl Chem 37, 2249-51(1975), Mossbauer Effect of 119Sn in Thiostannates, Selenostannates and Selenothiostannates

75I008   FE-57   P A Ioffe, L Sh Aleksandrova, L Sh Tsemekhman, S E Vaisburd, and B F Verner, Zh Fiz Khim 49,792(1975)/Russ J Phys Chem 49,465-6(1975), The Mossbauer Spectra of Iron in Troilite

75I009   FE-57   A Ito and S Morimoto, J Phys Soc Jpn 39,884-91(1975), Mossbauer Study of Magnetic Properties of $KFeF_3$

75I010   FE-57   V K Imshennik and I P Suzdalev, Study of Electronic $Fe3+$ Relaxation Processes in Some Polymeric Compounds, in "Int Conf Mossbauer Spectrosc, Proc," Vol 1, pp 419-20 (See 75H027)

75I011   SN-119   P A Ioffe, A A Baklagin, and V A Kozlova, Zh Neorg Khim 20, 1712(1975)/Russ J Inorg Chem 20,960(1975), Mossbauer Spectra and State of Tin in the Compound $Zn2SnO4$

75I012   THEORY   E M Iolin and E V Zolotoyabko, Zh Eksp Teor Fiz 68,1331-6 (1975)/Sov Phys-JETP 41,662-4(1976), Inelastic Scattering of Mossbauer Gamma Rays by Crystals near the Temperature of a Structural Phase Transiton

75I013   REVIEW   S P Ionov, G V Ionova, V S Lubimov, and E F Makarov, Phys Status Solidi B 71,11-57(1975), Instability of Crystal Lattices with Respect to Electron Density Redistributions

75I014   FE-57   V V Isaev-Ivanov, N M Kolchanova, V F Masterov, D N Nasledov, G N Talalakin, and V K Yarmarkin, Possible States of an Iron Impurity in Gallium Arsenide, in "Legir Poluprovodn," edited by N Kh Abriksov and V S Zemskov ("Nauka," Moscow, 1975), pp 32-7 (In Russian)

75I015   SN-119   Y Ishida and T Ozawa, Scr Metall 9,1103-6(1975), Grain Boundary Segregation of Tin and the Electronic and Vibrational State in Zn-Al Eutectoid

75I016   FE-57   V P Ivanitskii, A M Kalinichenko, I V Matyash, and T P Khomyak, Geokhimiya 1864-71(1975), Study of the Processes of Biotite Oxidation and Dehydroxylation by the PMR Method and Mossbauer (In Russian) (Possibly translated in Geochem Int)

75I017   FE-57   V P Ivanitskii, A V Smirnova, and K N Alekseeva, Geol Zh (Russ Ed) 35(4),143-7(1975), Phase Composition of the Krymka Meteorite According to Mossbauer Spectra (In Russian)

75I018   FE-57   V P Ivanitskii, I V Matyash, and F I Rakovich, Dopov Akad Nauk Ukr RSR, Ser B (3),206-9(1975), Mossbauer Spectra of Iron in Biotite near Ore as a Prospecting Criterion for Radioactive Elements (In Russian)

75I019   THEORY   E M Iolin and V V Martyshchenko, Latv PSR Zinat Akad Vestis, Fiz Teh Ser (3),29-34(1975), Mossbauer Irradiation from Thick Crystal near Kossel Cone Intersection (In Russian)

CODE	TOPIC	REFERENCE

75I020   THEORY   L N Ivanov and V S Letokhov, Zh Eksp Teor Fiz 68,1748-56 (1975)/Sov Phys-JETP 41,877-81(1976), Spectrum of Electron-nuclear Gamma Transitions in Atomic Nuclei

75J001   HF-178   C Jeandey and P Peretto, Phys Status Solidi A 28,529-37 (1975), Effet Mossbauer en Ligne et Defauts de Recul dans le Hafnium Metallique et dans l'Oxyde de Hafnium

75J002   FE-57   C Janot and P Delcroix, Electronic Properties of Iron Impurity in HCP Metals from Mossbauer Studies, in "5th Int Conf Mossbauer Spec, Proc," Part 1, pp 19-25 (See 75H017)

75J003   FE-57   L Y Johansson, R Larsson, J Blomquist, C Cederstrom, S Grangiesser, U Helgeson, L C Moberg, and M Sundbom, Mossbauer and ESCA Studies of a Series of Iron Compounds, in "5th Int Conf Mossbauer Spec, Proc," Part 1, pp 214-5 (See 75H017)

75J004   FE-57   V Jacopian, B Philipp, H Mehner, J Schulze, and H Dautzenberg, Faserforsch Textiltech 26,153-8(1975), Untersuchungen zur Kationenbindung an hydrolytisch abgebauten Cellulosepulvern

75J005   REVIEW   C E Johnson, Mossbauer Spectroscopy in Biology, in "Topics in Applied Physics, Volume 5 - Mossbauer Spectroscopy," edited by U Gonser (Springer-Verlag, Berlin, 1975), pp 139-66

75J006   I-129   C H W Jones, J Chem Phys 62,4343-9(1975), 129I Mossbauer Studies of Iodine in the +1 and +3 Oxidation States

75J007   FE-57   B M Jankowski, J E Frackowiak, J Kansy, J W Moron, T J Panek, A Salamon, and M R Uhlig, Some Remarks on the Structure of Fe-Ni Invar Alloys, in "Int Conf Mossbauer Spectrosc, Proc," Vol 1, pp 137-8   (See 75H027)

75J008   FE-57   V G Jadhao, R M Singru, G Rama Rao, D Bahadur, and C N R Rao, J Chem Soc, Faraday Trans II 71,1885-91(1975), Effect of the Rare Earth Ion on the Spin State Equilibria in Perovskite Rare Earth Metal Cobaltates

75J009   FE-57   C Jeandey and P Peretto, J Phys (Paris) 36,1103-14(1975), Effet Mossbauer en Ligne et Effet de Recul dans les Alliages de Substitution Fe(1-x)Alx(b.c.c.) et Fe-Ni(c.f.c.)

75J010   FE-57   D A Jefferson, M J Tricker, and A P Winterbottom, Clays Clay Miner 23,355-60(1975), Electron-microscopic and Mossbauer Spectroscopic Studies of Iron-stained Kaolinte Minerals

75J011   REVIEW   C E Johnson, Mossbauer Spectroscopy, in "Amino-acids, Peptides, and Proteins, Volume 6" (A Specialist Periodical Report: A Review of the Literature Published during 1973), edited by R C Sheppard (The Chemical Society, Burlington House, London, 1975), pp 256-61

75J012   FE-57   C Jeandey and P Peretto, Phys Status Solidi A 30,71-80 (1975), Etude par Effet Mossbauer en Ligne des Effets de Recul dans les Alliages Equiatomiques Fe-Ge

75K001   REVIEW   B Keisch, Mossbauer Effect Spectroscopy Without Sampling: Application to Art and Archaeology, in "Advances in Chemistry Series, Number 138: Archaeological Chemistry" (American Chemical Society, New York, 1975), pp 186-206

CODE	TOPIC	REFERENCE
75K002	FE-57	H Keller and W Kundig, Solid State Commun 16,253-6(1975), Mossbauer Studies of Brownian Motion
75K003	FE-57	E Kreber, U Gonser, and A Trautwein, J Phys Chem Solids 36, 263-5(1975), Mossbauer Measurements of the Bipyramidal Lattice Site in BaFe12019
75K004	THEORY	O Knop, E M Palmer, and R W Robinson, Acta Crystallogr, Sec A 31,19-31(1975), Arrangements of Point Charges Having Zero Electric-field Gradient (Errata: personal communication)
75K005	FE-57	M A Kobeissi, R Suter, A M Gottlieb, and C Hohenemser, Phys Rev B 11,2455-9(1975), Critical Fluctuations in Ni Observed with the Mossbauer Effect
75K006	FE-57	E Konig, G Ritter, and H A Goodwin, Chem Phys Lett 31, 543-6(1975), 57Fe Mossbauer Effect Study of a Presumed Singlet-triplet Transition in (Fe(P4)X)BPh4 Complexes, X = Br, I
75K007	INSTRM	M N Kamthe, K V Bankar, and S K Date, J Phys Educ 2,15-20 (1975), Design and Construction of Versatile Mossbauer Spectrometer
75K008	FE-57	W Keune and A Sette Camara, Phys Status Solidi A 27,181-90 (1975), Mossbauer Effect and X-ray Observation of Magnetic-field-induced Crystallographic Texture in Antiferromagnetic FeS Powder
75K009	IR-193	M Kanashiro, M Nishi, N Kunitomi, and H Sakai, J Phys Soc Jpn 38,897(1975), Mossbauer Effect of Ir193 in Fe-Pt-Ir Alloys
75K010	SN-119	M Katada, J Sci Hiroshima Univ, Ser A 39,45-72(1975), Mossbauer Effect of 119Sn in Tin Sulfides and Their Related Compounds
75K011	AU-197	C W Kimball, A E Dwight, G M Kalvius, B D Dunlap, and M V Nevitt, Phys Rev B 12,819-23(1975), Low-temperature Phase Transition and Isomer-shift Systematics in Intermediate Phases of Rare-earth-gold Compounds
75K011	ER-166	C W Kimball, A E Dwight, G M Kalvius, B D Dunlap, and M V Nevitt, Phys Rev B 12,819-23(1975), Low-temperature Phase Transition and Isomer-shift Systematics in Intermediate Phases of Rare-earth-gold Compounds
75K011	TM-169	C W Kimball, A E Dwight, G M Kalvius, B D Dunlap, and M V Nevitt, Phys Rev B 12,819-23(1975), Low-temperature Phase Transition and Isomer-shift Systematics in Intermediate Phases of Rare-earth-gold Compounds
75K011	YB-171	C W Kimball, A E Dwight, G M Kalvius, B D Dunlap, and M V Nevitt, Phys Rev B 12,819-23(1975), Low-temperature Phase Transition and Isomer-shift Systematics in Intermediate Phases of Rare-earth-gold Compounds
75K012	GD-155	R Kmiec, K Latka, T Matlak, K Ruebenbauer, and K Tomala, Phys Status Solidi B 68,K125-8(1975), Quadrupole Moment Ratio for the 86.5 keV Transition in 155Gd and the Goldanskii-Karyagin Effect for Sm2Sn207
75K013	FE-57	M Kopcewicz, A Kotlicki, and M Szefer, Phys Status Solidi A 29,635-43(1975), Mossbauer Study of Proton Irradiation Effects in Hydrated Iron(II) Ammonium Sulphate and Iron(III) Ammonium Alum
75K014	FE-57	M Kopcewicz, A Kotlicki, and M Szefer, Phys Status Solidi A 29,K125-8(1975), Evaluation of the Size of the Spike Regions Formed in Dielectric Iron Compounds Due to Proton Irradiation

CODE	TOPIC	REFERENCE
75K015	FE-57	M Kopcewicz and A Kotlicki, Radiat Eff 24,267-71(1975), Study of Proton Radiation Effects in Several Iron Compounds Using the Mossbauer Technique
75K016	FE-57	B Krishnamurthy and K P Sinha, J Magn Res 17,189-92(1975), Mossbauer Lineshape in the Presence of NMR Transitions
75K016	THEORY	B Krishnamurthy and K P Sinha, J Magn Res 17,189-92(1975), Mossbauer Lineshape in the Presence of NMR Transitions
75K017	SN-119	H J Kroth, H Schumann, H G Kuivila, C D Schaeffer, Jr, and J J Zuckerman, J Am Chem Soc 97,1754-60(1975), Tin-119 Chemical Shifts of Ortho, Meta, Para, 2,6- and Polysubstituted Aryltrimethyltin Derivatives and Related Organotin Compounds
75K018	FE-57	E Kreber, U Gonser, A Trautwein, and F E Harris, Peculiar Properties of the (Bi)pyramidal Lattice Site in BaFe12O19, in "5th Int Conf Mossbauer Spec, Proc," Part 1, pp 130-3 (See 75H017)
75K019	THEORY	L Korecz, On Comparison of Molecular Orbital and Isomeric Shift Data of Some Low-spin Iron Compounds, in "5th Int Conf Mossbauer Spec, Proc," Part 1, pp 215-7 (See 75H017)
75K020	FE-57	A Kostikas, D Petridis, and A Simopoulos, Mossbauer Studies of Magnetic and Crystal Field Properties in Fe(III) Mixed Chelates, in "5th Int Conf Mossbauer Spec, Proc," Part 1, pp 220-2 (See 75H017)
75K021	FE-57	H Kubsch, E Fritzsch, and H A Schneider, Some Problems of Phase Identification of Iron Corrosion Products, in "5th Int Conf Mossbauer Spec, Proc," Part 1, pp 233-40 (See 75H017)
75K022	FE-57	J Kinrade, G B Skippen, and D R Wiles, Geochim Cosmochim Acta 39,1325-7(1975), A Mossbauer Observation of Hedenbergite Synthesis
75K023	FE-57	Yu F Krupanskii and I P Suzdalev, Fiz Tverd Tela (Leningrad) 17,588-90(1975)/Sov Phys-Solid State 17,375-6 (1975), Some Features of the Magnetic Properties of Small alpha-Fe2O3 Particles
75K024	FE-57	K Kauffman and F Hazel, J Inorg Nucl Chem 37,1139-48(1975), Infrared and Mossbauer Spectroscopy. Electron Microscopy and Chemical Reactivity of Ferric Chloride Hydrolysis Products
75K025	FE-57	K Kauffman and F Hazel, J Colloid Interface Sci 51,422-6 (1975), Mossbauer Spectroscopy of Aged Ferric Oxide Gels
75K026	KR-83	B Kolk, Phys Rev B 12,1620-5(1975), Mossbauer Investigations of the Quadrupole Interaction of 83Kr in Various Hosts
75K027	SN-119	E J Kupchik and J A Feiccabrino, J Organomet Chem 93,325-9 (1975), Preparation of Some N-substituted N-(triphenylstannyl)-cyanamides
75K028	FE-57	W Keune, S K Date, I Dezsi, and U Gonser, J Appl Phys 46, 914-24(1975), Mossbauer-effect Study of Co57 and Fe57 Impurities in Ferroelectric LiNbO3
75K029	SN-119	N Karnezos, L B Welsh, and M W Shafer, Phys Rev B 11,1808-17(1975), Structural and NMR Properties of Niobium Dichalcogenides Intercalated with Post Transition Metals
75K030	FE-57	B Karvaly, Cs Badinka, L Keszthelyi, and L Erdei, A Mossbauer Study on the Interaction Between Biomolecular Lipid Membranes and Ferric/Ferrous Ions, in "5th Int Conf Mossbauer Spec, Proc," Part 2, pp 422-5 (See 75H017)

CODE	TOPIC	REFERENCE
75K031	FE-57	C Kellershohn, F Soubirou, and C Hubert, Quantitative Study by Mossbauer Effect of Radiolysis of Dihydrate Iron Oxide FeC2O4-2H2O, in "5th Int Conf Mossbauer Spec, Proc," Part 2, pp 432-7 (See 75H017)
75K032	FE-57	K Krop, J Korecki, J Zukrowski, W Karas, and J M Williams, The Relaxation Times of Superparamagnetic Particles as Determined from Field Dependent Mossbauer Spectra, in "5th Int Conf Mossbauer Spec, Proc," Part 2, pp 462-71 (See 75H017)
75K033	FE-57	P P Kirichok, L G Kustovskii, Yu S Avraamov, L M Letyuk, I N Bibolov, Sh S Madumarov, and V Ya Garmash, Izv Vyssh Uchebn Zaved, Fiz 18(3),141-3(1975)/Sov Phys J, Hyperfine Interactions in Yttrium Cobalt Titanium Ferrites (Y3Co(0.05x)Ti(0.05x)Fe(5-x)O12
75K034	FE-57	S Kumar, Phys Status Solidi B 69,K145-7(1975), On the Usability of Lattice Dynamical Theory in Mossbauer Spectroscopic Studies
75K034	THEORY	S Kumar, Phys Status Solidi B 69,K145-7(1975), On the Usability of Lattice Dynamical Theory in Mossbauer Spectroscopic Studies
75K035	SN-119	I Ya Kuramshin, Sh Sh Bashkirov, A A Muratova, R A Manapov, A S Khramov, and A N Pudovik, Zh Obshch Khim 45, 701-2(1975)/J Gen Chem (USSR) 45,684-5(1975), Localization Centers of Donor-acceptor Bond in Complexes of Tin Halides with Phosphorous Acid Amides
75K036	REVIEW	G M Kalvius, Electronic Structure of the Light Actinides from Mossbauer Spectroscopy, in "5th Int Conf Mossbauer Spec, Proc," Part 3, pp 485-98 (See 75H017)
75K037	FE-57	H Kubsch and H A Schneider, Phase Transition of Iron Oxyhydroxides by Mechanical Activation, in "5th Int Conf Mossbauer Spec, Proc," Part 3, pp 718-22 (See 75H017)
75K038	FE-57	M Kopcewicz and A Kotlicki, Application of the Mossbauer Spectroscopy to the Investigations of the Proton Irradiation Effects in Several Iron Compounds, in "5th Int Conf Mossbauer Spec, Proc," Part 3, pp 740-8 (See 75H017)
75K039	REVIEW	W Keune, U Gonser, and H Vollmar, Mossbauer-Spektroskopie als zerstorungsfreie Materialprufmethode, in "Handbuch der zerstorungsfreien Materialprufung," edited by E A W Muller (R Oldenbourg Verlag, Munich, 1975), U 3, 31, 32, 321, 322
75K040	FE-57	T Kobayashi and S Shimizu, Phys Lett 54A,311-2(1975), Time-filtering Effect in the Mossbauer Spectrum
75K041	KR-83	B Kolk, Phys Rev B 12,4695-701(1975), Mossbauer Studies on 83Kr Aftereffects in Solid Krypton
75K042	ANALYS	J Kansy, J E Frackowiak, B M Jankowski, and T J Panek, The Recurrent Formula for Transmission Integral Evaluation, in "Int Conf Mossbauer Spectrosc, Proc," Vol 1, pp 27-8 (See 75H027)
75K043	W-182	D K Kaipov, U M Makhanov, A V Kuz'minov, D N Smirin, and Zh I Adymov, The Study of Mossbauer Effect for the 100.1 keV State of 182W with the Current Registration Method, in "Int Conf Mossbauer Spectrosc, Proc," Vol 1, pp 35-6 (See 75H027) (In Russian)
75K044	FE-57	W Keune, D L Williamson, and J Lauer, Mossbauer Spectroscopy of Very Thin Fe Films, in "Int Conf Mossbauer Spectrosc, Proc," Vol 1, pp 39-40 (See 75H027)

CODE	TOPIC	REFERENCE

75K045   FE-57   M Kopcewicz, A Kotlicki, and M Szefer, Modulation of Moss-bauer Gamma Rays by rf Magnetic Field, in "Int Conf Moss-bauer Spectrosc, Proc," Vol 1, pp 49-50 (See 75H027)

75K046   FE-57   M Kopcewicz, A Kotlicki, and M Szefer, Influence of the Foil Coating on the Formation of Mossbauer rf Sidebands, in "Int Conf Mossbauer Spectrosc, Proc," Vol 1, pp 51-2 (See 75H027)

75K047   FE-57   M Kopcewicz, Mossbauer Study of rf Collapse in Permalloy, in "Int Conf Mossbauer Spectrosc, Proc," Vol 1, pp 53-4 (See 75H027)

75K048   FE-57   E Kuzman, S Nagy, A Sztarek, F Nagy, L Domonkos, Z Hegedus, and A Vertes, Mossbauer Study of Fe-Cr Carbides, in "Int Conf Mossbauer Spectrosc, Proc," Vol 1, pp 123-4 (See 75H027)

75K049   FE-57   J Kjeldgaard, G Trumpy, N Thrane, and S Morup, Influence of a Uniaxial Stress on the Mossbauer Spectrum of alpha-Iron, in "Int Conf Mossbauer Spectrosc, Proc," Vol 1, pp 127-8 (See 75H027)

75K050   FE-57   D S Kulgawczuk, E Mazanek, M Wyderko, and J Kraczka, Studies of Iron-Manganese Ore Agglomerates by Mossbauer Spectroscopy, in "Int Conf Mossbauer Spectrosc, Proc," Vol 1, p 139 (See 75H027)

75K051   FE-57   N M Kolchanova, D N Nasledov, Yu S Smetannikova, G N Talalakin, K I Vinogradova, and V K Yarmarkin, Mossbauer Effect in A3B5 Crystals Containing Magnetic Order Regions, in "Int Conf Mossbauer Spectrosc, Proc," Vol 1, pp 185-6 (See 75H027)

75K052   FE-57   K Konig, G Wortmann, and G M Kalvius, Search for Magnetic Ordering in the High-pressure Phase of Iron (epsilon-Fe) down to 2.2 K, in "Int Conf Mossbauer Spectrosc, Proc," Vol 1, p 189 (See 75H027)

75K053   FE-57   M Kopcewicz, I Sosnowska, M Szefer, and J Tatarkiewicz, Mossbauer Study of Proton Radiation Effects in FeCl2-4H2O, in "Int Conf Mossbauer Spectrosc, Proc," Vol 1 227-8 (See 75H027)

75K054   FE-57   A Kotlicki and N Boye Olsen, Proton Radiation Damage in Biotite, in "Int Conf Mossbauer Spectrosc, Proc," Vol 1, pp 229-30 (See 75H027)

75K055   FE-57   S Koch, R H Holm, and R B Frankel, J Am Chem Soc 97,6714-23(1975), Synthesis and Ground State Electronic Properties of Tetraaza Macrocyclic Iron(II, III) Complexes Containing (14)-, (15)-, and (16)-Membered Rings

75K056   FE-57   H Kubsch and H A Schneider, Investigations of the Initial Stage of Corrosion by Mossbauer Spectrometry, in "Int Conf Mossbauer Spectrosc, Proc," Vol 1, pp 315-6 (See 75H027)

75K057   FE-57   H Keller, W Kundig, and H Arend, Mossbauer Study of the Two-dimensional Antiferromagnet (CH3NH3)2FeCl4, in "Int Conf Mossbauer Spectrosc, Proc," Vol 1, pp 363-4 (See 75H027)

75K058   FE-57   Yu F Krupanskii and I P Suzdalev, Magnetic Phase Transi-tions of First Order in Fine Particles gamma-Fe203, delta-Fe203, in "Int Conf Mossbauer Spectrosc, Proc," Vol 1, pp 381-2 (See 75H027)

75K059   FE-57   D S Kulgawczuk, H Przywarska-Boniecka, L Trynda, and K Dolegowska, Catalase Model Complexes Investigated by Mossbauer Effect, in "Int Conf Mossbauer Spectrosc, Proc," Vol 1, pp 441-2 (See 75H027)

CODE	TOPIC	REFERENCE

75K060 FE-57 L Keszthelyi and I Vincze, Absorption of Circularly Polarized Gamma Radiation in L- and D-amino Acids, in "Int Conf Mossbauer Spectrosc, Proc," Vol 1, pp 447-8 (See 75H027)

75K061 PROPSL E Kankeleit, Z Phys A 275,119-21(1975), The Effect of Decaying Atomic States on Integral and Time Differential Mossbauer Spectra

75K062 FE-57 J M Knudsen, E Larsen, J E Moreira, and O Faurskov Nielsen, Acta Chem Scand, Ser A 29,833-9(1975), Characterization of Decaaqua-mu-oxodi-iron(III) by Mossbauer and Vibrational Spectroscopy

75K063 FE-57 L Keszthelyi and I Vincze, Radiat Environ Biophys 12,181-5 (1975), Absorption of Circularly Polarized Gamma-radiation in L- and D-Amino Acids

75K064 SN-119 P Kamenov, E Vapirev, B Slavov, K Bourin, and Cv Bontschev, Time Measurements of Resonance Scattered Gamma-rays, in "Int Conf Mossbauer Spectrosc, Proc," Vol 1, pp 503-4 (See 75H027)

75K065 REVIEW Yu M Kagan, Coherent Phenomenon on Interaction of Mossbauer Radiation with Crystals (Progress in Theory), in "Int Conf Mossbauer Spectrosc, Proc," Vol 2, pp 17-41 (See 75H038) (In Russian)

75K065 THEORY Yu M Kagan, Coherent Phenomenon on Interaction of Mossbauer Radiation with Crystals (Progress in Theory), in "Int Conf Mossbauer Spectrosc, Proc," Vol 2, pp 17-41 (See 75H038) (In Russian)

75K066 INSTRM E Kankeleit, Some Technical Developments in Mossbauer Spectroscopy, in "Int Conf Mossbauer Spectrosc, Proc," Vol 2, pp 43-58 (See 75H038)

75K066 REVIEW E Kankeleit, Some Technical Developments in Mossbauer Spectroscopy, in "Int Conf Mossbauer Spectrosc, Proc," Vol 2, pp 43-58 (See 75H038)

75K067 REVIEW W Kundig, Calibration of the Isomer Shift, in "Int Conf Mossbauer Spectrosc, Proc," Vol 2, pp 355-67 (See 75H038)

75K068 FE-57 P Kritidis, Cv Bontschev, and S Micheva, Investigation of Line Broadening and Recoil in the Mossbauer Experiments with Solutions, in "5th Int Conf Mossbauer Spectrosc, Proc," Vol 3, pp 643-5 (See 75H017) (In Russian)

75K069 FE-57 Yu F Krupanskii and I P Suzdalev, Magnetic Properties of Ultra-small Particles of the Ferric Oxide, in "5th Int Conf Mossbauer Spectrosc, Proc," Vol 3, pp 728-31 (See 75H017) (In Russian)

75K070 FE-57 N I Kaletina and B N Stepanenko, Dokl Akad Nauk SSSR 223, 494-6(1975)/Dokl Biochem 223,288-9(1975), Production of Anomeric Pairs of Certain N-arylglycosylamines and the Gamma-resonance Spectra of Their Complexes with Ferric Chloride

75K071 FE-57 A S Kamzin, Yu G Yuskevich, and S I Yushchuk, Fiz Tverd Tela 17,2419-21(1975)/Sov Phys-Solid State 17,1599-600 (1976), Connection Between Local Distortions of the Crystal Lattice and the Rectangularity Coefficient of the Hysteresis Loop of Ferrites Having the Spinel Structure

75K072 THEORY S V Karyagin, Fiz Tverd Tela 17,1856-8(1975)/Sov Phys-Solid State 17,1220-1(1975), Influence of Dimensions of Jump Diffusion on the Broadening of a Mossbauer Line of a Polycrystalline Sample

CODE	TOPIC	REFERENCE

75K073   THEORY   V N Kashcheev, Fiz Tverd Tela 17,3552-8(1975)/Sov Phys-Solid State 17,2194-8(1976), Critical Anomalies of the Mossbauer Line of a Ferromagnet

75K074   FE-57   A S Khlystov, N I Marusich, N F Noskova, and D V Sokol'skii, Koord Khim 1,942-4(1975), Gamma Resonance Spectroscopic Study of the Iron(III) Acetylacetonate-cobalt (III) Acetylacetonate-trialkylaluminum and the Iron(III) Acetylacetonate-nickel(II) Acetylacetonate-trialkylaluminum Catalytic Systems (In Russian)

75K075   SN-119   H Kohler, L Neef, L Korecz, and K Burger, J Organomet Chem 90,159-71(1975), Pseudochalkogenvergindungen VII. Synthese und Struktur von Pseudochalkogenooxoacyl-organozinnverbindungen

75K076   FE-57   A V Kolpakov, E N Ovchinnikova, and R N Kuz'min, Kristallografiya 20,221-5(1975)/Sov Phys-Crystallogr 20,135-8 (1975), Symmetry of Electric Field Gradients in Crystals

75K076   TE-125   A V Kolpakov, E N Ovchinnikova, and R N Kuz'min, Kristallografiya 20,221-5(1975)/Sov Phys-Crystallogr 20,135-8 (1975), Symmetry of Electric Field Gradients in Crystals

75K077   FE-57   M Kopcewicz, A Kotlicki, and M Szefer, Phys Status Solidi B 72,701-8(1975), Mossbauer Study of FM Sidebands Produced by the RF Magnetic Field Applied to an Iron Foil

75K078   FE-57   E V Korneev, A F Belov, T A Khimich, V F Belov, and I M Kolesnikov, Zh Fiz Khim 49,2640-4(1975)/Russ J Phys Chem 49,1547-50(1975), Crystallochemical and Magnetic Properties of Manganese-Zinc Ferrites

75K079   SN-119   V M Koshkin, E E Ovechkina, and V P Romanov, Zh Eksp Teor Fiz 69,2218-21(1975)/Sov Phys-JETP, Nuclear Gamma Resonance in Neutral Tin Atoms in the In2Te3 Crystal Lattice

75K080   FE-57   M Yu Kosygin, Yu N Novikov, and R A Stukan, Fiz Tverd Tela 17,1803-5(1975)/Sov Phys-Solid State 17,1176-7(1975), Use of Shock Waves in Obtaining Information on the Structure of Layered Compounds of Graphite

75K081   SN-119   H G Kurvila, J E Dixon, P L Maxfield, N M Scarpa, T M Topka, K H Tsai, and K R Wursthorn, J Organomet Chem 86, 89-107(1975), Preparation of Some Ketoorganostannanes and Ketoorganochlorostannanes. Intramolecular Coordination in Ketoorganochlorostannanes

75K082   FE-57   E N Kuznetsov, V D Checherskii, and V P Romanov, Ukr Fiz Zh 20,1781-6(1975)/Ukr Phys J, Nuclear Gamma Resonance Investigation of the Inversion Degree in Some Simple Spinel-ferrites at High Temperatures

75K083   SN-119   T S Khodashova, V A Varnek, E N Yurchenko, and M A Porai-Koshits, Dokl Akad Nauk SSSR 224,1323-6(1975)/ Dokl Chem 224,617-20(1975), The Crystal Structure and Mossbauer Parameters of a Complex of Rhodium with Tin(II) Fluoride, Cs4(Rh(SnF2(H2O)2)2(SnF15))-4H2O

75K084   FE-57   M R Knapp, K Li, and W O Philbrook, Metallurg Trans B 6, 513-21(1975), Kinetics of Reactions Between Magnetite and Carbon Monoxide Between 723 and 823 K

75K085   FE-57   V V Korovushkin, E V Korneev, A F Belov, T A Khimich, and I M Kolesnikov, Zh Fiz Khim 49,2810-5(1975)/Russ J Phys Chem 49,1656-9(1975), The Problem of the Ageing of Ferrite-spinels

75K086   FE-57   V F Kumeishin and O A Ivanov, Fiz Met Metalloved 40,1295-7 (1975)/Phys Met Metallogr (USSR), Study of Relaxation Processes in Nickel near the Curie Point Using Nuclear Gamma Resonance

CODE	TOPIC	REFERENCE
75K087	PROPSL	P Kamenov and Cv Bontschev, Dokl Bolg Akad Nauk 28,1175-7 (1975), On the Possibility of Realizing a Gamma Laser with Long-living Isomer Nuclei
75L001	FE-57	G Longworth and I R Harris, J Less-Common Met 41,175-85 (1975), Mossbauer Effect Study of the Pseudo-binary System Ce(Fe(1-x)Co(x)2
75L002	EU-151	S J Lyle and A D Witts, J Chem Soc, Dalton Trans 185-8 (1975), Mossbauer Spectroscopic Investigation of Some Europium(III) Diketonates
75L003	FE-57	T Lohner, I Dezsi, D L Nagy, and A M Afanas'ev, Phys Lett 53A,446-8(1975), On the Lineshape of Spin-relaxation Broadened Mossbauer Spectra of Some Fe3+ Paramagnets
75L004	FE-57	M E Lines and M Eibschutz, Phys Rev B 11,4583-94(1975), Magnetism in Orbitally Unquenched Chainar Compounds. I. The Antiferromagnetic Case: RbFeBr3
75L005	REVIEW	G Longworth, Nature 256,367(1975), Mossbauer Effect and Order-disorder Transitions in Alloys
75L006	FE-57	G Longworth and S E Warren, Nature 255,625-7(1975), Mossbauer Spectroscopy of Greek 'Etruscan' Pottery
75L007	SN-119	I S Lyubutin and T V Dmitrieva, Pis'ma Zh Eksp Teor Fiz 21, 132-5(1975)/JETP Lett 21,59-60(1975), Induction of Strong Magnetic Fields at the Nuclei of Diamagnetic Tin Atoms in Chalcogenide Spinels
75L008	FE-57	J Lipka, The Hyperfine Interaction in Ferrimagnetic Materials, in "5th Int Conf Mossbauer Spec, Proc," Part 1, pp 87-9 (See 75H017)
75L009	GD-155	S J Lyle, P T Walsh, A D Witts, and J W Ross, J Chem Soc, Dalton Trans 1406-9(1975), Mossbauer Spectroscopic Study of the Gadolinium-Hydrogen System
75L010	FE-57	J Lauermannova, Effective Magnetic Fields in the Fe-Ni-C Martensite, in "5th Int Conf Mossbauer Spec, Proc," Part 2, pp 302-4 (See 75H017)
75L011	I-129	G Langouche, M Van Rossum, K P Schmidt, and R Coussement, The Quadrupole Interaction of 125Te and 129I in Polycrystalline Te and in Te Single Crystals, in "5th Int Conf Mossbauer Spec, Proc," Part 3, pp 531-6 (See 75H017)
75L011	TE-125	G Langouche, M Van Rossum, K P Schmidt, and R Coussement, The Quadrupole Interaction of 125Te and 129I in Polycrystalline Te and in Te Single Crystals, in "5th Int Conf Mossbauer Spec, Proc," Part 3, pp 531-6 (See 75H017)
75L012	FE-57	J Lauer, W Keune, and T Shinjo, Surface Study of Iron by Mossbauer Spectroscopy, in "Int Conf Mossbauer Spectrosc, Proc," Vol 1, pp 41-2 (See 75H027)
75L013	SN-119	I S Lyubutin and T V Dmitrieva, Strong Magnetic Fields at Nuclei of Diamagnetic Tin Atoms in Chromium Chalcogenide Spinels, in "Int Conf Mossbauer Spectrosc, Proc," Vol 1, pp 147-8 (See 75H027)
75L014	I-129	G Langouche, P Boolchand, M Van Rossum, and R Coussement, Nuclear Quadrupole Interaction of 129I in Crystalline Te, Se and S, in "Int Conf Mossbauer Spectrosc, Proc," Vol 1 pp 203-4 (See 75H027)
75L015	FE-57	G H Loew and R K Kirchner, J Am Chem Soc 97,7388-90 (1975), Electronic Structure and Electric Field Gradients in Oxyhemoglobin and -cytochrome P-450 Model Compounds

276

CODE	TOPIC	REFERENCE

75L016   FE-57   F K Lotgering, A M Van Diepen, and J F Olijhoek, Solid State Commun 17,1149-53(1975), Mossbauer Spectra of Iron-Chromium Sulphospinels with Varying Metal Ratio

75L017   FE-57   V N Luzgina, E I Filippovich, R P Evstigneeva, and A F Vanin, Zh Obshch Khim 45,212-6(1975)/J Gen Chem (USSR) 45,196-9(1975), Physicochemical Properties of Complexes of Protohemin with Substituted Histidine-containing Peptides

75L018   FE-57   J Lipka, J Rijacek, and I Toth, Mossbauer and X-ray Study of Iron Nitrides near Fe2N, in "Int Conf Mossbauer Spectrosc, Proc," Vol 1, pp 279-80   (See 75H027)

75L019   FE-57   F J Litterst, R Ramisch, and G M Kalvius, Structural Relaxation due to Stabilization at the Glass Transition of Very Dilute Solutions of Fe2+ in Propane-(1,2)diol+), in "Int Conf Mossbauer Spectrosc, Proc," Vol 1, pp 355-6 (See 75H027)

75L020   FE-57   K B Lal, S Mendiratta, and G N Rao, Phys Status Solidi A 32,K79-83(1975), Mossbauer Studies of Fe-Se, Fe-Sb, Fe-Pr, and Fe-Nd Alloys

75L021   FE-57   C Le Caer, G Le Caer, and B Roques, C R Acad Sci (Paris), Ser C 281,527-9(1975), Domaines d'Existence et Proprietes Magnetiques des Germaniures Ternaires Cubiques, Fe(1-x)CrxGe et Fe(1-x)CoxGe

75L022   FE-57   W Levason, C A McAuliffe, M M Khan, and S M Nelson, J Chem Soc, Dalton Trans 1778-83(1975), Magnetic Cross-over in Six-co-ordinate Iron(II) Complexes of cis-1,2- Bis(diphenylphosphino)ethylene

75L023   REVIEW   F J Litterst and G M Kalvius, Investigations of Non-crystalline Materials and Liquid Crystals by the Mossbauer Effect, in "Int Conf Mossbauer Spectrosc, Proc," Vol 2, pp 189-220   (See 75H038)

75L024   FE-57   R A Lebedev, Yu D Perfil'ev, L A Kulikov, M I Afanasov, A M Babeshkin, and A N Nesmeyanov, Gamma-resonance Spectroscopy Investigation of the Chemical Subsequences of the Isomeric Transition, Electron Capture, Beta-minus-decay in Solid Substances, in "5th Int Conf Mossbauer Spectrosc, Proc," Vol 3, pp 537-45   (See 75H017)   (In Russian)

75L024   I-129   R A Lebedev, Yu D Perfil'ev, L A Kulikov, M I Afanasov, A M Babeshkin, and A N Nesmeyanov, Gamma-resonance Spectroscopy Investigation of the Chemical Subsequences of the Isomeric Transition, Electron Capture, Beta-minus-decay in Solid Substances, in "5th Int Conf Mossbauer Spectrosc, Proc," Vol 3, pp 537-45   (See 75H017)   (In Russian)

75L024   SB-121   R A Lebedev, Yu D Perfil'ev, L A Kulikov, M I Afanasov, A M Babeshkin, and A N Nesmeyanov, Gamma-resonance Spectroscopy Investigation of the Chemical Consequences of the Isomeric Transition, Electron Capture, Beta-minus-decay in Solid Substances, in "5th Int Conf Mossbauer Spectrosc, Proc," Vol 3, pp 537-45   (See 75H017)   (In Russian)

75L024   TE-125   R A Lebedev, Yu D Perfil'ev, L A Kulikov, M I Afanasov, A M Babeshkin, and A N Nesmeyanov, Gamma-resonance Spectroscopy Investigation of the Chemical Subsequences of the Isomeric Transition, Electron Capture, Beta-minus-decay in Solid Substances, in "5th Int Conf Mossbauer Spectrosc, Proc," Vol 3, pp 537-45   (See 75H017)   (In Russian)

75L025   FE-57   R W Lane, J A Ibers, R B Frankel, and R H Holm, Proc Nat Acad Sci USA 72,2868-72(1975), Synthetic Analogs of Active Sites of Iron-Sulfur Proteins: Bis(o-xylyldithiolato) ferrate(III) Monoanion, a Structurally Unconstrained Model for the Rubredoxin Fe-S4 Unit

CODE	TOPIC	REFERENCE
75L026	FE-57	D H Leech and D J Machin, J Inorg Nucl Chem 37,2279-82 (1975), The Preparation and Properties and Alkali-metal Salts of Iron(II) Chloro-complexes
75L027	FE-57	L M Levinson, I S Jacobs, C Greskovich, and G H Glover, Single-phase Inhomogeneity in Ceramic Garnets, in "AIP Conference Proceedings-No 24, Magnetism and Magnetic Materials-1974" (20th Annual Conference, San Francisco), edited by C D Graham, Jr, G H Lander, and J J Rhyne (American Institute of Physics, New York, 1975), pp 489-90
75L028	INSTRM	R A Levy, P A Flinn, and R A Hartzell, Nucl Instrum Methods 131,559-62(1975), A Proportional Counter for Efficient Mossbauer Scattering Experiments
75L029	FE-57	R A Levy and J A Rayne, Spin-glass Behaviour in the $(Pd(1-x)Agx)0.99Fe0.01$ System, in "14th International Conference on Low Temperature Physics, Proceedings: Volume 3. Low Temperature Properties of Solids" (Otaniemi, Finland, 1975), edited by M Krusius and M Vuorio (North-Holland Publishing Co, Amsterdam/American Elsevier Publishing Co, New York, 1975), pp 262-5
75L030	FE-57	B Loegel, J M Friedt, and R Poinsot, J Phys F 5,L54-7 (1975), Superparamagnetic Behaviour in FexCr(1-x) Alloys $(0.20 < x < 0.30)$
75L031	FE-57	I S Lyubutin, A P Dodokin, and E N Ageeva, Zh Eksp Teor Fiz 68,1363-7(1975)/Sov Phys-JETP 41,678-80(1976), Anomalies in the Mossbauer Spectra of Sn119 Nuclei in the Region of the Magnetic Phase Transition in Yttrium Iron Garnets
75L031	SN-119	I S Lyubutin, A P Dodokin, and E N Ageeva, Zh Eksp Teor Fiz 68,1363-7(1975)/Sov Phys-JETP 41,678-80(1976), Anomalies in the Mossbauer Spectra of Sn119 Nuclei in the Region of the Magnetic Phase Transition in Yttrium Iron Garnets
75M001	FE-57	C T Meyer, Y Gros, and H Vincent, Solid State Commun 16, 625-9(1975), Etude par Effet Mossbauer des Proprietes Cristallographiques et Magnetiques de Fe2GeS4
75M002	FE-57	A H Morrish and P E Clark, Phys Rev 11,278-86(1975), High-field Mossbauer Study of Manganese-Zinc Ferrites
75M003	SN-119	J P Motte and N N Greenwood, J Solid State Chem 13,41-8 (1975), Etude par Effect Mossbauer de la Structure et des Proprietes de Diffusion de la Phase Antifluorine Nonstoechimetrique: Li8SnP4
75M004	FE-57	P B Merrithew, C C Lo, and A J Modestino, Inorg Chem 14, 242-7(1975), t2g (pi) Electron Distribution in Some Low-spin Mixed-ligand Complexes of Iron(III)
75M005	FE-57	E V Mielczarek and W P Winfree, Phys Rev B 11,1026-9(1975), Isomer Shift in TiFe and a Calibration of the 57Fe Isomer Shift
75M006	FE-57	S Mitra and C Bansal, Chem Phys Lett 30,403-5(1975), 57Fe Mossbauer Study of Volcanic Hornblende
75M007	THEORY	A V Mitin and G P Chugunova, Phys Status Solidi A 28,39-48 (1975), Polarization Detection Methods of Double Gamma-ray Nuclear Resonances
75M008	FE-57	P A Montano, Z Shanfield, and P H Barrett, Phys Rev B 11, 3302-4(1975), Nuclear Orientation and Mossbauer Studies of Alloys of CoSi-FeSi
75M009	ANALYS	S Morup and E Both, Nucl Instrum Methods 124,445-8(1975), Interpretation of Mossbauer Spectra with Broadened Lines

CODE	TOPIC	REFERENCE

75M010   SN-119   K Matsui, R R Hasiguti, T Shoji, and A Ohkawa, Tin-vacancy
Interaction in Silicon Monitored by 119mSn
Mossbauer Probe, in "Lattice Defects in Semiconductors-
1974: Institute of Physics Conference Series-No 23"
(Freiburg, 1975), edited by F A Huntley (The Institute of
Physics, Bristol, 1975), pp 572-8

75M011   FE-57   R C Mercader and T E Cranshaw, J Phys F 5,L124-7(1975),
The Mossbauer Spectrum of Strained Iron and the Quadrupole
Coupling in Alpha Iron

75M012   FE-57   C T Meyer and M Pineri, J Polym Sci 13,1057-61(1975),
Crosslinking of Elastomers by Coordination Bonding. Study
of Clustering of the Coordination Complexes by Mossbauer
Spectroscopy

75M013   EU-151   M Meyer, J M Friedt, L Iannarella, and J Danon, Solid State
Commun 17,585-7(1975), Mossbauer Study of 151Eu in Hydrogen-loaded Palladium

75M014   FE-57   A Migliori, R J Bartlett, and R D Taylor, IEEE Trans Magn
Mag-11,347-9(1975), Magnetic Field Microprobe in Supercon-
ductors Carrying a Current

75M015   FE-57   W H Morrison, Jr, E Y Ho, and D N Hendrickson, Inorg Chem
14,500-6(1975), Reaction of Ferrocene with Polyaromatic
Molecules. pi-Arenebis(pi-cyclopentadienyliron) Dica-
tions. II. Electronic Structure

75M016   ANALYS   T Mukoyama, Nucl Instrum Methods 126,153-4(1975), Fitting
of Lorentzian to Mossbauer Spectra by Non-iterative Method

75M017   FE-57   A V Morornyi, B A Tallershik, S P Teslenko, and B A
Shustrov, Fiz Tverd Tela (Leningrad) 17,324-5(1975)/Sov
Phys-Solid State 17,198-9(1975), Magnetic and Electric
Phase Transitions in V2O3-Ti2O3 Solid Solutions

75M018   FE-57   Yu V Maksimov, I P Suzdalev, V I Gol'danskii, O V Krylov,
L Ya Margolis, and A E Nechitailo, Chem Phys Lett 34,172-4
(1975), Observation of Mossbauer Spectra of Heterogeneous
Catalysts during the Catalytic Reaction

75M019   FE-57   K Melzer and C Michalk, Detection of alpha-Fe2O3 Precipita-
tions in a Ca-V-Fe Garnet, in "5th Int Conf Mossbauer
Spec, Proc," Part 1, pp 94-6 (See 75H017)

75M020   FE-57   M Macrin, P Telnic, A Gelberg, and G Filoti, Mossbauer
Studies of the Barium Hexagonal Ferrite, in "5th Int Conf
Mossbauer Spec, Proc," Part 1, pp 133-6 (See 75H017)

75M021   REVIEW   W Meisel, Some Analytical Aspects of the Mossbauer Spectrometry, in "5th Int Conf Mossbauer Spec, Proc," Part 1,
pp 200-13 (See 75H017)

75M022   FE-57   V A Makarov, I M Puzei, and T V Sakharova, Phys Status
Solidi A 30,K21-4(1975), Magnetic Moment Distribution
of Fe Atoms in Ferromagnetic F.C.C. Fe-Ni Alloys

75M023   FE-57   R N Mullinger, R Cammack, K K K Rao, D O Hall, D P E
Dickson, C E Johnson, J D Rush, and A Simopoulos, Biochem
J 151,75-83(1975), Physicochemical Characterization of
Four-iron-Four-sulphide Ferredoxin from Bacillus
stearothermophilus

75M024   FE-57   W H Morrison, Jr and D N Hendrickson, Inorg Chem 14,2331-
46(1975), Electron Transfer in Oxidized Biferrocene, Bifer-
rocenylene, and (1.1) Ferrocenophane Systems

75M025   FE-57   Yu V Maksimov, I P Suzdalev, R A Arents, and M Ya
Kushnerev, Mossbauer and X-ray Investigation of the Thermal
Decomposition of the Iron Eta-nitride, in "5th Int Conf
Mossbauer Spec, Proc," Part 2, pp 332-9 (See 75H017)
(In Russian)

CODE	TOPIC	REFERENCE

75M026   FE-57   T V Malysheva, V D Shvagerev, and A N Yermakov, Camacite and Taenite Ratio in the Iron Meteorite Santa Catharina, in "5th Int Conf Mossbauer Spec, Proc," Part 2, pp 385-8 (See 75H017)

75M027   FE-57   W Manzel, G Vogl, and W Koch, Trapping of Low-temperature Irradiation Produced Interstitial Atoms at Co57 in Al, in "5th Int Conf Mossbauer Spec, Proc," Part 2, pp 455-8 (See 75H017)

75M028   FE-57   S Morup, Magnetic Field Dependence of the Spin-spin Relaxation Time of Fe(NO3)3-9H2O, in "5th Int Conf Mossbauer Spec, Proc," Part 2, pp 472-3 (See 75H017)

75M029   TE-125   G R Mackay, C Blaauw, and W Leiper, J Phys F 5,L166-70 (1975), The Hyperfine Field at Tellurium Impurity Sites in the Heusler Alloy Pd2MnSb

75M030   THEORY   G Meissner and K Binder, Phys Rev B 12,3948-55(1975), Debye-Waller Factor, Compressibility Sum Rule, and Central Peak at Structural Phase Transitions

75M031   FE-57   A Malliaris and A Simopoulos, J Chem Phys 63,595-6(1975), Magnetic Ordering in Solid Solutions of Bis(N,N-diethyl-dithiocarbamato)Fe+3 Halogenides

75M032   FE-57   A Malliaris and D Petridis, Chem Phys Lett 36,117-9(1975), Study of Hyperfine Interactions by Mossbauer Spectroscopy in Mixed Crystals

75M033   FE-57   Yu V Maksimov, I P Suzdalev, R A Arents, E F Makarov, M Ya Kushnerev, and L D Kuznetsov, Dokl Akad Nauk SSSR 222, 392-5(1975)/Dokl Phys Chem 222,483-5(1975), Investigation by Gamma-resonance Spectroscopy of the Interstitial Phase Arising on Defect Sections of a Promoted Iron Ammonia-synthesis Catalyst

75M034   FE-57   P G Manning and M J Tricker, Can Mineral 13,259-65(1975), Optical-absorption and Mossbauer Spectral Studies of Iron and Titanium Site-populations in Vesuvianites

75M035   FE-57   C Max, G Le Caer, and B Roques, J Solid State Chem 14,172-80(1975), Etude des Proprietes Magnetiques de la Variete Monoclinique du Moncgermaniure de Fer

75M036   FE-57   R I Mints and V A Semenkin, Ukr Fiz Zh 20,594-7(1975), Mossbauer Study of Deformed Martensite in an Iron-Nickel-Chromium-Carbon Alloy (In Russian)

75M037   SN-119   A Minkova, B Slavov, and Cv Bontschev, On the Application of the Diffusion Theory in the Depth-selective Mossbauer Spectroscopy, in "Int Conf Mossbauer Spectrosc, Proc," Vol 1, pp 45-6 (See 75H027)

75M037   THEORY   A Minkova, B Slavov, and Cv Bontschev, On the Application of the Diffusion Theory in the Depth-selective Mossbauer Spectroscopy, in "Int Conf Mossbauer Spectrosc, Proc," Vol 1, pp 45-6 (See 75H027)

75M038   TE-125   G R Mackay, C Blaauw, and W Leiper, The Hyperfine Field at Tellurium Impurity Sites in the Heusler Alloys Pd2MnSb125, in "Int Conf Mossbauer Spectrosc, Proc," Vol 1, pp 89-90 (See 75H027)

75M039   THEORY   A S Moskvin, N S Ovanesyan, and V A Trukhtanov, Angular Dependence of Supertransferred Hyperfine Interaction, in "Int Conf Mossbauer Spectrosc, Proc," Vol 1, pp 145-6 (See 75H027)

75M040   FE-57   W Mansel and G Vogl, Application of Mossbauer Spectroscopy to Defect Kinetics in Metals, in "Int Conf Mossbauer Spectrosc," Vol 1, pp 219-20 (See 75H027)

CODE	TOPIC	REFERENCE
75M041	FE-57	W Mansel and G Vogl, Direct Observation of Interstitial Atom Trapping at Co57-impurities in Silver, in "Int Conf Mossbauer Spectrosc, Proc," Vol 1, pp 221-2 (See 75H027)
75M042	EU-151	H Maletta and G Crecelius, Lattice Defects and Its Influence on Hyperfine Interactions on EuS Evaporated Thin Films, in "Int Conf Mossbauer Spectrosc, Proc," Vol 1, pp 235-6 (See 75H027)
75M043	EU-151	H Maletta and G Crecelius, Appl Phys 8,241-4(1975), Lattice Defects and Its Influence on Hyperfine Interactions on EuS Evaporated Thin Films
75M044	FE-57	K Melzer, G Dehe, B Seidel, and C Michalk, Hermsdorfer Tech Mitt 41,1296-300(1975), Der Mossbauereffekt und seine Anwendung zur Bestimmung der Kationen-verteilung in Ferriten
75M045	FE-57	J Mada and S Iida, J Phys Soc Jpn 39,1627-8(1975), Detailed Mossbauer Hyperfine Parameters of Low Temperature Phase of Fe3O4 (Addendum: personal communication)
75M046	SN-119	F P Mullins and C Curran, Can J Chem 53,3200-5(1975), Mossbauer and Related Studies cf Complexes of R2Sn(NCS)2 with Neutral Ligands Containing Oxygen Donor Atoms
75M047	FE-57	J F Myers, G W Rayner Canham, and A B P Lever, Inorg Chem 14,461-8(1975), Higher Oxidation Level Phthalocyanine Complexes of Chromium, Iron, Cobalt, and Zinc. Phthalocyanine Radical Species
75M048	TE-125	H Micklitz, P H Barrett, and P A Montano, Mossbauer Studies of Rare-gas Matrix Isolated 125Te Compounds, in "Int Conf Mossbauer Spectrosc, Proc," Vol 1, pp 241-2 (See 75H027)
75M049	FE-57	A Malliaris and C Niarchos, Mossbauer Studies of Paramagnetic Behavior of Disolved Bis(N, N-dialkyldithiocarbamato)Fe Halides, in "Int Conf Mossbauer Spectrosc, Proc," Vol 1, pp 253-4 (See 75H027)
75M050	FE-57	A Malliaris and D Petridis, Study of Magnetic Hyperfine Interactions by Mossbauer Spectroscopy in Mixed Crystals, in "Int Conf Mossbauer Spectrosc, Proc," Vol 1, pp 255-6 (See 75H027)
75M051	FE-57	J P Mathieu and J Chappert, Mossbauer Spectroscopy of Iron(II) Bisacetylacetonate Dihydrate, in "Int Conf Mossbauer Spectrosc, Proc," Vol 1, pp 267-8 (See 75H027)
75M052	FE-57	Y Maeda, H E Marcolin, A Trautwein, P Wende, and F Seel, Mossbauer Study of the Polymeric System Fe-imidazole II. Substituted Fe-imidazole, in "Int Conf Mossbauer Spectrosc, Proc," Vol 1, pp 273-4 (See 75H027)
75M053	FE-57	W Meisel, H J Jacobasch, H Mehner, and V Jacopian, The Bonding of Iron to Synthetic Fibres and Cellulose, in "Int Conf Mossbauer Spectrosc, Proc," Vol 1, pp 275-6 (See 75H027)
75M054	EU-153	D Mihaila-Tarabasanu, U Wagner, F E Wagner, and G M Kalvius, Production and Stability of 153Eu Species Observed after the beta- Decay of 153Sm in Anhydrous Rare Earth Chlorides, in "Int Conf Mossbauer Spectrosc, Proc," Vol 1, pp 291-2 (See 75H027)
75M055	FE-57	E F Makarov, A S Plachinda, and S I Alekseeva, Study of Spatial Distribution of Fe3+ Ions in Frozen Solutions and in Ion-exchangers, in "Int Conf Mossbauer Spectrosc, Proc," Vol 1, pp 297-8 (See 75H027) (In Russian)

CODE	TOPIC	REFERENCE

75M056   FE-57   Yu V Maksimov, I P Suzdalev, A I Matveev, E F Makarov, and A V Kravtsov, On Distribution of Cations in Structures of Catalytically Active Imperfect and Alloyed Magnetites, in "Int Conf Mossbauer Spectrosc, Proc," Vol 1, pp 307-8 (See 75H027)

75M057   FE-57   Yu V Maksimov, I P Suzdalev, V I Gol'danskii, A I Matveev, and E F Makarov, Investigation of the Surface State of the Iron-molybdenum Catalyst during the Reaction of Partial Methanol Oxidation, in "Int Conf Mossbauer Spectrosc, Proc," Vol 1, pp 309-10 (See 75H027)

75M058   FE-57   S Morup, M K Nielsen, J M Knudsen, and G Trumpy, Mossbauer Studies of Amorphous and Crystalline States in Frozen Aqueous Solutions of FeCl3, in "Int Conf Mossbauer Spectrosc, Proc," Vol 1, pp 353-4 (See 75H027)

75M059   FE-57   Y Maeda, T Harami, A Trautwein, and U Gonser, EFG Tensor in Deoxymyoglobin Single Crystals, in "Int Conf Mossbauer Spectrosc, Proc," Vol 1, pp 435-6 (See 75H027)

75M060   FE-57   K Mahesh, Mossbauer Study of Medicinal Calcinated Iron, in "Int Conf Mossbauer Spectrosc, Proc," Vol 1, pp 449-51 (See 75H027)

75M061   FE-57   K Mahesh, Phys Status Solidi B 71, K177-80 (1975), Effect of Pressure on the Recoilless Fraction of Fe57 in Vanadium

75M062   FE-57   T V Malysheva and G A Kazakov, Admixture Atoms in Natural Minerals, in "Int Conf Mossbauer Spectrosc, Proc," Vol 1, pp 475-6 (See 75H027)

75M063   THEORY   K V Makaryunas, The Problem of Scale Calibration in Mossbauer Isomer Shift and Chemical Influence on Electron Capture, in "Int Conf Mossbauer Spectrosc, Proc," Vol 1, pp 495-6 (See 75H027) (In Russian)

75M064   REVIEW   E F Makarov and R A Stukan, Emission Gamma-resonance Spectroscopy, in "Int Conf Mossbauer Spectrosc, Proc," Vol 2, pp 133-61 (See 75H038)

75M065   I-127   K V Makaryunas, E K Makaryuene, A K Dragunas, and M L Bal'chyuene, Electric Field Gradients at the Nuclei of Tellurium and Impurity Iodine in the A(1)B(2)Te2 and A(2)B(3)Te4 Crystals, in "5th Int Conf Mossbauer Spectrosc, Proc," Vol 3, pp 529-30 (See 75H017) (In Russian)

75M065   TE-125   K V Makaryunas, E K Makaryuene, A K Dragunas, and M L Bal'chyuene, Electric Field Gradients at the Nuclei of Tellurium and Impurity Iodine in the A(1)B(2)Te2 and A(2)B(3)Te4 Crystals, in "5th Int Conf Mossbauer Spectrosc, Proc," Vol 3, pp 529-30 (See 75H017) (In Russian)

75M066   I-127   L I Molkanov, I M Band, M B Trzhaskovskaya, Yu S Grushko, and A V Oleynik, Electron Valence Configurations of Iodine from the Chemical Shift of the X-ray Emission Lines and Mossbauer Effect, in "5th Int Conf Mossbauer Spectrosc, Proc," Vol 3 pp 546-9 (See 75H017) (In Russian)

75M067   THEORY   A V Mitin, Influence of the Alternating Magnetic Field on the Polarization of the Gamma-radiation, in "5th Int Conf Mossbauer Spectrosc, Proc," Vol 3 pp 615-8 (See 75H017) (In Russian)

75M068   REVIEW   B Manuschev, Technological Applications of the Mossbauer Effect, in "5th Int Conf Mossbauer Spectrosc, Proc," Vol 3, pp 687-98 (See 75H017) (In Russian)

75M069   FE-57   Y Maeda, Y Sasaki, and Y Takashima, Inorg Chim Acta 13, 141-4 (1975), Spectral Properties of Some 4-substituted Pyridine N-oxide Complexes with Iron Perchlorate

CODE	TOPIC	REFERENCE
75M070	FE-57	K Mahesh and D C Gupta, Lett Nuovo Cimento Soc Ital Fis 13, 561-2(1975), Evidence of the Force Constant Change with Pressure on the 57Fe:Cu System
75M071	FE-57	Yu V Maksimov, I P Suzdalev, V I Gol'danskii, O V Krylov, L Ya Margolis, and A E Nechitailo, Dokl Akad Nauk SSSR 221, 880-3(1975)/Dokl Phys Chem 221,332-4(1975), Investigation by Gamma Resonance Spectroscopy of the Process of Charge Transfer by an Fe-Co-Mo Catalyst during the Catalytic Reaction of Mild Oxidation of Propylene
75M072	FE-57	Yu V Maksimov, I P Suzdalev, A I Matveev, E F Makarov, and A V Kravtsov, Izv Akad Nauk SSSR, Ser Khim 2665-70(1975)/ Bull Acad Sci USSR, Div Chem Sci, Study of the Structural Modification of Oxide Catalysts in the Synthesis of Organic Compounds from Carbon Monoxide and Water by Gamma-resonance Spectroscopy
75M073	EU-151	H Maletta and G K Shenoy, Low Temperatrue Mossbauer Thermometry, in "14th International Conference on Low Temperature Physics, Proceedings: Volume 4. Techniques and Topics" (Otaniemi, Finland, 1975), edited by M Krusius and M Vuorio (North-Holland Publishing Co, Amsterdam/American Elsevier Publishing Co, New York, 1975), pp 68-71
75M074	GENRAL	T V Malysheva, Mossbauer Effect in Geochemistry and Cosmo-Chemistry ("Nauka," Moscow, 1975), 166 pages (In Russian)
75M075	SN-119	R A Manapov, I Ya Kuramshin, A A Muratova, and A N Pudovik, Zh Obshch Khim 45,1975-9(1975)/J Gen Chem (USSR) 45,1940-3(1975), Mossbauer Spectra of Complexes of Organo-phosphorus Compounds with Tin Halides
75M076	NP-237	K Matsui, A Ohkawa, and T Shoji, Kakuriken Kenkyu Hokoku 7,141-54(1975)8 Annealing Effects in Recoil-damaged Uranium Dioxide Studied by Np237 59.54 keV Mossbauer Effect
75M077	FE-57	N M Matveeva, A A Kolyada, R N Kuz'min, and I M Sharshakov, Metal Term Obrab Met (10),15-9(1975)/Met Sci Heat Treat Met (USSR) 830-2(1975), Structural Transformations in Fe + 30 at.% Ni
75M078	FE-57	F Menil, M Pezat, and B Tanguy, C R Acad Sci (Paris), Ser C 281,849-52(1975), Etude par Effet Mossbauer du Fluoroti-trure de Fer Fe4N3F3
75M079	FE-57	P A Montano, A Study of the Magnetic Behavior of Iron Impurities in the Linear Antiferromagnet CsNiCl3, in "AIP Conference Proceedings-No 24, Magnetism and Magnetic Materials-1974" (20th Annual Conference, San Francisco), edited by C D Graham, Jr, G H Lander, and J J Rhyne (American Institute of Physics, New York, 1975), pp 359-60
75M080	FE-57	H Mosbaek, Acta Chem Scand, Ser A 29,957-8(1975), A Mossbauer Investigation of Some Iron Carbonyl Compounds
75M081	FE-57	A S Moskvin, N S Ovanesyan, and V A Trukhtanov, Hyperfine Interac 1,265-81(1975), Angular Dependence of the Super-exchange Interaction Fe3+-O2-Cr3+
75M082	FE-57	L N Mulay and G H Ziegenfuss, Exchange Interactions in Isolated Trinuclear Clusters of Fe3+: Magnetic and Moss-bauer Studies, in "AIP Conference Proceedings-No 24, Magnetism and Magnetic Materials-1974" (20th Annual Conference, San Francisco), edited by C D Graham, Jr, G H Lander, and J J Rhyne (American Institute of Physics, New York, 1975), pp 213-4
75M083	FE-57	E Munck, H Rhodes, W H Orme-Johnson, L C Davis, W J Brill, and V K Shah, Biochim Biophys Acta 400,32-53(1975), Nitrogenase. VIII. Mossbauer and EPR Spectroscopy. The MoFe Protein Component from Azotobacter Vinelandii Op

CODE	TOPIC	REFERENCE
75M084	FE-57	W U Malik and K D Sharma, Curr Sci 44,661(1975), Evidence for Existence of +2 Oxidation State of Iron in KFeCr(CN)6 by Mossbauer Spectroscopy
75M085	FE-57	A Minkova and J P Schunck, Dokl Bolg Akad Nauk 28,1171-3 (1975), Study of Superfine gamma-FeOOH Layers by Means of Mossbauer Effect
75M086	FE-57	E Munck and P M Champion, Ann N Y Acad Sci 244,142-62 (1975), Heme Proteins and Model Compounds: Mossbauer Absorption and Emission Spectroscopy
75M086	REVIEW	E Munck and P M Champion, Ann N Y Acad Sci 244,142-62 (1975), Heme Proteins and Model Compounds: Mossbauer Absorption and Emission Spectroscopy
75M087	FE-57	E I Mal'tsev, V I Goman'kov, I M Puzei, V A Makarov, and E V Kozis, Fiz Met Metalloved 39,543-52(1975)/Phys Met Metallogr (USSR), Neutron Diffraction and Mossbauer Investigations of the Ordering and Decomposition in Iron-Cobalt-Nickel Alloys
75M088	SN-119	K P Mitrofanov, L P Benderskaya, S I Reiman, and V I Kongauz, Vestn Mosk Univ, Fiz 30,487-9(1975)/ Moscow Univ Phys Bull 30,82-4(1975), The Mossbauer Effect on Sn119 Nuclei in a System of mMgSnO4-(1-m)-Mg3(BO3)2 Solid Solutions
75N001	FE-57	D L Nagy, I Dezsi, and U Gonser, Neues Jahrb Mineral Monatsh 3,101-14(1975), Mossbauer Studies of FeCO3 (Siderite)
75N002	FE-57	D L Nagy, K Kulcsar, G Ritter, H Spiering, H Vogel, R Zimmermann, I Dezsi, and M Pardavi-Horvath, J Phys Chem Solids 36,759-67(1975), Magnetic Field Induced Texture in Mossbauer Absorbers
75N003	FE-57	Y Nishihara, J Phys Soc Jpn 38,710-7(1975), Effect of Nearest Neighbor Ions on the Hyperfine Fields at 57Fe Nuclei in TbFe(1-x)Cr(x)O3
75N004	SN-119	D L Nagy, G J Zimmer, T Lohner, J P Senateur, and I Bibicu, Mossbauer Study of the Magnetic Phase Transformations in SnMn3N, in "5th Int Conf Mossbauer Spec, Proc," Part 1, pp 75-8 (See 75H017)
75N005	FE-57	C I Nistor, C Boekema, F Van der Woude, and G A Sawatzky, A Mossbauer Study of Doped Magnetite, in "5th Int Conf Mossbauer Spec, Proc," Part 1, pp 99-103 (See 75H017)
75N006	THEORY	C I Nistor and T Beica, The Multiplicity of the Quadrupole Interaction in the Mixed Spinelic Ferrites Fe(Me(x)Fe(2-x))O4, in "5th Int Conf Mossbauer Spec, Proc," Part 1, pp 104-6 (See 75H017)
75N007	SN-119	F S Nasredinov, B T Melekh, L N Vasil'ev, and L N Seregina, Fiz Tverd Tela (Leningrad) 17,633-5(1975)/ Sov Phys-Solid State 17,413(1975), Crystal-glass Transition in Ge(0.15)Te(0.85) and Its Influence on the Local Environment of Germanium Atoms
75N007	TE-125	F S Nasredinov, B T Melekh, L N Vasil'ev, and L N Seregina, Fiz Tverd Tela (Leningrad) 17,633-5(1975)/ Sov Phys-Solid State 17,413(1975), Crystal-glass Transition in Ge(0.15)Te(0.85) and Its Influence on the Local Environment of Germanium Atoms

CODE	TOPIC	REFERENCE
75N008	SN-119	I N Nikolaev, A P Shotov, A F Volkov, and V P Mar'in, Pis'ma Zh Eksp Teor Fiz 21,144-7(1975)/JETP Lett 21,65-6 (1975), "Softening" of Phonon Spectrum in Semiconductors of the $Pb_{(1-x)}Sn_{(x)}Te$ System on Going to the Gapless State
75N009	FE-57	V I Nikolaev and V S Rusakov, Fiz Tverd Tela (Leningrad) 17,326-7(1975)/Sov Phys-Solid State 17,200-1(1975), "Magnetic Anomalies" of the Parameters of the Mossbauer Spectra of the Nuclei Fe57 and Sn119 in Antiferromagnet FeSn2
75N009	SN-119	V I Nikolaev and V S Rusakov, Fiz Tverd Tela (Leningrad) 17,326-7(1975)/Sov Phys-Solid State 17,200-1(1975), "Magnetic Anomalies" of the Parameters of the Mossbauer Spectra of the Nuclei Fe57 and Sn119 in Antiferromagnet FeSn2
75N010	FE-57	S Nasu and U Gonser, 57Fe in Al, in "5th Int Conf Mossbauer Spec, Proc," Part 2, pp 311-3 (See 75H017)
75N011	FE-57	I N Nikolaev, L S Pavlyukov, and V P Mar'in, Fiz Tverd Tela (Leningrad) 17,1548-50(1975)/Sov Phys-Solid State 17,1016-7 (1975), Effect of Pressure on Effective Magnetic Fields and Isomer Shifts at Fe57 Nuclei in Yttrium Iron Garnet
75N012	FE-57	V I Nikolaev, V S Rusakov, and S S Yakimov, Fiz Tverd Tela (Leningrad) 17,1405-7(1975)/Sov Phys-Solid State 17,903-4 (1975), Crystal-lattice Dynamics and Shift of the Mossbauer Line in Variable-composition Phases
75N013	SN-119	I V Nistiryuk and P P Seregin, Fiz Tverd Tela (Leningrad) 17,1192-4(1975)/Sov Phys-Solid State 17,768-9(1975), State of Tin Impurity Atoms in Silicon
75N014	ANALYS	C I Nistor, Several Remarks Concerning the Fitting of Mossbauer Spectra, in "5th Int Conf Mossbauer Spec, Proc," Part 3, pp 637-9 (See 75H017)
75N015	FE-57	T Notermann, G W Keulks, A Skliarov, Yu Maximov, L Ya Margolis, and O V Krylov, J Catal 39,286-93(1975), The Physicochemical Properties of the Bismuth Iron Molybdate System
75N016	GD-155	D J Newman and D C Price, J Phys C 8,2985-91(1975), Determination of the Electrostatic Contributions to Lanthanide Quadrupolar Crystal Fields
75N017	FE-57	H P Nissen and K Nagorny, Z Phys Chem (Frankfurt am Main) 95,301-4(1975), Mossbauerspektroskopische und magnetische Untersuchungen an Alkalithioferraten(III)
75N018	REVIEW	S Nasu and U Gonser, Zairyo 24,1083-91(1975), Mossbauer Effect and Its Applications in Material Science (In Japanese)
75N019	ANALYS	D L Nagy, I Dezsi, and K Kulcsar, Can We Prepare Texture-free Mossbauer Samples? in "Int Conf Mossbauer Spectrosc, Proc," Vol 1, pp 25-6 (See 75H027)
75N020	FE-57	Y Nakamura and R Tahara, Mossbauer Effect in Ordered fcc Fe3Ge, in "Int Conf Mossbauer Spectrosc, Proc," Vol 1, pp 173-4 (See 75H027)
75N021	DY-161	L Niesen, H P Wit, P J Kikkert, and H De Waard, Lattice Location of Rare-earth Impurities Implanted in Ferromagnetic Metals Derived from Mossbauer Experiments, in "Int Conf Mossbauer Spectrosc, Proc," Vol 1, pp 207-8 (See 75H027)

CODE	TOPIC	REFERENCE

75N021   ER-166   L Niesen, H P Wit, P J Kikkert, and H De Waard, Lattice Location of Rare-earth Impurities Implanted in Ferromagnetic Metals Derived from Mossbauer Experiments, in "Int Conf Mossbauer Spectrosc, Proc," Vol 1, pp 207-8 (See 75H027)

75N021   TM-169   L Niesen, H P Wit, P J Kikkert, and H De Waard, Lattice Location of Rare-earth Impurities Implanted in Ferromagnetic Metals Derived from Mossbauer Experiments, in "Int Conf Mossbauer Spectrosc, Proc," Vol 1, pp 207-8 (See 75H027)

75N022   FE-57   J A Nasser and F Varret, Shear Strains and Magnetic Structure in FeCl2 under Magnetic Field, H, Perpendicular to the Ternary Axis, c, of the Crystal. Theoretical Model and Mossbauer Experiments, in "AIP Conference Proceedings-No 24-Magnetism and Magnetic Materials-1974" (20th Annual Conference, San Francisco), edited by C D Graham, Jr, G H Lander, and J J Rhyne (American Institute of Physics, New York, 1975), pp 59-60

75N023   FE-57   G V Novikov, V K Egorov, V I Popov, and L B Sipavina, Alpha Transformation in Iron Sulfide $Fe(1-x)S$, in "Int Conf Mossbauer Spectosc, Proc," Vol 1, pp 391 (See 75H027)

75N024   FE-57   G V Novikov, V K Egorov, V I Popov, and V M Polovov, Low Temperature Magnetic Transition in Iron Sulfides $Fe(1-x)S$, in "Int Conf Mossbauer Spectrosc, Proc," Vol 1, pp 393 (See 75H027)

75N025   ANALYS   S Nagy, B Levay, and A Vertes, Acta Chim Acad Sci Hung 85, 273-88(1975), Investigations on the Correlation Between the Optimum Thickness of Mossbauer Absorbents and Certain Experimental Parameters

75N026   REVIEW   I Nowik, Magnetic Structure and Transferred Hyperfine Interactions, in "Int Conf Mossbauer Spectrosc, Proc," Vol 2, pp 83-93 (See 75H038)

75N026   SN-119   I Nowik, Magnetic Structure and Transferred Hyperfine Interactions, in "Int Conf Mossbauer Spectrosc, Proc," Vol 2, pp 83-98 (See 75H038)

75N027   FE-57   E N Nikitin, P P Seregin, V I Tarasov, and E Yu Turaev, Fiz Tverd Tela 17,2176-8(1975)/Sov Phys-Solid State 17, 1441(1976), Electronic Structure of MnSi1.7 from Mossbauer Spectroscopy Data

75N027   SN-119   E N Nikitin, P P Seregin, V I Tarasov, and E Yu Turaev, Fiz Tverd Tela 17,2176-8(1975)/Sov Phys-Solid State 17, 1441(1976), Electronic Structure of MnSi1.7 from Mossbauer Spectroscopy Data

75N028   FE-57   I N Nikolaev, L S Pavlyukov, and V P Mar'in, Fiz Tverd Tela 17,3389-91(1975)/Sov Phys-Solid State 17,2217-8(1976), Pressure-induced Shift of the Neel Temperature of FeF3

75N029   FE-57   I N Nikolaev, L S Pavlyukov, and V P Mar'in, Zh Eksp Teor Fiz 69,1844-52(1975)/Sov Phys-JETP, Effect of Pressure on the Effective Magnetic Fields at Fe57 Nuclei in the Dielectric Magnetic Substances Y3Fe5O12 and FeF3

75N030   FE-57   V I Nikolaev and V S Rusakov, Kristallografiya 20,845-7 (1975)/Sov Phys-Crystallogr 20,519-21(1976), The Nature of the "Magnetic Anomaly" in the Parameters of the Mossbauer Line in Rare Earth Metal Orthoferrites

75N031   FE-57   V I Nikolaev, V S Rosakov, and A B Anfisov, Zh Eksp Teor Fiz 68,287-94(1975)/Sov Phys-JETP 41,140-3(1975), Cluster Formations of FeSn2 in White Tin Single Crystals

CODE	TOPIC	REFERENCE

75N032   FE-57   Y Nishihara, S Ogawa, and S Waki, J Phys Soc Jpn 39,63-9 (1975), Mossbauer Study of Ni0.995Fe0.005S2 - Magnetic Structure of NiS2

75N033   FE-57   G V Novikov, V K Egorov, V I Popov, L V Sipavina, and G Yu Odinets, Geokhimiya 1776-85(1975), Investigation of Phase Transformations in Highly Ferruginous Pyrrhotites by Mossbauer Spectroscopy and Powder Diffractometry (In Russian) (Possibly translated in Geochem Int)

75N034   ANALYS   Y Nishihara, Denshi Gijutsu Sogo Kenkyujo Iho 12,13-28 (1975), Least-squares Fitting of Mossbauer Spectrum (In Japanese)

750001   SN-119   S Onaka and H Sano, Bull Chem Soc Jpn 48,258-61(1975), The Syntheses of R3Sn-Mn(CO)(5-n)L(n)(n=0 or 1) Compounds and Their 119Sn-Mossbauer and 1H-NMR Studies

750002   FE-57   I Ortalli and C Lamborizio, Lett Nuovo Cimento Soc Ital Fis 12,147-50(1975), Mossbauer Study of Magnetic Fields in Fe(2-x)Ge Compounds

750003   FE-57   N S Ovanesyan and V A Trukhtanov, Angular Dependence of the Superexchange Interactions Cr3+-O2-Fe3+ in the Ortho-chromites, in "5th Int Conf Mossbauer Spec, Proc," Part 1, pp 157-61 (See 75H017) (In Russian)

750003   SN-119   N S Ovanesyan and V A Trukhtanov, Angular Dependence of the Superexchange Interactions Cr3+-O2-Fe3+ in the Ortho-chromites, in "5th Int Conf Mossbauer Spec, Proc," Part 1, pp 157-61 (See 75H017) (In Russian)

750004   SN-119   I Ortalli and V Fano, The Mossbauer Effect in Binary Tin Chalcogenides of Tin-129, in "5th Int Conf Mossbauer Spec, Proc," Part 1, pp 263-6 (See 75H017)

750005   FE-57   K Ono, K Kimura, T Yagi, and H Inokuchi, J Chem Phys 63, 1640-2(1975), Mossbauer Study of Cytochrome c3

750006   THEORY   L G Onoprienko, Fiz Tverd Tela (Leningrad) 17,609-11 (1975)/Sov Phys-Solid State 17,394(1975), Mossbauer Effect in Uniaxial Ferromagnets with Domain Structure

750007   I-127   L W Oberley and J C Ehrhardt, J Chem Phys 63,2329-33 (1975), 127I Mossbauer Studies of Thyroid Compounds

750008   FE-57   H N Ok and C S Kim, Nuovo Cimento 28B, 138-41(1975), Mossbauer Study of Antiferromagnetic CuFeS2

750009   FE-57   T Ohtani, K Kosuge, S Kachi, and M Takano, Mater Res Bull 10,709-16(1975), Interpretation of Mossbauer Spectra of 57Fe Doped Ni(1-x)S

750010   FE-57   S Okamoto, H Sekizawa, and S I Okamoto, J Phys Chem Solids 36,591-5(1975), Hydrothermal Synthesis, Structure and Magnetic Properties of Barium Differite

750011   FE-57   A N Ozernoy, V P Novikov, V I Plotnikov, and A K Zhetbaev, The Study of Sorption of 57Co by Iron Hydrooxide in Complex-forming Conditions by NGR Method, in "Int Conf Mossbauer Spectrosc, Proc," Vol 1, pp 293-4 (See 75H027) (In Russian)

750012   FE-57   H Obelhack and F H Wittmann, Debye-Waller Factor of Colloidal Particles in Hydro- and Xerogels, in "Int Conf Mossbauer Spectrosc, Proc," Vol 1, pp 349-50 (See 75H027)

750013   REVIEW   D A O'Connor, Crystallography Using the Rayleigh Scattering of Gamma Rays, in "Int Conf Mossbauer Spectrosc, Proc," Vol 2, pp 369-77 (See 75H038)

CODE	TOPIC	REFERENCE
750014	SN-119	H J Odenthal, T Kruck, and K Ehlert, Z Naturforsch, Teil B 30,696-8(1975), Metallkomplexe mit anionischen Liganden von Elementen der 4. Hauptgruppe, X(1). Diskussion der Bindungsverhaltnisse in Trihalogenstannidometallat(0)-Komplexen anhand der Ergebnisse von 119mSn-Mossbauer-Untersuchungen
750015	TE-125	A A Opalenko, I A Avenarius, R P Vardapetyan, and R N Kuz'min, Phys Status Solidi B 72,K125-30(1975), The Anisotropy of Atomic Vibrations in Te and TeO2 Crystals
750016	FE-57	L G Onoprienko, Fiz Met Metalloved 39,751-6(1975)/Phys Met Metallogr (USSR), Mossbauer Effect in Magnetically Uniaxial Ferromagnets
750016	THEORY	L G Onoprienko, Fiz Met Metalloved 39,751-6(1975)/Phys Met Metallogr (USSR), Mossbauer Effect in Magnetically Uniaxial Ferromagnets
75P001	FE-57	G C Papaefthymiou, B H Huynh, C S Yen, J L Groves, and C S Wu, J Chem Phys 62,2995-3001(1975), Mossbauer Studies of Fe2+ in Anhydrous Hemoglobin and Its Isolated Subunits
75P002	TE-125	A K Prabhakaran, S B Raju, and R G Mendiratta, Solid State Commun 16,407-8(1975), Mossbauer f Factor for 35.5 keV-Te125 Transition in Zinc-Blende Type Crystals
75P003	FE-57	J Piekoszewski, J Suwalski, L Dabrowski, and S Makolagwa, Phys Status Solidi A 28,K143-5(1975), Mossbauer Effect of a Singly Substituted YIG System in an Applied Magnetic Field
75P004	FE-57	J Piekoszewski, L Dabrowski, and J Suwalski, Solid State Commun 16,75-7(1975), Variation of the B-site Spin Orientation in Zn-Ni Ferrite in External Magnetic Fields Observed with Mossbauer Effect
75P005	SN-119	L Pellerito, N Bertazzi, G C Stocco, A Silvestri, and R Barbieri, Spectrochim Acta, Part A 31,303-8(1975), Infrared and Mossbauer Spectroscopic Studies on N, N'ethylenebis (salicylideneiminato) Sn(VI)hal2
75P006	EU-151	A Polaczek, Rocz Chem 49,1191-6(1975), Isomer Shift in the Mossbauer Spectrum of 151Eu in the Cubic Eu(x)WO3
75P007	FE-57	J Piekoszewski, L Dabrowski, J Suwalski, and S Makolagwa, Study of Temperature Dependence of Spin Order in the Si-substituted YIG System by Mossbauer Effect and Magnetization Measurements, in "5th Int Conf Mossbauer Spec, Proc," Part 1, pp 83-6 (See 75H017)
75P008	SN-119	J Pebler, Study of Phase Transitions in Cd2Nb(2-2x)Sn(2x)O(7-2x)F(2x), in "5th Int Conf Mossbauer Spec, Proc," Part 1, pp 145-51 (See 75H017)
75P009	FE-57	T M Peev, L S Bozadhiev, and M V Kostova, Mossbauer Investigation of Crystallized and Amorphous Sodium-Iron-Silicate Melts, in "5th Int Conf Mossbauer Spec, Proc," Part 1, pp 244-7 (See 75H017) (In Russian)
75P010	FE-57	Yu D Perfil'ev, N V Gorelikova, and A M Babeshkin, Vestn Mosk Univ, Khim 30,117-8(1975)/Moscow Univ Chem Bull 30(1), 83-4(1975), Mossbauer Determination of Iron Content in Tourmalines
75P011	SB-121	F Petillon and J E Guerchais, J Inorg Nucl Chem 37,1863-70 (1975), Complexes Soufres (Partie VII) du Titane(III), de l'Antimoine(V) et (III), du Bismuth(III) et de l'Etain(IV) Etude Mossbauer

CODE	TOPIC	REFERENCE

75P011   SN-119   F Petillon and J E Guerchais, J Inorg Nucl Chem 37,1863-70 (1975), Complexes Soufres (Partie VII) du Titane(III), de l'Antimoine(V) et (III), du Bismuth(III) et de l'Etain(IV). Etude Mossbauer

75P012   FE-57   J Polakova, J Lauermannova, and T Zemcik, Mossbauer Study of Atomic Order in gamma-Ni-Fe Alloys, in "5th Int Conf Mossbauer Spec, Proc," Part 2, pp 298-301 (See 75H017)

75P013   FE-57   M Prejsa, J Cirak, and I Toth, Determination of the Influence of Metal Working on the Mossbauer Spectra and Measurement of Mechanical Stresses Beneath the Insulation Coating in the Iron Alloys, in "5th Int Conf Mossbauer Spec, Proc," Part 2, pp 340-5 (See 75H017) (In Russian)

75P014   FE-57   C Pietzsch, E Fritzsch, K Fritzsche, and H A Schneider, Mossbauer and X-ray Studies on the Existence of a Cubic Stannite Phase in Natural Ores, in "5th Int Conf Mossbauer Spec, Proc," Part 2, pp 379-84 (See 75H017)

75P015   FE-57   F Parak, W Zgorzalla, H Eicher, A Mayer, G M Kalvius, K Gersonde, M Breitenbach, and H E Schlaak, Mossbauer and EPR Spectroscopy of 57Fe in Bacterial Ferredoxin, in "5th Int Conf Mossbauer Spec, Proc," Part 2, pp 419-22 (See 75H017)

75P016   FE-57   V E Prusakov, V P Alekseev, R A Stukan, and V I Gol'danskii, Investigation of Certain Cobalt Complexes by the Method of Emission Gamma-resonance Spectroscopy, in "5th Int Conf Mossbauer Spec, Proc," Part 2, pp 429-32 (See 75H017) (In Russian)

75P017   FE-57   R K Puri and D A O'Connor, A Mossbauer Study of the Lattice Dynamics of Iron and Tin Impurities in Titanium and Palladium, in "5th Int Conf Mossbauer Spec, Proc," Part 2, pp 448-50 (See 75H017)

75P017   SN-119   R K Puri and D A O'Connor, A Mossbauer Study of the Lattice Dynamics of Iron and Tin Impurities in Titanium and Palladium, in "5th Int Conf Mossbauer Spec, Proc," Part 2, pp 448-50 (See 75H017)

75P018   I-129   M Pasternak, S Bukshpan, and T Sonnino, Solid State Commun 16,871-2(1975), Magnetic Structure Determination of NiI2 by 129I Mossbauer Effect

75P019   SN-119   P A Pella and J R DeVoe, J Radioanal Chem 25,185-8(1975), Systematic Error in Tin Ore Assay by Mossbauer Spectrometry

75P020   FE-57   T J A Popma and A M Van Diepen, Non-crystalline Y3Fe5O12 Studied by Mossbauer Effect and Magnetization, in "AIP Conf Proc No 24-Magnetism and Magnetic Materials-1974" (20th Annual Conference, San Francisco), edited by C D Graham, Jr, G H Lander, and J J Rhyne (American Institute of Physics, New York, 1975), pp 123-4

75P021   AU-197   H Prosser, F E Wagner, G Wortmann, G M Kalvius, and R Wappling, Hyperfine Interac 1,25-32(1975), Mossbauer Determination of the E2/M1 Mixing Ratio of the 77 keV Transition in 197Au and of the Sign of the Electric Field Gradient in KAu(CN)2

75P022   REVIEW   H D Pfannes and U Gonser, Texture Problems in Mossbauer Spectroscopy, in "5th Int Conf Mossbauer Spec, Proc," Part 3, pp 596-9 (See 75H017)

75P023   SN-119   R K Puri and L R Gupta, Phys Status Solidi B 70,785-92 (1975), An Estimation of Force Constant Change of Mossbauer 119Sn Impurity Nuclei in Palladium Host

75P024   FE-57   B C Parakkat, P J Ouseph, R L Vonnahme, and D H Gibson, J Inorg Nucl Chem 37,2340-2(1975), A Mossbauer Study of pi-Allyliron Carbonyl Complexes

CODE	TOPIC	REFERENCE

75P025   FE-57   E Popiel and J A Sawicki, Evaluations of ME-PAC Experiments in 57Fe, in "Int Conf Mossbauer Spectrosc, Proc," Vol 1, pp 13-4 (See 75H027)

75P025   PROPSL   E Popiel and J A Sawicki, Evaluations of ME-PAC Experiments in 57Fe, in "Int Conf Mossbauer Spectrosc, Proc," Vol 1, pp 13-4 (See 75H027)

75P026   SN-119   V N Panyushkin, L Bogner, and G Wortmann, Quadrupole Splitting of the 23,8 keV Mossbauer Line of 119Sn in Beta-Tin, in "Int Conf Mossbauer Spectrosc, Proc," Vol 1, pp 99 (See 75H027)

75P027   FE-57   J Polakova and T Zemcik, Mossbauer Study of Ni3Fe Short-range Order, in "Int Conf Mossbauer Spectrosc, Proc," Vol 1, pp 109-10 (See 75H027)

75P028   FE-57   J Piekoszewski, L Dabrowski, and J Suwalski, Analysis of Localized Canting Model Applied to Spin Structure of Zn-Ni Ferrite, in "Int Conf Mossbauer Spectrosc, Proc," Vol 1, pp 155-6 (See 75H027)

75P029   FE-57   J Piekoszewski, L Dabrowski, J Suwalski, and S Makolagwa, Mossbauer Effect of Singly Substituted YIG System in Applied Magnetic Field, in "Int Conf Mossbauer Spectrosc, Proc," Vol 1, pp 157-8 (See 75H027)

75P030   GE-73   L N Pfeiffer, R S Raghavan, C P Lichtenwalner, and A G Cullis, Phys Rev B 12,4793-804(1975), Mossbauer Effect of the 13.3-keV Transition in 73Ge

75P031   TE-125   B M Powell and P Martel, J Phys Chem Solids 36,1287-98 (1975), The Lattice Dynamics of Tellurium

75P032   FE-57   V Petrouleas, A Kostikas, and A Simopoulos, Mossbauer Study of Fine and Hyperfine Parameters of Fe2+ in KFeCl3, in "Int Conf Mossbauer Spectrosc, Proc," Vol 1, pp 249-250 (See 75H027)

75P033   FE-57   V Petrouleas, A Kostikas, A Simopoulos, and D Coucouvanis, Mossbauer Study of a Six-coordinate Fe(IV) Complex with Dithiolate Ligands, in "Int Conf Mossbauer Spectrosc, Proc," Vol 1, pp 251-2 (See 75H027)

75P034   FE-57   H Pollak and W Bruyneel, On the Iron Distribution Between Sites in Epidote, in "Int Conf Mossbauer Spectrosc, Proc," Vol 1, pp 259-60 (See 75H027)

75P035   FE-57   V G Pyl'nev, F M Chukhovskii, and I S Zheludev, The Energy Level Splitting of the Fe2+ Ion in Boracites as Derived from the Mossbauer Measurements, in "Int Conf Mossbauer Spectrosc, Proc," Vol 1, pp 265-6 (See 75H027)

75P036   FE-57   A S Plachinda, E F Makarov, E V Egorov, and S I Alekseeva, About the Dependence of Chemical State of Fe(III) Ions on Their Concentration in Ion-exchange Resin, in "Int Conf Mossbauer Spectrosc, Proc," Vol 1, pp 295-6 (See 75H027) (In Russian)

75P037   FE-57   T M Peev, Mossbauer Spectra of Iron-chromous Oxide Catalyst, in "Int Conf Mossbauer Spectrosc, Proc," Vol 1, pp 313-4 (See 75H027)

75P038   FE-57   V A Povitskii, A N Salugin, and E F Makarov, Morin Transitions in Hematite. Influence of Impurities and Defects, in "Int Conf Mossbauer Spectrosc, Proc," Vol 1, pp 377-8 (See 75H027) (In Russian)

75P039   FE-57   T M Peev, Relaxation Mossbauer Spectra of Iron Anionic Complexes, Sorbed in Cation Exchange Resin, in "Int Conf Mossbauer Spectrosc, Proc," Vol 1, pp 421-2 (See 75H027)

CODE	TOPIC	REFERENCE
75P040	FE-57	H Pollak and C Herinckx, Relation Between Pleochroism, Conductivity and Mossbauer Effect in Iron Oxygen O(h) Coordination, in "Int Conf Mossbauer Spectrosc, Proc," Vol 1, pp 425-6 (See 75H027)
75P041	FE-57	H Pollak and W Bruyneel, Fast Electron Hopping in Tourmaline, in "Int Conf Mossbauer Spectrosc, Proc," Vol 1, pp 427-8 (See 75H027)
75P042	FE-57	S Papp, P Kvintovics, F Nagy, and A Vertes, Magy Kem Foly 81,211-6(1975), Mossbauer, Infrared, and NMR Spectroscopic Investigation on Solutions of Some Alkali and Phosphonium Cyanoferrates
75P043	FE-57	R S Preston and B J Zabransky, Phys Lett 55A,179-80 (1975), Use of 57Mn in Mossbauer Radiation Damage Studies
75P044	FE-57	V Petrouleas, A Simopoulos, and A Kostikas, Phys Rev B 12, 4675-81(1975), Mossbauer Study of KFeCl3. II. Relaxation Effects and One-dimensional Ordering
75P045	FE-57	V Petrouleas, A Kostikas, and A Simopoulos, Phys Rev B 12, 4666-74(1975), Mossbauer Study of KFeCl3. I. Fine and Hyperfine Structure of Fe2+
75P046	FE-57	J Pebler and K Schmidt, Mossbauer Study of the Quaterny Fluoride Na2MnFeF7, in "Int Conf Mossbauer Spectrosc, Proc," Vol 1, p 529 (See 75H027)
75P047	FE-57	J Pebler and K Schmidt, Phase Transitions in Fe(alpha)V(1-alpha)O(2+epsilon), in "Int Conf Mossbauer Spectrosc, Proc," Vol 1, p 531 (See 75H027)
75P048	REVIEW	F Parak, Spatial Arrangement and Electronic Structure of the Active Center of Iron Containing Biomolecules, in "Int Conf Mossbauer Spectrosc, Proc," Vol 2, pp 285-303 (See 75H038)
75P049	THEORY	M Piecuch and C Janot, J Phys Chem Solids 36,1135-45 (1975), Contribution Electronique au Gradient de Champ Electrique sur une Impurete de Transition dans un Environnement Metallique Anisotrope
75P050	ANALYS	J Piekoszewski, K Kisynska, and L Dabrowski, Nukleonika 20, 947-52(1975), Determination of the Optimum Absorber Thickness in Mossbauer Effect Experiments (In Russian)
75P051	FE-57	V A Pletyushkin, V I Chechernikov, V V Muzaleva, V A Semenov, R N Kuz'min, and V K Slovyanskikh, Fiz Metal Metalloved 39,217-20(1975)/Phys Met Metallogr (USSR) 39, 200-3(1975), Magnetic Properties of Solid Solutions of the System UFe2-UCo2
75P052	FE-57	R S Preston, A E Dwight, A J Fedro, and C W Kimball, Mossbauer and X-ray Study of the Self-induced Magnetostatic Distortion of TbxY(1-x)Fe2 Laves Phases, in "AIP Conference Proceedings-No 24, Magnetism and Magnetic Materials-1974" (20th Annual Conference, San Francisco), edited by C D Graham, Jr, G H Lander, and J J Rhyne (American Institute of Physics, New York, 1975), pp 660-1
75P053	FE-57	V A Povitskii, A N Salugin, and E F Makarov, Fiz Tverd Tela 17,3649-51(1975)/Sov Phys-Solid State, Defect Nature of the Structure of Hematite and Morin Transition
75P054	SN-119	O Kh Poleshchuk, Yu K Maksyutin, and I G Orlov, Koord Khim 1,666-9(1975), Charge Transfer in Tin Chloride Complexes (In Russian)

CODE	TOPIC	REFERENCE

75R001   FE-57   W M Reiff, I E Grey, A Fan, Z Eliezer, and H Steinfink, J Solid State Chem 13,32-40(1975), The Oxidation State of Iron in Some Ba-Fe-S Phases: A Mossbauer and Electrical Resistivity Investigation of Ba2FeS3, Ba7Fe6S14, Ba6Fe8S15, BaFe2S3, and Ba9Fe16S32

75R002   FE-57   W M Reiff, B Dockum, M A Weber, and R B Frankel, Inorg Chem 14,800-6(1975), Magnetic Ordering of Mono(diimine)iron(II) Chlorides. Fe(2,2'-bipy)Cl2 and Fe(5,5'-(CH3)2-2,2'-bipy)Cl2

75R003   CS-133   S R Reintsema, S A Drentje, P J Schurer, and H De Waard, Radiat Eff 24,145-54(1975), Lattice Location of Xenon Impurities Implanted in Iron Derived from Mossbauer Effect Measurements

75R004   AU-197   L D Roberts, The Mossbauer Effect for 197Au, in "Mossbauer Effect Data Index, Covering the 1973 Literature," edited by J G Stevens and V E Stevens (Plenum Publishing Corp, New York, 1975), pp 349-91

75R004   REVIEW   L D Roberts, The Mossbauer Effect for 197Au, in "Mossbauer Effect Data Index, Covering the 1973 Literature," edited by J G Stevens and V E Stevens (Plenum Publishing Corp, New York, 1975), pp 349-91

75R005   ANALYS   K Ruebenbauer, Acta Phys Pol A 47,11-5(1975), Some Remarks on the Fourier Analysis of Mossbauer Spectra

75R005   FE-57   K Ruebenbauer, Acta Phys Pol A 47,11-5(1975), Some Remarks on the Fourier Analysis of Mossbauer Spectra

75R006   FE-57   D P Riley, P H Merrell, J A Stone, and D H Busch, Inorg Chem 14,490-4(1975), Five- and Six-coordinate Complexes of Iron(II) and -(III) with a Macrocycli Tetradentate Ligand

75R007   FE-57   J D Rush, D P E Dickson, C E Johnson, P J Hewitt, and H F Lam, Phys Med Biol 20,128-30(1975), The Use of Mossbauer Spectroscopy to Investigate Lung Samples Containing Iron

75R008   FE-57   T Raman, V K Nagesh, D Chakravorty, and G N Rao, J Appl Phys 46,972-3(1975), Mossbauer Studies in the Glass System Na2O-B2O3-Fe2O3

75R009   SB-121   J N R Ruddick and J R Sams, Inorg Nucl Chem Lett 11,229-31 (1975), Verification of the -1:2 Quadrupole Coupling Constant Ratio in cis- and trans-Octahedral Diorganoantimony(I) Complexes

75R010   SN-119   P Roggwiller and W Kundig, Phys Rev B 11,4179-83(1975), Isomer Shift and Influence of the Chemical Environment on the Lifetime of the Mossbauer Nucleus 119Sn

75R011   FE-57   E Realo and A Lijn, Mossbauer Investigation of alpha-Fe2O3-Sn, in "5th Int Conf Mossbauer Spec, Proc," Part 1, pp 151-6 (See 75H017) (In Russian)

75R011   SN-119   E Realo and A Lijn, Mossbauer Investigation of alpha-Fe2O3-Sn, in "5th Int Conf Mossbauer Spec, Proc," Part 1, pp 151-6 (See 75H017) (In Russian)

75R012   FE-57   S N Ray, T Lee, and T P Das, Phys Rev B 12,58-63(1975), Effect of Many-body Interactions on Isomer Shift in Iron Compounds

75R012   THEORY   S N Ray, T Lee, and T P Das, Phys Rev B 12,58-63(1975), Effect of Many-body Interactions on Isomer Shift in Iron Compounds

CODE	TOPIC	REFERENCE

75R013   REVIEW   K Raclavsky, Mossbauer Effect Technique in Mineral Science, in "5th Int Conf Mossbauer Spec, Proc," Part 2, pp 347-55 (See 75H017)

75R014   FE-57   K Raclavsky, J Sitek, and J Lipka, Mossbauer Spectroscopy of Iron in Clay Minerals, in "5th Int Conf Mossbauer Spec, Proc," Part 2, pp 368-71 (See 75H017)

75R015   FE-57   K R P M Rao and P K Iyengar, Phys Status Solidi A 30,397-401(1975), Mossbauer Spectroscopy Study of Short-range Magnetic Ordering in Co-Ga (Fe57) Intermetallic Compounds

75R016   SN-119   J N R Ruddick and J R Sams, J Inorg Nucl Chem 37,564-6 (1975), A Mossbauer Spectroscopic Study of Some Schiff Base Complexes of Tin(IV) Halides

75R017   FE-57   D Raj and J Danon, J Inorg Nucl Chem 37,2039-45(1975), Mossbauer Spectroscopic Studies of Thermal Decomposition of Alkali Ferricyanides

75R018   SN-119   A J Rein and R H Herber, J Chem Phys 63,1021-9(1975), Molecular Spectroscopy of Organometallic Compounds: Organotin(IV) Tropolonates

75R019   FE-57   Yu L Rodionov and V N Zambrzhitskii, Dokl Akad Nauk SSSR 221,825-7(1975)/Sov Phys-Dokl 20,285-6(1975), Effect of Deformation Temperature on the Distribution of the Atoms in Submicroscopic Regions and on the Martensite Transformation in an Iron-32.4% Nickel Alloy

75R020   FE-57   Yu L Rodionov, B S Machurin, and P L Gruzin, Dokl Akad Nauk SSSR 222,64-7(1975)/Sov Phys-Dokl 20,355-6(1975), Effect of Crystal Lattice Defects on Distribution of Atoms and Martensite Transformation of Iron-Nickel Alloys

75R021   SN-119   E Realo, Mossbauer Spectroscopy of Optical Sn-centers in Crystals of NaCl and CaS, in "Int Conf Mossbauer Spectrosc, Proc," Vol 1, pp 199-200 (See 75H027) (In Russian)

75R022   FE-57   S M Richardson, Am Mineral 60,73-8(1975), A Pink Muscovite with Reverse Pleochroism from Archer's Post, Kenya

75R023   GD-155   J W Ross and J Sigalas, J Phys F 5,1973-80(1975), Isomer Shifts in Gadolinium Compounds Having the Caesium Chloride Structure

75R024   FE-57   R Raudsepp, Mossbauer Study of Interaction Between Iron Ions and alpha-Chimotripsin, in "Int Conf Mossbauer Spectrosc, Proc," Vol 1, pp 299-300 (See 75H027) (In Russian)

75R025   SN-119   E Realo and S I Reiman, The Influence of the Electron Arrangement in Characteristic Admixture of Ionic Tin in Fe3O4, in "Int Conf Mossbauer Spectrosc, Proc," Vol 1, pp 373-4 (See 75H027) (In Russian)

75R026   FE-57   J R Regnard, Magnetically Induced Quadrupole Interactions and Vibronic Coupling for Fe2+ Impurities in Cubic CaO and KMgF3, in "Int Conf Mossbauer Spectrosc, Proc," Vol 1, pp 409-10 (See 75H027)

75R027   FE-57   F W Richter and K Schmidt, Z Naturforsch, Teil A 30,1621-6 (1975), Mossbauer-spektropische Untersuchungen am System Fe(1+x)Sb mit NiAs-Struktur

75R028   I-129   S R Reintsema, H De Waard, and S A Drentje, Mossbauer Effect Studies on Sources of 129mTe Implanted in Iron and Nickel, in "Int Conf Mossbauer Spectrosc, Proc," Vol 1, pp 525-6 (See 75H027)

75R029   FE-57   R Raudsepp, Eesti NSV Tead Akad Toim, Fuus, Mat 24,312-7 (1975), Investigation of the Interaction Between Iron and Glutathione by Mossbauer Spectroscopy (NGR) (In Russian)

CODE	TOPIC	REFERENCE

75R030   FE-57   R Raudsepp, Eesti NSV Tead Akad Toim, Fuus, Mat 24,438-41 (1975), Study of the Reaction Between Iron Ions and Erythrocyte Membranes by Mossbauer Spectroscopy (In Russian)

75R031   FE-57   K V Reddy and S C Chetty, Phys Status Solidi A 32,585-92 (1975), Mossbauer Studies on the Fe-Se System

75R032   SN-119   V O Reikhsfel'd, V A Ivanov, and I E Saratov, Zh Obshch Khim 45,2243-5(1975)/J Gen Chem (USSR) 45,2202-4(1975), Infrared- and Mossbauer-spectral Study of Hydride Organo-Silanes, -Germanes, and -Stannanes

75R033   FE-57   C N R Rao, V G Bhide, and N F Mott, Philos Mag 32,1277-82 (1975), Hopping Conduction in La(1-x)SrxCoO3 and Nd(1-x)SrxCoO3

75R034   SN-119   P F Rodesiler, T Auel, and E L Amma, J Am Chem Soc 97, 7405-10(1975), Metal Ion-aromatic Complexes. XXII. The Preparation, Structure, and Stereochemistry of Tin(II) in pi-C6H6Sn(AlCl4)2-C6H6

75R035   THEORY   S K Roy, M Singh, and B P Srivastava, Indian J Pure Appl Phys 13,217-20(1975), Study of Mossbauer Effect of a Lattice Containing a Pair of Point Defects at High Temperature Limit

75R036   SN-119   N M Rubinina, V B Shagdarov, V K Yanovskii, and R N Kuz'min, Kvan Elektron (Moscow) 2,1024-9(1975)/Sov J Quantum Electron, Study of Impurity Centers in Iron-doped Lithium Metaniobate by Mossbauer Spectroscopy

75R037   FE-57   T Ruskov and T Tomov, Iron Steel Int 48,405-7(1975), Rapid Phase Analysis of Iron Ore Treated by Magnetic Roasting

75R038   FE-57   S I Reiman and K P Mitrofanov, Eesti NSV Tead Akad Toim, Fuus, Mat 24,428-32(1975), The Preparation of the KMeF3 (Me = Mn,Fe,C,Ni) Crystals for NGR Investigations (In Russian)

75R039   FE-57   S L Ruby, J C Love, P A Flinn, and B J Zabransky, Appl Phys Lett 27,320-2(1975), Diffusion of Iron Ions in a Cold Liquid: Evidence Against a "Jump" Model

75S001   SN-119   H Sano and Y Mekata, Chem Lett 155-60(1975), Mossbauer Spectroscopic Studies of Bis(tri-n-butyltin) Sulfate, Selenate, and Chromate

75S002   FE-57   H Sano and T Ohnuma, Bull Chem Soc Jpn 48,266-9(1975), Mossbauer Spectroscopic Studies of the Effect of Anions in the Second Coordination Sphere in the EC-decay

75S003   FE-57   H F Steger, J Inorg Nucl Chem 37,39-43(1975), Stability of Metal Oxinates IV. Iron Oxinates

75S004   FE-57   V Subramanian, Phys Status Solidi A 27,303-8(1975), Mossbauer Effects in Fe-Mn Precipitates

75S005   FE-57   J P Schunck, J M Friedt, and Y Llabador, Rev Phys Appl 10, 121-6(1975), Spectroscopie Mossbauer de 57Fe et 119Sn par Detection des Electrons de Conversion et Auger Application a des Etudes de Surface

75S005   SN-119   J P Schunck, J M Friedt, and Y Llabador, Rev Phys Appl 10, 121-6(1975), Spectroscopie Mossbauer de 57Fe et 119Sn par Detection des Electrons de Conversion et Auger Application a des Etudes de Surface

75S006   GENRAL   N J Seeley, Nature 254,479(1975), Mossbauer Spectroscopy in Archaeology

CODE	TOPIC	REFERENCE
75S007	FE-57	R P Singh and B N Srivastava, Phys Status Solidi A 27, K29-33(1975), Mossbauer Studies of ZnMn(1-x)Cr(x)FeO4 Spinels
75S008	FE-57	R P Singh and B N Srivastava, Chem Phys Lett 30,300-2 (1975), Mossbauer Effect of Ferrous Zirconium Double Sulphate
75S009	FE-57	K Spartalian, W T Oosterhuis, and J B Neilands, J Chem Phys 62,3538-43(1975), Electronic State of Iron in Enterobactin Using Mossbauer Spectroscopy
75S010	FE-57	T Sakai and T Tominaga, Radiochem Radioanal Lett 22,11-7 (1975), Mossbauer and NMR Studies of Mixed-valence Fe2F5-7H2O and Related Compounds
75S011	FE-57	I Sakamoto, T Kinoshita, N Hayashi, and B Furubayashi, Jpn J Appl Phys 14,715-6(1975), Preparation of 57Fe-enriched alpha-Fe2O3 Film for the Mossbauer Magnetic Diffraction
75S012	FE-57	R Y Saleh and D K Straub, Inorg Chim Acta 13,105-8(1975), Halobis(dicyclohexyldithiophosphinato)iron(III) Complexes
75S013	FE-57	J R Sams and T B Tsin, J Chem Phys 62,734-5(1975), Para-magnetic Hyperfine Structure in an Octahedral High-spin Fe2+ Complex
75S014	FE-57	J R Sams and T B Tsin, Inorg Chem 14,1573-9(1975), Orbital Ground States and Crystal Field Splittings in Some Octahe-drally Coordinated High-spin Ferrous Complexes
75S015	FE-57	H Sato and T Tominaga, Radiochem Radioanal Lett 22,3-10 (1975), A Mossbauer Study on the Thermal Decomposition of Tris(2,2'-bipyridine)iron(II) Chloride
75S016	FE-57	B Seidel and G Dehe, Phys Status Solidi A 28,K117-20 (1975), Mossbauer Study of Cation Distribution in Fe(3-x)Cu(x)O4
75S017	FE-57	R R Sharma and P Moutsos, Phys Rev B 11,1840-6(1975), Mossbauer Quadrupole Splitting in Ferric Hemin
75S017	THEORY	R R Sharma and P Moutsos, Phys Rev B 11,1840-6(1975), Mossbauer Quadrupole Splitting in Ferric Hemin
75S018	FE-57	H Shechter and D Bukshpan-Ash, Phys Rev B 11,2673-7 (1975), Small-angle Scattering of Mossbauer Gamma Rays near Magnetic Phase Transitions
75S019	SN-119	S Solacolu, E Barbulescu, D Barb, and M Morariu, Rev Roum Chim 20,69-73(1975), Polymorphic Transformations in the BaTiO3-BaSnO3 System Revealed by Mossbauer Effect
75S020	FE-57	K Spartalian, W T Oosterhuis, and N Smarra, Biochim Biophys Acta 399,203-12(1975), Mossbauer Effect Studies in the Fungus Phycomyces
75S021	FE-57	P Steiner and S Hufner, Phys Rev B 12,842-6(1975), Local Magnetization of Fe in Ag
75S022	YB-170	J Stohr, Phys Rev B 11,3559-72(1975), Mossbauer Relaxation Studies of 170Yb in Dilute Au: 170Tm Sources in External Magnetic Fields up to 55 kG
75S023	GENRAL	I P Suzdalev, Vestn Akad Nauk SSSR (3),93-5(1975), Conference on Applications of Mossbauer Spectroscopy (In Russian)
75S024	FE-57	W Steiner and R Haferl, Concentration Dependence of Hyper-fine and Quadrupole Splitting in Y2(Fe(1-x)Co(x))17 Com-pounds, in "5th Int Conf Mossbauer Spec, Proc," Part 1, pp 49-54 (See 75H017)

CODE	TOPIC	REFERENCE
75S025	FE-57	M R Spender and A H Morrish, Mossbauer Study of Some Sulfur Spinels, in "5th Int Conf Mossbauer Spec, Proc," Part 1, pp 125-9 (See 75H017)
75S026	FE-57	W Siebke, S Hosl, H Spiering, and G Ritter, Nuclear Electric Field Gradient Determination in Single Crystals of FeCl2-4H2O Using the Mossbauer Effect in Applied Magnetic Fields, in "5th Int Conf Mossbauer Spec, Proc," Part 1, pp 176-8 (See 75H017)
75S027	FE-57	N I Shapiro, I P Suzdalev, V I Gol'danskii, A I Sherle, and A A Berlin, Mossbauer Investigations of Some Iron Polyazoporphines, in "5th Int Conf Mossbauer Spec, Proc," Part 1, pp 223-7 (See 75H017) (In Russian)
75S028	FE-57	M Seberini and K Volenik, Application of Mossbauer Spectroscopy in the Study of High-temperature Oxidation of Steel, in "5th Int Conf Mossbauer Spec, Proc," Part 1, pp 241-4 (See 75H017)
75S029	FE-57	C Song and J G Mullen, Solid State Commun 17,549-52(1975), Mossbauer Studies of Cobaltous Oxide at High Temperatures: Evidence for Relaxation Valence Averaging and Phase Transitions about Iron Impurities
75S030	SN-119	T K Sham and G M Bancroft, Inorg Chem 14,2281(1975), Tin-119 Mossbauer Quadrupole Splittings for Distorted Me2Sn(IV) Structures
75S031	SN-119	J Sitek, Mossbauer Study of the Lattice Dynamics of Tin Atoms in Antimony, in "5th Int Conf Mossbauer Spec, Proc," Part 2, pp 451-3 (See 75H017)
75S032	SN-119	Yu A Samarskii, N E Alekseevskii, and A P Kiryanov, Zh Eksp Teor Fiz 68,2330-4(1975)/Sov Phys-JETP, "Magnetic Anomaly" of the Probability of the Mossbauer Effect in Dilute Pd-Co Alloys
75S033	SN-119	P P Seregin, M A Sagatov, T F Mazets, and L N Vasil'ev, Phys Status Solidi A 28,127-32(1975), The Influence of the Crystal-glass Transition on the State of Impurity Tin Atoms in Chalcogenide Semiconductors
75S034	FE-57	A Sette Camara and W Keune, Corros Sci 15,441-53(1975), Oxidation Study of Iron by Mossbauer Conversion Electron and Gamma-ray Scattering
75S035	TE-125	K V Smith, J S Thayer, and B J Zabransky, Inorg Nucl Chem Lett 11,441-6(1975), Mossbauer Spectra of Some Alkyltellurium(IV) Compounds
75S036	SN-119	V I Shtanov, V P Zlomanov, and A V Novoselova, Izv Akad Nauk SSSR, Neorg Mater 11,358-60(1975)/Inorg Mater (USSR) 11,301-3(1975), Physicochemical Investigation of the System PbSe-SnSe2
75S037	FE-57	T Suzuki, Y Maeda, H Sakai, S Fujimoto, and Y Morita, J Biochem 78,555-60(1975), Mossbauer Effect and Electron Paramagnetic Resonance Studies on Yeast Aconitase
75S038	THEORY	A Szczepanski, Bull Acad Pol Sci, Ser Sci Tech 23,47-52 (1975), Coincidence Mossbauer Spectroscopy and the Dynamics of Point Defects. I. The Classical Theory
75S039	SB-121	G K Shenoy and J M Friedt, Methodology of the 121Sb Mossbauer Quadrupole Spectra, in "5th Int Conf Mossbauer Spec, Proc," Part 3, pp 510-3 (See 75H017)
75S040	TE-125	J Suwalski, L Dabrowski, J Leciejewicz, J Piekoszewski, and W Suski, The Mossbauer Effect of 125Te in Uranium-Tellurium Compounds, in "5th Int Conf Mossbauer Spec, Proc," Part 3, pp 514-6 (See 75H017)

CODE	TOPIC	REFERENCE
75S041	FE-57	K P Schmidt, J De Raedt, G Langouche, M Van Rossum, and R Coussement, A Computerized Piezoelectric Mossbauer Spectrometer, in "5th Int Conf Mossbauer Spec, Proc," Part 3, pp 649-52 (See 75H017)
75S041	INSTRM	K P Schmidt, J De Raedt, G Langouche, M Van Rossum, and R Coussement, A Computerized Piezoelectric Mossbauer Spectrometer, in "5th Int Conf Mossbauer Spec, Proc," Part 3, pp 649-52 (See 75H017)
75S042	FE-57	E P Stepanov, A N Artem'ev, I P Perstnev, V V Sklyarevskii, and G V Smirnov, The Interference of Nuclear Transitions in the Purely Nuclear Diffraction of 14.4 keV Gamma-rays in Hematite, in "5th Int Conf Mossbauer Spec, Proc," Part 3, pp 709-12 (See 75H017)
75S043	FE-57	B D Sawicka and J A Sawicki, Mossbauer Absorption by 57Fe Implanted into Solids, in "5th Int Conf Mossbauer Spec, Proc," Part 3, pp 731-5 (See 75H017)
75S044	FE-57	A Simopoulos, A Kostikas, and N H Gangas, Mossbauer Studies of Iron Oxide Transformations in Fired Clays, in "5th Int Conf Mossbauer Spec, Proc," Part 3, pp 759-62 (See 75H017)
75S045	FE-57	J W Schindler, J R Luoma, and J P Cusick, Ind Eng Chem, Prod Res Dev 14,212-6(1975), Photoelectron Spectra of Halogeno(tolyl isocyanide)iron(II) Complexes
75S046	FE-57	Z Shanfield, P H Barrett, and P A Montano, Properties of the Amorphous Magnet: MgF2/Fe, in "AIP Conference Proceedings No 24, Magnetism and Magnetic Materials-1974" (20th Annual Conference, San Francisco), edited by C D Graham, Jr, G H Lander, and J J Rhyne (American Institute of Physics, New York, 1975), pp 129-30
75S047	FE-57	T Shigematsu, J Phys Soc Jpn 39,1233-8(1975), Mossbauer and Structural Studies on (Fe(1-x)Mn(x))2B
75S048	FE-57	K Skeff Neto and V K Garg, J Inorg Nucl Chem 37,2287-90 (1975), Mossbauer Thermal Decomposition Studies of Fe(II) Sulphate
75S049	FE-57	T Sakai and T Tominaga, Radiochem Radioanal Lett 23,329-36 (1975), NMR and Mossbauer Spectra of Mixed-valence Iron Fluoride Dihydrate and Related Compounds
75S050	EU-151	E A Samuel and W N Delgass, J Chem Phys 62,1590-2(1975), The Hyperfine Structure of Eu2+ Ions in Zeolites
75S051	FE-57	A Singh and K N Shrivastava, Solid State Commun 17,1123-4 (1975), Crystal Field Contribution to Mossbauer Isomer Shift
75S051	THEORY	A Singh and K N Shrivastava, Solid State Commun 17,1123-4 (1975), Crystal Field Contribution to Mossbauer Isomer Shift
75S052	FE-57	V I Sorokin, G V Novikov, V K Egorov, V I Popov, and L V Sipavina, Geokhimiya 1329-36(1975), Investigation of Iron-containing Sphalerites by the Mossbauer Method (In Russian) (Possible Translation in Geochem Int)
75S053	I-129	M Sneh and B Dayal, Phys Status Solidi B 70,341-6(1975), Recoilless Fraction of Iodine Ion in RbI, NaI, and KI by the Shell Model
75S054	THEORY	E V Smirnov and V A Belyakov, On the Dynamic Theory of the Diffraction of Resonant Gamma Rays on Magnetically Ordered Crystals, in "Int Conf Mossbauer Spectrosc, Proc," Vol 1, pp 11-2 (See 75H027) (In Russian)

CODE	TOPIC	REFERENCE

75S055   THEORY   S O Svensson, S Morup, and G Trumpy, Anisotropic Diffusional Line Broadening in Single Crystals of fcc Metals in "Int Conf Mossbauer Spectrosc, Proc," Vol 1, pp 19-20 (See 75H027)

75S056   INSTRM   A Kh Sherif, A A Opalenko, and R N Kuz'min, Unit for Measuring of the Mossbauer Effect under High Pressure, in "Int Conf Mossbauer Spectrosc, Proc," Vol 1, pp 33-4 (See 75H027)

75S057   FE-57   R J Semper, C L Chien, and J C Walker, Observation of the Temperature Dependence of Magnetic Hyperfine Fields in Very Thin Iron Films, in "Int Conf Mossbauer Spectrosc, Proc," Vol 1, pp 37-8 (See 75H027)

75S058   FE-57   J Stanek, J A Sawicki, and B D Sawicka, Conversion Electron Mossbauer Spectroscopy of 57Fe Implanted into Solids, in "Int Conf Mossbauer Spectrosc, Proc," Vol 1, pp 43-4 (See 75H027)

75S059   TA-181   D Salomon, W Wallner, and P J West, High Temperature Behavior of the 6.2 keV Mossbauer Transition of 181Ta in Tantalum Metal, in "Int Conf Mossbauer Spectrosc, Proc," Vol 1, pp 105-6 (See 75H027)

75S060   FE-57   M Shiga and Y Nakamura, Collapse of Localized Moments in bcc Fe Alloys near the Critical Concentration, in "Int Conf Mossbauer Spectrosc, Proc," Vol 1, pp 175-6 (See 75H027)

75S061   FE-57   J Stanek, J A Sawicki, B D Sawicka, and M Drwiega, Properties of 57Fe Implanted in Al, in "Int Conf Mossbauer Spectrosc, Proc," Vol 1, pp 209-10 (See 75H027)

75S062   FE-57   B D Sawicka, J Stanek, and J A Sawicki, Isomer Shifts of 57Fe Implanted into Transition Elements, in "Int Conf Mossbauer Spectrosc, Proc," Vol 1, pp 211-2 (See 75H027)

75S063   FE-57   T Sakai and T Tominaga, Bull Chem Soc Jpn 48,3168-70(1975), An NMR and Mossbauer Spectroscopic Study of Mixed-valence Iron Fluoride Heptahydrate and Related Compounds

75S064   FE-57   J L Schurter, R G Barnes, and R D Willett, Hyperfine Fields and Critical Point Exponents of Some Two-dimensionally Layered Fe(2+) Salts, in "AIP Conference Proceedings-No 24-Magnetism and Magnetic Materials-1974" (20th Annual Conference, San Francisco), edited by C D Graham, Jr, G H Lander, and J J Rhyne (American Institute of Physics, New York, 1975), pp 307-8

75S065   FE-57   E R Seidel, F J Litterst, W Gierisch, and G M Kalvius, J Magn Magn Mat 1,19-22(1975), Moment Compensation in NixRh(1-x)(Fe) from Mossbauer Spectroscopy

75S066   SN-119   A Schichl, F J Litterst, H Micklitz, J P Devort, and J M Friedt, 119Sn Resonance in Matrix Isolated Sn(II) and Sn(IV) Halides, in "Int Conf Mossbauer Spectrosc, Proc," Vol 1, pp 239-40 (See 75H027)

75S067   FE-57   H Spiering, R Zimmermann, and D L Nagy, Ligand Field Theory and Hyperfine Interaction in Siderite, in "Int Conf Mossbauer Spectrosc, Proc," Vol 1, pp 261-2 (See 75H027)

75S068   FE-57   I P Saraswat, A C Vajpei, and V K Garg, Mossbauer Studies of Thermally Treated Iron(III) Coprecipitated with Ln(OH)3-(Ln = La, Pr or Nd), in "Int Conf Mossbauer Spectrosc, Proc," Vol 1, pp 269-70 (See 75H027)

75S069   FE-57   B A Shustrov, S P Teslenko, and V K Yarmarkin, Mossbauer Study of V2O3-Cr2O3 and V2O3-Ti2O3 Solid Solutions and VO2 in the Metal-semiconductor Transition Region, in "Int Conf Mossbauer Spectrosc, Proc," Vol 1, pp 323-4 (See 75H027)

CODE	TOPIC	REFERENCE
75S070	FE-57	J Sitek, M Prejsa, P Duhaj, M Hucl, and J Cirak, Mossbauer Spectroscopy on the Amorphous System Pd(80-x)Fe(x)Si20, in "Int Conf Mossbauer Spectrosc, Proc," Vol 1, pp 345-6 (See 75H027)
75S071	THEORY	G Ya Selyutin, Mossbauer Spectrum near the Critical Point in Ferromagnet, in "Int Conf Mossbauer Spectrosc, Proc," Vol 1, pp 361-2 (See 75H027) (In Russian)
75S072	FE-57	B Seidel, G Dehe, K Melzer, and C Michalk, On the Possibility of a Low Temperature Phase Transition in Sn-substituted Magnetite, in "Int Conf Mossbauer Spectrosc, Proc," Vol 1, pp 375-6 (See 75H027)
75S073	FE-57	F Sontheimer, D Seyboth, and H H F Wegener, The Dependence of Spin Relaxation in Frozen Solutions of Fe(NO3)3 on Spin Concentration and Strong Magnetic Fields, in "Int Conf Mossbauer Spectrosc, Proc," Vol 1, pp 399-400 (See 75H027)
75S074	FE-57	G K Shenoy and B D Dunlap, Calculation of Relaxation Spectra of K3(Co,Fe)(CN)6, in "Int Conf Mossbauer Spectrosc, Proc," Vol 1, pp 407-8 (See 75H027)
75S075	FE-57	M Soltwisch and D Quitmann, Viscous Motion of Glycerol Molecules, in "Int Conf Mossbauer Spectrosc, Proc," Vol 1, pp 429-30 (See 75H027)
75S076	FE-57	T Shinjo, H D Pfannes, and U Gonser, Mossbauer Study of Magnetic Recording Tapes, in "Int Conf Mossbauer Spectrosc, Proc," Vol 1, pp 465-6 (See 75H027)
75S077	FE-57	Z Shanfield, P A Montano, and P H Barrett, Phys Rev Lett 35,1789-92(1975), Resistivity and Mossbauer Measurements for Solid Xe-Fe Mixtures
75S078	FE-57	A Simopoulos, A Kostikas, I Sigalas, N H Gangas, and A Moukarika, Clays Clay Miner 23,393-9(1975), Mossbauer Study of Transformations Induced in Clay by Firing
75S079	ER-166	J Stohr and W Wagner, J Phys F 5,812-21(1975), Hyperfine Interactions and Relaxation Effects of Er Impurities in Y Studied by Mossbauer Spectroscopy
75S080	ER-166	J Stohr, J D Cashion, and W Wagner, J Phys F 5,1417-25 (1975), Mossbauer Spectroscopy of Rare Earth Impurities in Hexagonal Host Metals: ZrEr
75S081	FE-57	K Spartalian, G Lang, J P Collman, R R Gagne, and C A Reed, J Chem Phys 63,5375-82(1975), Mossbauer Spectroscopy of Hemoglobin Model Compounds: Evidence for Conformational Excitation
75S082	DY-160	J Stohr and J D Cashion, Phys Rev B 12,4805-11(1975), 160Dy and 166Er Mossbauer Studies of Concentrated and Diluted Rare-earth Dihydrides: Single-line Compounds and Crystal-field Effects
75S082	ER-166	J Stohr and J D Cashion, Phys Rev B 12,4805-11(1975), 160Dy and 166Er Mossbauer Studies of Concentrated and Diluted Rare-earth Dihydrides: Single-line Compounds and Crystal-field Effects
75S083	FE-57	D Samaras and A Collomb, Solid State Commun 16,1279-84 (1975), Rotation des Moments Magnetiques dans BaLa2Fe207
75S084	FE-57	Yu K Shubn'i, Mossbauer Effect in the Study of Ancient Pottery, in "Int Conf Mossbauer Spectrosc, Proc," Vol 1, pp 489-90 (See 75H027) (In Russian)
75S085	FE-57	O Schneeweiss and T Zemcik, Influence of Plastic Deformation on Phase Transformation in Manganese Steels, in "Int Conf Mossbauer Spectrosc, Proc," Vol 1, pp 509-10 (See 75H027)

CODE	TOPIC	REFERENCE
75S086	REVIEW	G K Shenoy and B D Dunlap, Mossbauer Relaxation Lineshapes in the Presence of Hyperfine Interactions Containing Off-diagonal Terms, in "Int Conf Mossbauer Spectrosc, Proc," Vol 2, pp 275-84 (See 75H038)
75S087	REVIEW	J G Stevens and V E Stevens, Mossbauer Effect Data Center, in "Int Conf Mossbauer Spectrosc, Proc," Vol 2, pp 467-73 (See 75H038)
75S088	FE-57	B D Sawicka, J A Sawicki, and J Stanek, Computer Simulations of EFG in Lattices Implanted with 57Fe, in "5th Int Conf Mossbauer Spectrosc, Proc," Vol 3, pp 632-5 (See 75H017) (In Russian)
75S089	FE-57	A N Salugin, V A Povitskii, M V Pilin, N V Elistratov, and A F Pis'marov, Fiz Tverd Tela 17,1806-8(1975)/Sov Phys-Solid State 17,1179-80(1975), Mossbauer Spectroscopic Investigation of Spin Reorientation in the System (1-c)Fe2O3-cAl2O3
75S090	FE-57	R H Sands and W R Dunham, Quart Rev Biophys 7,443-504 (1975), Spectroscopic Studies on Two-iron Ferredoxins
75S090	REVIEW	R H Sands and W R Dunham, Quart Rev Biophys 7,443-504 (1975), Spectroscopic Studies of Two-iron Ferredoxins
75S091	FE-57	F Schmidt, W Gunsser, and A Knappwost, Z Naturforsch, Teil A 30,1627-32(1975), Magnetische und Mossbauer-spektroskopische Untersuchungen von zeolithischen Eisenkontakten
75S092	INSTRM	M Seberini and J Cirak, Jad Energ 21,263-4(1975), Determination of Zero Velocity of a Mossbauer Spectrometer Using the Method of Electronic Magnifying Lens (In Slovak)
75S093	SN-119	P P Seregin, I V Nistiryuk, and F S Nasredinov, Fiz Tverd Tela 17,2330-4(1975)/Sov Phys-Solid State 17,1540-2(1976), Tin as an Isotopic Impurity in Silicon and Germanium
75S094	FE-57	N I Shapiro, I P Suzdalev, V I Gol'danskii, A I Sherle, and A A Berlin, Teor Eksp Khim 11,330-6(1975)/Theor Exp Chem (USSR), Use of Gamma Resonance Spectroscopy for Studying Iron Polyazaporphines
75S095	ER-166	J Sivardiere, M Blume, and M J Clauser, Hyperfine Interac 1,227-50(1975), Magnetic Relaxation and Paramagentic Mossbauer Spectra: Influence of the Off-diagonal Hyperfine Coupling
75S095	FE-57	J Sivardiere, M Blume, and M J Clauser, Hyperfine Interac 1,227-50(1975), Magnetic Relaxation and Paramagentic Mossbauer Spectra: Influence of the Off-diagonal Hyperfine Coupling
75S095	TM-169	J Sivardiere, M Blume, and M J Clauser, Hyperfine Interac 1,227-50(1975), Magnetic Relaxation and Paramagentic Mossbauer Spectra: Influence of the Off-diagonal Hyperfine Coupling
75S096	DY-161	J Sivardiere and M Blume, Hyperfine Interac 1,283-94(1975), Paramganetic Mossbauer Spectra of 161Dy(Gamma 6 or Gamma 7) and 166Er(Gamma 8) in Cubic Symmetry: Influence of Relaxation
75S096	ER-166	J Sivardiere and M Blume, Hyperfine Interac 1,283-94(1975), Paramganetic Mossbauer Spectra of 161Dy(Gamma 6 or Gamma 7) and 166Er(Gamma 8) in Cubic Symmetry: Influence of Relaxation
75S097	SN-119	F E Smith and B V Liengme, J Organomet Chem 91,C31-2 (1975), A Novel Series of Triorganotin Compounds

CODE	TOPIC	REFERENCE
75S098	FE-57	V K Sokolova, V A Varnek, N F Yudanov, and I I Tychinskaya, Izv Sib Otd Akad Nauk SSSR, Ser Khim Nauk (6),13-7(1975), Mossbauer Effect in Iron Hexafluoro-stannates (In Russian)
75S098	SN-119	V K Sokolova, V A Varnek, N F Yudanov, and I I Tychinskaya, Izv Sib Otd Akad Nauk SSSR, Ser Khim Nauk (6),13-7(1975), Mossbauer Effect in Iron Hexafluoro-stannates (In Russian)
75S099	FE-57	G E Stein and J A Marinsky, J Inorg Nucl Chem 37,2421-8 (1975), A Mossbauer Spectroscopic Study of Some Ferric Polyaminocarboxylates
75S100	FE-57	P Steiner, D Gumprecht, W Von Zdrojewski, and S Hufner, Mossbauer Experiments on Dilute 57Fe in Cu, Ag, and Au, in "AIP Conference Proceedings-No 24, Magnetism and Magnetic Materials-1974" (20th Annual Conference, San Francisco), edited by C D Graham, Jr, G H Lander, and J J Rhyne (American Institute of Physics, New York, 1975), pp 479-80
75S101	FE-57	J Stanek, J A Sawicki, and B D Sawicka, Nucl Instrum Methods 130,613-4(1975), Conversion Electron Mossbauer Spectroscopy of 57Fe Implanted into Solids
75S102	SN-119	G C Stocco, G Alonzo, N Bertazzi, and F Di Bianca, Gazz Chim Ital 105,355-60(1975), 36/Mossbauer, Infrared and Other Studies on Tin(II) Halide Derivatives of Multidentate Ligands
75S103	FE-57	G D Sultanov, N G Guseinov, I Ismailzade, R M Mizababaev, and L A Aliev, Fiz Tverd Tela 17,1940-3(1975)/Sov Phys-Solid State 17,1271-2(1976), Magnetic Properties of Layered Perovskite-like Ferroelectric-magnets
75S104	SN-119	T N Sumarokova, R A Slavinskaya, and T A Tember, Zh Obshch Khim 45,2687-92(1975)/J Gen Chem (USSR) 45,2651-4(1975), Coordination Compounds of Sn(IV) with N,N'-Diacyldicar-boxamides
75S105	SB-121	T N Sumarokova, E F Makarov, D Kh Kamysbaev, A Yu Aleksandrov, I I Amelin, and M I Usanovich, Izv Akad Nauk Kaz SSR, Ser Khim 25(5),9-13(1975), Mossbauer Effect in Some Complexes of Antimony(III) with Organic Substances (In Russian)
75S106	FE-57	V A Starodub, M S Novakovskii, V G Kirichenko, V V Chekin, and A I Velikodnyi, Koord Khim 1,1706-10(1975), Study of Iron(II) Chloride Complexes with Phenylhydrazine and Its Derivatives Using Nuclear Gamma Resonance Spectroscopy (In Russian)
75T001	FE-57	L Takacs, M C Cadeville, and I Vincze, J Phys F 5,800-11 (1975), Mossbauer Study of the Intermetallic Compounds (Fe(1-x)Co(x))2B and (Fe(1-x)Co(x))B
75T002	AU-197	J O Thomson, F E Obenshain, P G Huray, J C Love, and J Burton, Phys Rev B 11,1835-9(1975), Mossbauer Measurements with 197Au in AuCl2, AuGa2, AuIn2, and AuSb2
75T003	FE-57	A Trautwein, E Kreber, U Gonser, and F E Harris, J Phys Chem Solids 36,325-8(1975), Molecular Orbital and Mossbauer Study of Iron-Oxygen Compounds
75T004	FE-57	C A Taft, D Raj, and J Danon, J Phys Chem Solids 36,283-7 (1975), Charge Distribution and Covalency Effects on Mossbauer Parameters in KFeS2

CODE	TOPIC	REFERENCE

75T005　FE-57　C P Tsang, A J F Boyle, and E H Morgan, Biochim Biophys Acta 386,32-40(1975), Mossbauer Spectra of Bicarbonate-free Ferric-transferrin Complex

75T006　REVIEW　T Tominaga, Gendai Kagaku (46),32-9(1975), Mossbauer Spectroscopy (In Japanese)

75T007　FE-57　M Tanaka, T Tokoro, and T Mori, Mossbauer Studies of Fe2+ in Some Spinel Type Compounds, in "5th Int Conf Mossbauer Spec, Proc," Part 1, pp 118-24 (See 75H017)

75T008　FE-57　M Teodorescu and G Filoti, Mossbauer Spectra of Some Complex Compounds of Fe(II) with Pyridine, in "5th Int Conf Mossbauer Spec, Proc," Part 1, pp 218-9 (See 75H017)

75T009　FE-57　H E Toma, E Giesbrecht, J M Malin, and E Fluck, Inorg Chim Acta 14,11-5(1975), Correlations of Mossbauer and Visible-UV Spectra with the Aqueous Substitution Reactivity of Several Substituted Pentacyanoferrate(II) Complexes

75T010　FE-57　J R Thompson and J O Thomson, Phys Rev B 12,2572-8(1975), Investigation of Dilute Magnetic Impurities Via the Mossbauer Effect: Ag57Fe and Ag57Co

75T011　FE-57　A Trautwein and F E Harris, Theor Chim Acta 38,65-9(1975), Calculated Electron Densitites and Experimental Isomer Shifts of Fe57 in the Deoxy- and CO- Compounds of Myoglobin and Hemoglobin

75T012　FE-57　A Trautwein, F E Harris, A J Freeman, and J P Desclaux, Phys Rev B 11,4101-5(1975), Relativistic Electron Densities and Isomer Shifts of 57Fe in Iron-Oxygen and Iron-Fluorine Clusters and of Iron in Solid Noble Gases

75T012　THEORY　A Trautwein, F E Harris, A J Freeman, and J P Desclaux, Phys Rev B 11,4101-5(1975), Relativistic Electron Densities and Isomer Shifts of 57Fe in Iron-Oxygen and Iron-Fluorine Clusters and of Iron in Solid Noble Gases

75T013　SN-119　E W Thornton and P G Harrison, J Chem Soc, Faraday Trans 1 71,461-72(1975), Tin Oxide Surfaces. Part 1, Surface Hydroxyl Groups and the Chemisorption of Carbon Dioxide and Carbon Monoxide on Tin(IV) Oixde

75T014　FE-57　A Trautwein, Y Maeda, U Gonser, F Parak, and H Formanek, Mossbauer Spectroscopy on Fe57 Enriched Deoxygenated Myoglobin Single Crystals, in "5th Int Conf Mossbauer Spec, Proc," Part 2, pp 412-4 (See 75H017)

75T015　FE-57　K I Turta, A V Ablov, N V Gerbeleu, R A Stukan, and Ch V Dyatlova, Zh Neorg Khim 20,150-4(1975)/Russ J Inorg Chem, 20,82-4(1975), Mossbauer Spectroscopic Study of the Magnettic State of Complexes of Iron(III) with Thiosemicarbazones of Substituted o-Hydroxybenzaldehydes

75T016　SN-119　M Takano, Y Takeda, M Shimada, T Matsuzawa, T Shinjo, and T Takada, J Phys Soc Jpn 39,656-60(1975), Mossbauer Study of Supertransferred Hyperfine Field of 119Sn (Sn4+) in Ca(1-x)Sn(x)MnO3

75T017　FE-57　J R Thompson and J O Thomson, Mossbauer Studies of Dilute Ag:Fe and Ag:Co Alloys, in "AIP Conference Proceedings-No 24, Magnetism and Magnetic Materials-1974" (20th Annual Conference, San Francisco), edited by C D Graham, Jr, G H Lander, and J J Rhyne (American Institute of Physics, New York, 1975), pp 477-8

75T018　FE-57　J M Thomas, M J Tricker, and A P Winterbottom, J Chem Soc, Faraday Trans 2 71,1708-19(1975), Conversion Electron Mossbauer Spectroscopic Study of Iron Containing Surfaces. Monitoring the Early Stages of Oxidation of a Low-carbon Steel by a Non-destructive Procedure

CODE	TOPIC	REFERENCE
75T019	FE-57	V A Tsurin, E E Yurchikov, and A Z Men'shikov, Fiz Tverd Tela 17,2915-21(1975)/Sov Phys-Solid State 17,1942-5 (1975), Mossbauer Investigations of the Hyperfine Magnetic Field in FePd Alloys
75T020	FE-57	M J Tricker and A G Freeman, Surf Sci 52,549-52(1975), Application of 57Fe Conversion Electron Mossbauer Spectroscopy to the Study of Single Crystal Specimens
75T021	FE-57	W F Tucker, R O Asplund, and S L Holt, Arch Biochem Biophys 166,433-8(1975), Preparation and Properties of Fe3+-Amino Acid Complexes. Crystalline Complexes with Aliphatic Amino Acids
75T022	FE-57	R Tahara, Y Nakamura, M Inagaki, and Y Iwama, Mossbauer Study of Spinodal Decomposition in Fe-Co-Cr Alloys, in "Int Conf Mossbauer Spectrosc, Proc," Vol 1, 107-8 (See 75H027)
75T023	FE-57	M Takeda and N N Greenwood, J Chem Soc, Dalton Trans 2207-12(1975), Tellurium-125 Mossbauer Spectra of Some Mixed Oxides of Tellurium(IV) and Some Mixed-valence Oxides of Tellurium(IV, VI)
75T023	SN-119	M Takeda and N N Greenwood, J Chem Soc, Dalton Trans 2207-12(1975), Tellurium-125 Mossbauer Spectra of Some Mixed Oxides of Tellurium(IV) and Some Mixed-valence Oxides of Tellurium(IV, VI)
75T023	TE-125	M Takeda and N N Greenwood, J Chem Soc, Dalton Trans 2207-12(1975), Tellurium-125 Mossbauer Spectra of Some Mixed Oxides of Tellurium(IV) and Some Mixed-valence Oxides of Tellurium(IV, VI)
75T024	PROPSL	G T Trammell and J P Hannon, Opt Commun 15,325-9(1975), Threshold Conditions for Pulsed Gamma-ray Lasers
75T025	FE-57	A Trautwein and R Zimmermann, Electronic and Magnetic Structure of alpha-FeSO4, in "Int Conf Mossbauer Spectrosc, Proc," Vol 1, pp 243-4 (See 75H027)
75T026	FE-57	C A Taft, Mossbauer Studies of Chemical Structure and Bonding Effects in Hyperfine Interactions in AFeO2 Type Compounds, in "Int Conf Mossbauer Spectrosc, Proc," Vol 1, pp 245-6 (See 75H027)
75T027	FE-57	H Topsoe and S Morup, Mossbauer Effect Study of Hydrodesulfurization Catalysts, in "Int Conf Mossbauer Spectrosc, Proc," Vol 1, pp 305-6 (See 75H027)
75T028	FE-57	H Topsoe and S Morup, Mossbauer Study of the Verwey Transition in Microcrystals of Fe3O4, in "Int Conf Mossbauer Spectrosc, Proc," Vol 1, pp 321-2 (See 75H027)
75T029	FE-57	R C Thiel, H T Le Fever, and F J Van Steenwijk, Very Slow Paramagnetic Relaxation of Ferrous Ions in a Linear Chain System, in "Int Conf Mossbauer Spectrosc, Proc," Vol 1, pp 405-6 (See 75H027)
75T030	INSTRM	T Tomov, T Ruskov, S Georgiev, and N Pavlov, Design of an Electro-mechanical Vibrator and Its Temperature Stabilization, in "5th Int Conf Mossbauer Spectrosc, Proc," Vol 3, pp 653-6 (See 75H017) (In Russian)
75U001	SN-119	D L Uhrich, V O Aimiuwu, P I Ktorides, and W J LaPrice, Phys Rev A 12,211-8(1975), Smectic B Liquid-crystalline Glass (at 77 K) as Seen by the Mossbauer Effect of Sn-bearing Solute Molecules

CODE	TOPIC	REFERENCE

75U002   FE-57   V A Uskov and V I Prudovskii, Izv Akad Nauk SSSR, Neorg Mater 11,158-9(1975)/Inorg Mater (USSR) 11,131-2(1975), Use of the Mossbauer Method to Investigate the States of Iron in Diffusion Layers of Silicon

75U003   FE-57   M R Uhlig, B M Jankowski, J Kansy, T J Panek, and A Salamon, Mossbauer Study of Age Hardening of Fe-Co-Cr-Mo Maraging Steel, in "Int Conf Mossbauer Spectrosc, Proc," Vol 1, pp 117-8 (See 75H027)

75V001   FE-57   F Varret, P Imbert, G Jehanno, and R Saint-James, Phys Status Solidi A 27,K99-101(1975), Compact Linearly Polarized Source for Mossbauer 57Fe Studies

75V002   FE-57   H J Von Bardeleben, A Goltzene, C Schwab, J M Friedt, and R Poinsot, J Appl Phys 46,1736-8(1975), Stoichiometry of the Ternary Semiconductor CuGaS2: 57Fe as Determined by Mossbauer Spectroscopy

75V003   SN-119   A Vertes, S Nagy, I Czako-Nagy, and E Csakvari, J Phys Chem 79,149-51(1975), Mossbauer Study of Equilibrium Constants of Solvates. I. Determination of Equilibrium Constants of Tetraiodotin-trimethylisopropoxysilane and Tetrabromotin-acetic Anhydride Solvates

75V004   ANALYS   E D Von Meerwall, Comput Phys Commun 9,117-28(1975), A Least-squares Spectral Curve Fitting Routine for Strongly Overlapping Lorentzians or Gaussians

75V005   FE-57   F Van der Woude, P J Schurer, and G A Sawatzky, Mossbauer Effect Studies of Magnetic Interactions in Iron and Dilute Iron Alloys, in "5th Int Conf Mossbauer Spec, Proc," Part 1, pp 1-9 (See 75H017)

75V006   FE-57   A M Van der Kraan, J J Van Loef, and W Tolksdorf, Mossbauer Spectra of Gallium Substituted Yttrium-Iron Garnets in the Vicinity of Tc, in "5th Int Conf Mossbauer Spec, Proc," Part 1, pp 79-82 (See 75H017)

75V007   FE-57   M F Vereshchak, M M Goldman, and A K Zhetbaev, Study of the Phase Transitions in NaFeO2 by the Mossbauer Method, in "5th Int Conf Mossbauer Spec, Proc," Part 1, pp 173-5 (See 75H017) (In Russian)

75V008   REVIEW   A Vertes, Mossbauer Effect Studies of Chemical Reactions, in "5th Int Conf Mossbauer Spec, Proc," Part 1, pp 179-99 (See 75H017)

75V009   SN-119   V A Varnek, L I Strugova, and E G Avvakumov, Fiz Tverd Tela (Leningrad) 17,561-4(1975)/Sov Phys-Solid State 17,355-6 (1975), Magnetic Structure of Particles Formed as a Result of a Solid State Reaction Between Tin and Manganese

75V010   FE-57   I Vincze and A T Aldred, Solid State Commun 17,639-41 (1975), Mossbauer Measurements in Iron Base Alloys with Be, Cu and Au Impurities

75V011   FE-57   M F Vereshchak, A K Zhetbaev, and D K Kaipov, Study of the Phase Transformations in the Binary Fe-Al and Ternary Fe-Al-C Alloys by the Mossbauer Effect, in "5th Int Conf Mossbauer Spec, Proc," Part 2, pp 314-7 (See 75H017) (In Russian)

75V012   FE-57   G Vogl and W Manzel, Influence of Interstitial-atom Trapping on the Debye-temperature of Co57 in Al, in "5th Int Conf Mossbauer Spec, Proc," Part 2, pp 453-4 (See 75H017)

75V013   EU-151   F J Van Steenwijk, H T Le Fever, R C Thiel, and K H J Buschow, Physica 79B,604-9(1975), Mossbauer Effect and Magnetic Properties of EuCu5

CODE	TOPIC	REFERENCE
75V014	FE-57	Yu B Voitkovskii, Yu S Yusfin, and M N Shatalov, Izv Vyssh Uchebn Zaved, Chern Metall (7),30-1(1975), Use of the Mossbauer Effect to Study the Calcium Oxide-ferric Oxide System (In Russian)
75V015	SN-119	W Vogl and G Vogl, Solid State Commun 17,1029-33(1975), Defect Cascades and Point Defects in Low-temperature Neutron Irradiated alpha-Tin Monitored by Mossbauer Spectroscopy
75V016	I-129	M Van Rossum, G Langouche, K P Schmidt, and R Coussement, Evidence of a Quadrupole Interaction of 129I Implanted in Iron Foils, in "5th Int Conf Mossbauer Spec, Proc," Part 3, pp 552-4 (See 75H017)
75V017	FE-57	A M Van der Kraan, Mossbauer Effect Studies of Surface Ions of Ultrafine alpha-Fe2O3 Particles, in "5th Int Conf Mossbauer Spec, Proc," Part 3, pp 723-8 (See 75H017)
75V018	FE-57	P O Voznyuk, V N Dubinin, V V Kuz'movich, and N A Ivkina, Ukr Khim Zh 41,706-9(1975)/Ukr J Chem, Nuclear Gamma-resonance Study of the Stability of Silicomolybdic and Silicotungstic Blues and Their Barium Salts
75V019	SN-119	Kh Kh Valiev, K A Duldina, and R N Kuz'min, Mossbauer Effect in Heusler Alloys (Co,Ni)2(Zr,Hf)Sn, in "Int Conf Mossbauer Spectrosc, Proc," Vol 1, pp 87-8 (See 75H027)
75V020	FE-57	A M Van der Kraan, J N J Van der Velden, P C M Gubbens, and K H J Buschow, Mossbauer Effect Study and Magnetization Measurements of Some RFe3 Intermetallic Compounds, in "Int Conf Mossbauer Spectrosc, Proc," Vol 1, pp 179-80 (See 75H027)
75V021	FE-57	J N J Van der Velden, A M Van der Kraan, P C M Gubbens, and K H J Buschow, Mossbauer Effect Study of GdFe2, in "Int Conf Mossbauer Spectrosc, Proc," Vol 1, pp 181-2 (See 75H027)
75V022	AU-197	M P A Viegers, J C H Van Eijkeren, M M Van Deventer, and J M Trooster, Mossbauer Fraction of Gold Microcrystals, in "Int Conf Mossbauer Spectrosc, Proc," Vol 1, p 201 (See 75H027)
75V023	SN-119	M Van Rossum, G Langouche, P Boolchand, M Rots, F Namavar, and R Coussement, Electric Field Gradient and Lattice Location of Tin in Tellurium, in "Int Conf Mossbauer Spectrosc, Proc," Vol 1, pp 205-6 (See 75H027)
75V024	FE-57	A M Van der Kraan, P C M Gubbens, and K H J Buschow, Phys Status Solidi A 31,495-501(1975), Mossbauer Effect Investigation of ErFe3 and YFe3
75V025	ANALYS	J Van Dongen Torman, R Jagannathan, and J M Trooster, Hyperfine Interac 1,135-44(1975), Analysis of 57Fe Mossbauer Hyperfine Spectra
75V025	FE-57	J Van Dongen Torman, R Jagannathan, and J M Trooster, Hyperfine Interac 1,135-44(1975), Analysis of 57Fe Mossbauer Hyperfine Spectra
75V026	FE-57	K Volenik, M Seberini, and J Neid, Czech J Phys B 25,1063-71(1975), A Mossbauer and X-ray Diffraction Study of Non-stoichiometry in Magnetite
75V027	FE-57	H J Von Bardeleben, A Goltzene, and C Schwab, J Phys (Paris), Colloq C3 36,47-51(1975), Transition Metal Ions as Stoichiometry Sensors of CuGaS2
75V028	AU-197	M P A Viegers, A W P G Peters Rit, and J M Trooster, Mossbauer Fraction of a Au(I)-Au(III) Complex, in "Int Conf Mossbauer Spectrosc, Proc," Vol 1, pp 287-8 (See 75H027)

CODE	TOPIC	REFERENCE

75V029   FE-57   A Vertes and I Nagy-Czako, Mossbauer Study of the Parameters of Solvation of Tin Tetrahalides, in "Int Conf Mossbauer Spectrosc, Proc," Vol 1, pp 301-2 (See 75H027)

75V030   FE-57   A Vertes, E Kuzman, S Nagy, L Domonkos, and Z Hegedus, Mossbauer Study of the Corrosion Products of Iron Alloys, in "Int Conf Mossbauer Spectrosc, Proc," Vol 1, pp 317-8 (See 75H027)

75V031   FE-57   Yu B Voitkovskii, T S Gendler, L G Dainyak, and R N Kuz'min, Phase Transformations under Oxidation and Decompotion of Biotite (K0.95Na0.05Ca0.05)(Mg1.05Fe1.18Fe0.23 Mn0.02Al0.41Ti0.11)(Si2.75Al1.25)O10.8(OH)1.2, in "Int Conf Mossbauer Spectrosc, Proc," Vol 1, pp 387-8 (See 75H027)

75V032   FE-57   M E Vol'pin, Yu N Novikov, N D Lapkina, V I Kasatochkin, Yu T Struchkov, M E Kazakov, R A Stukan, V A Povitskii, Yu S Karimov, and A V Zvarikina, J Am Chem Soc 97,3366-73 (1975), Lamellar Compounds of Graphite with Transition Metals. Graphite as a Ligand

75V033   FE-57   Yu B Voitkovskii, O N Generalov, M N Shatalov, M A Butyugin, and Yu S Yusfin, The Study of Reaction Kinetics during Ferrite Synthesis in Systems Fe2O3-MgO and Fe3O4-CaO, in "Int Conf Mossbauer Spectrosc, Proc," Vol 1, pp 327-8 (See 75H027) (In Russian)

75V034   TE-125   R P Vardapetyan, R N Kuz'min, A A Opalenko, and W Fischer, Mossbauer Effect in the Polycrystals of Paratellurite TeO2, in "5th Int Conf Mossbauer Spectrosc, Proc," Vol 3, pp 525-8 (See 75H017) (In Russian)

75V035   AU-197   H Van Kempen, J A A J Pereboom, and M P A Viegers, Susceptibility of the Magnetic Gold Complex (Au(II)(mnt)2) Between 9 mK and 100 K, in "14th International Conference on Low Temperature Physics, Proceedings: Volume 4. Techniques and Special Topics" (Otaniemi, Finland, 1975), edited by M Krusius and M Vuorio (North-Holland Publishing Co, Amsterdam/American Elsevier Publishing Co, New York, 1975), pp 372-5

75V036   XE-129   M Van Rossum, G Langouche, H Pattyn, G Dumont, J Odeurs, A Meykens, R Coussement, and P Boolchand, Lattice Location of Xe in Iron, in "AIP Conference Proceedings-No 24, Magnetism and Magnetic Materials-1974" (20th Annual Conference, San Francisco), edited by C D Graham, Jr, G H Lander, and J J Rhyne (American Institute of Physics, New York, 1975), pp 460-1

75V037   THEORY   R P Vardapetyan, Fiz Tverd Tela 17,1850-2(1975)/Sov Phys-Solid State 17,1215(1975), Inhomogeneous Broadening of the Components of a Mossbauer Doublet of Single-crystal and Polycrystalline Samples

75V038   SB-121   M B Varfolomeev, M N Sotnikova, F Kh Chibirova, and V S Shpinel', Zh Neorg Khim 20,1163-6(1975)/Russ J Inorg Chem 20,655-7(1975), Some Structural Differences in Binary Oxides Formed by Indium and Gallium with Sb(V)

75V039   SN-119   M B Varfolomeev, A S Mironova, F Kh Chibirova, and V E Plyushchev, Izv Akad Nauk SSSR, Neorg Mater 11,2242-4 (1975)/Inorg Mater (USSR), Interaction of Indium Sesquioxide with Stannic Oxide

75V040   SN-119   V A Varnek, E N Yurchenko, V A Kogan, L N Mazalov, Yu K Maksyutin, O Kh Poleshchuk, A S Egorov, and O A Osipov, Zh Strukt Khim 16,359-66(1975)/J Struct Chem (USSR) 16, 337-43(1975), Temperature Dependence of the Resonance Absorption of Gamma Quanta in Complexes of Tin(IV) Chloride with Organic Ligands

CODE	TOPIC	REFERENCE

75V041 FE-57 F Varret and G Jehanno, J Phys (Paris) 36,415-26(1975), Etude Mossbauer des Proprietes Electroniques, Magnetiques et Hyperfines de Fe2+ dans les Fluosilicates

75V042 FE-57 G P Vdovykin, V I Grachev, T V Malysheva, and L M Satarova, Geokhimiya 1872-84(1975), Study of Iron Forms in Carbonaceous Meteorites by the Method of Mossbauer Effect. I. Iron Phases in Carbonaceous Chondrites and Ureilites (In Russian) (Possibly translated in Geochem Int)

75V043 FE-57 Yu S Vishnyakov, L G Lyubutina, and I S Lyubutin, Fiz Tverd Tela 17,629-31(1975)/Sov Phys-Solid State 17,409-10 (1975), Spin Reorientation in Orthoferrite YFe0.9976Co0.002403

75V044 FE-57 Yu B Voitkovskii, M N Shatalov, and Yu S Yusfin, Izv Vyssh Uchebn Zaved, Chern Metall (11),27-9(1975), Analysis of the Chemical Reaction in the Magnesium Oxide-ferric Oxide System (In Russian)

75V045 FE-57 K Volenik, J Cirak, and M Seberini, Br Corros J 10,196-200 (1975), Mossbauer Transmission Spectroscopy of Oxide Films Stripped from Low-carbon Steel

75V046 THEORY V I Vorontsov and V I Vysotskii, Fiz Tverd Tela 17,2944-52 (1975)/Sov Phys-Solid State 17,1959-64(1976), Bragg Diffraction of Mossbauer Radiation in the Case of Hyperfine Splitting

75V047 SN-119 V A Varnek, E N Yurchenko, G L Elizarova, A I Shan'ko, L G Matvienko, P G Antonov, and Yu N Kukushkin, Koord Khim 1 161-4(1975), Mossbauer Effect in Pentacoordinated Complexes of the Platinum Metals with Tin-containing Ligands (In Russian)

75V048 SN-119 L N Vasil'ev and A Sh Bakhtyarov, Izv Akad Nauk SSSR, Neorg Mater 11,2074-6(1975)/Inorg Mater (USSR) 11,1780-1 (1975), Mossbauer Spectra of Alloys Ti2Se-As2Se3-SnSe

75V049 FE-57 A P Vinnikov, L V Zubenko, V P Zubenko, D V Balashov, and E G Semin, Izv Akad Nauk SSSR, Neorg Mater 11,1860-3 (1975)/Inorg Mater (USSR) 11,1594-6(1975), Solubility of alpha-Fe2O3 in BeO, Al2O3, and Synthetic Chrysoberyl

75V050 ANALYS E D Von Meerwall, Comput Phys Commun 9,351-9(1975), A Fortran Code for Automatic Spectrum Analysis on Medium Scale Computers

75V051 FE-57 I M V'yunnik, Fiz Met Metalloved 39,1284-6(1975)/Phys Met Metallogr (USSR), Nuclear Gamma Resonance Study of Carbide Phases Formed as a Result of Tempering under Gamma Irradiation of Steel EI69

75W001 SN-119 W Wilder, H D Pfannes, and U Gonser, Z Metallkd 66,161-4 (1975), Mossbauer Spektroskopie am Schmelzpunkt von Zinn

75W002 SN-119 G Weyer, J U Andersen, B I Deutch, J A Golovchenko, and A Nylandsted-Larsen, Radiat Eff 24,117-21(1975), Direct Comparison of Mossbauer and Channeling Studies of Implanted 119Sn in Silicon Single Crystals

75W003 FE-57 R Wappling, L Haggstrom, T Ericsson, S Devanarayanan, E Karlsson, B Carlsson, and S Rundqvist, J Solid State Chem 13,258-71(1975), First Order Magnetic Transition, Magnetic Structure, and Vacancy Distribution in Fe2P

75W004 FE-57 H P Wit, Rev Sci Instrum 46,927-8(1975), Simple Moire Calibrator for Velocity Transducers Used in Mossbauer Effect Measurements

CODE	TOPIC	REFERENCE

75W004   INSTRM   H P Wit, Rev Sci Instrum 46,927-8(1975), Simple Moire Calibrator for Velocity Transducers Used in Mossbauer Effect Measurements

75W005   FE-57   D L Williamson, Gamma-Fe Thermal Shift, in "5th Int Conf Mossbauer Spec, Proc," Part 1, pp 15-8   (See 75H017)

75W006   FE-57   M Wautelet and A Gerard, Magnetic Properties of the Compound $Zn(Mn(0.98)Fe(0.02))2O4$, in "5th Int Conf Mossbauer Spec, Proc," Part 1, pp 114-8   (See 75H017)

75W007   SN-119   E Wenschuh, W D Riedmann, L Korecz, and K Burger, Z Anorg Allg Chem 413,143-9(1975), N-Triorganostannyl-sulfinamide

75W008   FE-57   H L Wehner and H H F Wegener, Anomalous Spectra of alpha-FeSO4 below and above the Neel Temperature, in "5th Int Conf Mossbauer Spec, Proc," Part 2, pp 474-7 (See 75H017)

75W009   THEORY   C Wissel, Solid State Commun 17,1011-2(1975), Anomalies in the Mossbauer Spectrum due to Soft Modes

75W010   ER-166   K Weber, Magnetization Measurements and Mossbauer 166Er Polarimetry on the Metamagnetic Phase Transitions in Hexagonal ErAl3 Single Crystals, in "5th Int Conf Mossbauer Spec, Proc," Part 3, pp 504-10   (See 75H017)

75W011   REVIEW   J M Williams, Cryogenics 15,307-22(1975), The Mossbauer Effect and Its Applications at Very Low Temperatures

75W012   INSTRM   M Wroblewski, J Mirkowski, and J Wroblewska, Camac Mossbauer Spectrometer System with Small On-line Computer for Multi-spectra Recording and Processing, in "Int Conf Mossbauer Spectrosc, Proc," Vol 1, pp 31-2   (See 75H027)

75W013   FE-57   M Wautelet, A Gerard, and F Grandjean, Band Model for Transition-metal Phosphides and Arseno-phosphides, in "Int Conf Mossbauer Spectrosc, Proc," Vol 1, pp 63-4 (See 75H027)

75W014   FE-57   J M Williams and J B Dunlop, 57Fe Isomer Shift and Quadrupole Splitting Measurements in the Intermetallic Compound Zn13Fe, in "Int Conf Mossbauer Spectrosc, Proc," Vol 1, pp 121-2   (See 75H027)

75W015   FE-57   D L Williamson and W Keune, Magnetic Properties of alpha-Fe Precipitates in Cu, in "Int Conf Mossbauer Spectrosc, Proc," Vol 1, pp 133-4   (See 75H027)

75W016   FE-57   G Weyer, B I Deutch, A Nylandsted-Larsen, and O Holck, Mossbauer Studies on Isotope Separator Implanted 57Co in Group IV Elements, in "Int Conf Mossbauer Spectrosc, Proc," Vol 1, pp 213-4   (See 75H027)

75W017   FE-57   J B Ward, Scr Metall 9,1211-8(1975), An 57Fe and 57Co Mossbauer Study of Vacancy-impurity Binding in Aluminium

75W018   FE-57   P Wende, F Seel, Y Maeda, H E Marcolin, and A Trautwein, Mossbauer Study of the Polymeric System Fe-imidazole I. Unsubstituted Fe-imidazole, in "Int Conf Mossbauer Spectrosc, Proc," Vol 1, pp 271-2   (See 75H027)

75W019   TA-181   P J West, E Matthias, D Salomon, W Wallner, and G Weyer, Mossbauer Conversion Electron Studies of the Interference Between Photoeffect and Internal Conversion in the Absorption of 6.2 keV -E1 Gamma Radiation of 181Ta, in "Int Conf Mossbauer Spectrosc, Proc," Vol 1, pp 457-8   (See 75H027)

75W020   IR-193   A F Williams, S Bhaduri, and A G Maddock, J Chem Soc, Dalton Trans 1958-62(1975), Mossbauer Spectroscopy of Iridium Compounds.   Part II. Some Iridium(I) Compounds

75W021   IR-193   A P Williams, G C H Jones, and A G Maddock, J Chem Soc,
                  Dalton Trans 1952-7(1975), Mossbauer Spectroscopy of Iri-
                  dium Compounds.   Part I. Some Iridium(III) Complexes

75W022   SN-119   J L Wardell, J Chem Soc, Dalton Trans 1786-93(1975),
                  Sulphur-substituted Organometallic Compounds.   Part II.
                  Reaction of Vinyltin Compounds with Arenesulphenyl Halides
                  and Thiocyanates and Some Properties and Reactions of the
                  (2-Arylthio-1-halogenoethyl)-triphenyltin Addition Products

75W023   FE-57   W E Wallace, A S Ilyushin, and D Lopez, Study of Ternary
                  Alloy Systems Containing Fe-Mn and Fe-Co, in "Int Conf
                  Mossbauer Spectrosc, Proc," Vol 2, pp 99-112
                  (See 75H038)

75W024   REVIEW   H H F Wegener, The Study of Relaxation Mechanisms by Means
                  of the Mossbauer Spectroscopy, in "Int Conf Mossbauer
                  Spectrosc, Proc," Vol 2, pp 257-74 (See 75H038)

75W025   REVIEW   W K Wojtowiecki and S B Sazonov, The Amplitude and Phase
                  Modulation of Mossbauer Gamma-quanta, in "Int Conf Moss-
                  bauer Spectrosc, Proc," Vol 2, pp 399-411 (See 75H038)

75W026   SN-119   G Weyer, A Nylandsted-Larsen, B I Deutch, J U Andersen,
                  and E Antoncik, Hyperfine Interac 1,93-112(1975), Covalency
                  Effects on Implanted 119Sn in Group IV Semiconductors
                  Studied by Mossbauer and Channeling Experiments

75W027   FE-57   J M Williams and D I C Pearson, Anomalous Behaviour of the
                  Low Temperature Electric Quadrupole Splitting in Hexagonal
                  Fe-Ru Alloys in an External Magnetic Field, in "14th In-
                  ternational Conference on Low Temperature Physics, Proceed-
                  ings: Volume 3. Low Temperature Properties of Solids"
                  (Otaniemi, Finland, 1975), edited by M Krusius and M Vuorio
                  (North-Holland Publishing Co, Amsterdam/American Elsevier
                  Publishing Co, New York, 1975), pp 266-9

75W028   ANALYS   J M Williams and J S Brooks, Nucl Instrum Methods 128,
                  363-72(1975), The Thickness Dependence of Mossbauer Absorp-
                  tion Line Areas in Unpolarized and Polarized Absorbers

75W028   FE-57   J M Williams and J S Brooks, Nucl Instrum Methods 128,
                  363-72(1975), The Thickness Dependence of Mossbauer Absorp-
                  tion Line Areas in Unpolarized and Polarized Absorbers

75W029   FE-57   G Wortmann and D L Williamson, Hyperfine Interac 1,167-76
                  (1975), Electric Field Gradient at Fe in H.C.P. Transition
                  Metals from Mossbauer Measurements

75W030   FE-57   G Wortmann, N S Ovanesyan, V A Trukhtanov, and N I
                  Bezmen, Zh Eksp Teor Fiz 69,2093-2100(1975)/Sov Phys-JETP,
                  Magnetic Hyperfine Field at Nuclei of the Nonmagnetic $Fe2+$
                  Ions in Ferromagnetic $Co0.97Fe0.03S2$

75Y001   INSTRM   T Yoshimura, M Fujiwara, and N Wakabayashi, Jpn J Appl Phys
                  14,691-5(1975), A Derivative Mossbauer Spectrometer Stabil-
                  ized by Digital Method

75Y002   FE-57   T Yoshihashi and H Sano, Chem Phys Lett 34,289-91(1975),
                  Mossbauer Spectroscopic Studies of the Crystallographic
                  Phase Transition in 57Fe-doped $Mn(py)2Cl2$ and $Ni(py)2Cl2$

75Y003   DY-161   J K Yakinthos and J Chappert, Solid State Commun 17,979-81
                  (1975), Crystal Field and Mossbauer Spectroscopy of the
                  Intermetallic Compound DyCo3

75Y004   DY-161   V K Yarmarkin, B A Shustrov, and A V Motorny, Mossbauer
                  Effect on Dy161, Sb121, and Fe57 Nuclei in BaTiO3 Ceramics,
                  in "Int Conf Mossbauer Spectrosc, Proc," Vol 1, pp 325-6
                  (See 75H027)

CODE	TOPIC	REFERENCE

75Y004   FE-57   V K Yarmarkin, B A Shustrov, and A V Motorny, Mossbauer Effect on Dy161, Sb121, and Fe57 Nuclei in BaTiO3 Ceramics, in "Int Conf Mossbauer Spectrosc, Proc," Vol 1, pp 325-6 (See 75H027)

75Y004   SB-121   V K Yarmarkin, B A Shustrov, and A V Motorny, Mossbauer Effect on Dy161, Sb121, and Fe57 Nuclei in BaTiO3 Ceramics, in "Int Conf Mossbauer Spectrosc, Proc," Vol 1, pp 325-6 (See 75H027)

75Y005   YB-170   R Yanovsky, E R Bauminger, D Levron, I Nowik, and S Ofer, Solid State Commun 17,1511-4(1975), Crystalline Fields, Exchange Interactions and Spin Reorientations in TmxHo(1-x)Fe2 Systems, Studied by a Yb170 Mossbauer Probe

75Y006   SN-119   E N Yurchenko, V A Varnek, G L Elizarova, and L G Matvienko, Koord Khim 1,1406-14(1975), Study of Tin-containing Ligands in Complexes of Platinum Metals by Nuclear Gamma Resonance, X-ray Photoelectron and IR Spectroscopic, and MO LCAO Methods (In Russian)

75Z001   TE-125   V S Zasimov and R N Kuz'min, Phys Status Solidi B 70,K55-7 (1975), Diffraction of Resonant Gamma-quanta in Tellurium Single Crystals for Three Orders of Reflection

75Z002   FE-57   T Zemcik and J Vrestal, Temperature Dependence of Atomic Rearrangement in the Ni3Co Alloy, in "5th Int Conf Mossbauer Spec, Proc," Part 2, pp 305-6 (See 75H017)

75Z003   FE-57   A K Zhetbaev, D K Kaipov, and K K Satpaev, Mossbauer Effect of a Mixture of Iron Atoms in Monocrystalline Samples of Mica, Asbestos and Quartz, in "5th Int Conf Mossbauer Spec, Proc," Part 2, pp 364-7 (See 75H017) (In Russian)

75Z004   FE-57   T Zemcik, Refined Mossbauer Measurements of Lunar Silicates and Metallic Iron, in "5th Int Conf Mossbauer Spec, Proc," Part 2, pp 388-92 (See 75H017)

75Z005   FE-57   T Zemcik, J Lauermannova, and K Raclavsky, Mossbauer Micro-measurement of Fine Lunar Particles, in "5th Int Conf Mossbauer Spec, Proc," Part 2, pp 392-4 (See 75H017)

75Z006   FE-57   R Zimmermann, H Spiering, and G Ritter, Relaxation Phenomena in Mossbauer Spectra of Paramagnetic Iron(II) High-spin Complexes in External Magnetic Fields, in "5th Int Conf Mossbauer Spec, Proc," Part 2, pp 477-80 (See 75H017)

75Z007   FE-57   I S Zheludev, T M Perekalina, V G Pyl'nev, E M Smirnovskaya, V F Belov, A M Kostsov, and Yu N Yarmukhamedov, Izv Akad Nauk SSSR, Ser Fiz 39,724-7 (1975)/Bull Acad Sci USSR Phys Ser 39(4),72-5(1975), The Structure and Properties of Certain Boracites

75Z008   FE-57   R Zimmermann, A Trautwein, and F E Harris, Phys Rev B 12, 3902-7(1975), Electronic and Magnetic Structure of alpha-FeSO4

75Z009   THEORY   R Zimmermann, Chem Phys Lett 34,416-8(1975), Description of the Angular Dependence of Dipole Transitions by Intensity Tensors

75Z010   FE-57   J Zemcikova, Iron Magnetic Moments and Distribution of Ni Atoms in Quenched Fe-Ni Alloys, in "Int Conf Mossbauer Spectrosc, Proc," Vol 1, pp 75-6 (See 75H027)

75Z011   FE-57   T Zemcik, Ordering Transformation in an Fe-Si Alloy, in "Int Conf Mossbauer Spectrosc, Proc," Vol 1, pp 111-2 (See 75H027)

CODE	TOPIC	REFERENCE

75Z012   FE-57   R Zimmermann, Nucl Instrum Methods 128,537-43(1975), A Method for Evaluation of Single Crystal 57Fe Mossbauer Spectra (FeCl2-4H2O)

75Z013   FE-57   R Zimmermann, G Ritter, and H Spiering, Temperature Dependence of Spin-lattice Relaxation in Ferrous Compounds, in "Int Conf Mossbauer Spectrosc, Proc," Vol 1, pp 401-2 (See 75H027)

75Z014   REVIEW   T Zemcik, Cesk Cas Fyz 25,464-75(1975), Information on the Electronic Structure of Metals and Alloys from the Hyperfine Structure of Mossbauer Spectra (In Czech)

75Z015   FE-57   T Zemcik and K Baclavsky, Mossbauer Measurements on Soviet Luna-16 and Luna-20 Particles, in "Int Conf Mossbauer Spectrosc, Proc," Vol 1, pp 473-4 (See 75H027)

75Z016   FE-57   K Zapletal, Mossbauer Spectroscopy Study of Phases in Natural Troilite, in "Int Conf Mossbauer Spectrosc, Proc," Vol 1, pp 477-8 (See 75H027)

75Z017   FE-57   T Zak and T Zemcik, Hysteresis of Phase Transformation in an Fe-Mn Alloy, in "Int Conf Mossbauer Spectrosc, Proc," Vol 1, pp 511-2 (See 75H027)

75Z018   REVIEW   T Zemcik, Progress in Mossbauer Spectroscopy of Metallic Systems, in "Int Conf Mossbauer Spectrosc, Proc," Vol 2, pp 59-81 (See 75H038)

75Z019   FE-57   R R Zakiorv, G M Bartenev, and A D Tsyganov, Zh Strukt Khim 16,610-5(1975)/J Struct Chem (USSR) 16,568-72(1975), Temperature Dependence of the Mossbauer Effect in Garnet-like Ferric Molybdate

75Z020   TE-125   V S Zasimov, B N Kuz'min, and A Yu Aleksandrov, Fiz Tverd Tela 17,3083-6(1975)/Sov Phys-Solid State 17,2044-5(1976), Diffraction of 35.6 keV Resonance Gamma Rays in a Tellurium Single Crystal

75Z021   FE-57   A I Zakharov, B V Molotilov, V A Makarov, and L V Pastukhova, Fiz Met Metallov 40,668-70(1975)/Phys Met Metallogr (USSR), Nuclear Gamma Resonance Study of the Redistribution of Carbon Atoms in Invar

75Z022   SN-119   V S Zavgorodnii, E T Bogradovskii, V L Maksimov, V B Lebedev, B I Rogozev, and A A Petrov, Zh Obshch Khim 45, 2466-71(1975)/J Gen Chem (USSR) 45,2421-5(1975), Unsaturated Stanna Hydrocarbons. Application of 19F NMR Spectra in the Investigation of the Electron Structure of Pentafluorophenylethynl Derivatives of Group IV B Elements

75Z023   PROCES   E V Zolotoyabko and E M Iolin, Latv PSR Zinat Akad Vestis, Fiz Teh Zinat Ser (4),46-50(1975), Study of Structural Phase Transitions by Using the Coherent Scattering of Mossbauer Radiation (In Russian)

1975 Alphabetical Author Index

317

319

321

322

324

328

329

Donaldson, J D   (continued)
  J D Donaldson, A Kjekshus, D G Nicholson,
  and T Rakke
    SB-121 . . . . 75D013

  J D Donaldson, J Silver, S Hadjiminolis,
  and S D Ross
    SN-119 . . . . 75D049

  J D Donaldson, S D Ross, J Silver,
  and P J Watkiss
    SN-119 . . . . 75D056

  J D Donaldson, A Kjekshus, D G Nicholson,
  and T Rakke
    SB-121 . . . . 75D074

Dora, D
  D Dora, E Kisch, and L Varhaemi
    INSTRM 75D067

Dosmaganbetov, T
  T Dosmaganbetov, S A Bashanov, A K Zhetbaev,
  L S Sergeeva
    FE-57 . . . . 75D037

Douek, I
  D Cunningham, I Douek, M J Frazer,
  M McPartlin, and J D Matthews
    SN-119 . . . . 75C006

Drager, G
  O Brummer, G Drager, and D Katzer
    FE-57 . . . . 75B051

  O Brummer, K Berndt, G Dlubek, U Berg,
  G Drager, W Beier, and G Dworzak
    REVIEW 75B147

Drago, R S
  M A Hoselton, L J Wilson, and R S Drago
    FE-57 . . . . 75H031

Drago, V
  V Drago, E Baggio-Saitovitch, and J Danon
    FE-57 . . . . 75D038

Dragunas, A K
  A K Dragunas and K V Makaryunas
    TE-125 . . . . 75D063

  K V Makaryunas, E K Makaryuene,
  A K Dragunas, and M L Bal'chyuene
    I-127,TE-125 . . . . 75M065

Drentje, S A
  S R Reintsema, S A Drentje, and H De Waard
    CS-133 . . . . 74R042

  S R Reintsema, S A Drentje, P J Schurer,
  and H De Waard
    CS-133 . . . . 75R003

  S R Reintsema, H De Waard, and S A Drentje
    I-129 . . . . 75R028

Drijver, J W
  J W Drijver, F Van der Woude, and S Radelaar
    FE-57 . . . . 75D009

  J W Drijver and F Van der Woude
    FE-57 . . . . 75D014

Drokin, A I
  O A Bayukov, G I Gashimov, A I Drokin,
  V P Klonnikov, M I Petrov, A G Rustamov,
  and E A Eivazov
    FE-57 . . . . 75B034

Drost, H
  H Drost, H Von Lojewski, K Palow,
  R Wallenstein, and G Weyer
    FE-57 . . . . 75D027

Drwiega, M
  J Stanek, J A Sawicki, B D Sawicka,
  and M Drwiega
    FE-57 . . . . 75S061

Dubiel, S M
  S M Dubiel, J Zukrowski, J Korecki,
  and K Krop
    FE-57 . . . . 75D031

  S M Dubiel and J Zukrowski
    FE-57 . . . . 75D032

  S M Dubiel, J Zukrowski, and L Kozlowski
    FE-57 . . . . 75D033

  S M Dubiel, J Zukrowski, and I Vincze
    FE-57 . . . . 75D035

Dubinin, V N
  P O Voznyuk, V N Dubinin, V V Kuz'movich,
  and N A Ivkina
    FE-57 . . . . 74V040

  P O Voznyuk and V N Dubinin
    FE-57 . . . . 74V041

  V N Dubinin, P O Voznyuk, V V Kuz'movich,
  and N A Ivkina
    FE-57 . . . . 75D029

  P O Voznyuk, V N Dubinin, V V Kuz'movich,
  and N A Ivkina
    FE-57 . . . . 75V018

Dublon, G
  G Dublon, U Atzmony, M P Dariel,
  and H Shaked
    FE-57 . . . . 75D075

Dudas, J
  J Dudas and T Zemcik
    FE-57 . . . . 75D077

Dudreva, B Ts
  B Ts Dudreva and R K Pirinchieva
    FE-57 . . . . 74D058

Duff, K J
  K J Duff
    FE-57 . . . . 75D073

Duhaj, P
  J Sitek, M Prejsa, P Duhaj, M Hucl,
  and J Cirak
    FE-57 . . . . 75S070

Duldina, K A
  Kh Kh Valiev, K A Duldina, and R N Kuz'min
    SN-119 . . . . 75V019

Dumesic, J A
  M Boudart, A Delbouille, J A Dumesic,
  S Khammouma, and H Topsoe
    FE-57 . . . . 75B058

  J A Dumesic, H Topsoe, and M Boudart
    FE-57 . . . . 75D051

  J A Dumesic, H Topsoe, and M Boudart
    FE-57 . . . . 75D057

Dumont, G
  R Coussement, G Dumont, G Langouche,
  H Pattyn, M Rots, K P Schmidt,
  and M Van Rossum
    I-129 . . . . 75C023

  M Van Rossum, G Langouche, H Pattyn,
  G Dumont, J Odeurs, A Meykens, R Coussement,
  and P Boolchand
    XE-129 . . . . 75V036

Duncken, H H
  B Mehliss and H H Duncken
    FE-57,SN-119 . . . . 74M060

Dunham, W R
  R E Anderson, W R Dunham, R H Sands,
  A J Bearden, and H L Crespi
    FE-57 . . . . 75A045

  R H Sands and W R Dunham
    FE-57,REVIEW 75S090

333

Erkin, V M
  G N Belozerskii, V M Erkin,
  V A Malyshevskii, O G Sokolov,
  and Yu P Khimich
    FE-57 . . . . 75B078

Espinosa, G P
  R M Housley, S Geller, and G P Espinosa
    FE-57 . . . . 75H043

Ettwig, H H
  H H Ettwig and W Pepperhoff
    FE-57 . . . . 75E005

Evans, B J
  B J Evans
    FE-57 . . . . 75E016

  B J Evans and L J Swartzendruber
    FE-57 . . . . 75E027

  B J Evans and E W Sergent, Jr
    FE-57 . . . . 75E028

Evdokimov, M D
  M D Evdokimov and S B Tomilov
    FE-57 . . . . 74E022

Evstigneeva, R P
  V N Luzgina, E I Filippovich,
  R P Evstigneeva, and A F Vanin
    FE-57 . . . . 75L017

Evstyukhina, I A
  P L Gruzin, Yu P Bychkov, I A Evstyukhina,
  and L A Alekseev
    SN-119 . . . . 74G065

Ewings, P F R
  P F R Ewings, P G Harrison, and D E Fenton
    SN-119 . . . . 75E017

  P F R Ewings and P G Harrison
    SN-119 . . . . 75E018

  P F R Ewings and P G Harrison
    SN-119 . . . . 75E019

Faleev, D S
  D S Faleev
    SN-119 . . . . 73F033

Fan, A
  W M Reiff, I E Grey, A Fan, Z Eliezer,
  and H Steinfink
    FE-57 . . . . 75R001

Fano, V
  I Ortalli and V Fano
    SN-119 . . . . 75O004

Faurskov Nielsen, O
  J M Knudsen, E Larsen, J E Moreira,
  and O Faurskov Nielsen
    FE-57 . . . . 75K062

Fayek, M K
  M K Fayek, A A Bahgat, and Y M Abass
    FE-57 . . . . 75F030

Fedorin, V L
  V I Gudov, V I Stepanenko, V L Fedorin,
  and V S Shkalikov
    SN-119,THEORY 74G067

Fedorov, V E
  L M Belyaev, T V Dmitrieva, I S Lyubutin,
  A P Mazhara, and V E Fedorov
    FE-57 . . . . 75B125

Fedro, A J
  R S Preston, A E Dwight, A J Fedro,
  and C W Kimball
    FE-57 . . . . 75P052

Feiccabrino, J A
  J A Feiccabrino and E J Kupchik
    SN-119 . . . . 75F006

Feiccabrino, J A   (continued)
  E J Kupchik and J A Feiccabrino
    SN-119 . . . . 75K027

Felner, I
  E R Bauminger, A Diamant, I Felner, I Nowik,
  and S Ofer
    GD-155 . . . . 75B022

  E R Bauminger, I Felner, D Levron, I Nowik,
  and S Ofer
    EU-151 . . . . 75B074

  I Felner, I Mayer, A Grill, and M Schieber
    FE-57 . . . . 75F020

Fenger, J
  E B Andersen, J Fenger, and J Rose-Hansen
    FE-57 . . . . 75A026

Fenton, D E
  P F R Ewings, P G Harrison, and D E Fenton
    SN-119 . . . . 75E017

Figueras, F
  R Bacaud, P Bussiere, F Figueras,
  and J P Mathieu
    SN-119 . . . . 75B067

  R Bacaud, P Bussiere, R Dutartre,
  F Figueras, G A Martin, and J P Mathieu
    FE-57,SN-119 . . . . 75B106

Filin, M V
  A N Salugin, V A Povitskii, M V Filin,
  N V Elistratov, and A F Pis'marov
    FE-57 . . . . 75S089

Filin, V M
  V M Filin, V I Gol'danskii, and V S Nefedov
    NP-237 . . . . 75F028

  V M Filin, V I Gol'danskii, and V S Nefedov
    NP-237 . . . . 75F029

Filippovich, E I
  V N Luzgina, E I Filippovich,
  R P Evstigneeva, and A F Vanin
    FE-57 . . . . 75L017

Filoti, G
  G Filoti, A Gelberg, V Spanu, P Telnic,
  and H Niculescu-Majewska
    FE-57 . . . . 75F004

  V Florescu, G Filoti, and D Barb
    FE-57 . . . . 75F018

  G Filoti and C N Turcanu
    FE-57 . . . . 75F021

  G Filoti, C N Turcanu, and C Oproiu
    FE-57 . . . . 75F022

  G Filoti
    FE-57 . . . . 75F031

  M Macrin, P Telnic, A Gelberg, and G Filoti
    FE-57 . . . . 75M020

  M Teodorescu and G Filoti
    FE-57 . . . . 75T008

Fink, J
  G Czjzek, J Fink, H Schmidt, F Gautier,
  G Krill, M F Lapierre, P Panissod,
  and C Robert
    NI-61 . . . . 75C030

  J Fink, G Czjzek, and H Schmidt
    NI-61 . . . . 75F016

  F Gautier, G Krill, M F Lapierre,
  P Panissod, C Robert, G Czjzek, J Fink,
  and H Schmidt
    NI-61 . . . . 75G064

Firov, A I
  V S Zasimov, R N Kuz'min, and A I Firov
    TE-125 . . . . 74Z009

Ivanitskii, V P   (continued)
  V P Ivanitskii, A M Kalinichenko,
  I V Matyash, and T P Khomyak
    FE-57 . . . . 75I016

  V P Ivanitskii, A V Smirnova,
  and K N Alekseeva
    FE-57 . . . . 75I017

  V P Ivanitskii, I V Matyash,
  and P I Rakovich
    FE-57 . . . . 75I018

Ivanov, L I
  I Ya Dekhtyar, L I Ivanov, N A Litvinova,
  and M M Nishchenko
    FE-57 . . . . 75D040

Ivanov, L N
  L N Ivanov and V S Letokhov
    THEORY 75I020

Ivanov, O A
  V F Kumeishin and O A Ivanov
    FE-57 . . . . 75K086

Ivanov, V A
  V O Reikhsfel'd, V A Ivanov, and I E Saratov
    SN-119 . . . . 75R032

Ivanyushkin, E M
  S S Budagovskii, V N Bykov, M I Gavrilyuk,
  E M Ivanyushkin, and I I Rudnev
    FE-57 . . . . 75B060

Ivkina, N A
  P O Voznyuk, V N Dubinin, V V Kuz'movich,
  and N A Ivkina
    FE-57 . . . . 74V040

  V N Dubinin, P O Voznyuk, V V Kuz'movich,
  and N A Ivkina
    FE-57 . . . . 75D029

  P O Voznyuk, V N Dubinin, V V Kuz'movich,
  and N A Ivkina
    FE-57 . . . . 75V018

Ivoilov, N G
  Sh Sh Bashkirov, N G Ivoilov, I K Kosterina,
  and V A Chistyakov
    FE-57 . . . . 74B096

  Sh Sh Bashkirov, R K Gubaidullin,
  N G Ivoilov, and V A Chistyakov
    FE-57 . . . . 75B079

Iwama, Y
  R Tahara, Y Nakamura, M Inagaki, and Y Iwama
    FE-57 . . . . 75T022

Iyengar, P K
  K R P M Rao and P K Iyengar
    FE-57 . . . . 74R034

  K R P M Rao and P K Iyengar
    FE-57 . . . . 75R015

Izawa, M
  M Mishima, M Nakamura, M Izawa,
  and M Idogaki
    SN-119 . . . . 74M062

Izumi, M
  H Inoue, M Izumi, and E Imoto
    FE-57 . . . . 74I016

Jach, J
  J Jach
    FE-57 . . . . 74J016

Jacobasch, H J
  W Meisel, H J Jacobasch, H Mehner,
  and V Jacopian
    FE-57 . . . . 75M053

Jacobs, I S
  A E Berkowitz, J A Lahut, I S Jacobs,
  L M Levinson, and D W Forester
    FE-57 . . . . 75B048

  L M Levinson, I S Jacobs, C Greskovich,
  and G H Glover
    FE-57 . . . . 75L027

Jacopian, V
  V Jacopian, B Philipp, H Mehner, J Schulze,
  and H Dautzenberg
    FE-57 . . . . 75J004

  W Meisel, H J Jacobasch, H Mehner,
  and V Jacopian
    FE-57 . . . . 75M053

Jadhao, V G
  V G Jadhao, G N Rao, R M Singru, D Bahadur,
  and C N R Rao
    EU-151,FE-57 . . . . 74J017

  V G Bhide, D S Rajoria, C N R Rao,
  G Rama Rao, and V G Jadhao
    FE-57 . . . . 75B105

  V G Jadhao, R M Singru, G Rama Rao,
  D Bahadur, and C N R Rao
    FE-57 . . . . 75J008

Jagannathan, R
  J Van Dongen Torman, R Jagannathan,
  and J M Trooster
    ANALYS,FE-57 . . . . 75V025

Jankowski, B M
  J E Frackowiak, B M Jankowski, and T Panek
    FE-57 . . . . 75F008

  B M Jankowski, J E Frackowiak, J Kansy,
  J W Moron, T J Panek, A Salamon,
  and M R Uhlig
    FE-57 . . . . 75J007

  J Kansy, J E Frackowiak, B M Jankowski,
  and T J Panek
    ANALYS 75K042

  M R Uhlig, B M Jankowski, J Kansy,
  T J Panek, and A Salamon
    FE-57 . . . . 75U003

Janot, C
  C Janot and P Delcroix
    REVIEW 74J018

  C Janot and P Delcroix
    FE-57 . . . . 75J002

  M Piecuch and C Janot
    THEORY 75P049

Jauho, A P
  M T Hirvonen, A P Jauho, T E Katila,
  and K J Riski
    FE-57 . . . . 75H042

Jeandey, C
  C Jeandey and P Peretto
    HF-178 . . . . 75J001

  C Jeandey and P Peretto
    FE-57 . . . . 75J009

  C Jeandey and P Peretto
    FE-57 . . . . 75J012

Jefferson, D A
  D A Jefferson, M J Tricker,
  and A P Winterbottom
    FE-57 . . . . 75J010

Jehanno, G
  J Chappert, G Jehanno, and F Varret
    FE-57 . . . . 75C034

  F Varret, P Imbert, G Jehanno,
  and R Saint-James
    FE-57 . . . . 75V001

Kozlowski, L
  S M Dubiel, J Zukrowski, and L Kozlowski
    FE-57 . . . . 75D033

Kraczka, J
  D S Kulgawczuk, E Mazanek, M Wyderko,
  and J Kraczka
    FE-57 . . . . 75K050

Kravtsov, A V
  Yu V Maksimov, I P Suzdalev, A I Matveev,
  E F Makarov, and A V Kravtsov
    FE-57 . . . . 75M056

  Yu V Maksimov, I P Suzdalev, A I Matveev,
  E F Makarov, and A V Kravtsov
    FE-57 . . . . 75M072

Kreber, E
  E Kreber, U Gonser, and A Trautwein
    FE-57 . . . . 75K003

  E Kreber, U Gonser, A Trautwein,
  and F E Harris
    FE-57 . . . . 75K018

  A Trautwein, E Kreber, U Gonser,
  and F E Harris
    FE-57 . . . . 75T003

Krill, G
  G Czjzek, J Fink, H Schmidt, F Gautier,
  G Krill, M F Lapierre, P Panissod,
  and C Robert
    NI-61 . . . . 75C030

  F Gautier, G Krill, M F Lapierre,
  P Panissod, C Robert, G Czjzek, J Fink,
  and H Schmidt
    NI-61 . . . . 75G064

Krishnamurthy, B
  B Krishnamurthy and K P Sinha
    FE-57,THEORY 75K016

Kritidis, P
  P Kritidis, Cv Bontschev, and S Micheva
    FE-57 . . . . 75K068

Krizhanskii, L M
  K K Khristoforov, L P Nikitina,
  L M Krizhanskii, and S P Eknmov
    FE-57 . . . . 73K053

  L P Nikitina, S P Ekimov, L M Krizhanskii,
  R G Grebenshchikov, and V V Motsartov
    FE-57 . . . . 74N021

  S P Ekimov, L P Nikitina,
  R G Grebenshchikov, L M Krizhanskii,
  O G Chigareva, and V V Motsartov
    FE-57 . . . . 75E013

Krop, K
  S M Dubiel, J Zukrowski, J Korecki,
  and K Krop
    FE-57 . . . . 75D031

  K Krop, J Korecki, J Zukrowski, W Karas,
  and J M Williams
    FE-57 . . . . 75K032

Kroth, H J
  H J Kroth, H Schumann, H G Kuivila,
  C D Schaeffer, Jr, and J J Zuckerman
    SN-119 . . . . 75K017

Kruck, T
  H J Odenthal, T Kruck, and K Ehlert
    SN-119 . . . . 75O014

Krupanskii, Yu F
  Yu F Krupanskii and I P Suzdalev
    FE-57 . . . . 74K069

  Yu F Krupanskii and I P Suzdalev
    FE-57 . . . . 75K023

  Yu F Krupanskii and I P Suzdalev
    FE-57 . . . . 75K058

Krupanskii, Yu F   (continued)
  Yu F Krupanskii and I P Suzdalev
    FE-57 . . . . 75K069

Krylov, O V
  Yu V Maksimov, I P Suzdalev,
  V I Gol'danskii, O V Krylov, L Ya Margolis,
  and A E Nechitailo
    FE-57 . . . . 75M018

  Yu V Maksimov, I P Suzdalev,
  V I Gol'danskii, O V Krylov, L Ya Margolis,
  and A E Nechitailo
    FE-57 . . . . 75M071

  T Notermann, G W Keulks, A Skliarov,
  Yu Maximov, L Ya Margolis, and O V Krylov
    FE-57 . . . . 75N015

Ktorides, P I
  D L Uhrich, V O Aimiuwu, P I Ktorides,
  and W J LaPrice
    SN-119 . . . . 75U001

Kubsch, H
  H Kubsch, E Fritzsch, and H A Schneider
    FE-57 . . . . 75K021

  H Kubsch and H A Schneider
    FE-57 . . . . 75K037

  H Kubsch and H A Schneider
    FE-57 . . . . 75K056

Kudryashova, V A
  N D Devyatkov, V V Khrapov, R E Garibov,
  V A Kudryashova, V I Gaiduk, G P Bakaushina,
  A M Khrapko, A A Levina, and A P Andreeva
    FE-57 . . . . 75D076

Kuivila, H G
  H J Kroth, H Schumann, H G Kuivila,
  C D Schaeffer, Jr, and J J Zuckerman
    SN-119 . . . . 75K017

Kukushkin, Yu N
  V A Varnek, E N Yurchenko, G L Elizarova,
  A I Shan'ko, L G Matvienko, P G Antonov,
  and Yu N Kukushkin
    SN-119 . . . . 75V047

Kulcsar, K
  D L Nagy, K Kulcsar, G Ritter, H Spiering,
  H Vogel, R Zimmermann, I Dezsi,
  and M Pardavi-Horvath
    FE-57 . . . . 75N002

  D L Nagy, I Dezsi, and K Kulcsar
    ANALYS 75N019

Kulgawczuk, D S
  A Antonow, A Z Hrynkiewicz, D S Kulgawczuk,
  S Lasocki, and B Jezowska-Trzebiatowska
    FE-57 . . . . 75A020

  D S Kulgawczuk, E Mazanek, M Wyderko,
  and J Kraczka
    FE-57 . . . . 75K050

  D S Kulgawczuk, H Przywarska-Boniecka,
  L Trynda, and K Dolegowska
    FE-57 . . . . 75K059

Kulikov, L A
  Yu D Perfil'ev, M I Afanasov, L A Kulikov,
  and A M Babeshkin
    FE-57 . . . . 74P053

  R A Lebedev, Yu D Perfil'ev, L A Kulikov,
  M I Afanasov, A M Babeshkin,
  and A N Nesmeyanov
    FE-57,I-129,SB-121,TE-125 . . . . 75L024

Kumar, S
  S Kumar
    FE-57,THEORY 75K034

Kumeishin, V F
  V F Kumeishin and O A Ivanov
    FE-57 . . . . 75K086

355

360

Makarov, V A   (continued)
  A I Zakharov, B V Molotilov, V A Makarov,
  and L V Pastukhova
    FE-57 . . . . 75Z021

Makaryuene, E K
  K V Makaryunas, E K Makaryuene,
  A K Dragunas, and M L Bal'chyuene
    I-127,TE-125 . . . . 75M065

Makaryunas, K V
  A K Dragunas and K V Makaryunas
    TE-125 . . . . 75D063

  K V Makaryunas
    THEORY 75M063

  K V Makaryunas, E K Makaryuene,
  A K Dragunas, and M L Bal'chyuene
    I-127,TE-125 . . . . 75M065

Makhanov, U M
  D K Kaipov, U M Makhanov, A V Kuz'minov,
  D N Smirin, and Zh I Adymov
    W-182 . . . . 75K043

Makhnev, E S
  Sh Sh Bashkirov, G D Kurbatov, E S Makhnev,
  and V A Chistyakov
    FE-57 . . . . 75B116

Makolagwa, S
  L Dabrowski, J Piekoszewski, J Suwalski,
  and S Makolagwa
    FE-57 . . . . 75D016

  J Piekoszewski, J Suwalski, L Dabrowski,
  and S Makolagwa
    FE-57 . . . . 75P003

  J Piekoszewski, L Dabrowski, J Suwalski,
  and S Makolagwa
    FE-57 . . . . 75P007

  J Piekoszewski, L Dabrowski, J Suwalski,
  and S Makolagwa
    FE-57 . . . . 75P029

Maksimov, V L
  V S Zavgorodnii, E T Bogoradovskii,
  V L Maksimov, V B Lebedev, B I Rogozev,
  and A A Petrov
    SN-119 . . . . 75Z022

Maksimov, Yu V
  R A Arents, Yu V Maksimov, and I P Suzdalev
    FE-57 . . . . 75A019

  R A Arents, Yu V Maksimov, I P Suzdalev,
  and E F Makarov
    FE-57 . . . . 75A036

  V I Gol'danskii, Yu V Maksimov,
  and I P Suzdalev
    REVIEW 75G059

  Yu V Maksimov, I P Suzdalev,
  V I Gol'danskii, O V Krylov, L Ya Margolis,
  and A E Nechitailo
    FE-57 . . . . 75M018

  Yu V Maksimov, I P Suzdalev, R A Arents,
  and M Ya Kushnerev
    FE-57 . . . . 75M025

  Yu V Maksimov, I P Suzdalev, R A Arents,
  E F Makarov, M Ya Kushnerev,
  and L D Kuznetsov
    FE-57 . . . . 75M033

  Yu V Maksimov, I P Suzdalev, A I Matveev,
  E F Makarov, and A V Kravtsov
    FE-57 . . . . 75M056

  Yu V Maksimov, I P Suzdalev,
  V I Gol'danskii, A I Matveev,
  and E F Makarov
    FE-57 . . . . 75M057

Maksimov, Yu V   (continued)
  Yu V Maksimov, I P Suzdalev,
  V I Gol'danskii, O V Krylov, L Ya Margolis,
  and A E Nechitailo
    FE-57 . . . . 75M071

  Yu V Maksimov, I P Suzdalev, A I Matveev,
  E F Makarov, and A V Kravtsov
    FE-57 . . . . 75M072

Maksyutin, Yu K
  O Kh Poleshchuk, Yu K Maksyutin,
  and I G Orlov
    SN-119 . . . . 75P054

  V A Varnek, E N Yurchenko, V A Kogan,
  L N Mazalov, Yu K Maksyutin,
  O Kh Poleshchuk, A S Egorov, and O A Osipov
    SN-119 . . . . 75V040

Mal'tsev, E I
  E I Mal'tsev, V I Goman'kov, I M Puzei,
  V A Makarov, and E V Kozis
    FE-57 . . . . 75M087

Malden, P J
  C S Hogg, P J Malden, and R E Meads
    FE-57 . . . . 75H014

Maletta, H
  G Crecelius and H Maletta
    DY-161 . . . . 75C033

  H Maletta and G Crecelius
    EU-151 . . . . 75M042

  H Maletta and G Crecelius
    EU-151 . . . . 75M043

  H Maletta and G K Shenoy
    EU-151 . . . . 75M073

Malik, W U
  W U Malik and K D Sharma
    FE-57 . . . . 75M084

Malin, J M
  H E Toma, E Giesbrecht, J M Malin,
  and E Fluck
    FE-57 . . . . 75T009

Malliaris, A
  A Malliaris and A Simopoulos
    FE-57 . . . . 75M031

  A Malliaris and D Petridis
    FE-57 . . . . 75M032

  A Malliaris and C Niarchos
    FE-57 . . . . 75M049

  A Malliaris and D Petridis
    FE-57 . . . . 75M050

Malysheva, T V
  T V Malysheva, V D Shvagerev,
  and A N Yermakov
    FE-57 . . . . 75M026

  T V Malysheva and G A Kazakov
    FE-57 . . . . 75M062

  T V Malysheva
    GENRAL 75M074

  G P Vdovykin, V I Grachev, T V Malysheva,
  and L M Satarova
    FE-57 . . . . 75V042

Malyshevskii, V A
  G N Belozerskii, V M Erkin,
  V A Malyshevskii, O G Sokolov,
  and Yu P Khimich
    FE-57 . . . . 75B078

Mamaluy, Yu A
  Sh Sh Bashkirov, A B Liberman,
  V I Sinyavskii, N N Yefimova,
  and Yu A Mamaluy
    FE-57 . . . . 74B095

362

Montano, P A   (continued)
  Z Shanfield, P A Montano, and P H Barrett
    FE-57 . . . . 75S077

Moore, F E
  J M Daniels, F E Moore, and S K Panda
    FE-57 . . . . 75D044

Morariu, M
  D Barb, M Morariu, L Diamandescu,
  and I Bibicu
    FE-57 . . . . 74B093

  D Barb, E Burzo, and M Morariu
    FE-57 . . . . 75B027

  S Solacolu, E Barbulescu, D Barb,
  and M Morariu
    SN-119 . . . . 75S019

Morawiec, H
  J E Frackowiak, H Morawiec, and T J Panek
    FE-57 . . . . 75F023

Moreira, J E
  J M Knudsen, E Larsen, J E Moreira,
  and O Faurskov Nielsen
    FE-57 . . . . 75K062

Morgan, E H
  C P Tsang, A J F Boyle, and E H Morgan
    FE-57 . . . . 75T005

Mori, T
  M Tanaka, T Tokoro, and T Mori
    FE-57 . . . . 75T007

Morimoto, S
  A Ito and S Morimoto
    FE-57 . . . . 75I002

  A Ito and S Morimoto
    FE-57 . . . . 75I003

  A Ito and S Morimoto
    FE-57 . . . . 75I009

Morinaga, K
  M Nakamura, N Maeda, and K Morinaga
    FE-57 . . . . 73N027

Morita, Y
  T Suzuki, Y Maeda, H Sakai, S Fujimoto,
  and Y Morita
    FE-57 . . . . 75S037

Moron, J W
  B M Jankowski, J E Frackowiak, J Kansy,
  J W Moron, T J Panek, A Salamon,
  and M R Uhlig
    FE-57 . . . . 75J007

Morornyi, A V
  A V Morornyi, B A Tallershik, S P Teslenko,
  and B A Shustrov
    FE-57 . . . . 75M017

Moroz, V F
  B N Maimur and V F Moroz
    REVIEW 74M058

Morozumi, T
  H Ohashi, M Koizumi, and T Morozumi
    FE-57 . . . . 74O014

Morrish, A H
  A H Morrish and P E Clark
    FE-57 . . . . 75M002

  M R Spender and A H Morrish
    FE-57 . . . . 75S025

Morrison, Jr, W H
  W H Morrison, Jr, E Y Ho,
  and D N Hendrickson
    FE-57 . . . . 75M015

  W H Morrison, Jr and D N Hendrickson
    FE-57 . . . . 75M024

Morup, S
  J Kjeldgaard, G Trumpy, N Thrane,
  and S Morup
    FE-57 . . . . 75K049

  S Morup and E Both
    ANALYS 75M009

  S Morup
    FE-57 . . . . 75M028

  S Morup, M K Nielsen, J M Knudsen,
  and G Trumpy
    FE-57 . . . . 75M058

  S O Svensson, S Morup, and G Trumpy
    THEORY 75S055

  H Topsoe and S Morup
    FE-57 . . . . 75T027

  H Topsoe and S Morup
    FE-57 . . . . 75T028

Mosbaek, H
  H Mosbaek
    FE-57 . . . . 75M080

Moskvin, A S
  A S Moskvin, N S Ovanesyan,
  and V A Trukhtanov
    THEORY 75M039

  A S Moskvin, N S Ovanesyan,
  and V A Trukhtanov
    FE-57 . . . . 75M081

Mostafa, A G
  N A Eissa, A L Hussein, and A G Mostafa
    FE-57 . . . . 75E014

Motorny, A V
  V K Yarmarkin, B A Shustrov, and A V Motorny
    DY-161,FE-57,SB-121 . . . . 75Y004

Motsartov, V V
  L P Nikitina, S P Ekimov, L M Krizhanskii,
  R G Grebenshchikov, and V V Motsartov
    FE-57 . . . . 74N021

  S P Ekimov, L P Nikitina,
  R G Grebenshchikov, L M Krizhanskii,
  O G Chigareva, and V V Motsartov
    FE-57 . . . . 75E013

Mott, N F
  C N R Rao, V G Bhide, and N F Mott
    FE-57 . . . . 75R033

Motte, J P
  J P Motte and N N Greenwood
    SN-119 . . . . 75M003

Moukarika, A
  A Simopoulos, A Kostikas, I Sigalas,
  N H Gangas, and A Moukarika
    FE-57 . . . . 75S078

Moutsos, P
  R R Sharma and P Moutsos
    FE-57,THEORY 75S017

Mueller, M H
  D J Lam, B D Dunlap, A R Harvey,
  M H Mueller, A T Aldred, I Nowik,
  and G H Lander
    NP-237 . . . . 74L045

  A T Aldred, B D Dunlap, D J Lam, G H Lander,
  M H Mueller, and I Nowik
    NP-237 . . . . 75A004

  A T Aldred, B D Dunlap, D J Lam, G H Lander,
  M H Mueller, and I Nowik
    NP-237 . . . . 75A012

Mukerji, Z
  Y W Chow and Z Mukerji
    FE-57 . . . . 75C003

383

Review in
Mössbauer Spectroscopy

^{170}Yb MÖSSBAUER SPECTROSCOPY

I. Nowik and E. R. Bauminger

Racah Institute of Physics
The Hebrew University
Jerusalem, Israel

I. INTRODUCTION

A number of 2+ → 0+ gamma ray transitions are available for Mössbauer studies in even rare earth isotopes. Mössbauer spectra obtained for gamma rays from 2+ → 0+ transitions are usually comparatively simple and easy to analyze. Among these transitions, the 84.3 keV gamma ray transition in ^{170}Yb is the most convenient and promising. It is distinguished by the following favorable properties:

(a) Strong sources are easily prepared by neutron irradiation of ^{169}Tm. ^{169}Tm is the only stable isotope of Tm.

(b) The half life of the radioactive source, ^{170}Tm, is comparatively long ($T_{\frac{1}{2}} = 130$ d).

(c) Single emission line sources (or absorbers) with relatively large recoil free fractions are easily obtained.

(d) The gamma ray spectrum of ^{170}Yb is composed of a single line at 84 keV, and any kind of gamma ray detector can be used.

(e) The nuclear properties of the 2+, 84 keV level (half life, change in nuclear radius, magnetic moment, and quadrupole moment) are favorable and well resolved recoilless absorption spectra can be obtained, from which exact values of hyperfine parameters can be deduced.

Ytterbium ions appear either in the divalent or in the trivalent state in solids. The magnetic properties of the Yb ions in the two valence states are very different. The ionic ground state of $Yb^{2+}(4f^{14})$ is $^1S_0$ which is diamagnetic. The study of hyperfine interactions in Yb^{2+} compounds yields information on the electric field gradient at the Yb nucleus, produced by the charge distribution in the lattice, and on the magnetic hyperfine field caused by super-transfer or conduction electron polarization mechanisms. The ionic ground state of $Yb^{3+}(4f^{13})$ is $^2F_{\frac{7}{2}}$. Hyperfine interaction studies on Yb^{3+} compounds may yield information on crystalline and exchange fields, magnetic spin structure and spin relaxation phenomena. Finally, other phenomena connected with the Kondo state or with spin and charge fluctuations in compounds in which the ion is in an intermediate valence state can also be studied by the Mössbauer effect in ^{170}Yb.

II. INSTRUMENTATION

A. SOURCES

The radioactive source used in the recoilless absorption measurements of the 84 keV gamma rays of ^{170}Yb is ^{170}Tm which decays by β^- emission to ^{170}Yb with a half life of $t_{\frac{1}{2}} = 130$ d. The source is produced by neutron irradiation of natural thulium. In the early experiments the source was produced by irradiating Tm metal [62W083]. However, since Tm metal has a hexagonal structure, the 84 keV emission line was somewhat broadened by quadrupole interactions. In most cases a source of $TmAl_2$, $TmAl_3$, or 10% Tm in Al is used. In these cases the Yb ions produced in the source are divalent, and thus, diamagnetic. They are located in a cubic symmetry site. No hyperfine structure is present and the width of the emission line is close to the natural line width. The recoil-free fraction of these sources is higher than that of Tm metal. Several milligrams of ^{169}Tm irradiated for several days (~ 100 mg day) in a neutron flux of 10^{14} n/cm^{2}s produce a very reasonable source (~ 20 mC of 84 keV gamma rays) for ^{170}Yb experiments. It is evident that almost any Mössbauer group has the capability to obtain such a source. It was suggested recently [73G032] that an even better source for Mössbauer experiments is TmB_{12}. It has a higher recoil-free fraction and an emission line width as good as or better than that of $TmAl_2$. In the future, particularly in experiments involving absorbers containing dilute Yb, the advantage of using TmB_{12} sources will be more pronounced.

In order to study hyperfine fields acting on ^{170}Yb in sources, Tm metal or oxide may be irradiated and the compound of interest prepared afterwards with the radioactive material. The desired compound may also be prepared first and irradiated afterwards. Each way has its advantages and disadvantages. The first procedure guarantees that the only radioactivity present is that of ^{170}Tm and very dilute systems can be studied. However, the compounds can be used only for Mössbauer measurements, and X-ray analysis or susceptibility measurements cannot be made on them. The second procedure has the advantage that the sample can be analyzed, before irradiation, by any technique. However, additional unwanted radioactivity may be produced during the irradiation, especially in Tm dilute samples. The neutron irradiation will, in this case, also cause radiation damage to the sample, and sources prepared in this way must be annealed for some time [74B055].

B. ABSORBERS

The natural abundance of ^{170}Yb is only 3%, but since the 84 keV gamma ray emitted from the source is the only gamma ray in the spectrum (see Figure 1) and the recoil-free fraction of the source can be as high as ~30% (for TmB_{12}), absorbers of natural abundance of ^{170}Yb can be used in most cases. For example, with a nonenriched $YbAl_2$ absorber, 40 mg/cm^2 thick, and a $TmAl_2$ source, a single absorption line with a width of 2.3 mm/s and an effect of 0.3% is obtained (the natural line width, W_0, is 2.03 mm/s). Any compound of divalent Yb, in which the local symmetry around the Yb is cubic, can serve as a single narrow line absorber for the study of hyperfine structures in the source. $YbAl_2$ [73G032], $YbAl_3$ [72W001], YbB_6 [74G013], and Yb metal itself (which is generally cubic, unless very pure when it becomes hexagonal) [70H032] have been used as single line absorbers. Yb metal has a low recoil-free fraction and is, therefore, less favorable.

In studies of hyperfine structures in absorbers, the absorbers used may be about 200 mg/cm^2 thick and still give unbroadened absorption lines. Absorbers containing divalent Yb show little or no hyperfine structure and thus, thin nonenriched absorbers can be used. For absorption studies of trivalent Yb compounds, particularly for studies of dilute systems, thick absorbers enriched in ^{170}Yb are needed.

Figure 1. Gamma ray spectrum emitted from a ^{170}TmAl2 source in a 3 mm thick Ge(Li) detector, with and without a 1.5 mm thick Cu absorber.

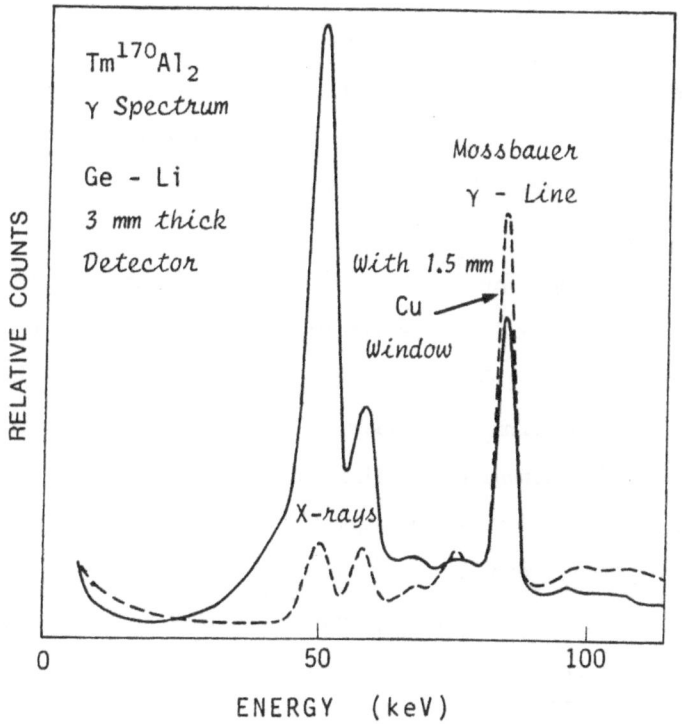

C. EXPERIMENTAL PROCEDURES

Because of the relatively high energy of the gamma ray (84 keV), Mössbauer studies of ^{170}Yb usually have to be performed with both source and absorber at temperatures below 70 K. Since it is easy to obtain single line sources or absorbers even at 4.2 K (in cubic Yb^{2+} compounds), most experiments have been performed with either source or absorber at 4.2 K. In most cases, the source and the absorber have been held in a conventional helium cryostat in vertical geometry. In some cases, a horizontal arrangement has been used [74G013]. The variable temperature of the absorber (or source) has been obtained by conventional ovens or by lucite ovens.[1] Using a single line absorber (or source) at 4.2 K, the source (or absorber) can be held at 60-70 K and still a reasonable effect ($\sim 0.1\%$) can be obtained, even if the spectrum is composed of five resolved lines (a magnetic hyperfine interaction will yield a Mössbauer spectrum composed of five lines of equal intensity).

The Mössbauer spectrum of the 84 keV $2^+ \to 0^+$ ^{170}Yb transition extends in velocity units in the most extreme case of ^{170}Yb in $TmFe_2$ (in which a large magnetic splitting is seen, $H_{eff} = 4.5$ MOe), from -40 mm/s to +30 mm/s [75Y005]. Velocity calibration is usually performed with a ^{57}Fe in Fe_2O_3 or Fe metal Mössbauer spectrum. The linearity of the Mössbauer drive system when changing the amplitude and, therefore, the velocity, has to be perfect in this case. By now the recoilless absorption spectrum of the 84 keV gamma ray of ^{170}Yb in $YbCl_3 \cdot 6H_2O$ could serve as a calibration standard for ^{170}Yb measurements. It gives at 4.2 K a beautiful magnetically split spectrum, composed of five equally spaced lines of equal intensity. The line positions are well established by several measurements [65H005, 66M013, 66M014, 67E001, 70H032, 71H008, 71K058] (Figure 2c).

Single line ^{170}Yb Mössbauer spectra extend in velocity from about -5 to +5 mm/s. For this range of velocities, the recoilless absorption spectra of the 14 keV gamma ray of ^{57}Fe in either Fe metal or Fe_2O_3 can very well serve for calibration.

The gamma ray spectrum of ^{170}Tm is composed only of the 84 keV gamma ray and Yb X-rays. The gamma-ray spectrum, as observed with a 3 mm thick Ge(Li) detector, is shown in Figure 1. For most purposes a NaI(Tl) detector may be used. With such a detector usually better geometry and larger efficiency is obtained and thus higher counting rates may be achieved.

In Figure 2 some typical ^{170}Yb Mössbauer spectra are shown. Figure 2a shows the single absorption line obtained with a $TmAl_2$ source and an $YbAl_2$ absorber, both at 4.2 K.

---

[1] J. Gal and J. Hess, Rev. Sci. Instrum. 42, 543 (1971).

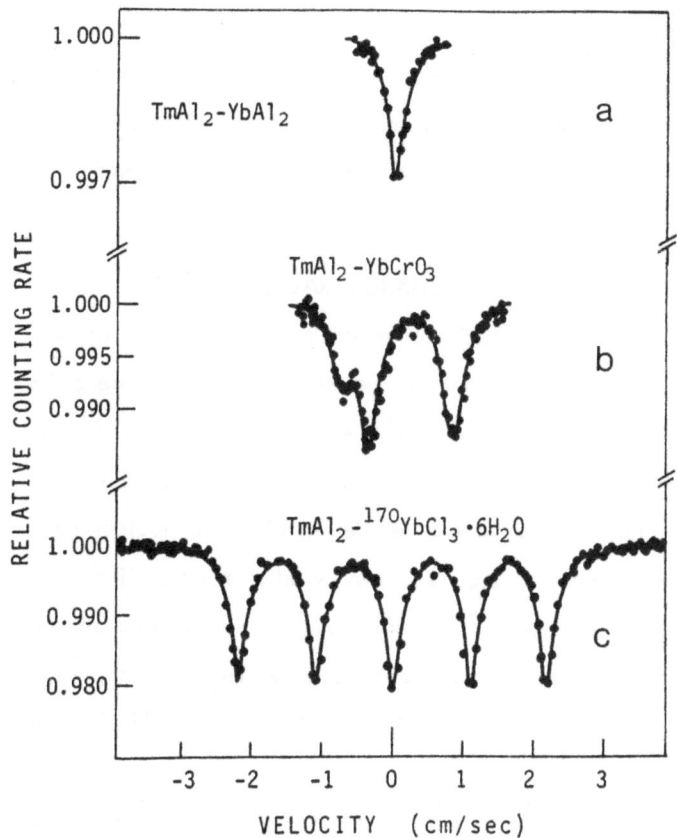

Figure 2. Recoilless absorption spectra of the 84.3 keV gamma ray emitted from a ^{170}TmAl$_2$ source in a) YbAl$_2$, b) YbCrO$_3$, c) YbCl$_3 \cdot 6H_2O$.

Figure 2b shows the recently observed absorption spectrum in YbCrO$_3$ [74D014]. Such a pure three line quadrupole spectrum is obtained if the electric field gradient acting on the ^{170}Yb nuclei is axial ($\eta = 0$), and the magnetic hyperfine field is is zero. In a nonaxial case a five line spectrum could be obtained (the case of YbCrO$_3$ is almost axial, $\eta = 0.15$). Figure 2c shows the absorption spectrum obtained in YbCl$_3 \cdot 6H_2O$. Here the five line pattern is due to magnetic hyperfine structure [66M013].

III. NUCLEAR PARAMETERS

All the nuclear parameters of the 2+ state and the 2+ → 0+, 84 keV, E2 transition of ^{170}Yb are now well known. The lifetime of the 2+ state was measured to be 1.60 ns, which gives a natural line width for Mössbauer studies of $W_o = 2.03$ mm/s. The narrowest experimental line width so far was obtained using a TmAl$_2$ source and a thin YbAl$_2$ absorber, both at 20 K ($W_{exp} = 2.1$ mm/s) [67N012]. The conversion coefficient α_{tot} for the 84 keV, E2 transition is 6.2. Many Mössbauer studies were devoted to the determination of the radius and moments of the 2+ state. The magnetic moment is now very accurately known as $\mu = 0.668(4)$ nm [75B038]. Because of limited

411

knowledge on the electric field gradients acting on the ^{170}Yb nucleus, the quadrupole moment cannot be determined directly from Mössbauer studies. The accepted value, $Q = -2.12$ barn, relies on Coulomb excitation measurements [2] and general systematics. The most recent and reliable value of the change in the mean square charge radius between the 2+ state and the 0+ ground state, $\Delta<r^2>$, obtained from Mössbauer studies is $1.2(0.4) \times 10^{-3}$ fm^2 [73R024].

A. THE CHANGE IN THE MEAN SQUARE CHARGE RADIUS

As is well known, the isomer shift measured in a Mössbauer spectrum is proportional to the quantity $\Delta\rho \cdot \Delta<r^2>$, where $\Delta\rho = \rho_a - \rho_s$ is the difference in the electronic charge densities in the nuclear volume in the absorber and in the source, and $\Delta<r^2> = <r^2>_e - <r^2>_g$ is the difference between the mean-square nuclear charge radii in the excited and the ground state of the nucleus. A measurement of the isomer shift thus provides a way of determining the nuclear parameter $\Delta<r^2>$, provided the electron density difference $\Delta\rho$ is known by other means. Since there is particular interest in $\Delta<r^2>$ for highly deformed axially symmetric nuclei, which have energy level schemes similar to that of a quantum mechanical rotor, many authors have tried to determine $\Delta<r^2>$ for ^{170}Yb and other even-even Yb isotopes [67A017, 68A027, 68H025, 70H032, 71H008, 71H059, 73R024].

It is generally difficult to obtain a reliable value of $\Delta\rho$, and therefore, there is a great spread in the values cited for $\Delta<r^2>$ in the various publications. In most cases the value of $\Delta\rho$ (Yb^{3+} - Yb^{2+}) was obtained from free ion Hartree-Fock calculations or from optical isotope shift data for free ions. It was shown lately that optical isotope shifts are the same for ions in the solid state and for the free ions [73R024]. One may use, therefore, to a good approximation, the electron density differences derived for free ions without correction, for the analysis of isomer shifts obtained in the Mössbauer effect. The estimated accuracy of the values of $\Delta\rho$ so derived is about 30%.

The dispersion term arising from interference effects between conversion electrons emitted after nuclear resonance absorption and photoelectrons has been taken into account in the analysis of the experimental data only by Russell et al. [73R024]. The use of Lorentzian line shapes in the fitting of the Mössbauer spectra, as done by all other authors, leads to non negligible pseudoshifts, but, as stated by Russell et al., the relative isomer shifts between different single line absorbers are almost unaffected by this approximation.

---

[2] H. Ryde, G. D. Symons, and Z. Szymanski, Nucl. Phys. **80**, 241 (1966).

Table I. Isomer Shifts Between Trivalent and Divalent Yb Compounds, Difference in Charge Densities and Estimates of $\Delta <r^2>$

Yb^{3+} compound	Yb^{2+} compound	δ mm/s	$\Delta\ (Yb^{3+}-Yb^{2+})$ (a_o^{-3})	$\Delta <r^2>$ $10^{-3}fm^2$	Reference
$Yb_3Ga_5O_{12}$	$YbSO_4$	0.63(14)	40(10)	3.2(1)	[68A027]
$YbCl_3$	$YbCl_{2.3}$	0.20(4)	40	1.1	[68K007]
Average Yb^{3+}	$YbSO_4$	0.41(8)	70(10) 50(20)	1.2(3) 1.7(6)	[68H025] [71H008]
Yb_2S_3	YbF_2	0.362(14)	62(20)	1.2(4)	[73R024]

In Table I we cite values of isomer shifts between trivalent and divalent Yb compounds, measured by various groups, their estimates for $\Delta\rho\ (Yb^{3+}-Yb^{2+})$ and their results for $\Delta <r^2>$. In Figure 3 some of the isomer shifts measured are shown.

$\Delta <r^2>$ was also determined from isomer shift measurements between Yb^{3+} compounds and Yb metal or other Yb metallic compounds [67A017, 68K007]. The conduction electron densities at the Yb nuclei, $\rho(CE)$, were calibrated using the electron densities in Eu metal determined by isomer shift measurements in Eu. As the shifts between Yb metal and Yb^{3+} compounds are even smaller than the shifts between Yb^{2+} and Yb^{3+} compounds, the results for $\Delta <r^2>$ obtained from these measurements are less accurate.

The systematic variation of $\Delta <r^2>$ over a family of isotopes can be determined even without the knowledge of $\Delta\rho$. The ratios of $\Delta <r^2>$ of two different Mössbauer transitions of the same element in different isotopes may be extracted from the isomer shifts in identical absorbers: $\Delta <r^2>_1/ \Delta <r^2>_2 = \delta_1/\delta_2$. The observation of non zero changes in the mean square charge radius in rotational transitions in the nuclei of even Yb isotopes provides information about non adiabatic effects in the intrinsic nuclear structure, induced by rotational excitations. Nuclear models which describe the collective and single particle motions of nucleons giving rise to such non adiabatic effects may thus be tested by the determination of $\Delta <r^2>$ for analogous rotational transitions in $^{170,172,174,176}Yb$.

An extensive study on $\Delta <r^2>$ in even isotopes of Yb has been reported by Russell et al. [73R024]. The ratios of $\Delta <r^2>$ of even Yb isotopes to $\Delta <r^2>$ of ^{170}Yb as derived in this work are $\Delta <r^2>_{172}/ \Delta <r^2>_{170} = +0.341\ (82)$, $\Delta <r^2>_{174}/ \Delta <r^2>_{170} = -0.366(71)$, and

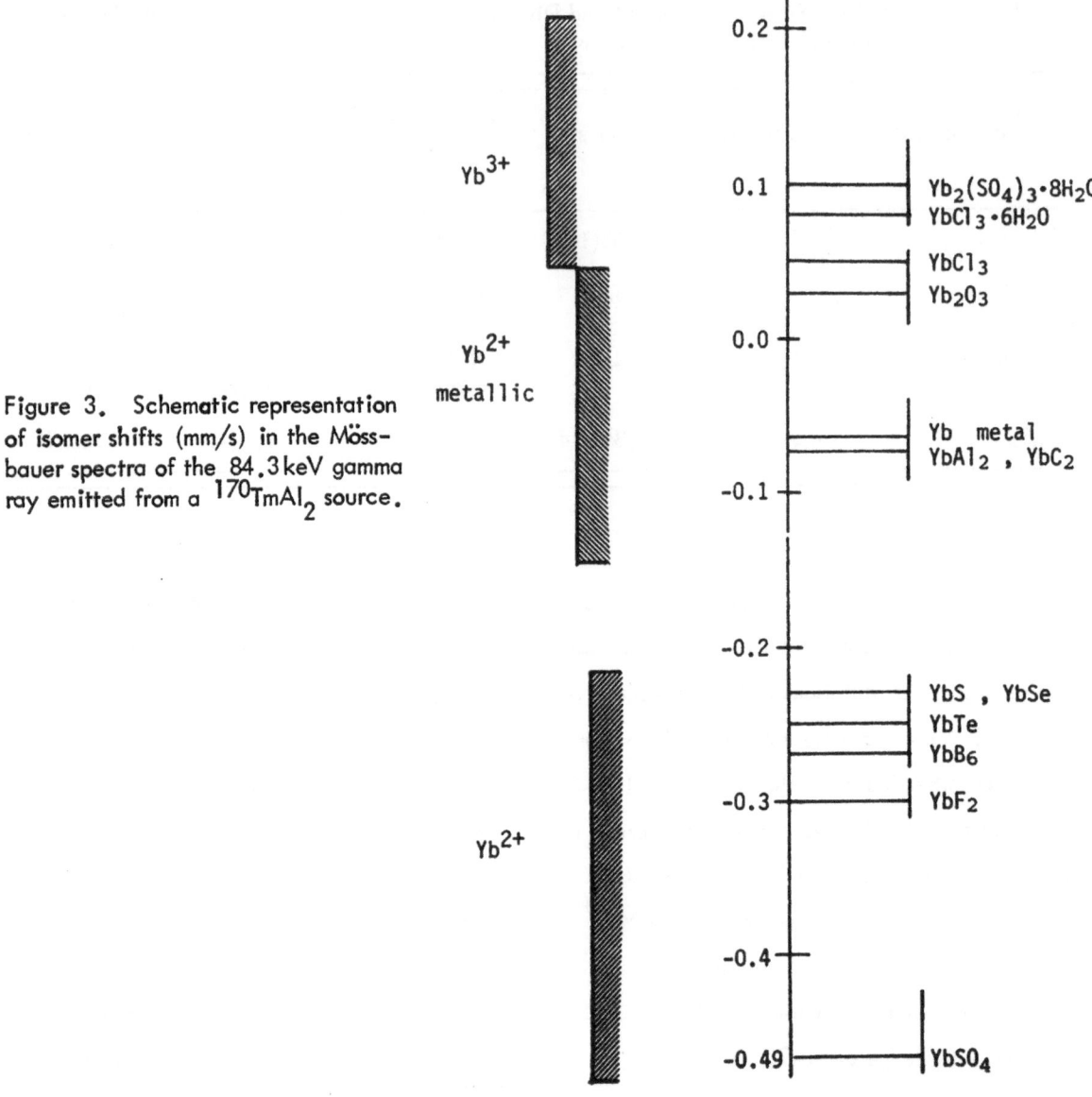

Figure 3. Schematic representation of isomer shifts (mm/s) in the Mössbauer spectra of the 84.3 keV gamma ray emitted from a ^{170}TmAl$_2$ source.

$\Delta < r^2 >_{176} / \Delta < r^2 >_{170} = 0.141(43)$. All absorbers used in this work were single line absorbers. The shifts measured in ^{176,174,172}Yb isotopes are less than 1/40 of the experimental line width obtained in the various absorbers. It is therefore not too surprising that an entirely different ratio of $\Delta < r^2 >_{174} / \Delta < r^2 >_{170} = +0.52(15)$ was obtained in [71H008], where the isomer shifts were derived from broad, partially split, spectra. According to Russell et al. [73R024], the value of $\Delta < r^2 >$ depends on the neutron number, even though the nuclear deformation in this series of Yb isotopes is constant. The authors show that a microscopic theory due to Meyer and Speth[3] is capable to explain,

[3] J. Meyer and J. Speth, Nucl. Phys. A203, 17(1973).

at least qualitatively, the observed behavior of $\Delta <r^2>$ in the Yb isotopes. The predictions of the centrifugal stretching model[4] and of Marshalek's theory[5] for $\Delta <r^2>$ are much too large when compared to the observed values of $\Delta <r^2>$.

B. THE QUADRUPOLE MOMENT

The quadrupole moment of the 2+ state of ^{170}Yb has been determined as -2.12 b by Coulomb excitation measurements[2]. Quadrupole interaction parameters, which were deduced from ^{170}Yb Mössbauer spectra, served therefore to determine the electric field gradient (EFG) acting on the ^{170}Yb nuclei and no attempt was made to calculate the EFG's and to deduce the quadrupole moment of the 2+ state of ^{170}Yb.

The quadrupole interaction parameter of the 84 keV level of ^{170}Yb was measured in a number of compounds by Henning et al. [71H008], in order to compare the quadrupole moment of the 84 keV level of ^{170}Yb with the quadrupole moment of the 76.5 keV level of ^{171}Yb. Measurements performed with even Yb isotopes in Yb_2O_3, YbOOH, $YbSO_4$, and $Yb_2(SO_4)_3 \cdot 8H_2O$ absorbers showed that upon addition of two neutron pairs to ^{170}Yb, the quadrupole moment does not change by more than 2%. Pure quadrupole split spectra were also observed in Yb_2O_2S, in Yb_2O_2Se [72G004] and in $YbCrO_3$ [74D014]. An attempt to deduce ratios of quadrupole moments of the 2+ levels of ^{172,174}Yb and the $\frac{3}{2}$-level of ^{171}Yb to the quadrupole moment of the 2+ level of ^{170}Yb was performed, by studying the Mössbauer spectra of Yb_2O_3 in all these isotopes [71P009]. The exact analysis of these spectra seems somewhat doubtful, as it was wrongly assumed $n \neq 0$ for the ions occupying the C_{3i} symmetry site.

C. THE MAGNETIC MOMENT

The magnetic moment of the 2+ state of ^{170}Yb has been measured by the Mössbauer effect as early as 1965 [65H005]. Since then the measurement has been repeated several times [66M013, 66M014, 67E001, 68M011]. In all these measurements ^{170}YbCl$_3 \cdot$6H$_2$O absorbers at 4.2 K have been used. In YbCl$_3 \cdot$H$_2$O, the spin relaxation time is long. A magnetic hyperfine spectrum, shown in Figure 2c, is obtained. This magnetic hyperfine structure can be treated in the effective magnetic field approximation (see below discussion on relaxation spectra). From the analysis of the spectrum $g(2+)\mu_n H_{eff}$ was obtained. The magnetic splitting of the I = 1/2- ground state of ^{171}Yb in YbCl$_3 \cdot$6H$_2$O was also measured by the Mössbauer technique[6,7] [66H002]. Since $g_{171}(1/2-)$ is

---

[4] W. Henning, Z. Phys. _217_, 438 (1968).

[5] E. R. Marshalek, Phys. Rev. Lett. _20_, 214 (1968).

[6] C. Günther and E. Kankeleit, Phys. Lett. _22_, 443 (1966).

[7] W. Henning, P. Kienle, and H. I. Körner, Z. Phys. _199_, 207 (1967).

known with great accuracy[8] to be 0.983778(8), the hyperfine field (H_{eff}) in this compound could be accurately determined and thus the value of the g factor of the 2+, 84 keV level of ^{170}Yb was determined as 0.335(4). The most precise determination of the $g_{170}(2+)$ as 0.334(2) was obtained recently from the slow spin relaxation spectrum of Yb^{3+} in a single crystal of 0.025% Tm in $\underline{CaF}_2$ and a previous ENDOR measurement of $^{171}Yb^{3+}$ in the same compound [75B038].

In [66M013, 66M014, 67E001, and 68M011] the nuclear moments of the 2+ levels of the other even Yb isotopes were measured, by comparing the hyperfine splittings obtained in an $YbCl_3 \cdot 6H_2O$ absorber to the splitting obtained for ^{170}Yb in the same absorber. In [68M011] the values of $g_{170}(2+) = 0.335(6)$, $g_{172}(2+) = 0.332(8)$, and $g_{174}(2+) = 0.337(8)$ were obtained and compared with theoretical predictions of Nilsson and Prior[9] and Greiner[10]. Very good agreement with Greiner's calculated values were found.

D. MOSSBAUER ABSORPTION LINE SHAPE IN PRESENCE OF INTERFERENCE EFFECTS

Several authors[11] [72W001] have pointed out the influence of interference effects on the Mössbauer absorption line. The interference between conversion electrons emitted after nuclear resonance absorption and photoelectrons leads to a dispersion term in the Mössbauer absorption cross section and, therefore, the shape of a single transmission line in a Mössbauer spectrum is given by

$$\frac{N(v)}{N(\infty)} = 1 - \epsilon \, (1 - 2\xi x)/(1 + x^2)$$

where $x = 2(v - S)/\Gamma$. Here $N(v)$ is the intensity transmitted at a Doppler velocity v, S is the velocity corresponding to the position of the absorption line, Γ is the full line width at half maximum, and ϵ is the depth of the absorption dip at resonance. ξ determines the relative magnitude of the dispersion term.

Since the interaction of low energy gamma rays with atomic electrons has predominantly electric dipole nature, it was believed for a while that only E1 transitions would show this effect. However, it was shown by Wagner et al. [72W001] that dispersion terms are also present in absorption spectra of gamma rays with E2 and mixed E2/M1 character. For ^{170}Yb the value $2\xi = -0.034(8)$ was obtained. Subsequently Russell et al. [73R024] have remeasured the line shape of various ^{170}Yb single line spectra and obtained a weighted mean of $2\xi = -0.030$. Hence, in the

---

[8] L. Olschewski and E. W. Otten, Z. Phys. 200, 224 (1967).

[9] S. G. Nilsson and O. Prior, Kgl. Danske Videnskab, Selskab, Mat. Phys. Medd. 32, 616 (1961).

[10] W. Greiner, Nucl. Phys. 80, 417 (1966).

[11] G. T. Trammell and J. P. Hannon, Phys. Rev. 180, 337 (1969).

analyses of Mössbauer spectra of ^{170}Yb, instead of Lorentzian lines, line shapes of the form:

$$L(v) \;=\; \frac{\frac{\Gamma}{2} + 0.03\,(v-S)}{(\frac{\Gamma}{2})^2 + (v-S)^2}$$

have to be used.

IV. APPLICATIONS

A. VALENCE STATES OF Yb

A Mössbauer study of ^{170}Yb may yield the valence state of Yb in the compound investigated, by either of the three parameters determined in a Mössbauer study: the isomer shift, the electric field gradient, or the magnetic hyperfine field.

Rare earth ions in solids generally appear as tripositive ions by giving up the $5d6s^2$ or $4f6s^2$ external electrons. In the ionized state all have $5s^2 5p^6$ outer electrons and differ only in the number of the inner 4f electrons. This is, of course, the reason why the rare earth elements and compounds have similar chemical properties. Some of the rare earth ions appear also in other valence states. These ions achieve the spectroscopically stable configuration of the closed or half closed shell, as e.g. $Ce^{4+}(4f^0$ like $La^{3+})$, Tb^{4+} and Eu^{2+} ($4f^7$ like Gd^{3+}) and $Yb^{2+}(4f^{14}$ like $Lu^{3+})$.

Though the chemical properties of rare earth ions are similar, their magnetic properties, which are mainly determined by the number of electrons in the 4f shell, are very different. In Table II the atomic and magnetic properties of the Tm^{3+}, Yb^{3+}, Yb^{2+} and Lu^{3+} ions are presented. One observes extreme differences in the magnetic properties of Yb in the two valence states. Yb^{3+} is strongly paramagnetic, Yb^{2+} is diamagnetic. Yb^{2+} is magnetically equivalent to Lu^{3+}. The two charge states of Yb are stable and appear in both insulators and metals. Only recently unstable or intermediate Yb valence states in metals have been observed[12] and studied also by the Mössbauer technique [74B055].

The isomer shift measured in a Mössbauer study is a measure of the charge density in the nucleus. Since in $Yb^{2+}(4f^{14})$ there are more 4f electrons than in $Yb^{3+}(4f^{13})$, they shield more effectively the nucleus from the penetration of the 4s, 5s and 6s electrons to the nucleus, and thus reduce the charge density at the nucleus. Since for ^{170}Yb $\Delta <r^2>$ is positive, absorption lines due to Yb^{2+} will appear in a Mössbauer absorption spectrum at more negative velocities than absorption lines due to Yb^{3+} (Figure 3).

$\overline{}$
[12] E. E. Havinga, K.H.J. Buschow, and J. H. van Daal, Solid State Commun. 13, 621 (1973).

Until now, there was no generally accepted standard for isomer shift measurements. Most authors cite the shifts relative to an $Yb:TmAl_2$ source. Some refer to an $YbAl_2$ absorber [74P024].

In all divalent nonmetallic Yb compounds, small negative isomer shifts relative to $Yb:TmAl_2$ are obtained. All trivalent compounds give small positive isomer shifts relative to this source.

The isomer shifts measured in the recoilless absorption spectra of the 84 keV gamma ray of ^{170}Yb between compounds in which the Yb ions are divalent and trivalent are quite small and thus the measurement of the isomer shift alone is not always enough to establish the valency of the Yb ionic state. However, the electric field gradient and magnetic hyperfine fields are usually sensitive to the charge state of the Yb ions and yield additional information needed.

As seen from Table II, the hyperfine interactions due to the 4f electrons vanish in the case of Yb^{2+}. Mössbauer spectra of Yb^{2+} may show, therefore, hyperfine interactions which are due to external sources only, such as an electric field gradient due to the lattice charges and a magnetic hyperfine field due to transferred exchange polarization of internal shells or due to conduction electron polarization. A single line spectrum is, therefore, obtained in cubic Yb^{2+} compounds (e.g. in

Table II. Atomic and Magnetic Properties of Tm, Yb, and Lu

	Tm^{3+}	Yb^{3+}	Yb^{2+}	Lu^{3+}
Atomic Configuration	$4f^{13}6s^2$	$4f^{14}6s^2$	$4f^{14}6s^2$	$4f^{14}5d6s^2$
Ionic Configuration	$4f^{12}$	$4f^{13}$	$4f^{14}$	$4f^{14}$
Ground Term	$^3H_6$	$^2F_{\frac{7}{2}}$	$^1S_0$	$^1S_0$
Landé Factor	7/6	8/7	0	0
Magnetic Moment	7.0	4.0	0	0
Effective Moment	7.56	4.53	0	0
Ionic Radius A°	0.95	0.94	1.13	0.93
$<1/r^3>_{4f}(a_o^{-3}$ units)	11.72	12.5	---	---
Magnetic Hyperfine Field MOe	6.93	4.23	0	0
Electric Field Gradient $(10^{18}V/cm^2)$	7.87	8.44	0	0

YbTe, YbSe, YbS, YbB$_6$ [73R024], Yb metal, YbAl$_2$, YbPb$_3$, YbIn$_3$ [74P024]) and a hardly resolved quadrupole structure, composed of three (for $\eta = 0$) or five (for $\eta \neq 0$) absorption lines in the non cubic Yb^{2+} compounds (e.g. YbSO$_4$ [71H008] and YbH$_2$ [74M018]). The observed quadrupole interactions are comparatively small. From quadrupole spectra in Yb^{2+} compounds the lattice electric field gradient, the second order crystalline field potential parameter A$^\circ_2$ and the asymmetry parameter η can be determined [70H032, 71H008, 74M018]. Hyperfine parameters of some Yb^{2+} compounds are given in Table III. The small quadrupole interactions and the absence of magnetic interactions may serve as a means to determine the valence of the Yb ion as 2+ in a newly investigated compound. In this way, for example, the Yb ion was found to be divalent in YbSi$_2$ [69S22] and in Yb$_{.22}$Ca$_{.78}$O [71A008]. In Yb$_4$Ge$_7$ [69S22] the observation of two different quadrupole interactions indicated the presence of ions of both valencies in this compound. The large quadrupole interaction found in Yb$_5$Si$_3$ indicated that the Yb is trivalent in this compound. In YbH$_2$ [74M018] the values of the quadrupole interaction parameters and the isomer shift are shown to favor the hydridic model for the rare earth hydrides. In some special cases, as e.g. in Yb$_6$Fe$_{23}$ [71G055], a small magnetic hyperfine field produced by the polarized conduction electrons was found. The small negative isomer shift confirms that the Yb ion is divalent in this compound.

Mössbauer spectra observed in Yb^{3+} compounds may be much more complex, and may show a large variety of phenomena. The rest of this review will be devoted to these.

Table III. Hyperfine Interaction Parameters of Several Yb^{2+} Compounds

Compound	Isomer Shift δ (mm/s) [a]	Quadrupole Interaction eqQ (MHz) [b]	Asymmetry Parameter η	Magnetic Interaction (MHz)	Reference
YbSO$_4$	−0.42(6)	520(15)	0.85(25)	−	[68H025] [71H008]
YbH$_2$	−0.11(1)	366(8)	0.89(5)	−	[74M018]
Yb in Tm metal	+0.04(3)[c]	330(30)	0	−	[70H032] [71H059]
Yb$_6$Fe$_{23}$	−0.31(3)	36(8)	−	66(2)	[71G055]

a   Relative to a TmAl$_2$ source.

b   The crystalline field parameter A$^\circ_2$ is given by:

eqQ $= -4$A$^\circ_2$ $(1-\gamma_\infty)$Q, where γ_∞ is the Sternheimer antishielding factor (-65 to -80 in rare earths).

c   The positive isomer shift of Yb^{2+} in Tm metal in comparison with the negative shifts of Yb^{2+} in most divalent compounds (Figure 3) is due to the atomic volume effect. The volume per Yb^{2+} ion in Tm metal is much smaller than in Yb metal ($\delta = -0.08$) and the conduction band contains three electrons per Tm atom, whereas it contains only two in Yb metal.

B. PARAMAGNETIC SPIN RELAXATION PHENOMENA

The ionic ground state of $Yb^{3+}(4f^{13})$ is $^2F_{\frac{7}{2}}$, which is split by crystalline fields in a solid into four Kramers' doublets in non cubic symmetry sites and into two doublets (Γ_6 and Γ_7) and one quartet (Γ_8) in a cubic site. At low temperatures, at which most recoilless absorption measurements in ^{170}Yb are performed, usually only the lowest Stark level is populated. In the limit of slow spin relaxation, recoilless absorption spectra in Yb^{3+} in a cubic symmetry site may show a two line spectrum with relative intensities 3:2, corresponding to the two states, $F = \frac{5}{2}$ and $F = \frac{3}{2}$, which arise from the coupling of the nuclear $I = 2$ state to a cubic Kramers' doublet (Γ_6 or Γ_7) of effective spin $S = \frac{1}{2}$. In the case of a Γ_8 ground state and slow spin relaxation, a seven line spectrum is expected. For medium spin relaxation rates (rates comparable in size to the hyperfine interactions), the spectra can obtain complicated shapes and will reduce to a single narrow line in the limit of fast relaxation. In sites of axial symmetry, Yb^{3+} will show a five line magnetic pattern for long spin relaxation times (even in the absence of magnetic order) and a three line pure quadrupole structure for short spin relaxation times.

The hyperfine spin Hamiltonian for the general doublet is given by

$$\mathcal{H}_{hf} = A_x S_x I_x + A_y S_y I_y + A_z S_z I_z + \frac{eqQ}{4I(2I-1)}[3I_z^2 - I(I+1) + \tfrac{1}{2}\eta(I_+^2 + I_-^2)]$$

where $S = \frac{1}{2}$ is the effective spin.

The spin Hamiltonian in the quartet (Γ_8) state is more complicated and given in reference [72N003]. The shape of the observed Mössbauer spectrum of the $2+ \rightarrow 0+$ transition will strongly depend on the tensor A, the parameters q and η, and the rate at which the effective spin components change in space (the relaxation times). A detailed theoretical analysis of the shape of the expected Mössbauer spectra as function of relaxation times is given by Hirst[13] and Blume and coworkers[14]. Gonzalez et al. give a closed form formula, which contains two relaxation rate parameters ($W_\perp$ and $W_{||}$) [74G013]. Two relatively simple cases can be seen: a) the case of a cubic doublet, where $A_x = A_y = A_z$, $q = 0$, and $W_\perp = W_{||}$. b) The case of extreme axial symmetry, where $\eta = 0$, $A_x = A_y = 0$ and $W_{||} = 0$; in this case the formula reduces to the ordinary Wickman formula[15].

[13] L.L. Hirst, J. Phys. Chem. Solids 31, 655 (1970).

[14] J. H. Clauser and M. Blume, Phys. Rev. B3, 583 (1971); J. Sivardiere, M. Blume, and M. J. Clauser, Hyperfine Interac. 1, 227 (1975); J. Sivardiere and M. Blume, Hyperfine Interac. 1, 283 (1975).

[15] H. H. Wickman, in Mössbauer Effect Methodology, Volume 2, edited by I. J. Gruverman (Plenum Press, New York, 1966), p. 39.

Among the Yb^{3+} paramagnetic compounds investigated, various relaxation rates may be found: in Yb_2O_3 e.g., extreme fast relaxation times are found ($\tau \leq 10^{-12}$s); in $YbCl_3 \cdot 6H_2O$, $Yb_2(SO_4)_3 \cdot 8H_2O$, Rb_2NaYbF_6, $Cs_2NaYbCl_6$ and Yb:Au very long spin relaxation times ($\tau > 10^{-9}$s) are seen. Many examples of intermediate relaxation times exist (YbF_3, Yb acetate e.g.).

The observed spin relaxation phenomena in Mössbauer spectra of Yb^{3+} lead to the following conclusions: (i) A not extremely anisotropic A tensor, which also means a not extremely anisotropic g tensor, generally leads to fast spin relaxation phenomena (Yb_2O_3, $YbCrO_3$). This is not the case when the interatomic Yb-Yb distances are very large as, e.g. in $Yb_2(SO_4)_3 \cdot 8H_2O$ [71K058], Rb_2NaYbF_6 [74D045], and $Cs_2NaYbCl_6$ [74S090]. (ii) When the A tensor is extremely anisotropic, $A_x = A_y = 0$, the spin relaxation times are long. Spectra showing magnetic hyperfine structure of the "effective field" type, similar to the spectra observed in magnetically ordered systems, are observed at low temperatures ($YbCl_3 \cdot 6H_2O$). (iii) In cases where the total relaxation time . τ_R, of the electronic spin is of the same order of magnitude as the nuclear Larmor precession period, Mössbauer studies of the temperature dependence of relaxation times in insulators may yield the size of the various contributions to the general spin relaxation process, which include the spin-spin (temperature independent) relaxation process, the spin-lattice Orbach process (with exponential temperature dependence) and the Raman process (with a T^9 temperature dependence). In metals, an additional term due to relaxation with the conduction electrons can be investigated.

Kalvius et al. studied the temperature dependence of the $YbCl_3 \cdot 6H_2O$ spectrum [71K058] and obtained the temperature dependence of the spin relaxation time as $\frac{1}{\tau_R} = 4.8 \times 10^{11} \exp(-197/T) + 1.5 \times 10^8$. The first term is a typical spin-lattice relaxation term due to an Orbach process and the second term is a temperature independent typical spin-spin relaxation term. In [75B018] Borely et al. deduced the relaxation rate of Yb^{3+} in YbES between 1.3 and 30K. The thermal variation of the relaxation rate hinted the existence of an additional exponential term which was attributed to an optical mode of phonons. In this recent work an enriched absorber and a TmB_{12} source were used. The shielding factors R = 0.24 and $\sigma_2 = 0.61$ and the hyperfine parameters were derived from the experiments ($A_{||} = 904(14)$ MH eqQ = 904(16) MHz).

The first systematic Mössbauer study of several Yb^{3+} metallic compounds [67N012] revealed the somewhat surprising fact that slow spin relaxation phenomena can be observed even in concentrated Yb^{3+} metallic systems. Since then, spin relaxation phenomena have been observed in many other rare earth metallic systems. The Mössbauer spectrum observed in $YbNi_5$ [67N012], in which clearly relaxation phenomena are seen, is shown in Figure 4. The quadrupole interaction parameters found in $YbNi_5$ at 4.2 K and at 20K indicate that at these temperatures only the $|\pm\frac{7}{2}\rangle$ ground Kramers' doublet

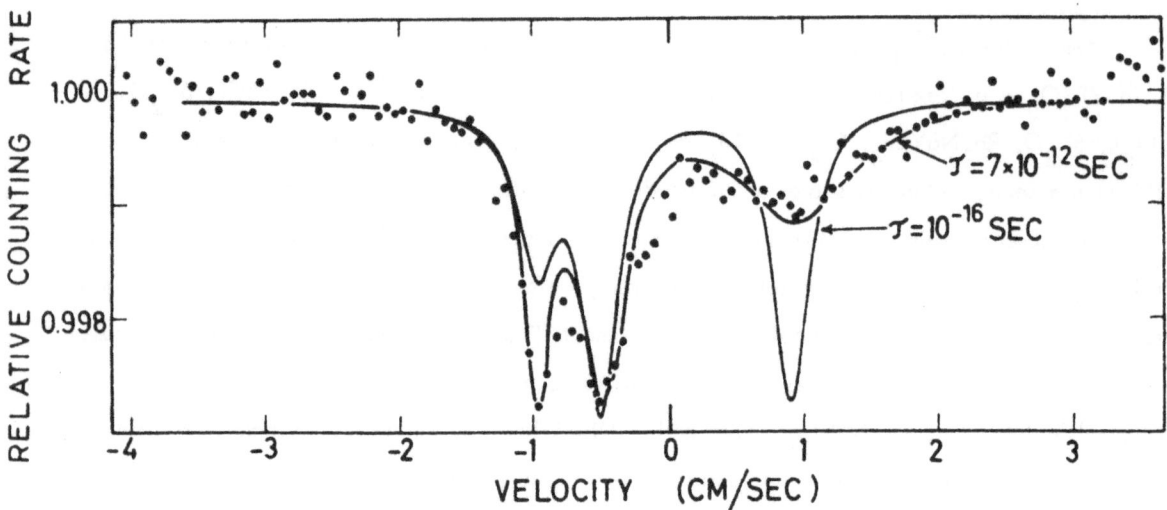

Figure 4.   Recoilless absorption spectrum of the 84.3 keV gamma ray emitted from a ^{170}TmAl$_2$ source in YbNi$_5$ at 4.2 K. The solid lines are theoretical spectra assuming the parameters $g\mu_n H_{eff}/h$ = 1200 MHz, eqQ/4h = 650 MHz, and the relaxation times indicated in the figure [67N012].

is populated. The spin relaxation times observed in this compound indicate that the relaxation is predominantly due to spin-spin interactions.

Very long spin relaxation times for a rare earth ion, submitted to an isotropic hyperfine interaction, were observed for the first time by Gonzalez et al. in the dilute metallic system of Yb:Au [72G028]. This system has been studied since then by the Mössbauer effect over a wide temperature range (between 0.3 and 50 K) [73G021], [74G013], and [74G029]. The samples studied were obtained by neutron irradiating Tm metal and melting it with gold. Samples with concentrations of Tm between 170 and 7500 ppm in gold were investigated. An YbB$_6$ absorber, enriched in ^{170}Yb, was used in these experiments. Some of the spectra observed are shown in Figure 5. The electronic ground state of Yb^{3+} in Yb:Au is the Γ_7 Kramers' doublet and its separation from the Γ_6 and Γ_8 levels is about 80 and 90 K. The Mössbauer spectra were analyzed, using the algebraic expression for the Mössbauer line shape in the presence of relaxation, developed in [74G013]. From the analysis of the temperature dependence of the spectra, the spin-conduction electron relaxation rate was derived and a Kondo anomaly was detected [73G021]. By comparing the experimental temperature dependence with the Kondo relaxation rate formula [73G021], the sign and size of J_{sf}, the exchange coupling constant between the localized moment and the conduction electrons, were deduced (J_{sf} = -0.55 eV). In a later publication Gonzalez et al. report the extension of this work to lower temperatures [74G029]. They show that from the relative intensities of the lines observed in the Mössbauer

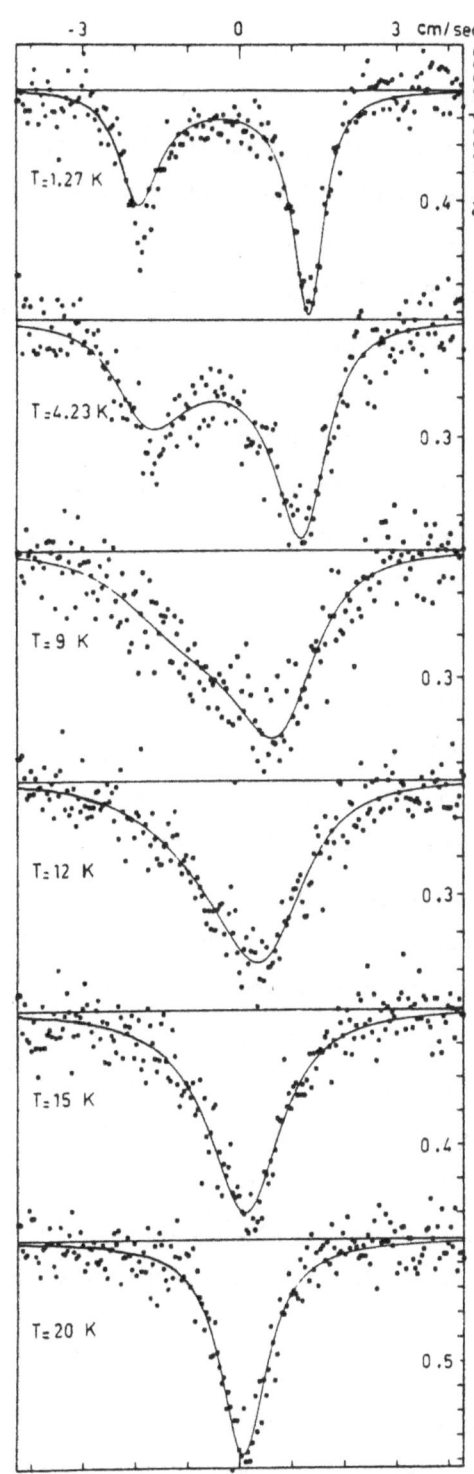

Figure 5. Relaxation spectra obtained between 1.27 and 20 K with a ^{170}Tm:Au source and an YbB$_6$ absorber. The solid lines represent least squares computer fits to the experimental points [74G013].

emission spectra relaxation rates can be obtained, if the spin relaxation rate is slow compared to the separation between the two hyperfine states, but comparable to the natural width of the 84 keV state. Measurements of the intensities thus give information on the electronic relaxation in a temperature range where the relaxation no longer distorts the line profiles. A very detailed analysis of the relaxation processes as obtained from the Mössbauer spectra measured between 0.6 and 26 K, taking into consideration the influence of the crystalline field on the Kondo effect, is given by Gonzalez et al. [75G016]. In [74S034] Shenoy et al. show that the relaxation rate is independent of small ($\leq$ 1 kG), externally applied magnetic fields. Very recently Mössbauer relaxation studies of ^{170}Yb in dilute ^{170}Tm:Au sources in external magnetic fields were performed by Stöhr [75S022]. In order to analyze the results, a general microscopic relaxation theory for an effective-field hyperfine interaction was derived. It was pointed out that in Mössbauer-source experiments, the preceeding nuclear beta decay may produce population effects in the lowest electronic energy states, due to slow relaxation rearrangement in the electronic shell. The experimental Mössbauer spectra of ^{170}Yb in dilute ^{170}Tm:Au sources were found to be well described by the relaxation theory, including the new rearrangement ideas for Mössbauer-source experiments.

In $Cs_2NaYbCl_6$ [74S055, 74S090] and Rb_2NaYbF_6 [74D045] the Yb-Yb distance is so large that the mutual interactions between the Yb ions are very weak and the Yb ions behave as isolated ions in a cubic crystalline field. At low temperatures the Mössbauer spectrum is similar to that observed for Yb:Au. In $Cs_2NaYbCl_6$ the derived spin relaxation time is temperature independent between 1.6 and 20 K ($\tau \sim 5 \times 10^{-9}$ s) and thus it is concluded to be mainly due to spin-spin interactions. Application of magnetic fields leads to spectra predicted by the Breit-Rabi diagrams. A detailed analysis of the spectra, using the closed form equations given by Gonzalez-Jimenez et at. [74G013], shows that the relaxation times increase with increasing magnetic fields, as expected if the relaxation mechanisms arise mainly through spin-spin coupling. Similar results were obtained with Rb_2NaYbF_6. Mössbauer spectra of $Cs_2Na(Yb_{1-x}Gd_x)Cl_6$ prove that the observed spin relaxation phenomena are indeed due to spin-spin interactions. The relaxation rate is linear in Gd concentration and its relative value for $x = 0$ to $x = 1$ agrees with theoretical predictions [75A037].

C. CRYSTALLINE FIELDS

In some Yb^{3+} metallic compounds, the crystal field parameters can be obtained from a study of the dependence of the hyperfine interactions in ^{170}Yb on temperature and on applied external magnetic fields [72N003]. In $YbPd_3$ the Yb^{3+} ion occupies a cubic site. In zero external field a broad absorption line is observed [67N012]. In external fields, an effective magnetic field and electric field gradient are observed to act on the Yb nucleus [72N003]. By comparing the experimental data to calculated values of H_{eff} and q_{eff}, taking into account the wave functions of the $^2F_{\frac{7}{2}}$ state in a cubic crystalline field, the crystal field parameters A_4 and A_6 could be derived. The values obtained for the crystal field constants imply that Yb^{3+} in $YbPd_3$ has a Γ_7 ground state, with Γ_8 and Γ_6 lying at approximately 29 and 39 cm^{-1}, respectively. This is roughly one order of magnitude lower than in cubic Yb^{3+} insulators. In $YbNi_2$ [73N003] the ordering temperature was found to be low (5.7K), but well defined magnetic hyperfine structure spectra were observed when an external field was applied. The comparison of the temperature and external field dependence of the magnetic hyperfine field with a simple molecular field model, taking into account crystalline fields, yielded values for the exchange field and crystalline field parameters (Table IV).

Recently the recoilless absorption spectra of the 84 keV gamma ray, emitted from an Yb^{3+}:$TmFe_2$ source in an enriched Yb metal absorber was measured as a function of temperature [75Y005]. The ordering temperature of $TmFe_2$ is high. The temperature dependence of the magnetic hyperfine field and electric field gradient yield values for the exchange and crystalline fields acting on the Yb^{3+} ion (Table IV).

Table IV. Hyperfine Interactions, Crystalline and Exchange Fields and Spin Relaxation Times in $^{170}Yb^{3+}$ Cubic Compounds

Compound	Ionic Ground State	$g\mu_n H_{eff}$ MHz	eqQ MHz	Spin Relaxation Time at 4.2 K (ns)	Crystalline Fields $A_4\langle r^4\rangle$ cm^{-1}	$A_6\langle r^6\rangle$ cm^{-1}	Exchange Field $\mu_B H_{exch}$ °K	Reference
YbPd$_3$	Γ_7	0	0	0.04	-12(1)	0.6(6)	-	[72N003]
YbNi$_2$	Γ_6	440(20)	375(50)	-	44(2)	<2	~15	[73N003]
Yb:TmFe$_2$	~$\lvert\tfrac{7}{2}\rangle$	1130(30)	2400(250)	0.70(15)	25(3)	-2(1)	116(4)	[75Y005]

Table V. Ground State Hyperfine Structure Parameters and Relaxation Times of $^{170}Yb^{3+}$ in Cubic Compounds

Compound	Ground State	Hyperfine Interaction Constant A (mm/s)	Spin Relaxation Time at 4.2 K ns	Reference
Yb:Au	Γ_7	13.3	1.11(4)	[72G028, 74S034]
Yb:CaF$_2$	Γ_7	13.18(7)	1.2	[75B038]
Cs$_2$NaYbF$_6$	Γ_6	-10.9(1)	4.7(5)	[74S055, 74S090]
Rb$_2$NaYbCl$_6$	Γ_6	-10.8(3)	0.98(5)	[74D045]
Yb:TmBe$_{13}$	Γ_6	-8.5(9)	1.07(6)	16
YbAuNi$_4$	Γ_6	-10.3(4)	0.96(4)	16

In Cs$_2$NaYbCl$_6$ [74S090], Rb$_2$NaYbF$_6$ [74D045], and Yb:Au [74S034], the nature of the electronic ground state of the Yb^{3+} ion (Γ_6 or Γ_7) and the magnetic hyperfine coupling constant A were derived from the Mössbauer spectra in a straight-forward way (Table V). Dramatic changes in the shape of the Mössbauer spectra were observed even when small magnetic fields ($\leq$ 1 kG) were applied (Figure 6). The dependence of the hyperfine structure on external magnetic fields was shown to follow a Breit-Rabi diagram (Figure 7) in the three compounds.

---

[16] I. Nowik, I. Felner, and R. Yanovsky, to be presented at the Corfu Conference, 1976.

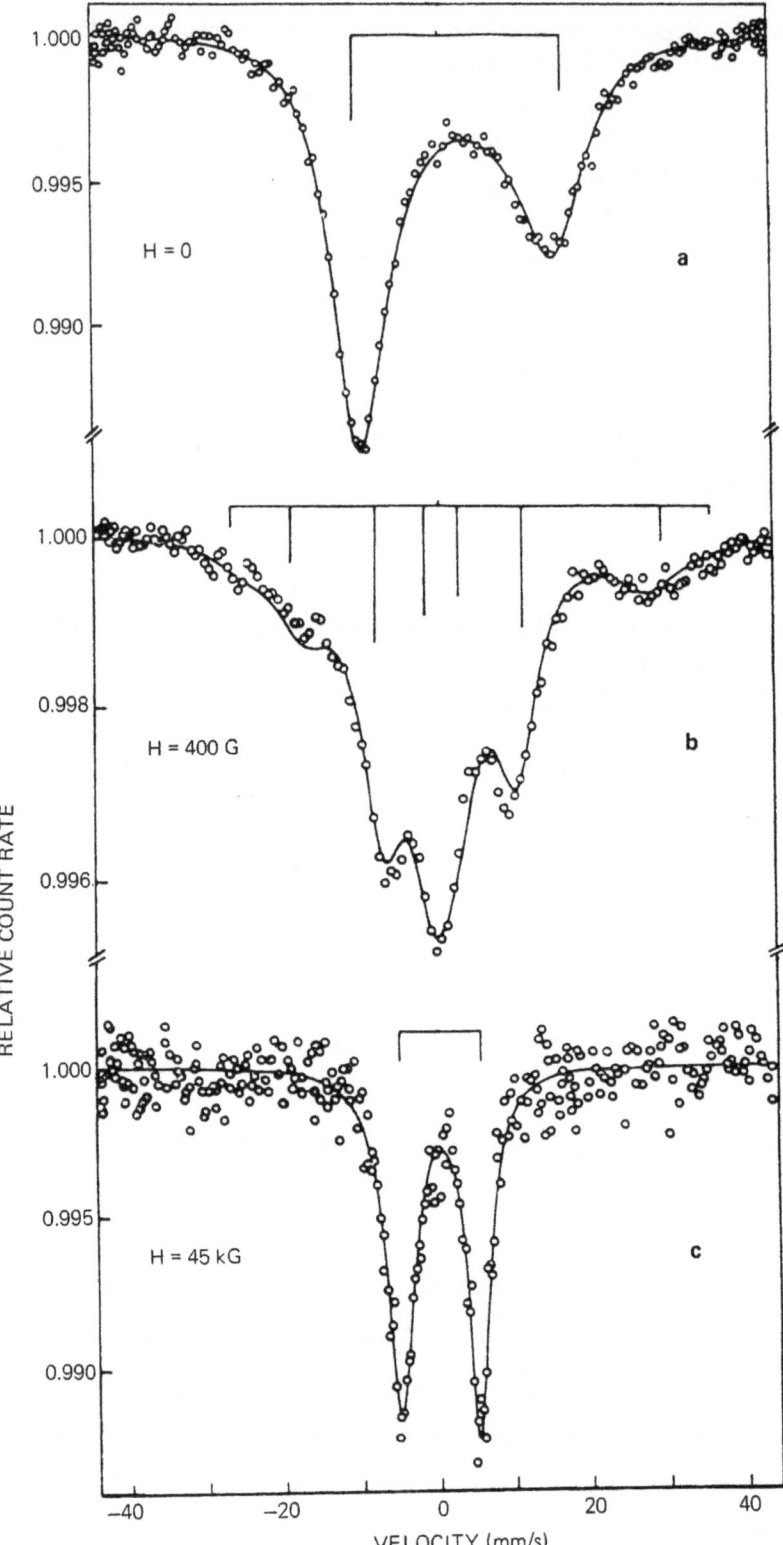

Figure 6. ^{170}Yb Mössbauer spectra in $Cs_2NaYbCl_6$ at 4.2 K for different external fields: a) 0kG; b) 0.4kG c) 45kG [74S090].

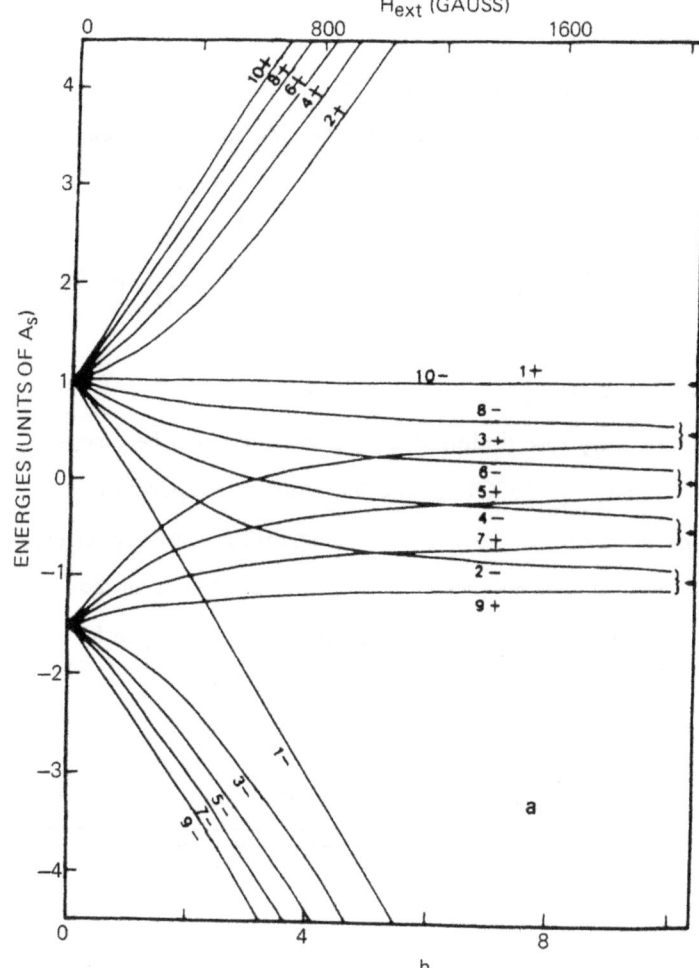

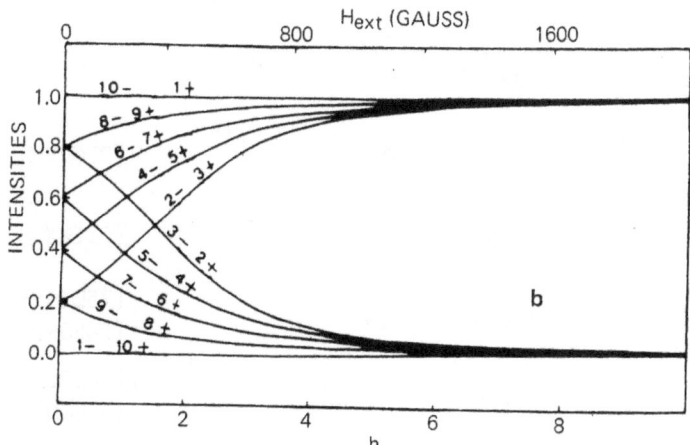

Figure 7. Breit-Rabi diagram for a 2+ → 0+ gamma ray transition: a) energies, and b) intensities of the Mössbauer transition as a function of external magnetic field [74S090].

427

Spectra obtained with absorbers of dilute ^{170}Yb in CaF_2 showed that about 2/3 of the Yb ions are divalent. The Yb^{2+} and about 1/3 of the Yb^{3+} ions are in cubic sites, the other Yb^{3+} ions are mainly in tetragonal sites. In dilute $Tm:CaF_2$ sources only the Yb^{3+} cubic hyperfine structure is observed. The spectra could be fitted, assuming a Γ_7 ground state and hyperfine parameters obtained previously from ENDOR measurements [75B038]. Recently[16] a well resolved Γ_6 Mössbauer spectrum has been observed for ^{170}Yb in concentrated paramagnetic metallic systems: in ^{170}Yb:$TmBe_{13}$ and in $YbAuNi_4$. The experimental results are summarized in Table V.

D. MAGNETIC STRUCTURES

The Curie points of Yb^{3+} salts are usually very low (below 4.2 K) and hence very few magnetically ordered systems have been studied. In $Yb_3Fe_5O_{12}$ [67O002] the Yb ordering temperature is high due to the large Fe-Fe and Fe-Yb exchange interaction. In this case a magnetically split spectrum is observed even at 20 K, though ferromagnetic spin relaxation phenomena complicate the observed spectra. A detailed analysis yields, however, all hyperfine parameters, including the ferromagnetic spin relaxation time (Table VI).

Table VI. Hyperfine Interaction Parameters in Yb Magnetically Ordered Compounds

Compound	Ordering Temperature K	Temperature K	$g\mu_n H_{eff}$ MHz	eqQ MHz	η	Relaxation Time s	Reference
$Yb_3Fe_5O_{12}$	548	4.2	467	140		$9(3)\cdot10^{-11}$ at 20 K	[67O002]
Yb_2O_3 site C_2 site C_{3i}	2.3	1.6	520(50) 280(50)	1600(80) 700(40)	0.12 0	$<10^{-12}$	[70K015]
Yb_2O_2S	2.65	1.3	427(8)	−545(60)	0	$<10^{-12}$	[72G004]
Yb_2O_2Se	2.65	1.3	405(8)	−1000(60)	0	$<10^{-12}$	[72G004]
$YbFeO_3$	630	4.1	270(3)	1660(60)	∼0.4		[75D019] [74B061]
$YbCrO_3$	118(1)	1.3	∼ 25	2060(30)	0.16(5)		[74B061] [74D014]
$YbVO_3$	94(1)	4.1	∼188	2020(60)	0.20(5)		[74B061]

A detailed study of the hyperfine interaction in Yb_2O_3 below $T_N = 2.3$ K was performed by Kalvius et al. [70K015]. The results indicate that the hyperfine magnetic field at the C_2 site is along the principal electric field gradient axis. This axis which is in the [110] direction, is perpendicular to the C_2 symmetry axis. This determination of the direction of the magnetic moment agrees with results obtained from neutron diffraction measurements[17]. Kalvius et al. stress the high extent to which the hyperfine fields in Yb^{3+} ions are proportional to the magnetic moment (Figure 8). Gonzalez-Jimenez et al. [72G004] have studied Yb_2O_2S and Yb_2O_2Se above and below their Néel temperatures. In these compounds the Yb^{3+} ions occupy an axial symmetry site C_{3v}. From a detailed analysis of the hyperfine interaction parameters, the authors manage to determine the composition of the ground Kramers' doublet wave function of Yb^{3+} in Yb_2O_2S. This determination is consistent with susceptibility measurements. The results obtained also show that the moments in both compounds are perpendicular to the high symmetry axis of the crystal. Davidson et al. [74D014] report an interesting study of Yb^{3+} in $YbCrO_3$ in external fields. $YbCrO_3$ is an orthorombic weak ferromagnet. The absorber consisted of loosely packed powder whose crystallites were oriented by an external field

---

[17] R. M. Moon, W. C. Koehler, H. R. Child, and L. J. Raybenheimer, Phys. Rev. 176, 722 (1968).

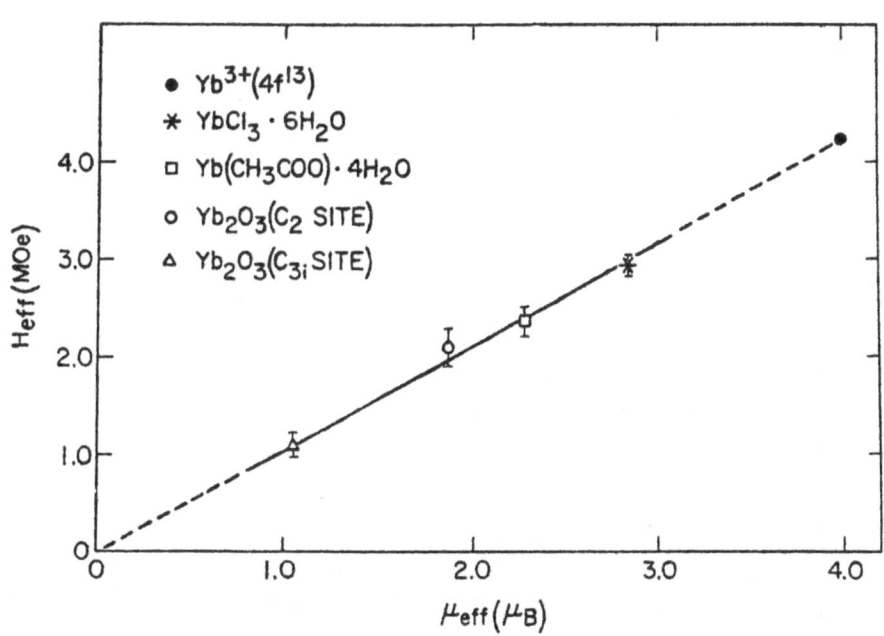

Figure 8. Effective magnetic field at the [170]Yb nucleus as function of the magnetic moment of the Yb^{3+} ion in several ytterbium compounds [70K015].

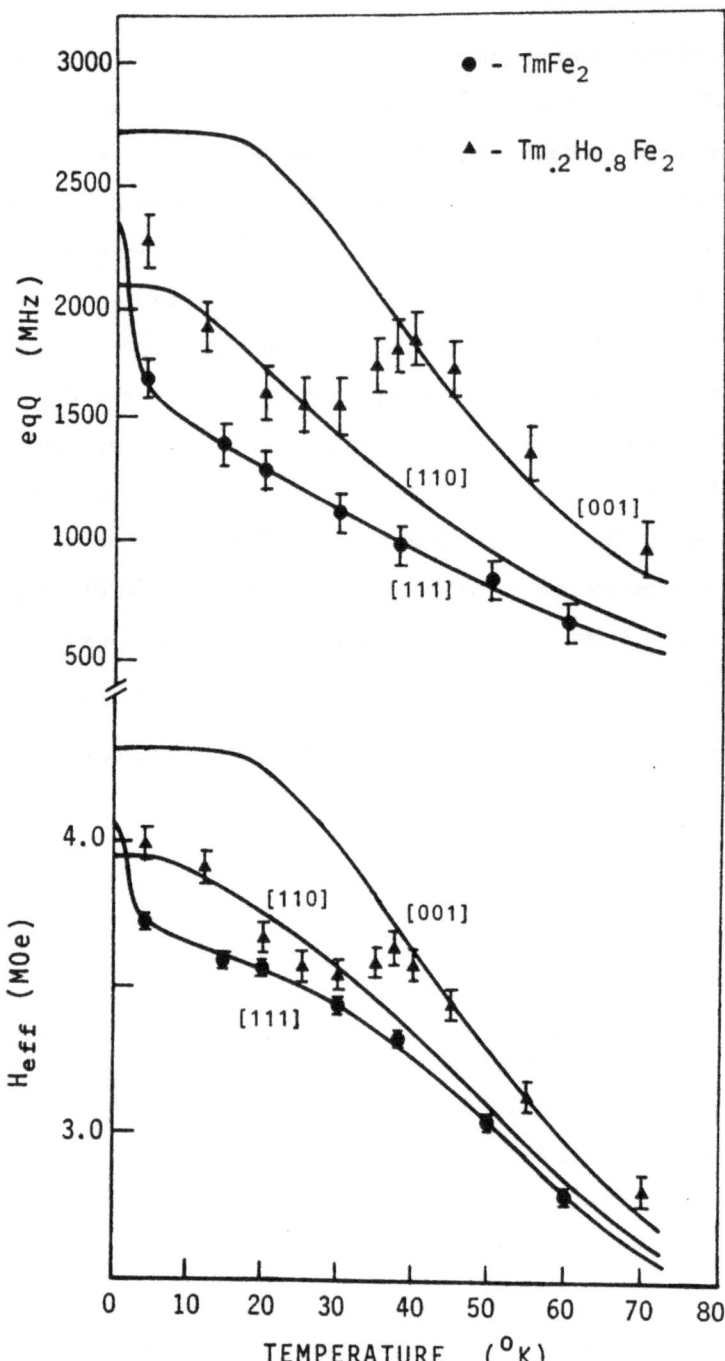

Figure 9. Temperature dependence of hyperfine fields acting on ^{170}Yb in TmFe$_2$ and Tm$_{0.2}$Ho$_{0.8}$Fe$_2$. The solid lines are theoretical curves for magnetizations in the [111], [110], and [001] directions, calculated using the parameters given in Table IV [75Y005].

so that each had its weak ferromagnetic a axis parallel to the gamma ray direction. In this way the angle θ between the a axis and the direction of the EFG principal axis could be determined as $(25 \pm 1)°$. The axis was found to lie in the ab plane. The component of the hyperfine field in the direction of the EFG axis was found to increase linearly with the externally applied field. The analysis of these results, together with results obtained before from susceptibility measurements, yields the full anisotropic local susceptibility matrix and indicates that the ground state of Yb^{3+} in $YbCrO_3$ is predominantly a $\left| +\frac{7}{2} \right>$ Kramers' doublet. Another study of $YbCrO_3$ and several other perovskite compounds ($YbVO_3$, $YbFeO_3$ and $Yb:YAlO_3$) is reported by Bonville et al. [74B061]. From the temperature dependence of the hyperfine fields in $YbVO_3$ and $YbFeO_3$, the hyperfine fields at saturation and the splittings of the ground state doublet Δ were determined. The values found for Δ are in agreement with values obtained from specific heat measurements. In 10% Yb in $YAlO_3$ slow relaxation was obtained and a magnetic hyperfine spectrum seen. It is explained as resulting from an anisotropic hyperfine A tensor ($A_\perp = 0$) and an axial electric field gradient. At 120K a pure quadrupole spectrum, with somewhat broadened lines, is obtained. The wave functions of the ground state of Yb^{3+} in $YAlO_3$ and $YbFeO_3$ were calculated. The hyperfine interaction parameters in Yb magnetically ordered compounds are summarized in Table VI.

The sensitivity of the ^{170}Yb Mössbauer spectra to the orientation of the magnetic moment relative to the local coordinate system is very well seen from the Mössbauer spectra observed with a $Tm_x Ho_{1-x} Fe_2$ source and an Yb metal absorber. In $TmFe_2$ the easy axis of magnetization is the cubic [111] axis. In $Tm_x Ho_{1-x} Fe_2$ the axis can be either along [111], [110] or [001], depending on x and may switch direction as function of temperature[18]. The value of the magnetic hyperfine field and the electric field gradient acting on the ^{170}Yb nucleus depends on the angle between the magnetization axis and the local symmetry axis. Mössbauer studies of the hyperfine fields in this system, as function of x and temperature, determine the reorientation transition temperature and the nature of this reorientation transition. It is found, e.g., that the transition from the [110] to the [100] axis takes place at 35K in $Tm_{.2} Ho_{.8} Fe_2$ and at 45K in $Tm_{.4} Ho_{.6} Fe_2$. The hyperfine fields and quadrupole interaction parameters obtained are plotted in Figure 9, as a function of temperature. The solid lines in the figure are the theoretical curves, calculated with the crystal field parameters and exchange field obtained from the Mössbauer study of Yb^{3+} in $TmFe_2$. It is seen that the reorientation transition is a continuous second order transition, though the general symmetry of the system is cubic and a first order transition could be expected [75Y005].

---

[18] U. Atzmony, M. P. Dariel, E. R. Bauminger, D. Lebenbaum, I. Nowik, and S. Ofer, Phys. Rev. 7B, 4220 (1973).

Another reorientation study utilizing ^{170}Yb was performed on YbFeO$_3$ [75D019, 74B061]. The ^{170}Yb Mössbauer spectra show a magnetic hyperfine splitting due to a small exchange field produced by the weak ferromagnetic moment of the iron. This moment is produced by a slight canting of the antiferromagnetically coupled iron spins. The observed spectra are consistent with the occurrence of a continuous rotation of the Yb spins over a temperature interval $6.5 < T < 7.83$K. The orientations of the principal axes of the EFG at the Yb nuclei were also determined and information on the arrangement of the Fe moments in these compounds was deduced.

E. FUTURE EXPERIMENTS

No doubt that the Mössbauer effect in ^{170}Yb will be used in future studies of crystalline fields, exchange fields, magnetic spin structure and spin relaxation phenomena. It will surely serve in the near future for more studies of the Kondo effect. Among the rare earth ions only Ce and Yb are candidates for the observation of Kondo anomalies. Ce has no Mössbauer isotope and thus ^{170}Yb is the only suitable isotope for Mössbauer studies of Kondo anomalies. More studies like those reported so far for Yb:Au are therefore expected.

Yb is one of the few rare earth ions which may exist in a metallic system in a mixed valence state. It is known, for example, from susceptibility and Mössbauer studies, that the Yb ion is divalent in YbAl$_3$ at 4.2K [68H025], whereas, its effective magnetic moment measured at room temperature shows that it is trivalent at 300K. In the temperature region between 100 and 200K it goes over continuously from one valence state to the other[12]. This intermediate state may be described as resulting from fast fluctuations between the $4f^{14}$ configuration and the $4f^{13}$ configuration with the extra electron in the conduction band. Intermediate valence states were studied quite extensively in $Eu^{2+} - Eu^{3+}$ systems, using the Mössbauer effect in ^{151}Eu. In Eu the different valence states are characterized by very different isomer shifts. In ^{170}Yb the isomer shifts obtained are not sensitive enough for such studies. The Yb^{2+} and Yb^{3+} configurations can, however, easily be distinguished by their magnetic properties and the Mössbauer spectra obtained in the two valence states in a magnetic field (external or exchange field) will be very different. Spectra being consistent with interconfiguration fluctuations have been reported for Yb in Tm$_x$Ho$_{1-x}$Fe$_2$ [74B055]. Progress in this field in the future may be expected.

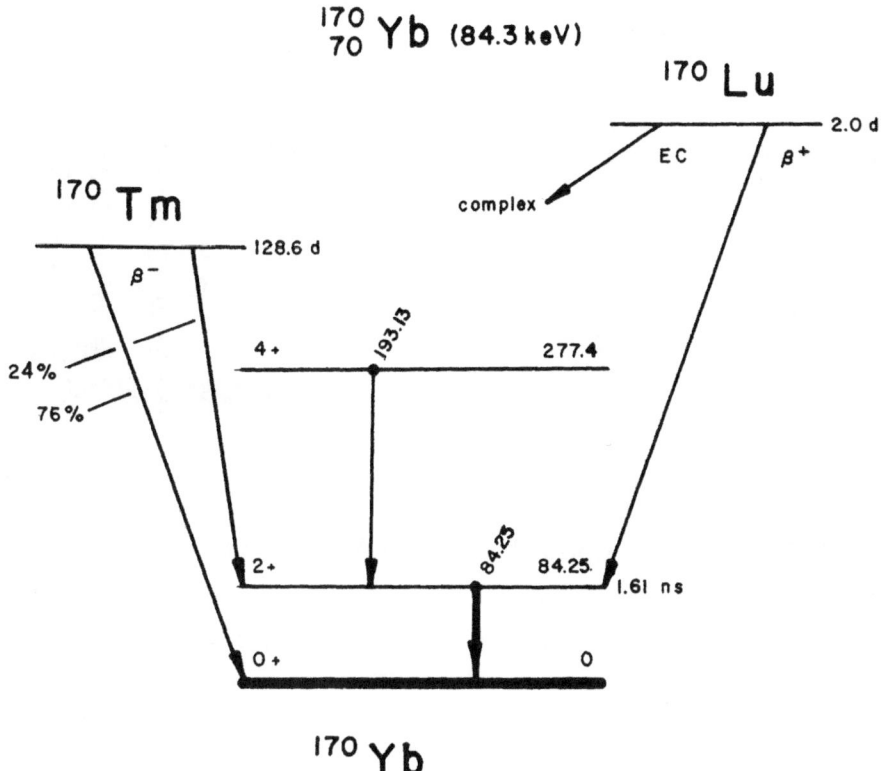

$^{170}_{70}$ Yb (84.3 keV)

^{170} Lu

^{170} Tm

^{170} Yb

Measured Properties

E_γ = 84.2529 ± 0.0012 keV (1),(2),(3)

$t_{\frac{1}{2}}$ = 1.608 ± 0.017 ns (4),(5)

α_T = 8.05 ± 0.20 (4)

IA = 3.03%

μ^* = 0.669 ± 0.008 nm (7),(8),(9)

Q^* = -2.11 ± 0.11 b (10)

Energies and Intensities

	E_γ	I_{Lu} (1)
γ_M :	84.26 keV	195
γ_1 :	152.6 keV	6
γ_2 :	194. keV	46
K_{α_1} :	54.4 keV	

Derived Parameters

σ_o = 1.90 (4) × 10^{-19} cm^2

Γ = ~ 2.84 (3) × 10^{-7} eV

W_o = 2.019 (21) mm/s

E_r = 2.241323 (19) × 10^{-2} eV

Energy Conversions

1 mm/s = 67.9529 (5) MHz

1 mm/s = 2.81033 (1) × 10^{-7} eV

(1) W. Beer and J. Kern, Nucl. Imstrum. Methods 117, 183 (1974).
(2) D. E. Raeside, Nucl. Instrum. Methods 87, 7 (1970).
(3) G. L. Borchert, W. Scheck, and K. P. Wieder, Z. Naturforsch, Teil A 30, 274 (1975).
(4) D. K. Gupta and G. N. Rao, Nucl. Phys. A182, 669 (1972).
(5) A. Bäcklin, S. G. Malmskog, and H. Solhod, Ark. Fys. 34, 495 (1967).
(6) J. Plch, J. Zderadicka, and L. Kokta, Czech. J. Phys. B 23, 1181 (1973).
(7) 65H05.
(8) 68M011.
(9) 75B038.
(10) H. Ryde, G. D. Symons, and Z. Szymanski, Nucl. Phys. 80, 249 (1966).

[Other data from M. R. Schmorak and R. L. Auble, Nucl. Data Sheets 15, 371 (1975)]

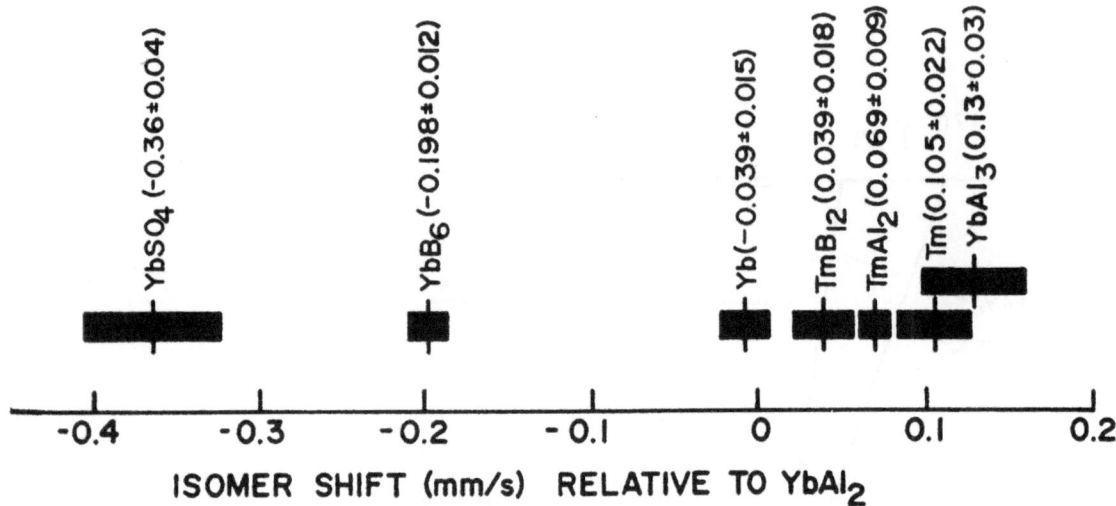

ISOMER SHIFT (mm/s) RELATIVE TO YbAl$_2$

Note: No correction is made for the second order Doppler shift. This correction is relatively small, e.g. a material having an effective Debye temperature of 200 K, the correction between 4.2K and 77K is 0.006 mm/s

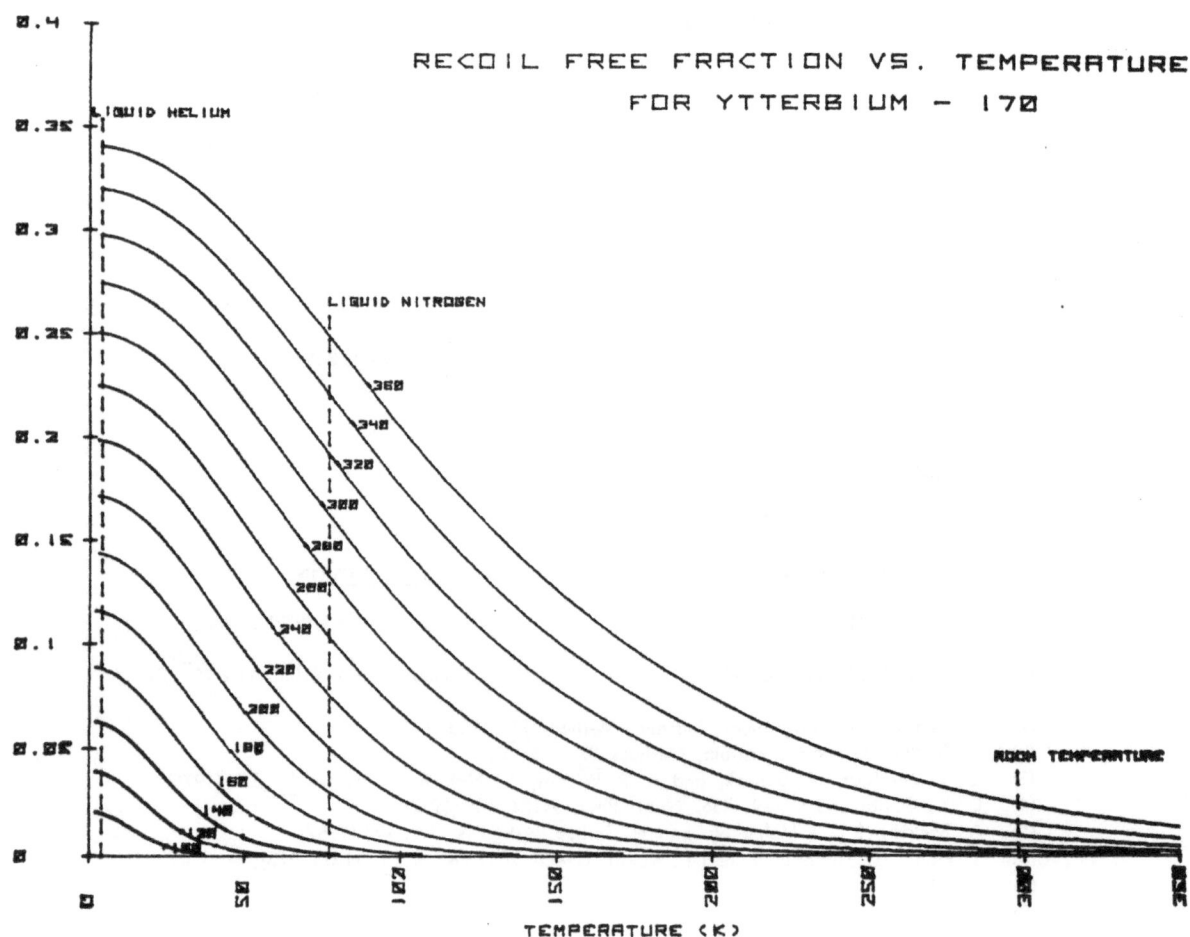

RECOIL FREE FRACTION VS. TEMPERATURE FOR YTTERBIUM – 170

YB-170, 84.3 KEV TRANSITION

INORGANIC HALIDES

SOURCE	S-TEMP	ABSORBER	A-TEMP	SHIFT	QS	REMARKS	REF
TmB12	4.2	CaF2	4.2	—		g1=.334 nm	75B038
XX		Cs2NaYbCl6	4.2			hfs=-10.9 mm/s	74S055
XX		Cs2NaYbCl6	v			spin relaxation	74S090
XX		Cs2NaYb(1-x) =				=GdxCl6, spin relaxation	75A037
XX		Rb2NaYbF6	4.2			relaxation effects	74D045
Al2Tm(IS/Al2Yb)		YbCl2	4.2	-.25		nuclear radius experiment	67A017
Al2Tm	4.2	YbCl2.5	4.2	-.10		nuclear radius experiment	68HC25
Al2Tm	4.2	YbCl2.5	4.2	-.10		nuclear radius experiment	68K007
Al2Tm	4.2	YbCl3	4.2	+.09		nuclear radius experiment	68HC25
Al2Tm	4.2	YbCl3	4.2	.085		nuclear radius experiment	68K007
Al2Tm	4.2	YbCl3	4.2		--	EQ<80 M Hz, spin relaxation	68NC11
Al2Yb	4.2	YbCl3	4.2	--		nuclear radius experiment	73R024
Al2Tm(IS/Al2Yb)		YbCl3-anhydrous	4.2	+.11		nuclear radius experiment	67A017
Tm	He	YbCl3-6H2O	He			MM=.656	67E001
xx(IS/YbSO4)	4.2	YbCl3-6H2O	4.2	--		nuclear radius experiment	70HC32
Al-Er alloy	4.2	YbCl3-6H2O	4.2		--		66HC02
Al2Tm	4.2	YbCl3-6H2O	4.2			g1=.332	68NC11
Al2Tm	4.2	YbCl3-6H2O	4.2	+.10		nuclear radius experiment	68H025
Al2Tm	4.2	YbCl3-6H2O	4.2			MM=.676	66NC14
Al2Tm	4.2	YbCl3-6H2O	4.2	+.15			71H008
Al2Tm	4.2	YbCl3-6H2O	4.2			MM(Yb170)/MM(Yb174)=.994	66NC13
XX		YbCl3-6H2O	v			temperature dependence of paramagnetic relaxation	71KC58
Al2Tm	4.2	YbF2	4.2	—		nuclear radius experiment	73R024
Al2Tm	4.2	YbF3	4.2		—	EQ=1720 M Hz,spin relaxation	68NC11

METALS AND ALLOYS

SOURCE	S-TEMP	ABSORBER	A-TEMP	SHIFT	QS	REMARKS	REF
Al-Tm alloy		Al-Yb alloy				W=2.29	71EC09
TmB12(IS/Al2Yb)		Al2Yb					74EC24
TmB12		Al2Yb		-.05			74EC61
Al2Tm(IS/Al2Yb)		Al2Yb	4.2	0.00		nuclear radius experiment	67A017
Al2Tm	4.2	Al2Yb	4.2	—		nuclear radius experiment	73R024
TmB12(IS/Al2Yb)	4.2	Al2Yb	4.2	.00		fa=.19, Debye temperature=233	73G032
Al2Tm	4.2	Al3Yb	4.2	+.06		nuclear radius experiment	68HC25
XX		Au				HA=400 G	74S131
Al2Tm		Fe23Yb6	4.2	-.08	--	HI=258 kOe, EQ=.52	71GC55
TmB12(IS/Al2Yb)		In3Yb	4.2	-.13			74EC24
Al2Tm		Ni2Yb				complex spectra	67N012
Al2Tm		Ni2Yb	v		--	HA data	73NC03
Al2Tm		Ni5Yb				spin relaxation=7x10(-12) s	67N012
TmB12(IS/Al2Yb)		Pb3Yb	4.2	-.12			74EC24
Al2Tm		Pd3Yb	4.2			spin relaxation=2x10(-11) s	67NC12
Al2Tm		Pd3Yd	4.2			HA=0,15,30 & 44 kOe	72N003
Al2Tm		Pd3Yd	v			no evidence of magnetic ordering down to 1.4 K	72N003
TmB12(IS/Al2Yb)		Sn3Yb	4.2	-.08			74PC24
Al2Tm		Yb					71GC55
Tm(Yb)		Yb		-.12		detn of conduction-electron densities	71HC59
Al-Er alloy	4.2	Yb	4.2		—		66HC02
Al2Tm	4.2	Yb	4.2	-.08		nuclear radius experiment	68EC25
Al2Tm	4.2	Yb	4.2	-.075		nuclear radius experiment	68KC07
Al2Tm(IS/Al2Yb)		Yb	4.2	-.03		nuclear radius experiment	67AC17
Al2Tm	4.2	Yb	4.2	—		nuclear radius experiment	73R024
TmB12(IS/Al2Yb)		Yb	4.2	-.10			74EC24

MISCELLANEOUS EXPERIMENTS

SOURCE	S-TEMP	ABSORBER	A-TEMP	SHIFT	QS	REMARKS	REF
Al3Tm	4.2	Al3Yb	4.2			obsvn of interference effects	72WC01
Al-Tm alloy(IS=		Yb+3	4.2	.39		=/YbSO4), nuclear radius exp	71S064

MISCELLANEOUS INORGANIC COMPOUNDS

SOURCE	S-TEMP	ABSORBER	A-TEMP	SHIFT	QS	REMARKS	REF
Al2Tm(IS/Al2Yb)		GaYb-garnet	4.2	+.24		nuclear radius experiment	67A017
TmB12		YbB6		-.24			74BC61
Al2Tm	4.2	YbB6	4.2	--		nuclear radius experiment	73RC24
TmB12(IS/Al2Yb)	4.2	YbB6	4.2	-.22		W=2.4, fa=.10, Debye T=171 K	73GC32
Al2Tm	4.2	YbC2	4.2	--		nuclear radius experiment	73RC24
Al2Tm		YbCl3-6H2O	v			Yb is divalent	65HC05
Al2Tm		YbGe1.75	4.2			Yb is divalent	69SC22
Al2Tm		YbH2	4.1	-.11	--	EQ=-91.5 MHz, eta=.89	74MC18
Al2Tm	4.2	Yb(MeCO2)3	4.2		--	EQ=980 M Hz, spin relaxation	68NC11
Al2Tm	4.2	Yb(NO3)3-6H2O	4.2		--	EQ=940 M Hz, spin relaxation	68NC11
xx(IS/YbSO4)	4.2	YbOCH	4.2	--		nuclear radius experiment	70HC32
Al2Tm	4.2	YbSe	4.2	--		nuclear radius experiment	73RC24
Al2Tm		YbSi2	Be			Yb is Yb+2	69SC22
Al2Tm(IS/Al2Yb)		YbSi2	4.2	+.02		nuclear radius experiment	67A017
Al2Tm	4.2	YbTe	4.2	--		nuclear radius experiment	73RC24
Al2Tm		Yb5Si3	4.2		--	Yb is trivalent	69S022

ORGANIC COMPOUNDS

SOURCE	S-TEMP	ABSORBER	A-TEMP	SHIFT	QS	REMARKS	REF
TmB12		YbEtSO4	v	--	--	1.3<T<30 relaxation	75E018

INORGANIC OXIDES

SOURCE	S-TEMP	ABSORBER	A-TEMP	SHIFT	QS	REMARKS	REF
Tm	20	Tb2O3	20	--	--	first reported observation of the Mossbauer effect for this transition, HL=1.5 ns	62WC03
TmB12		YAlO3	4.2	.16		electronic structure info	74BC61
Al2Tm	4.2	Yb	4.2	-.06			71BC08
Al2Tm(IS/YbA12)		Yb.22Ca.78C	5	-.14		Yb is divalent	71A008
Al2Tm	4.2	YbCl3	4.2	+.10			71BC08
Al2Tm		YbCrO3	4.2		--	EQ=2060 MHz,HI=60 kCe	74EC14
Al2Tm		YbCrO3	77		--	EQ=1860 MHz, eta=.16	74DC14
TmB12		YbCrO3	v	--		HI data	74EC61
Al2Tm		YbFeO3	v			obsvn of the spin reorienta- tion transition at Yb sites, 4.2<T<10 K	75DC69
Al2Tm		YbFeO3	v		--	spin reorientation, 4.2K<T<1CK	75DC19
TmB12		YbFeO3	v	--		electronic structure info	74BC61
Al2Tm		YbGaG	4.2			ferromagnetic relax=2x10(-11)s	67CC02
Al2Tm		YbGaG	20			ferromagnetic relax=9x10(-11)s	67CC02
Al2Tm		YbIG	4.2			ferromagnetic relax=2x10(-11)s	67CC02
Al2Tm		YbIG	20			ferromagnetic relax=9x10(-11)s	67CC02

SOURCE	S-TEMP	ABSORBER	A-TEMP	SHIFT	QS	REMARKS	REF
Al2Tm	4.2	YbOOH	4.2	+.12		nuclear radius experiment	68B025
Al2Tm	4.2	YbOOH	4.2	+.08	--	EQ=-10.2	71BC08
Al2Tm	4.2	YbOx	4.2	.12		nuclear radius experiment	68KC07
Al2Tm	4.2	YbSO4	4.2	-.42	--	EQ=+7.6, eta=.85	71BCC8
TmB12		YbVO3	1.2	0.11			74B061
TmB12		YbVO3	v	--		HI data	74EC61
Al2Tm	15	Yb2O2S	1.3		--	EQ=-8.08, HI=1683 kOe	72G004
Al2Tm	15	Yb2O2S	5.2	.07	--	EQ=-8.64	72G004
Al2Tm	15	Yb2O2S	150	.12	--	EQ=-7.92	72G004
Al2Tm	15	Yb2O2Se	1.3		--	EQ=-14.8, HI=1597 kOe	72GC04
Al2Tm	15	Yb2O2Se	4.2	.05	--	EQ=-14.7	72G004
Tm	He	Yb2O3	He		--		67EC02
xx(IS/YbSO4)	4.2	Yb2O3	4.2	--		nuclear radius experiment	70HC32
Al2Tm	4.2	Yb2O3	4.2	+.10	--	EQ=+24.8, eta=.12	71BC08
Al2Tm	4.2	Yb2O3	4.2		--	EQ, spin relaxation	68NC11
Tm	4.2	Yb2O3	4.2		--	QM(Yb174)/QM(Yb170)=1.1	66EC02
Tm	20	Yb2O3	20	--	--	HI=1.5 ns,	62KC16
Al-Tm alloy		Yb2O3	v		--	EQ & HI data for the 2 sites	70KC15
Al-Tm alloy		Yb2O3	v	--	--	QM and RQ data	71EC09
Al2Tm	4.2	Yb2S3	4.2	+.10			71BC08
Al2Tm	4.2	Yb3Ga5O12	4.2		--	EQ<80 M Hz, spin relaxation	68NC11

SOURCE EXPERIMENTS

SOURCE	S-TEMP	ABSORBER	A-TEMP	SHIFT	QS	REMARKS	REF
Al2Tm	4.2	xx(IS/Al2Yb)	4.2	+.09		fs=.18, Debye temperature=230	73G032
Au(Tm)	v					study of Kondo anomaly	74GC59
Au(Tm)	v	xx				.6 K<T<26 K, relaxation	75GC16
Au(Tm)	v	xx				presence of Kondo effect	74G029
Au(Tm)	v	xx				observation of a Kondo anomaly	73GC21
Au(Tm)	v	YbB6				source experiment, study of electronic relaxation	74GC13
Au(Tm)		Al3Yb	1.4			study of the affect of HA on the relaxation spectra	74S034
Au(Tm)		Al3Yb	4.2			study of the affect of HA on the relaxation spectra	74SC34
Au(Tm)		Al3Yb	4.2	--	--	4.2 K<T<40 K, relaxation	75S022
Au(Tm)	v	YbB6	15			source experiment, relaxation	72GC28
CaF2(Tm)	4.2	YbB6	4.2	--		source experiment, g1=.334 nm	75BC38
Fe2Ho(1-x)Tmx	v	Yb	4.1		--	4.1<T<80 K, study of crystalline fields, exchange interactions & spin reorientations	75Y005
Fe2Ho(1-x)Tmx	v	Yb	4.1				74E055
Fe2Tm	4.2	Al3Tb	--	--		study of electronic rearrangement effects	74HC52
Fe2Tm	4.2	Al3Yb				electronic rearrangement effects	74E050
TmB12	4.2	xx(IS/Al2Yb)	4.2	+.01		fs=.34, Debye temperature=358	73GC32

INORGANIC SULFATES

SOURCE	S-TEMP	ABSORBER	A-TEMP	SHIFT	QS	REMARKS	REF
Al2Tm	4.2	YbSO4	4.2	-.27		nuclear radius experiment	68B025
Al2Tm(IS/Al2Yb)		YbSO4	4.2	-.39		nuclear radius experiment	67AC17
Al2Tm	4.2	Yb2(SO4)3-8H2O	4.2		--	EQ=172 M Hz, spin relaxation	68NC11
Al2Tm	4.2	Yb2(SO4)3-8H2O	4.2	+.17		nuclear radius experiment	68B025
Al2Tm	4.2	Yb2(SO4)3-8H2O	4.2	+.17	--	EQ=2.8	71H008
xx		Yb2(SO4)3-8H2O	v			temperature dependence of paramagnetic relaxation	71KC58

INCRGANIC SULFIDES

SCURCE	S-TEMP	ABSORBER	A-TEMP	SHIFT	QS	REMARKS	REF
Al2Tm	4.2	YbS	4.2	-.296		nuclear radius experiment	73K024
Tm(Yb)		Yb2S3		+.06		detn of conduction-electron densities	71H059
xx(IS/YbSO4)	4.2	Yb2S3	4.2	--		nuclear radius experiment	70H032

CODE	TOPIC	REFERENCE
62K016	YB-170	P Kienle, G M Kalvius, F Stanek, F E Wagner, H Eicher, and W Wiedemann, Hyperfine Splitting cf Gamma Rays from Rare Earth Nuclides, in "Proceedings of the Second International Conference on the Mossbauer Effect" (Saclay, France, 1961), edited ty D M J Compton and A H Schoen (John Wiley and Sons Inc, New York, 1962), pp 185-9
62W003	YB-170	F E Wagner, F W Stanek, F Kienle, and H Eicher, Z Phys 166, 1-5 (1962), Hyperfeinstrukturaufspaltung von ruckstossfreien gamma-Linien III. Das 84 keV-Niveau in Yb170
65H005	YB-170	A Huller, W Wiedemann, F Kienle, and S Hufner, Phys Lett 15,269-71 (1965), Hyperfine Splitting and g Factor of the 84-keV State in Yb170
66E002	YB-170	J S Eck, Y K Lee, E T Ritter, R B Stevens, Jr, and J C Walker, Phys Rev Lett 17,120-2 (1966), Observation of New Mossbauer Effects in Rare-earth Isotopes Following Coulomb Excitation
66H002	YB-170	W Henning, P Kienle, E Steichele, and F E Wagner, Phys Lett 22,446-8 (1966), Magnetic Properties of the K = 1/2 Rotational Band of Yb171
66M013	YB-170	E Munck, S Hufner, H Prange, and D Quitmann, Z Naturforsch, A 21,1507-8 (1966), Mossbauer-Effekt am Yb174-- Der g(R)-Faktor des Tiefsten (2+)-Zustandes
66M014	YB-170	E Munck, D Quitmann, H Prange, and S Hufner, Z Naturforsch, A 21,1318-9 (1966), Messung des g(R)-Faktors fur den (2+)-Rotationszustand von Yb172 mit der Mossbauer-Methode
67A017	YB-170	U Atzmony, E R Bauminger, J Hess, A Mustachi, and S Ofer, Phys Rev Lett 18,1061-3 (1967), Iscmer Shifts in the 2+ 0+ Rotational Transition of Yb170
67E001	YB-170	J S Eck, Y K Lee, and J C Walker, Phys Rev 163,1295-8 (1967), Measurement of the g(R) Factors for the First Excited States of Yb174 and Yb176 Using the Mossbauer Effect
67E002	YB-170	J S Eck, Y K Lee, J C Walker, and R R Stevens, Jr, Phys Rev 156,246-50 (1967), Mossbauer Effect Following Coulomb Excitation in the Even-even Isotopes of Ytterbium
67N012	YB-170	I Nowik, S Ofer, and J H Wernick, Phys Lett 24A,89-90 (1967), Relaxation Phencmena in Mosssbauer Spectra cf 170Yb in Intermetallic Compounds of Ytterbium
67O002	YB-170	S Ofer and I Nowik, Phys Lett 24A,88-9 (1967), Mossbauer Effect Studies of Ferromagnetic Relaxaticn in YbIG
68A027	YB-170	U Atzmony, E R Bauminger, J Hess, and S Ofer, Analysis cf Isomer Shift Measurements of the 2+ 0+ Rotational Transition of 170Yb, in "Hyperfine Structure and Nuclear Radiations" (Proceedings of Conference, Asilomar, Pacific Grove, California, 1967), edited by E Matthias and D A Shirley (North-Holland Publishing Cc, Amsterdam, 1968), pp 71-6

CODE	TOPIC	REFERENCE

68HO25 YB-170 W Henning, Z Phys 217,438-56(1968), Anderung des quadra-
tischen Kernladungsradius bei Rotationanregung

68KO07 YB-170 P Kienle, W Henning, G Kaindl, H J Korner, H Schaller, and
F E Wagner, Change cf Nuclear Charge Radii by Collective
Excitations, in "Proc Int Conf Nucl Struct" (Tokyo, Japan,
1968), edited by J Sanada (Supplement to J Phys Scc Jap,
Vol 24), pp 207-16

68M011 YB-170 E Munck, Z Phys 208,184-207(1968), Messung der magnetischen
Momente der tiefsten 2+-Zustande fur einige Dy-, Er- und
Yb-Isotope

68NO11 YB-170 I Nowik and S Ofer, J Phys Chem Solids 29,2117-9(1968),
Mossbauer Studies of 170Yb in Several Paramagnetic Salts

69SO22 YB-170 I Shidlovsky and I Mayer, J Phys Chem Solids 30,1207-13
(1969), Mossbauer Spectra cf Rare Earth Silicides and
Germanides

70HO32 YB-170 W Henning, G M Kalvius, and G K Shenoy, Phys Rev C 2,2414-
21(1970), Isomer Shifts cf the First and Second Excited
Levels of the Ground-State Rotational Band in Yb171

70KO15 YB-170 G M Kalvius, G K Shenoy, and E D Dunlap, Hyperfine Inter-
actions in Er2O3 and Yb2O3 Between 20K and 1.5K, in "Colloq
Int Cent Nat Rech Sci, No 180," (Proceedings of the Confer-
ence, Les Elements des Terres Rares, Paris-Grenoble, 1969)
(Editions du Centre National de la Recherche Scientifique,
Paris, 1970), pp 477-84

71AO08 YB-170 J-C Achard, C Gorochov, F Gonzalez, and P Imbert, C R Acad
Sci, Ser C 272,868-71(1971), Caractere Divalent de l'Ytterbium dans l'Oxyde Mixte Yb(1-x)Ca(x)O

71GO55 YB-170 G Goretzki, G Crecelius, and S Hufner, Hyperfine Interactions in Yb6Fe23, in "Hyperfine Interactions in Excited
Nuclei" (Proceedings cf Conference, Rehovot and Jerusalem,
1971), edited by G Goldring and R Kalish (Gordon and Breach
Science Publishers, Inc, New York, 1971), vol 3, pp 878-80

71HO08 YB-170 W Henning, G Bahre, and P Kienle, Z Phys 241,138-49(1971),
Isomer Shift and Quadrupole Splitting of the 2+-Rotational
State in Yb174

71H059 YB-170 W Henning, G Bahre, and P Kienle, On the Magnetic Field and
Charge Density cf Conduction Electrons at Nuclei in Rare
Earth Metals, in "Hyperfine Interactions in Excited Nuclei"
(Proceedings cf Conference, Rehovot and Jerusalem, 1971),
edited by G Goldring and R Kalish (Gordon and Breach
Science Publishers, Inc, New York, 1971), vol 3, pp 795-801

71KO58 YB-170 G M Kalvius, G K Shenoy, and B D Dunlap, Temperature Dependence of the Paramagnetic Relaxation in YbCl3-6H2O and
Yb2(SO4)3-8H2O frcm Mossbauer Measurements, in "Magnetic Resonance and Related Phenomena, Proc Congress AMPERE,
16th" (Bucharest, 1970), edited by I Ursu (Publ House Acad
Soc Repub Rom, Bucharest, 1971), pp 584-8

71PO09 YB-170 K-G Plingen, B Wolbeck, and F-J Schroder, Nucl Phys A165,
97-117(1971), Hyperfine Interaction in Polycrystalline
Paramagnetic Ytterbium Oxide Studied by the Mossbauer
Effect

71SO64 YB-170 G K Shenoy, G M Kalvius, W Henning, G Bahre, and P
Kienle, Change cf Nuclear Charge Radius in Even and Odd Yb
Isotopes, in "Hyperfine Interactions in Excited Nuclei"
(Proceedings cf Conference, Rehovot and Jerusalem, 1971),
edited by G Goldring and F Kalish (Gordon and Breach
Science Publishers, Inc, New York, 1971), vol 2, pp 699-705

72GO04 YB-170 F Gonzalez Jimenez and P Imbert, Solid State Commun 10,9-13
(1972), Etude des Composes Antiferromagnetiques Yb2O2S et
Yb2O2Se par Effet Mossbauer sur Yb170

CODE	TOPIC	REFERENCE

72G028   YB-170   F Gonzalez Jimenez and P Imbert, Solid State Commun 11, 861-6(1972), Mossbauer Study of Localized Moments in a Cubic Site: Ytterbium in Gold

72M050   YB-170   J Meyer and J Speth, Phys Scr 6,283-4(1972), Isomer Shifts of Rare-earth Nuclei

72N003   YB-170   I Nowik, B D Dunlap, and G M Kalvius, Phys Rev B 6,1048-53 (1972), Magnetic Properties and Crystalline Fields in YbPd3

72W001   YB-170   F E Wagner, B D Dunlap, G M Kalvius, H Schaller, R Felscher, and H Spieler, Phys Rev Lett 28,530-3(1972), Observation of Interference Effects in the Mossbauer Absorption Spectra of E2 and E2/M1 Transitions

73G021   YB-170   F Gonzalez Jimenez and P Imbert, Solid State Commun 13,85-7 (1973), First Observation of a Kondo Anomaly on the Relaxation Rate: Mossbauer Study of Ytterbium in Gold

73G032   YB-170   F Gonzalez Jimenez, P Imbert, J C Achard, and A Percheron, Phys Status Solidi A 19,201-6(1973), Spectrometrie Moss-bauer sur l'Isotope 170Yb dans les Borures YbB6 et TmB12

73N003   YB-170   I Nowik and B D Dunlap, J Phys Chem Solids 34,465-71(1973), Crystalline Fields and Exchange Interactions in YbNi2

73R024   YB-170   P B Russell, G L Latshaw, S S Hanna, and G Kaindl, Nucl Phys A210,133-56(1973), Change of the Nuclear Charge Radius in Excitation of 2+ Levels in Ytterbium Isotopes

74B055   YB-170   E R Bauminger, I Felner, D Froindlich, D Levron, I Nowik, S Ofer, and R Yanovsky, J Phys (Paris), Colloq C6 35,43-4 (1974), Mossbauer Effect Studies of Interconfiguration Fluc tuations in Metallic Rare Earth Compounds

74B061   YB-170   P Bonville, F Gonzalez-Jimenez, P Imbert, and F Varret, J Phys (Paris), Colloq C6 35,575-9(1974), Etude par Effet Mossbauer de Quelques Perovskites d'Ytterbium

74C014   YB-170   J D Cashion, M A Coulthard, and D B Prowse, J Phys C 7, 3620-32(1974), Mossbauer Isomer Shifts in Rare Earth Compounds: I. General Trends

74D014   YB-170   G R Davidson, B D Dunlap, M Eibschutz, and L G Van Uitert, Mossbauer Study of Oriented YbCrO3 Powder, in "AIP Conf Proc No 18-Magnetism and Magnetic Materials-1973" (19th Annual Conference, Boston, 1973), edited by C D Graham, Jr and J J Rhyne (American Institute of Physics, New York, 1974), Part 1, pp 381-5

74D045   YB-170   B D Dunlap, G R Davidson, M Eibschutz, H J Guggenheim, and R C Sherwood, J Phys (Paris), Colloq C6 35,429-31(1974), Crystal Field and Electronic Relaxation Effects in Rb2NaYbF6

74G013   YB-170   F Gonzalez-Jimenez, P Imbert, and F Hartmann-Boutron, Phys Rev B 9,95-107(1974), Mossbauer Effect in the Presence of Electronic Relaxation: Application to the 170Yb Nucleus (Erratum, op cit 10,2134(1974))

74G029   YB-170   F Gonzalez-Jimenez, F Hartmann-Boutron, and P Imbert, Phys Rev B 10,2122-8(1974), Interpretation of Low-temperature Mossbauer Spectra in the Presence of Kondo Deviations. II. Population Effects in Au170Yb

74G042   YB-170   F Gonzalez-Jimenez, F Hartmann-Boutron, P Imbert, B Cornut, and B Coqblin, J Phys (Paris), Colloq C6 35,421-4 (1974), Etude par Effet Mossbauer du Comportement Dynamique de l'Ytterbium Dilue dans l'Or

74G059   YB-170   Kondo Anomaly in Au-Yb, in "Proceedings of the Interna-tional Conference of Magnetism ICM-73" (Moscow, 1973) ("Nauka," Moscow, 1974), Vol V, pp 73-8

CODE	TOPIC	REFERENCE

74H027   YB-170   F Hartmann-Boutron, Phys Rev B 10,2113-21(1974), Interpretation of Low-temperature Mossbauer Spectra in the Presence of Kondo Deviations. I. General Considerations

74H050   YB-170   L L Hirst, J Stohr, E Zech, G K Shenoy, and G M Kalvius, Mossbauer Studies of Magnetic Rare Earth Intermetallics: Electronic Re-arrangement Effects in Source Experiments, in "Magnetic Resonance and Related Phenomena, Proceedings of AMPERE Congress, 18th" (Nottingham, 1974), edited by P S Allen, E R Andrew, and C A Bates (North-Holland Publishing Co, Amsterdam, 1974), pp 67-8

74H052   YB-170   L L Hirst, J Stohr, E Zech, G K Shenoy, and G M Kalvius, Mossbauer Studies of Magnetic Rare Earth Intermetallics: Electronic Re-arrangement Effects in Source Experiments, in "Magnetic Resonance and Related Phenomena, Proceedings of AMPERE Congress, 18th" (Nottingham, 1974), edited by P S Allen, E R Andrew, and C A Bates (North-Holland Publishing Co, Amsterdam, 1974), pp 67-8

74M018   YB-170   A Mustachi, J Phys Chem Solids 35,1447-50(1974), Mossbauer Studies of Europium and Ytterbium Hydrides

74P024   YB-170   A Percheron-Guegan, J C Achard, O Gorochov, F Gonzalez-Jimenez, and P Imbert, J Less-Common Met 37,1-8 (1974), Valence de l'Ytterbium dans Certains Composes Cubiques

74S034   YB-170   G K Shenoy, J Stohr, W Wagner, G M Kalvius, and B D Dunlap, Solid State Commun 15,1485-9(1974), The Influence of Small Magnetic Fields on the Mossbauer Relaxation Spectra of Au:Yb Alloys

74S055   YB-170   G K Shenoy, B Boinsot, L Asch, J M Friedt, and B D Dunlap, Phys Lett 49A,429-30(1974), Hyperfine Structure of Gamma(6) Electronic Level in Cubic Yb Compound from Mossbauer Spectroscopy

74S076   YB-170   P Steiner, Arch Sci 27,343-56(1974), Mossbauer Effect Measurements on Magnetic Impurities in Metals

74S090   YB-170   G K Shenoy, L Asch, J M Friedt, and B D Dunlap, J Phys (Paris), Colloq C6 35,425-7(1974), Spin Relaxation and Breit-Rabi Spectra of Cubic Yb Salt

74S130   YB-170   J Stohr, G M Kalvius, G K Shenoy, and L L Hirst, Electronic Rearrangement and Relaxation Effects Following the Nuclear Transformation in Mossbauer Source Experiments, in "Hyperfine Interactions Studied in Nuclear Reactions and Decay: Contributed Papers" (Conference, Uppsala, Sweden, 1974), edited by E Karlsson and R Wappling (Upplands Grafiska, Uppsala, 1974), pp 80-1

74S131   YB-170   J Stohr, W Wagner, and G M Kalvius, Mossbauer Spectroscopy of Dilute Au:Yb Alloys in Small External Magnetic Fields, in "Hyperfine Interactions Studied in Nuclear Reactions and Decay: Contributed Papers" (Conference, Uppsala, Sweden, 1974), edited by E Karlsson and R Wappling (Upplands Grafiska, Uppsala, 1974), pp 188-9

75A037   YB-170   L Asch, G K Shenoy, B D Dunlap, and G M Kalvius, Spin Relaxation Studies in Cubic Cs2Na(Yb(1-x)Gdx)Cl6, in "Int Conf Mossbauer Spectrosc, Proc" (Cracow, Poland, 1975), edited by A Z Hrynkiewicz and J A Sawicki (Wykonano w Pow-ielarni Akademii Gorniczo-Hutniczej im S Staszica, Cracow, 1975), Vol 1, pp 403-4

75B018   YB-170   C Borely, F Gonzalez-Jimenez, P Imbert, and F Varret, J Phys Chem Solids 36,605-9(1975), Mossbauer Study of Ytterbium Ethylsulphate

75B038   YB-170   C Borely, F Gonzalez-Jimenez, P Imbert, and F Varret, J Phys Chem Solids 36,683-8(1975), Mossbauer Study of 170Yb in CaF2

CODE	TOPIC	REFERENCE

75C008   YB-170   J D Cashion, M A Coulthard, and D E Prowse, J Phys C 8,
1267-75(1975), Mossbauer Isomer Shifts in Rare Earth Compounds: II. Nuclear Charge Radius and Electron Density
Studies

75D019   YB-170   G R Davidson, B D Dunlap, M Eibschutz, and L G Van Uitert,
Phys Rev B 12,1681-8(1975), Mossbauer Study of Yb Spin
Reorientation and Low-temperature Magnetic Configuration
in YbFeO3

75D069   YB-170   G R Davidson, B D Dunlap, M Eibschutz, and L G Van Uitert,
Direct Observation of the Spin-reorientation Transition at
Yb Sites in YbFeO3, in "AIP Conference Proceedings-
No 24, Magnetism and Magnetic Materials-1974" (20th
Annual Conference, San Francisco), edited by C D Graham,
Jr, G H Lander, and J J Rhyne (American Institute of
Physics, New York 1975), pp 63-4

75G016   YB-170   F Gonzalez-Jimenez, B Cornut, and B Coqblin, Phys Rev B
11,4674-82(1975), Influence of the Crystalline Field and
Kondo Effects on the Relaxation Rate: Application to
Mossbauer Experiments of Ytterbium Diluted in Gold

75S022   YB-170   J Stohr, Phys Rev B 11,3559-72(1975), Mossbauer Relaxation
Studies of 170Yb in Dilute Au: 170Tm Sources in External
Magnetic Fields up to 55 kG

75Y005   YB-170   R Yanovsky, E R Bauminger, D Levron, I Nowik, and S Ofer,
Solid State Commun 17,1511-4(1975), Crystalline Fields,
Exchange Interactions and Spin Reorientations in
TmxHo(1-x)Fe2 Systems, Studied by a Yb170 Mossbauer Probe

Cumulative List of Reviews
and Useful Papers in Previous MEDI's

Table of the Location of Most Recent Isotope Page for Each Mössbauer Isotope

Isotope	Last Year of Literature Coverage*	Page	Isotope	Last Year of Literature Coverage*	Page	Isotope	Last Year of Literature Coverage*	Page
^{40}K	1966–1968	138	^{153}Sm	1970	174	^{176}Hf	1966–1968	116
^{57}Fe	1975	54	^{154}Sm	1970	176	^{177}Hf	1970	124
^{61}Ni	1975	118	^{151}Eu	1975	46	^{178}Hf	1975	96
^{67}Zn	1975	194	^{153}Eu	1975	50	^{180}Hf	1973	100
^{73}Ge	1975	92	^{154}Gd	1966–1968	96	^{181}Ta	1975	166
^{83}K	1975	110	^{155}Gd	1975	88	^{180}W	1975	180
^{99}Tc	1973	160	^{156}Gd	1974	98	^{182}W	1975	182
^{99}Ru	1975	128	^{157}Gd	1974	100	^{183}W	1974	194
^{101}Ru	1973	130	^{158}Gd	1966–1968	108	^{184}W	1974	198
^{117}Sn	1974	150	^{160}Gd	1966–1968	110	^{186}W	1973	178
^{119}Sn	1975	154	^{159}Tb	1975	170	^{187}Re	1958–1965	118
^{121}Sb	1975	130	^{160}Dy	1975	36	^{186}Os	1966–1968	152
^{125}Te	1975	172	^{161}Dy	1975	38	^{188}Os	1958–1965	112
^{127}I	1975	100	^{162}Dy	1966–1968	22	^{189}Os	1974	130
^{129}I	1975	102	^{164}Dy	1966–1968	24	^{190}Os	1972	258
^{129}Xe	1975	184	^{165}Ho	1975	98	^{191}Ir	1966–1968	130
^{131}Xe	1974	202	^{164}Er	1966–1968	26	^{193}Ir	1975	106
^{133}Cs	1975	34	^{166}Er	1975	42	^{195}Pt	1974	136
^{133}Ba	1970	52	^{167}Er	1975	44	^{197}Au	1975	30
^{139}La	1973	112	^{168}Er	1966–1968	30	^{201}Hg	1971	146
^{141}Pr	1975	126	^{170}Er	1966–1968	32	^{231}Pa	1966–1968	158
^{145}Nd	1975	114	^{169}Tm	1975	178	^{232}Th	1973	166
^{145}Pm	1975	124	^{170}Yb	1975	186	^{234}U	1974	186
^{147}Pm	1971	176	^{171}Yb	1975	190	^{236}U	1974	188
^{147}Sm	1971	194	^{172}Yb	1973	184	^{238}U	1974	190
^{149}Sm	1975	134	^{174}Yb	1974	212	^{237}Np	1975	120
^{151}Sm	1970	170	^{176}Yb	1973	188	^{239}Pu	1972	264
^{152}Sm	1966–1968	174	^{175}Lu	1975	112	^{243}Am	1969	2

*The 1966–1968 literature coverage was not completed until 1974; therefore, data appearing on the isotope pages in this coverage reflect the literature through 1973.